Problems in Probability Theory, Mathematical Statistics and Theory of Random Functions

Problems in Probability Theory, Mathematical Statistics and Theory of Random Functions

Edited by A. A. SVESHNIKOV

Translated by Scripta Technica, Inc.
Edited by Bernard R. Gelbaum

DOVER PUBLICATIONS, INC.
NEW YORK

Published in Canada by General Publishing Company, Ltd., 30 Lesmill Road, Don Mills, Toronto, Ontario.
Published in the United Kingdom by Constable and Company, Ltd.

This Dover edition, first published in 1978, is an unabridged and unaltered republication of the English translation originally published by W. B. Saunders Company in 1968.
The work was originally published by the Nauka Press, Moscow, in 1965 under the title *Sbornik zadach po teorii veroyatnostey, matematicheskoy statistike i teorii sluchaynykh funktsiy.*

International Standard Book Number: 0-486-63717-4
Library of Congress Catalog Card Number: 78-57171

Manufactured in the United States of America
Dover Publications, Inc.
180 Varick Street
New York, N.Y. 10014

Foreword

Students at all levels of study in the theory of probability and in the theory of statistics will find in this book a broad and deep cross-section of problems (and their solutions) ranging from the simplest combinatorial probability problems in finite sample spaces through information theory, limit theorems and the use of moments.

The introductions to the sections in each chapter establish the basic formulas and notation and give a general sketch of that part of the theory that is to be covered by the problems to follow. Preceding each group of problems, there are typical examples and their solutions carried out in great detail. Each of these is keyed to the problems themselves so that a student seeking guidance in the solution of a problem can, by checking through the examples, discover the appropriate technique required for the solution.

Bernard R. Gelbaum

Contents

I RANDOM EVENTS

1. RELATIONS AMONG RANDOM EVENTS

Basic Formulas

Random events are usually designated by the letters $A, B, C, \ldots, U, V$, where U denotes an event certain to occur and V an impossible event. The equality $A = B$ means that the occurrence of one of the events inevitably brings about the occurrence of the other. The intersection of two events A and B is defined as the event $C = AB$, said to occur if and only if both events A and B occur. The union of two events A and B is the event $C = A \cup B$, said to occur if and only if at least one of the events A and B occurs. The difference of two events A and B is defined as the event $C = A \setminus B$, said to occur if and only if A occurs and B does not occur. The complementary event is denoted by the same letter as the initial event, but with an over bar. For instance, $\bar{A}$ and A are complementary, $\bar{A}$ meaning that A does not occur. Two events are said to be mutually exclusive if $AB = V$. The events A_k $(k = 1, 2, \ldots, n)$ are said to form a complete set if the experiment results in at least one of these events so that $\bigcup_{k=1}^{n} A_k = U$.

SOLUTION FOR TYPICAL EXAMPLES

Example 1.1 What kind of events A and B will satisfy the equality $A \cup B = A$?

SOLUTION. The union $A \cup B$ means the occurrence of at least one of the events A and B. Then, for $A \cup B = A$, the event A must include the event B. For example, if A means falling into region S_A and B falling into region S_B, then S_B lies within S_A.

The solution to Problems 1.1 to 1.3 and 1.8 is similar.

Example 1.2 Two numbers at random are selected from a table of random numbers. If the event A means that at least one of these numbers is prime and the event B that at least one of them is an even number, what is the meaning of events AB and $A \cup B$?

SOLUTION. Event AB means that both events A and B occur. The event $A \cup B$ means that at least one of the two events occurs; that is, from two selected numbers at least one number is prime or one is even, or one number is prime and the other is even.

One can solve Problems 1.4 to 1.7 analogously.

Example 1.3 Prove that $\overline{\bar{A}\bar{B}} = A \cup B$ and $\overline{\bar{C} \cup \bar{D}} = CD$.

PROOF. If $C = \bar{A}$ and $D = \bar{B}$, the second equality can be written in the form $\overline{A \cup B} = \bar{A}\bar{B}$. Hence it suffices to prove the validity of the first equality.

The event $\bar{A}\bar{B}$ means that both events A and B do not occur. The complementary event $\overline{\bar{A}\bar{B}}$ means that at least one of these events occurs: the union $A \cup B$. Thus $\overline{\bar{A}\bar{B}} = A \cup B$. The proof of this equality can also be carried out geometrically, an event meaning that a point falls into a certain region.

One can solve Problem 1.9 similarly. The equalities proved in Example 1.3 are used in solving Problems 1.10 to 1.14.

Example 1.4 The scheme of an electric circuit between points M and N is represented in Figure 1. Let the event A be that the element a is out of order, and let the events B_k $(k = 1, 2, 3)$ be that an element b_k is out of order. Write the expressions for C and $\bar{C}$, where the event C means the circuit is broken between M and N.

SOLUTION. The circuit is broken between M and N if the element a or the three elements b_k $(k = 1, 2, 3)$ are out of order. The corresponding events are A and $B_1B_2B_3$. Hence $C = A \cup B_1B_2B_3$.

Using the equalities of Example 1.3, we find that

$$\bar{C} = \overline{A \cup B_1B_2B_3} = \bar{A}\overline{B_1B_2B_3} = \bar{A}(\bar{B}_1 \cup \bar{B}_2 \cup \bar{B}_3).$$

Similarly one can solve Problems 1.16 to 1.18.

PROBLEMS

1.1 What meaning can be assigned to the events $A \cup A$ and AA?

1.2 When does the equality $AB = A$ hold?

1.3 A target consists of 10 concentric circles of radius r_k $(k = 1, 2, 3, \ldots, 10)$. An event A_k means hitting the interior of a circle of radius r_k $(k = 1, 2, \ldots, 10)$. What do the following events mean:

$$B = \bigcup_{k=1}^{6} A_k, \qquad C = \prod_{k=5}^{10} A_k\ ?$$

1.4 Consider the following events: A that at least one of three devices checked is defective, and B that all devices are good. What is the meaning of the events (a) $A \cup B$, (b) AB?

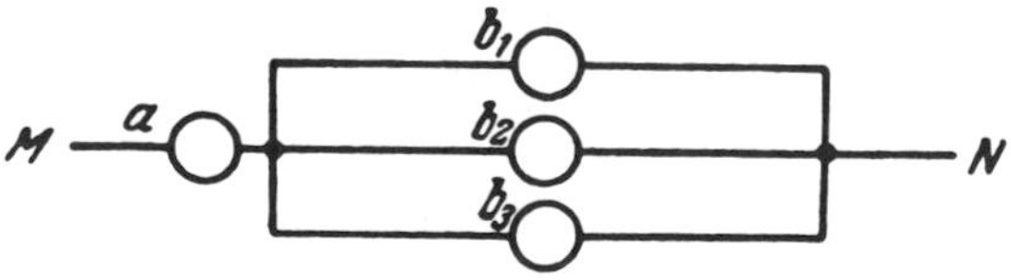

FIGURE 1

1.5 The events A, B and C mean selecting at least one book from three different collections of complete works; each collection consists of at least three volumes. The events A_s and B_k mean that s volumes are taken from the first collection and k volumes from the second collection. Find the meaning of the events (a) $A \cup B \cup C$, (b) ABC, (c) $A_1 \cup B_3$, (d) A_2B_2, (e) $(A_1B_3 \cup B_1A_3)C$.

1.6 A number is selected at random from a table of random numbers. Let the event A be that the chosen number is divisible by 5, and let the event B be that the chosen number ends with a zero. Find the meaning of the events $A \setminus B$ and $\overline{AB}$.

1.7 Let the event A be that at least one out of four items is defective, and let the event B be that at least two of them are defective. Find the complementary events $\bar{A}$ and $\bar{B}$.

1.8 Simplify the expression $A = (B \cup C)(B \cup \bar{C})(\bar{B} \cup C)$.

1.9 When do the following equalities hold true: (a) $A \cup B = \bar{A}$, (b) $AB = \bar{A}$, (c) $A \cup B = AB$?

1.10 From the following equality find the random event X:

$$\overline{X \cup A} \cup \overline{X \cup \bar{A}} = B.$$

1.11 Prove that $\bar{A}B \cup A\bar{B} = \overline{AB}$.

1.12 Prove that the following two equalities are equivalent:

$$\overline{\bigcup_{k=1}^{n} A_k} = \prod_{k=1}^{n} \bar{A}_k, \qquad \bigcup_{k=1}^{n} \bar{A}_k = \overline{\prod_{k=1}^{n} A_k}.$$

1.13 Can the events A and $\overline{A \cup B}$ be simultaneous?

1.14 Prove that A, $\bar{A}B$ and $\overline{A \cup B}$ form a complete set of events.

1.15 Two chess players play one game. Let the event A be that the first player wins, and let B be that the second player wins. What event should be added to these events to obtain a complete set?

1.16 An installation consists of two boilers and one engine. Let the event A be that the engine is in good condition, let B_k $(k = 1, 2)$ be that the kth boiler is in good condition, and let C be that the installation can operate if the engine and at least one of the boilers are in good condition. Express the events C and $\bar{C}$ in terms of A and B_k.

1.17 A vessel has a steering gear, four boilers and two turbines. Let the event A be that the steering gear is in good condition, let B_k $(k = 1, 2, 3, 4)$ be that the boiler labeled k is in good condition, let C_j $(j = 1, 2)$ be that the turbine labeled j is in good condition, and let D be that the vessel can sail if the engine, at least one of the boilers and at least one of the turbines are in good condition. Express D and $\bar{D}$ in terms of A and B_k.

1.18 A device is made of two units of the first type and three units of the second type. Let A_k $(k = 1, 2)$ be that the kth unit of the first type is in good condition, let B_j $(j = 1, 2, 3)$ be that the jth unit of the second type is in good condition, and let C be that the device can operate if at least one unit of the first type and at least two units of the second type are in good condition. Express the event C in terms of A_k and B_j.

2. A DIRECT METHOD FOR EVALUATING PROBABILITIES

Basic Formulas

If the outcomes of an experiment form a finite set of n elements, we shall say that the outcomes are equally probable if the probability of each outcome is $1/n$. Thus if an event consists of m outcomes, the probability of the event is $p = m/n$.

Solution for Typical Examples

Example 2.1 A cube whose faces are colored is split into 1000 small cubes of equal size. The cubes thus obtained are mixed thoroughly. Find the probability that a cube drawn at random will have two colored faces.

Solution. The total number of small cubes is $n = 1000$. A cube has 12 edges so that there are eight small cubes with two colored faces on each edge. Hence $m = 12 \cdot 8 = 96$, $p = m/n = 0.096$.

Similarly one can solve Problems 2.1 to 2.7.

Example 2.2 Find the probability that the last two digits of the cube of a random integer will be 1.[1]

Solution. Represent N in the form $N = a + 10b + \cdots$, where $a, b, \ldots$ are arbitrary numbers ranging from 0 to 9. Then $N^3 = a^3 + 30a^2b + \cdots$. From this we see that the last two digits of N^3 are affected only by the values of a and b. Therefore the number of possible values is $n = 100$. Since the last digit of N^3 is a 1, there is one favorable value $a = 1$. Moreover, the last digit of $(N^3 - 1)/10$ should be 1; i.e., the product $3b$ must end with a 1. This occurs only if $b = 7$. Thus the favorable value ($a = 1$, $b = 7$) is unique and, therefore, $p = 0.01$.

Similarly one can solve Problems 2.8 to 2.11.

Example 2.3 From a lot of n items, k are defective. Find the probability that l items out of a random sample of size m selected for inspection are defective.

Solution. The number of possible ways to choose m items out of n is C_n^m. The favorable cases are those in which l defective items among the k defective items are selected (this can be done in C_k^l ways), and the remaining $m - l$ items are nondefective, i.e., they are chosen from the total number $n - k$ (in C_{n-k}^{m-l} ways). Thus the number of favorable cases is $C_k^l C_{n-k}^{m-l}$. The required probability will be $p = (C_k^l C_{n-k}^{m-l})/C_n^m$.

One can solve Problems 2.12 to 2.20 similarly.

Example 2.4 Five pieces are drawn from a complete domino set. Find the probability that at least one of them will have six dots marked on it.

Solution. Find the probability q of the complementary event. Then $p = 1 - q$. The probability that all five pieces will not have a six (see Example 2.3) is $q = (C_7^0 C_{21}^5)/C_{28}^5$ and, hence,

$$p = 1 - \frac{C_{21}^5}{C_{28}^5} = 0.793.$$

[1] By a "random number" here we mean a k-digit number ($k > 1$) such that any of its digits may range from 0 to 9 with equal probability.

By a similar passage to the complementary event, one can solve Problems 2.21 and 2.22.

PROBLEMS

2.1 Lottery tickets for a total of n dollars are on sale. The cost of one ticket is r dollars, and m of all tickets carry valuable prizes. Find the probability that a single ticket will win a valuable prize.

2.2 A domino piece selected at random is not a double. Find the probability that the second piece also selected at random, will match the first.

2.3 There are four suits in a deck containing 36 cards. One card is drawn from the deck and returned to it. The deck is then shuffled thoroughly and another card is drawn. Find the probability that both cards drawn belong to the same suit.

2.4 A letter combination lock contains five disks on a common axis. Each disk is divided into six sectors with different letters on each sector. The lock can open only if each of the disks occupies a certain position with respect to the body of the lock. Find the probability that the lock will open for an arbitrary combination of the letters.

2.5 The black and white kings are on the first and third rows, respectively, of a chess board. The queen is placed at random in one of the free squares of the first or second row. Find the probability that the position for the black king becomes checkmate if the positions of the kings are equally probable in any squares of the indicated rows.

2.6 A wallet contains three quarters and seven dimes. One coin is drawn from the wallet and then a second coin, which happens to be a quarter. Find the probability that the first coin drawn is a quarter.

2.7 From a lot containing m defective items and n good ones, s items are chosen at random to be checked for quality. As a result of this inspection, one finds that the first k of s items are good. Determine the probability that the next item will be good.

2.8 Determine the probability that a randomly selected integer N gives as a result of (a) squaring, (b) raising to the fourth power, (c) multiplying by an arbitrary integer, a number ending with a 1.

2.9 On 10 identical cards are written different numbers from 0 to 9. Determine the probability that (a) a two-digit number formed at random with the given cards will be divisible by 18, (b) a random three-digit number will be divisible by 36.

2.10 Find the probability that the serial number of a randomly chosen bond contains no identical digits if the serial number may be any five-digit number starting with 00001.

2.11 Ten books are placed at random on one shelf. Find the probability that three given books will be placed one next to the other.

2.12 The numbers 2, 4, 6, 7, 8, 11, 12 and 13 are written, respectively, on eight indistinguishable cards. Two cards are selected at random from the eight. Find the probability that the fraction formed with these two random numbers is reducible.

2.13 Given five segments of lengths 1, 3, 5, 7 and 9 units, find the probability that three randomly selected segments of the five will be the sides of a triangle.

2.14 Two of 10 tickets are prizewinners. Find the probability that among five tickets taken at random (a) one is a prizewinner, (b) two are prizewinners, (c) at least one is a prizewinner.

2.15 This is a generalization of Problem 2.14. There are $n + m$ tickets of which n are prizewinners. Someone purchases k tickets at the same time. Find the probability that s of these tickets are winners.

2.16 In a lottery there are 90 numbers, of which five win. By agreement one can bet any sum on any one of the 90 numbers or any set of two, three, four or five numbers. What is the probability of winning in each of the indicated cases?

2.17 To decrease the total number of games, $2n$ teams have been divided into two subgroups. Find the probability that the two strongest teams will be (a) in different subgroups, (b) in the same subgroup.

2.18 A number of n persons are seated in an auditorium that can accommodate $n + k$ people. Find the probability that $m \leqslant n$ given seats are occupied.

2.19 Three cards are drawn at random from a deck of 52 cards. Find the probability that these three cards are a three, a seven and an ace.

2.20 Three cards are drawn at random from a deck of 36 cards. Find the probability that the sum of points of these cards is 21 if the jack counts as two points, the queen as three points, the king as four points, the ace as eleven points and the rest as six, seven, eight, nine and ten points.

2.21 Three tickets are selected at random from among five tickets worth one dollar each, three tickets worth three dollars each and two tickets worth five dollars each. Find the probability that (a) at least two of them have the same price, (b) all three of them cost seven dollars.

2.22 There are $2n$ children in line near a box office where tickets priced at a nickel each are sold. What is the probability that nobody will have to wait for change if, before a ticket is sold to the first customer, the cashier has $2m$ nickels and it is equally probable that the payments for each ticket are made by a nickel or by a dime.

3. GEOMETRIC PROBABILITIES

Basic Formulas

The geometric definition of probability can be used only if the probability of hitting any part of a certain domain is proportional to the size of this domain (length, area, volume, and so forth), and is independent of its position and shape.

If the geometric size of the whole domain equals S, the geometric size of a part of it equals S_B, and a favorable event means hitting S_B, then the probability of this event is defined to be

$$p = \frac{S_B}{S}.$$

The domains can have any number of dimensions.

SOLUTION FOR TYPICAL EXAMPLES

Example 3.1 The axes of indistinguishable vertical cylinders of radius r pass through an interval l of a straight line AB, which lies in a horizontal plane. A ball of radius R is thrown at an angle q to this line. Find the probability that this ball will hit one cylinder if any intersection point of the path described by the center of the ball with the line AB is equally probable.[2]

SOLUTION. Let x be the distance from the center of the ball to the nearest line that passes through the center of a cylinder parallel to the displacement direction of the center of the ball. The possible values of x are determined by the conditions (Figure 2)

$$0 \leqslant x \leqslant \frac{1}{2} l \sin q .$$

The collision of the ball with the cylinder may occur only if $0 \leqslant x \leqslant R + r$.

The required probability equals the ratio between the length of the segment on which lie the favorable values of x and the length of the segment on which lie all the values of x. Consequently,

$$p = \begin{cases} \dfrac{2(R+r)}{l \sin q} & \text{for } R + r \leqslant \dfrac{l}{2} \sin q; \\ 1 & \text{for } R + r \geqslant \dfrac{l}{2} \sin q . \end{cases}$$

One can solve problems 3.1 to 3.4 and 3.24 analogously.

Example 3.2 On one track of a magnetic tape 200 m. long some information is recorded on an interval of length 20 m., and on the second track similar information is recorded. Estimate the probability that from 60 to 85 m. there is no interval on the tape without recording if the origins of both recordings are located with equal probability at any point from 0 to 180 m.

SOLUTION. Let x and y be the coordinates of origin of the recordings, where $x \geqslant y$. Since $0 \leqslant x \leqslant 180$, $0 \leqslant y \leqslant 180$ and $x \geqslant y$, the domain of all the possible values of x and y is a right triangle with hypotenuse 180 m. The area of this triangle is $S = 1/2 \cdot 180^2$ sq. m. Find the domain of values of x and y

FIGURE 2

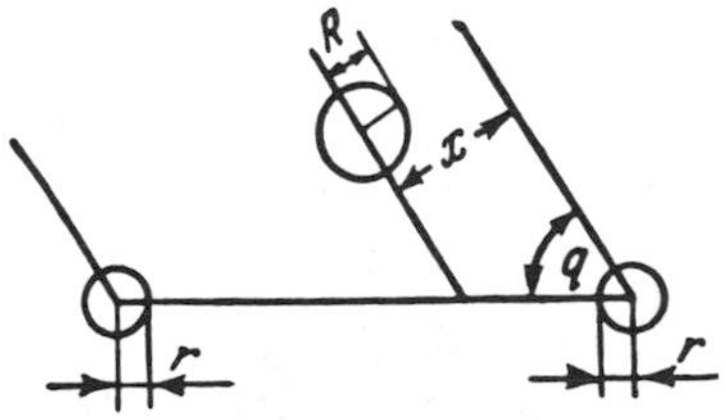

[2] The restriction of equal probability used in formulating several problems with a point that hits the interior of any part of a domain (linear, two-dimensional, and so forth) is understood only in connection with the notion of geometric probability.

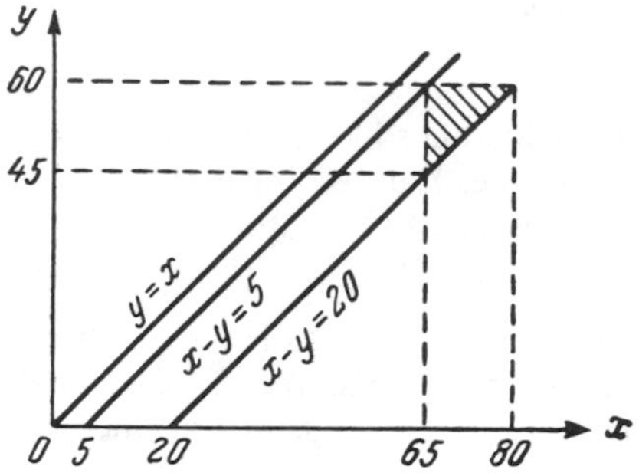

FIGURE 3

favorable to the given event. To obtain a continuous recording, it is necessary that the inequality $x - y \leqslant 20$ m. hold true. To obtain a recording interval longer than or equal to 25 m., we must have $x - y \geqslant 5$ m. Moreover, to obtain a continuous recording on the interval from 60 to 85 m., we must have

$$45 \text{ m.} \leqslant y \leqslant 60 \text{ m.},$$

$$65 \text{ m.} \leqslant x \leqslant 80 \text{ m}.$$

Drawing the boundaries of the indicated domains, we find that the favorable values of x and y are included in a triangle whose area $S_B = 1/2 \cdot 15^2$ sq. m. (Figure 3). The required probability equals the ratio of the area S_B favorable to the given event and the area of the domain S containing all possible values of x and y, namely,

$$p = \left(\frac{15}{180}\right)^2 = \frac{1}{144}.$$

One can solve Problems 3.5 to 3.15 similarly.

Example 3.3 It is equally probable that two signals reach a receiver at any instant of the time T. The receiver will be jammed if the time difference in the reception of the two signals is less than τ. Find the probability that the receiver will be jammed.

Solution. Let x and y be the instants when the two signals are received. The domain of all the possible values of x, y is a square of area T^2 (Figure 4).

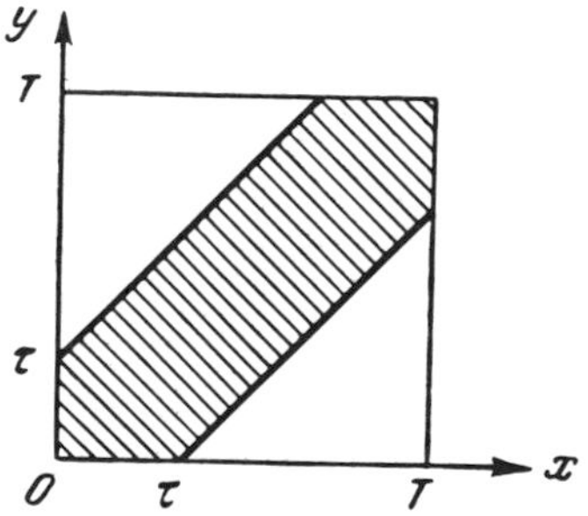

FIGURE 4

FIGURE 5

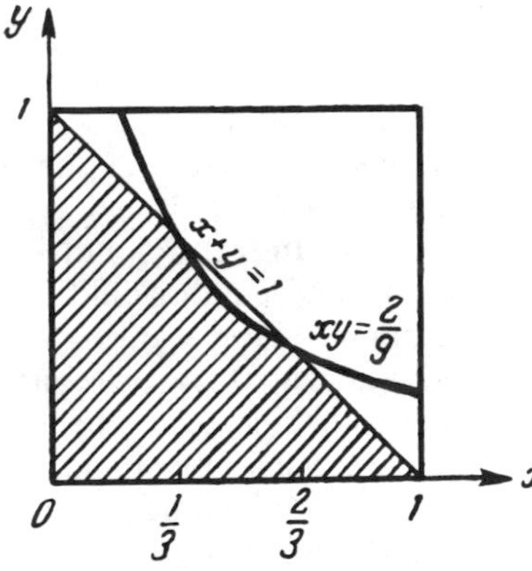

The receiver will be jammed if $|x - y| \leqslant \tau$. The given domain lies between the straight lines $x - y = \tau$ and $x - y = -\tau$. Its area equals

$$S_B = S - (T - \tau)^2,$$

and, therefore,

$$p = 1 - \left(1 - \frac{\tau}{T}\right)^2.$$

One can solve Problems 3.16 to 3.19 analogously.

Example 3.4 Find the probability that the sum of two random positive numbers, each of which does not exceed one, will not exceed one, and that their product will be at most 2/9.

SOLUTION. Let x and y be the chosen numbers. Their possible values are $0 \leqslant x \leqslant 1$, $0 \leqslant y \leqslant 1$, defining in the plane a square of area $S = 1$. The favorable values satisfy the conditions $x + y \leqslant 1$ and $xy \leqslant 2/9$. The boundary $x + y = 1$ divides the square in two so that the domain $x + y \leqslant 1$ represents the lower triangle (Figure 5). The second boundary $xy = 2/9$ is a hyperbola. The x's of the intersection points of these boundaries are: $x_1 = 1/3$ and $x_2 = 2/3$. The area of the favorable domain is

$$S_B = \frac{1}{3} + \int_{1/3}^{2/3} y\,dx = \frac{1}{3} + \frac{2}{9}\int_{1/3}^{2/3}\frac{dx}{x} = \frac{1}{3} + \frac{2}{9}\ln 2.$$

The desired probability is $p = S_B/S = 0.487$.

One can solve Problems 3.20 to 3.23 in a similar manner.

PROBLEMS

3.1 A break occurs at a random point on a telephone line AB of length L. Find the probability that the point C is at a distance not less than l from the point A.

3.2 Parallel lines are drawn in a plane at alternating distances of 1.5 and 8 cm. Estimate the probability that a circle of radius 2.5 cm. thrown at random on this plane will not intersect any line.

3.3 In a circle of radius R chords are drawn parallel to a given direction. What is the probability that the length of a chord selected at random will not exceed R if any positions of the intersection points of the chord with the diameter perpendicular to the given direction are equally probable?

3.4 In front of a disk rotating with a constant velocity we place a segment of length $2h$ in the plane of the disk so that the line joining the midpoint of the segment with the center of the disk is perpendicular to this segment. At an arbitrary instant a particle flies off the disk. Estimate the probability that the particle will hit the segment if the distance between the segment and the center of the disk is l.

3.5 A rectangular grid is made of cylindrical twigs of radius r. The distances between the axes of the twigs are a and b respectively. Find the probability that a ball of diameter d, thrown without aiming, will hit the grid in one trial if the flight trajectory of the ball is perpendicular to the plane of the grid.

3.6 A rectangle 3 cm. $\times$ 5 cm. is inscribed in an ellipse with the semi-axes $a = 100$ cm. and $b = 10$ cm. so that its larger side is parallel to a. Furthermore, one constructs four circles of diameter 4.3 cm. that do not intersect the ellipse, the rectangle and each other.

Determine the probability that (a) a random point whose position is equally probable inside the ellipse will turn out to be inside one of the circles, (b) the circle of radius 5 cm. constructed with the center at this point will intersect at least one side of the rectangle.

3.7 Sketch five concentric circles of radius kr, where $k = 1, 2, 3, 4, 5$, respectively. Shade the circle of radius r and two annuli with the corresponding exterior radii of $3r$ and $5r$. Then select at random a point in the circle of radius $5r$. Find the probability that this point will be in (a) the circle of radius $2r$, (b) the shaded region.

3.8 A boat, which carries freight from one shore of a bay to the other, crosses the bay in one hour. What is the probability that a ship moving along the bay will be noticed if the ship can be seen from the boat at least 20 minutes before the ship intersects the direction of the boat and at most 20 minutes after the ship intersects the direction of the boat? All times and places for intersection are equally likely.

3.9 Two points are chosen at random on a segment of length l. Find the probability that the distance between the points is less than kl, if $0 < k < l$.

3.10 Two points L and M are placed at random on a segment AB of length l. Find the probability that the point L is closer to M than to A.

3.11 On a segment of length l, two points are placed at random so that the segment is divided into three parts. Find the probability that these three parts of the segment are sides of a triangle.

3.12 Three points A, B, C are placed at random on a circle of radius R. What is the probability that the triangle ABC is acute-angled?

3.13 Three line segments, each of a length not exceeding l, are chosen at random. What is the probability that they can be used to form the sides of a triangle?

3.14 Two points M and N are placed on a segment AB of length l. Find the probability that the length of each of the three segments thus obtained does not exceed a given value a ($l \geqslant a \geqslant 1/3$).

3.15 A bus of line A arrives at a station every four minutes and a bus of line B every six minutes. The length of an interval between the arrival of a bus of line A and a bus of line B may be any number of minutes from zero to four, all equally likely.

Find the probability that (a) the first bus that arrives belongs to line A, (b) a bus of any line arrives within two minutes.

3.16 Two ships must arrive at the same moorings. The times of arrival for both ships are independent and equally probable during a given period of 24 hours. Estimate the probability that one of the ships will have to wait for the moorings to be free if the mooring time for the first ship is one hour and for the second ship two hours.

3.17 Two persons have the same probability of arriving at a certain place at any instant of the interval T. Find the probability that the time that a person has to wait for the other is at most t.

3.18 Two ships are sailing in a fog, one along a bay of width L and the other across the same bay. Their velocities are v_1 and v_2. The second ship emits sounds that can be heard at a distance $d < L$. Find the probability that the sounds will be heard on the first ship if the trajectories of the two ships may intersect with equal probabilities at any point.

3.19 A bar of length $l = 200$ mm. is broken at random into pieces. Find the probability that at least one piece between two break-points is at most 10 mm. if the number of break-points is (a) two, (b) three, and a break can occur with equal probability at any point of the bar.

3.20 Two arbitrary points are selected on the surface of a sphere of radius R. What is the probability that an arc of a great circle passing through these points will make an angle less than α, where $\alpha < \pi$?

3.21 A satellite moves on an orbit between 60 degrees northern latitude and 60 degrees southern latitude. Assuming that the satellite can splash down with equal probability at any point on the surface of the earth between the previously mentioned parallels, find the probability that the satellite will fall above 30 degrees northern latitude.

3.22 A plane is shaded by parallel lines at a distance L between adjacent lines. Find the probability that a needle of length l, where $l < L$, thrown at random will intersect some line (Buffon's problem).

3.23 Estimate the probability that the roots of (a) the quadratic equation $x^2 + 2ax + b = 0$, (b) the cubic equation $x^3 + 3ax + 2b = 0$ are real, if it is known that the coefficients are equally likely in the rectangle $|a| \leqslant n$, $|b| \leqslant m$. Find the probability that under the given conditions the roots of the quadratic equation will be positive.

3.24 A point A and the center B of a circle of radius R move independently in a plane. The velocities of these points are constant and equal u and v. At a given instant, the distance AB equals r ($r > R$), and the angle made by the line AB with the vector $\mathbf{v}$ equals β. Assuming that all directions for the point A are equally probable, estimate the probability that the point A will be inside the circle.

4. CONDITIONAL PROBABILITY. THE MULTIPLICATION THEOREM FOR PROBABILITIES

Basic Formulas

The conditional probability $\mathbf{P}(A \mid B)$ of the event A is the probability of A under the assumption that the event B has occurred. (It is assumed that the probability of B is positive.) The events A and B are independent if $\mathbf{P}(A \mid B) = \mathbf{P}(A)$. The probability for the product of two events is defined by the formula

$$\mathbf{P}(AB) = \mathbf{P}(A)\mathbf{P}(B \mid A) = \mathbf{P}(B)\mathbf{P}(A \mid B),$$

which, generalized for a product of n events, is

$$\mathbf{P}\left(\prod_{k=1}^{n} A_k\right) = \mathbf{P}(A_1)\mathbf{P}(A_2 \mid A_1)\mathbf{P}(A_3 \mid A_1A_2)\cdots\mathbf{P}\left(A_n \,\middle|\, \prod_{k=1}^{n-1} A_k\right).$$

The events $A_1, A_2, \ldots, A_n$ are said to be independent if for any m, where $m = 2, 3, \ldots, n$, and any k_j $(j = 1, 2, \ldots, n)$, $1 \leqslant k_1 < k_2 < \cdots < k_m \leqslant n$,

$$\mathbf{P}\left(\prod_{j=1}^{m} A_{k_j}\right) = \prod_{j=1}^{m} \mathbf{P}(A_{k_j}).$$

SOLUTION FOR TYPICAL EXAMPLES

Example 4.1 The break in an electric circuit occurs when at least one out of three elements connected in series is out of order. Compute the probability that the break in the circuit will not occur if the elements may be out of order with the respective probabilities 0.3, 0.4 and 0.6. How does the probability change if the first element is never out of order?

SOLUTION. The required probability equals the probability that all three elements are working. Let A_k $(k = 1, 2, 3)$ denote the event that the kth element functions. Then $p = \mathbf{P}(A_1A_2A_3)$. Since the events may be assumed independent,

$$p \geqslant \mathbf{P}(A_1)\mathbf{P}(A_2)\mathbf{P}(A_3) = 0.7 \cdot 0.6 \cdot 0.4 = 0.168.$$

If the first element is not out of order, then

$$p = \mathbf{P}(A_2A_3) = 0.24.$$

Similarly one can solve Problems 4.1 to 4.10.

Example 4.2 Compute the probability that a randomly selected item is of first grade if it is known that 4 per cent of the entire production is defective, and 75 per cent of the nondefective items satisfy the first grade requirements.

It is given that $\mathbf{P}(A) = 1 - 0.04 = 0.96$, $\mathbf{P}(B \mid A) = 0.75$.

The required probability $p = \mathbf{P}(AB) = (0.96)(0.75) = 0.72$.

Similarly one can solve Problems 4.11 to 4.19.

Example 4.3 A lot of 100 items undergoes a selective inspection. The entire lot is rejected if there is at least one defective item in five items checked. What is the probability that the given lot will be rejected if it contains 5 per cent defective items?

SOLUTION. Find the probability q of the complementary event A consisting of the situation in which the lot will be accepted. The given event is an intersection of five events $A = A_1A_2A_3A_4A_5$, where A_k ($k = 1, 2, 3, 4, 5$) means that the kth item checked is good.

The probability of the event A_1 is $\mathbf{P}(A_1) = 95/100$ since there are only 100 items, of which 95 are good.

After the occurrence of the event A_1, there remain 99 items, of which 94 are good and, therefore, $\mathbf{P}(A_2 \mid A_1) = 94/99$. Analogously, $\mathbf{P}(A_3 \mid A_1A_2) = 93/98$, $\mathbf{P}(A_4 \mid A_1A_2A_3) = 92/97$ and $\mathbf{P}(A_5 \mid A_1A_2A_3A_4) = 91/96$. According to the general formula, we find that

$$q = \frac{95}{100}\cdot\frac{94}{99}\cdot\frac{93}{98}\cdot\frac{92}{97}\cdot\frac{91}{96} = 0.77.$$

The required probability $p = 1 - q = 0.23$.

One can solve Problems 4.20 to 4.35 similarly.

PROBLEMS

4.1 Two marksmen whose probabilities of hitting a target are 0.7 and 0.8, respectively, fire one shot each. Find the probability that at least one of them will hit the target.

4.2 The probability that the kth unit of a computer is out of order during a time T equals p_k ($k = 1, 2, \ldots, n$). Find the probability that during the given interval of time at least one of n units of this computer will be out of order if all the units run independently.

4.3 The probability of the occurrence of an event in each performance of an experiment is 0.2. The experiments are carried out successively until the given event occurs. Find the probability that it will be necessary to perform a fourth experiment.

4.4 The probability that an item made on the first machine is of first grade is 0.7. The probability that an item made on the second machine is first grade is 0.8. The first machine makes two items and the second machine three items. Find the probability that all items made will be of first grade.

4.5 A break in an electric circuit may occur only if one element K or two independent elements K_1 and K_2 are out of order with respective probabilities 0.3, 0.2 and 0.2. Find the probability of a break in the circuit.

4.6 A device stops as a result of damage to one tube of a total of N. To locate this tube one successively replaces each tube with a new one. Find the probability that it will be necessary to check n tubes if the probability is p that a tube will be out of order.

4.7 How many numbers should be selected from a table of random numbers so that the probability of finding at least one even number among them is 0.9?

4.8 The probability that as a result of four independent trials the event A will occur at least once is 0.5. Find the probability that the event will occur in one trial if this probability is constant through all the other trials.

4.9 An equilateral triangle is inscribed in a circle of radius R. What is the probability that four points taken at random in the given circle are inside this triangle?

4.10 Find the probability that a randomly written fraction will be irreducible (Chebyshev's problem).[3]

4.11 If two mutually exclusive events A and B are such that $\mathbf{P}(A) \neq 0$ and $\mathbf{P}(B) \neq 0$, are these events independent?

4.12 The probability that the voltage of an electric circuit will exceed the rated value is p_1. For an increase in the voltage, the probability that an electric device will stop is p_2. Find the probability that the device will stop as a result of an increase in the voltage.

4.13 A motorcyclist in a race must pass through 12 obstacles placed along a course AB; he will stop at each of them with probability 0.1. Knowing the probability 0.7 with which the motorcyclist passes from B to the final point C without stops, find the probability that *no* stops will occur on the segment AC.

4.14 Three persons play a game under the following conditions: At the beginning, the second and third play in turns against the first. In this case, the first player does not win (but might not lose either) and the probabilities that the second and third win are both 0.3. If the first does not lose, he then makes one move against each of the other two players and wins from each of them with the probability 0.4. After this, the game ends. Find the probability that the first player wins from at least one of the other two.

4.15 A marksman hits a target with the probability 2/3. If he scores a hit on the first shot, he is allowed to fire another shot at another target. The probability of failing to hit both targets in three trials is 0.5. Find the probability of failing to hit the second target.

4.16 Some items are made by two technological procedures. In the first procedure, an item passes through three technical operations, and the probabilities of a defect occurring in these operations are 0.1, 0.2 and 0.3. In the second procedure, there are two operations, and the probability of a defect occurring in each of them is 0.3. Determine which technology ensures a greater probability of first grade production if in the first case, for a good item, the probability of first grade production is 0.9, and in the second case 0.8.

4.17 The probabilities that an item will be defective as a result of a mechanical and a thermal process are p_1 and p_2, respectively. The probabilities of eliminating defects are p_3 and p_4, respectively.

Find (a) how many items should be selected after the mechanical process in order to be able to claim that at least one of them can undergo the thermal process with a chance of eliminating the defect, (b) the probability that at least one of three items will have a nonremovable defect after passing through the mechanical and thermal processes.

[3] Consider that the numerator and denominator are randomly selected numbers from the sequence $1, 2, \ldots, k$, and set $k \to \infty$.

4.18 Show that if the conditional probability $\mathbf{P}(A \mid B)$ exceeds the unconditional probability $\mathbf{P}(A)$, then the conditional probability $\mathbf{P}(B \mid A)$ exceeds the unconditional probability $\mathbf{P}(B)$.

4.19 A target consists of two concentric circles of radius kr and nr, where $k < n$. If it is equally probable that one hits any part of the circle of radius nr, estimate the probability of hitting the circle of radius kr in two trials.

4.20 With six cards, each containing one letter, one forms the word *latent*. The cards are then shuffled and at random cards are drawn one at a time. What is the probability that the arrangement of letters will form the word *talent*?

4.21 A man has forgotten the last digit of a telephone number and, therefore, he dials it at random. Find the probability that he must dial at most three times. How does the probability change if one knows that the last digit is an odd number?

4.22 Some m lottery tickets out of a total of n are the winners. What is the probability of a winner in k purchased tickets?

4.23 Three lottery tickets out of a total of 40,000 are the big prizewinners. Find (a) the probability of getting at least one big prize-winner (ticket) per 1000 tickets, (b) how many tickets should be purchased so that the probability of one big winner is at least 0.5.

4.24 Six regular drawings of state bonds plus one supplementary drawing after the fifth regular one take place annually. From a total of 100,000 serial numbers, the winners are 170 in each regular drawing and 270 in each supplementary one. Find the probability that a bond wins after ten years in (a) any drawing, (b) a supplementary drawing, (c) a regular drawing.

4.25 Consider four defective items: one item has the paint damaged, the second has a dent, the third is notched and the fourth has all three defects mentioned. Consider also the event A that the first item selected at random has the paint damaged, the event B that the second item has a dent and the event C that the third item is notched. Are the given events independent in pairs or as a whole set?

4.26 Let $A_1, A_2, \ldots, A_n$ be a set of events independent in pairs. Is it true that the conditional probability that an event occurs, computed under the assumption that other events of the same set have occurred, is the unconditional probability of this event?

4.27 A square is divided by horizontal lines into n equal strips. Then a point whose positions are equally probable in the strip is taken in each strip. In the same way one draws $n - 1$ vertical lines. Find the probability that each vertical strip will contain only one point.

4.28 A dinner party of $2n$ persons has the same number of males and females. Find the probability that two persons of the same sex will not be seated next to each other.

4.29 A party consisting of five males and 10 females is divided at random into five groups of three persons each. Find the probability that each group will have one male member.

4.30 An urn contains $n + m$ identical balls, of which n are white and m black, where $m \geqslant n$. A person draws balls n times, two balls at a time, without returning them to the urn. Find the probability of drawing a pair of balls of different colors each time.

4.31 An urn contains n balls numbered from 1 to n. The balls are drawn one at a time without being replaced in the urn. What is the probability that in the first k draws the numbers on the balls will coincide with the numbers of the draws?

4.32 An urn contains two kinds of balls, white ones and black ones. The balls are drawn one at a time until a black ball appears, and each time when a white ball is drawn it is returned to the urn together with two additional balls. Find the probability that in the first 50 trials no black balls will be drawn.

4.33 There are $n + m$ men in line for tickets that are priced at five dollars each; n of these men have five-dollar bills and m, where $m \leqslant n + 1$, have ten-dollar bills. Each person buys only one ticket. The cashier has no money before the box office opens. What is the probability that no one in the line will have to wait for change?

4.34 The problem is the same as in 4.33, but now the ticket costs one dollar and n of the customers have one-dollar bills whereas m have five-dollar bills, where $2m \leqslant n + 1$.

4.35 Of two candidates, No. 1 receives n votes whereas No. 2 receives m $(n > m)$ votes. Estimate the probability that at all times during the vote count No. 1 will lead No. 2.

5. THE ADDITION THEOREM FOR PROBABILITIES

Basic Formulas

The probability of the union of two events is given by

$$\mathbf{P}(A \cup B) = \mathbf{P}(A) + \mathbf{P}(B) - \mathbf{P}(AB),$$

which can be extended to a union of any number of events:

$$\mathbf{P}\left(\bigcup_{k=1}^{n} A_k\right) = \sum_{k=1}^{n} \mathbf{P}(A_k) - \sum_{k=1}^{n-1} \sum_{j=k+1}^{n} \mathbf{P}(A_k A_j) + \sum_{k=1}^{n-2} \sum_{j=k+1}^{n-1} \sum_{i=j+1}^{n} \mathbf{P}(A_k A_j A_i) - \cdots + (-1)^{n-1} \mathbf{P}\left(\prod_{k=1}^{n} A_k\right).$$

For mutually exclusive events, the probability of a union of events is the sum of the probabilities of these events: that is,

$$\mathbf{P}\left(\bigcup_{k=1}^{n} A_k\right) = \sum_{k=1}^{n} \mathbf{P}(A_k).$$

SOLUTION FOR TYPICAL EXAMPLES

Example 5.1 Find the probability that a lot of 100 items, of which five are defective, will be accepted in a test of a randomly selected sample containing half the lot if, to be accepted, the number of defective items in a lot of 50 cannot exceed one.

SOLUTION. Let A be the event denoting that there is no defective item among those tested and B that there is only one defective item. The required probability is $p = \mathbf{P}(A) + \mathbf{P}(B)$. The events A and B are mutually exclusive. Thus $p = \mathbf{P}(A) + \mathbf{P}(B)$.

There are C_{100}^{50} ways of selecting 50 items from a total of 100. From 95 nondefective items one can select 50 items in C_{95}^{50} ways. Therefore, $\mathbf{P}(A) = C_{95}^{50}/C_{100}^{50}$. Analogously $\mathbf{P}(B) = C_5^1 C_{95}^{49}/C_{100}^{50}$. Then

$$p = \frac{C_{95}^{50}}{C_{100}^{50}} + \frac{C_5^1 C_{95}^{49}}{C_{100}^{50}} = \frac{47 \cdot 37}{99 \cdot 97} = 0.181.$$

Problems 5.1 to 5.12 are solved similarly.

Example 5.2 The scheme of the electric circuit between two points M and N is given in Figure 6. Malfunctions during an interval of time T of different elements of the circuit represent independent events with the following probabilities (Table 1).

TABLE 1

Element	K_1	K_2	L_1	L_2	L_3
Probability	0.6	0.5	0.4	0.7	0.9

Find the probability of a break in the circuit during the indicated interval of time.

SOLUTION. Denote by A_j $(j = 1, 2)$ the event meaning that an element K_j is out of order, by A that at least one element K_j is out of order and by B that all three elements L_i $(i = 1, 2, 3)$ are out of order.
Then, the required probability is

$$p = \mathbf{P}(A \cup B) = \mathbf{P}(A) + \mathbf{P}(B) - \mathbf{P}(A)\mathbf{P}(B).$$

Since

$$\mathbf{P}(A) = \mathbf{P}(A_1) + \mathbf{P}(A_2) - \mathbf{P}(A_1)\mathbf{P}(A_2) = 0.8,$$

$$\mathbf{P}(B) = \mathbf{P}(L_1)\mathbf{P}(L_2)\mathbf{P}(L_3) = 0.252,$$

we get $p \simeq 0.85$.

One can solve Problems 5.13 to 5.16 analogously.

Example 5.3 The occurrence of the event A is equally probable at any instant of the interval T. The probability that A occurs during this interval is p. It is known that during an interval $t < T$, the given event does not occur.

FIGURE 6

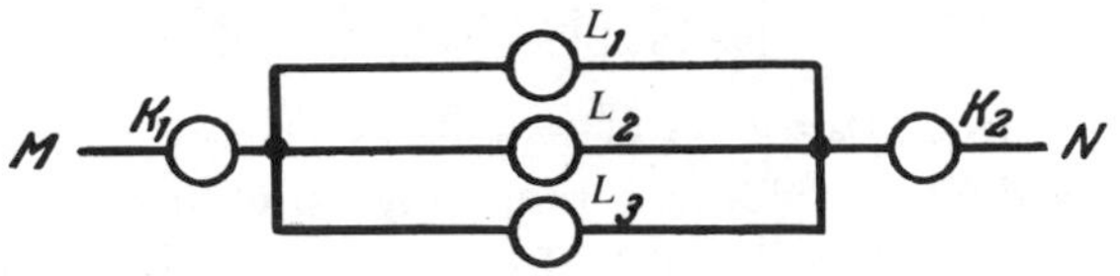

Find the probability P that the event A will occur during the remaining interval of time.

SOLUTION. The probability p that the event A occurs during the interval T is the probability $\frac{t}{T}p$ that the given event occurs during time t plus the product of the probability $\left(1 - \frac{t}{T}p\right)$ that A will not occur during t by the conditional probability that it will occur during the remaining time if it did not occur before. Thus, the following equality holds true:

$$p = \frac{t}{T}p + \left(1 - \frac{t}{T}p\right)P.$$

From this we find

$$P = \frac{p\left(1 - \frac{t}{T}\right)}{1 - \frac{t}{T}p}.$$

Example 5.4 An urn contains n white balls, m black balls and l red balls, which are drawn at random one at a time (a) without replacement, (b) with replacement of each ball to the urn after each draw. Find the probability that in both cases a white ball will be drawn before a black one.

SOLUTION. Let P_{I} be the probability for a white ball to be drawn before a black one, and P_{II} be the probability for a black ball to be drawn before a white ball.

The probability P_{I} is the sum of probabilities of drawing a white ball immediately after a red ball, two red balls, and so forth. Thus, in the case without replacement we have

$$P_{\mathrm{I}} = \frac{n}{n+m+l} + \frac{l}{n+m+l}\frac{n}{n+m+l-1} + \frac{l}{n+m+l}\frac{l-1}{n+m+l-1}\frac{n}{n+m+l-2} + \cdots,$$

and in the case with replacement,

$$P_{\mathrm{I}} = \frac{n}{n+m+l} + \frac{ln}{(n+m+l)^2} + \frac{l^2n}{(n+m+l)^3} + \cdots = \frac{n}{n+m}.$$

To obtain the probabilities P_{II}, replace n by m and m by n in the preceding formulas. From this it follows in both cases that $P_{\mathrm{I}}:P_{\mathrm{II}} = n:m$. Furthermore, since $P_{\mathrm{I}} + P_{\mathrm{II}} = 1$, the required probability in the case without replacement is also $P_{\mathrm{I}} = n/(n+m)$.

One can solve Problems 5.23 to 5.27 similarly.

Example 5.5 A person wrote n letters, sealed them in envelopes and wrote the different addresses randomly on each of them. Find the probability that at least one of the envelopes has the correct address.

SOLUTION. Let the event A_k mean that the kth envelope has the correct address, where $k = 1, 2, \ldots, n$. The desired probability is $p = \mathbf{P}(\bigcup_{k=1}^{n} A_k)$. The events A_k are simultaneous; for any $k, j, i, \ldots$ the following equalities obtain:

$$\mathbf{P}(A_k) = \frac{1}{n} = \frac{(n-1)!}{n!},$$

$$\mathbf{P}(A_k A_j) = \mathbf{P}(A_k)\mathbf{P}(A_j \mid A_k) = \frac{(n-2)!}{n!},$$

$$\mathbf{P}(A_k A_j A_i) = \frac{(n-3)!}{n!}$$

and finally,

$$\mathbf{P}\left(\prod_{k=1}^{n} A_k\right) = \frac{1}{n!}.$$

Using the formula for the probability of a sum of n events we obtain

$$p = C_n^1 \frac{(n-1)!}{n!} - C_n^2 \frac{(n-2)!}{n!} + C_n^3 \frac{(n-3)!}{n!} - \cdots + (-1)^{n-1} \frac{1}{n!}$$

or

$$p = 1 - \frac{1}{2!} + \frac{1}{3!} - \cdots + (-1)^{n-1} \frac{1}{n!}.$$

For large n, $p \approx 1 - e^{-1}$.
Similarly, one can solve Problems 5.32 to 5.38.

PROBLEMS

5.1 Any one of four mutually exclusive events may occur with the corresponding probabilities 0.012, 0.010, 0.006 and 0.002. Find the probability that the outcome of an experiment is at least one of these events.

5.2 A marksman fires one shot at a target consisting of a central circle and two concentric annuli. The probabilities of hitting the circle and the annuli are 0.20, 0.15 and 0.10, respectively. Find the probability of not hitting the target.

5.3 A ball is thrown at a square divided into n^2 identical squares. The probability that the ball will hit a small square of the horizontal strip i and vertical strip j is p_{ij} ($\sum_{i,j=1}^{n} p_{ij} = 1$). Find the probability that the ball will hit a horizontal strip.

5.4 Two identical coins of radius r are placed inside a circle of radius R at which a point is thrown at random. Find the probability that this point hits one of the coins if the coins do not overlap.

5.5 What is the probability of drawing from a deck of 52 cards a face card (jack, queen or king) of any suit or a queen of spades?

5.6 A box contains ten 20-cent stamps, five 15-cent stamps and two 10-cent stamps. One draws six stamps at random. What is the probability that their sum does not exceed one dollar (100 cents)?

5.7 Given the probabilities of the events A and AB, find the probability of the event $A\bar{B}$.

5.8 Prove that from the condition

$$\mathbf{P}(B \mid \bar{A}) = \mathbf{P}(B \mid A)$$

it follows that the events A and B are independent.

5.9 The event B includes the event A. Prove that $\mathbf{P}(A) \leqslant \mathbf{P}(B)$.

5.10 Two urns contain balls differing only in color. The first urn has five white, 11 black and eight red balls, the second has 10 white, eight black and six red balls. One ball at a time is drawn at random from both urns. What is the probability that both balls will be of the same color?

5.11 Two parallel strips 10 mm. wide are drawn in the plane at a distance of 155 mm. Along a perpendicular to these strips, at a distance of 120 mm. lie the centers of circles of radius 10 mm. Find the probability that at least one circle will cross one of the strips if the centers of the circles are situated along the line independent of the position of the strips.

5.12 The seeds of n plants are sown in a line along the road at equal distances from each other. The probability that a pedestrian crossing the road at any point will damage one plant is p ($p < 1/n$). Find the probability that the mth pedestrian who crosses the road at a nonpredetermined point will damage a plant if the pedestrians cross the road successively and independently.

5.13 Find the probability that a positive integer randomly selected will be nondivisible by (a) two and three (b) two or three.

5.14 The probability of purchasing a ticket in which the sums of the first and last three digits are equal is 0.05525. What is the probability of receiving such a ticket among two tickets selected at random if both tickets (a) have consecutive numbers, (b) are independent of each other.

5.15 Prove that if $\mathbf{P}(A) = a$ and $\mathbf{P}(B) = b$, then

$$\mathbf{P}(A \mid B) \geqslant \frac{a + b - 1}{b}.$$

5.16 Given that $\mathbf{P}(X \leqslant 10) = 0.9$, $\mathbf{P}(|Y| \leqslant 1) = 0.95$, prove that regardless of the independence of X and Y if $Z = X + Y$, then the following inequalities hold:

$$\mathbf{P}(Z \leqslant 11) \geqslant 0.85, \qquad \mathbf{P}(Z \leqslant 9) \leqslant 0.95.$$

5.17 A game between A and B is conducted under the following rules: as a result of the first move, always made by A, he can win with the probability 0.3; if A does not win in the first move, B plays next and can win with the probability 0.5; if in this move B does not win, A makes the next move, in which he can win with the probability 0.4. Find the probabilities of winning for A and B.

5.18 Given the probability p that a certain sportsman improves his previous score in one trial, find the probability that the sportsman will improve his score in a competition in which two trials are allowed.

5.19 Player A plays two games each in turn with players B and C. The probabilities that the first game is won by B and C are 0.1 and 0.2, respectively; the probability that the second game is won by B is 0.3, and by C, 0.4.

Find the probability that (a) B wins first, (b) C wins first.

5.20 From an urn containing n balls numbered from 1 to n two balls are drawn successively; the first ball is returned to the urn if its number is 1. Find the probability that the ball numbered 2 is drawn on the second trial.

5.21 Player A plays in turn with players B and C with the probability of winning in each game 0.25; he ends the game after the first loss or after two games played with each of the other players. Find the probabilities that B and C win.

5.22 The probability that a device breaks after it has been used k times is $G(k)$. Find the probability that the device is out of order after n consecutive uses if during the previous m operations it was not out of order.

5.23 Two persons alternately flip a coin. The one who gets heads first is the winner. Find the probabilities of winning for each player.

5.24 Three persons successively toss a coin. The one who gets heads first is the winner. Find the probabilities of winning for each player.

5.25 The probability of gaining a point without losing service in a game between two evenly matched volleyball teams is 0.5. Find the probability that the serving team will gain a point.

5.26 An urn contains n white and m black balls. Two players successively draw one ball at a time and each time return the ball to the urn. The game continues until one of them draws a white ball. Find the probability that the white ball will be first drawn by the player who starts the game.

5.27 Two marksmen shoot in turn until one of them hits the target. The probability of hitting the target is 0.2 for the first marksman and 0.3 for the second one. Find the probability that the first marksman fires more shots than the second.

5.28 Prove the validity of the equality

$$\mathbf{P}\left(\prod_{k=1}^{n} A_k\right) = 1 - \mathbf{P}\left(\bigcup_{k=1}^{n} \bar{A}_k\right).$$

5.29 Simplify the general formula for the probability of a union of events applicable to the case when the probabilities for products of equal numbers of events coincide.

5.30 Prove that

$$\mathbf{P}\left(\prod_{k=1}^{n} A_k\right) = \sum_{k=1}^{n} \mathbf{P}(A_k) - \sum_{k=1}^{n-1} \sum_{j=k+1}^{n} \mathbf{P}(A_k \cup A_j) + \sum_{k=1}^{n-2} \sum_{j=k+1}^{n-1} \sum_{i=j+1}^{n} \mathbf{P}(A_k \cup A_j \cup A_i) - \cdots + (-1)^{n-1}\mathbf{P}\left(\bigcup_{k=1}^{n} A_k\right).$$

5.31 Prove that for any events A_k ($k = 0, 1, \ldots, n$) the following equality holds true:

$$\mathbf{P}\left(\bar{A}_0 \prod_{k=1}^{n} A_k\right) = \mathbf{P}(\bar{A}_0) - \sum_{k=1}^{n} \mathbf{P}(\overline{A_0 \cup A_k}) + \sum_{k=1}^{n-1} \sum_{j=k+1}^{n} \mathbf{P}(\overline{A_0 \cup A_k \cup A_j}) - \cdots + (-1)^n \mathbf{P}\left(\overline{\bigcup_{k=0}^{n} A_k}\right).$$

5.32 An urn contains n balls numbered from 1 to n. The balls are drawn from the urn one at a time without replacement. Find the probability that in some draw the number on the ball coincides with the number of the trial.

5.33 An auditorium has n numbered seats; n tickets are distributed among n persons. What is the probability that m persons will be seated at seats that correspond to their ticket numbers if all the seats are occupied at random.

5.34 A train consists of n cars; k ($k \geqslant n$) passengers get on it and select their cars at random. Find the probability that there will be at least one passenger in each car.

5.35 Two persons play until there is a victory, which occurs when the first wins m games or the second n games. The probability that a game is won is p for the first player and $q = 1 - p$ for the second. Find the probability that the whole competition is won by the first player.

5.36 Two persons have agreed that a prize will go to the one who wins a given number of games. The game is interrupted when m games remain to be won by the first player and n by the second. How should the stakes be divided if the probability of winning a game is 0.5 for each player?

5.37 This is the problem of four liars. One person (a) out of four a, b, c and d receives information that he transmits in the form of a "yes" or "no" signal to the second person (b). The second person transmits to the third (c), the third to the fourth (d) and the fourth communicates the received information in the same manner as all the others. Given the fact that only one person in three tells the truth, find the probability that the first liar tells the truth if the fourth told the truth.

5.38 Some parallel lines separated by the distance L are drawn in a horizontal plane. A convex contour of perimeter s is randomly thrown at this plane. Find the probability that it will intersect one of the parallels if the diameter of the smallest circle circumscribed about the contour is less than L.

6. THE TOTAL PROBABILITY FORMULA

Basic Formulas

The probability $\mathbf{P}(A)$ that an event A will occur simultaneously with one of the events $H_1, H_2, \ldots, H_n$ forming a complete set of mutually exclusive events (hypotheses) is given by the total probability formula

$$\mathbf{P}(A) = \sum_{k=1}^{n} \mathbf{P}(H_k)\mathbf{P}(A \mid H_k),$$

where

$$\sum_{k=1}^{n} \mathbf{P}(H_k) = 1.$$

Solution for Typical Examples

Example 6.1 Among n persons, $m \leqslant n$ prizes are distributed by random drawing in turn from a box containing n tickets. Are the chances of winning equal for all participants? When is it best to draw a ticket?

Solution. Denote by A_k the event that consists of drawing a winning ticket in k draws from the box. According to the results of the preceding experiments, one can make $k + 1$ hypotheses. Let the hypothesis H_{ks} mean that among k drawn tickets, s are prizewinners. The probabilities of these hypotheses are

$$\mathbf{P}(H_{ks}) = \frac{C_m^s C_{n-m}^{k-s}}{C_n^k} \qquad (s = 0, 1, \ldots, k),$$

where

$$\mathbf{P}(H_{ks}) = 0, \quad \text{if } s > m.$$

Since there are $n - k$ tickets left, of which $m - s$ are winners, for $m \geqslant s$

$$\mathbf{P}(A_k \mid H_{ks}) = \frac{m - s}{n - k}.$$

By the total probability formula, we find

$$\mathbf{P}(A_k) = \sum_{s=0}^{k} \frac{C_m^s C_{n-m}^{k-s}}{C_n^k} \frac{m - s}{n - k},$$

where $C_m^s = 0$ for $s > m$.

This equality can also be written in the form

$$\mathbf{P}(A_k) = \frac{m}{n} \sum_{s=0}^{k} \frac{C_{m-1}^s C_{n-m}^{k-s}}{C_{n-1}^k}.$$

We have

$$\sum_{s=0}^{k} C_{m-1}^s C_{n-m}^{k-s} u^{n-k-1} = \frac{1}{k!} \sum_{s=0}^{k} C_k^s \frac{d^s u^{m-1}}{du^s} \frac{d^{k-s} u^{n-m}}{du^{k-s}}$$

$$= \frac{1}{k!} \frac{d^k}{du^k} (u^{m-1} u^{n-m}) = \frac{1}{k!} \frac{d^k u^{n-1}}{du^k} = C_{n-1}^k u^{n-k-1};$$

that is, the following equality holds true:

$$\sum_{s=0}^{k} C_{m-1}^s C_{n-m}^{k-s} = C_{n-1}^k.$$

The required probability $\mathbf{P}(A_k) = m/n$ for any k. Therefore, all participants have equal chances and the sequence in which the tickets are drawn is not important.

Analogously, one can solve Problems 6.1 to 6.17.

Example 6.2 A marked ball can be in the first or second of two urns with probabilities p and $1 - p$. The probability of drawing the marked ball from the urn in which it is located is $P(P \neq 1)$. What is the best way to use n draws of balls from any urn so that the probability of drawing the marked ball is largest if the ball is returned to its urn after each draw?

SOLUTION. Denote by A the event consisting of drawing the marked ball. The hypotheses are H_1 that the ball is in the first urn, H_2 that the ball is in the second urn. By assumption $\mathbf{P}(H_1) = p$, $\mathbf{P}(H_2) = 1 - p$. If m balls are drawn from the first urn and $n - m$ balls from the second urn, the conditional probabilities of drawing the marked ball are

$$\mathbf{P}(A \mid H_1) = 1 - (1 - P)^m, \qquad \mathbf{P}(A \mid H_2) = 1 - (1 - P)^{n-m}.$$

According to the total probability formula, the required probability is

$$\mathbf{P}(A) = p[1 - (1 - P)^m] + (1 - p)[1 - (1 - P)^{n-m}].$$

One should find m so that the probability $\mathbf{P}(A)$ is largest. Differentiating $\mathbf{P}(A)$ with respect to m (to find an approximate value of m, we assume that m is a continuous variable), we obtain

$$\frac{d\mathbf{P}(A)}{dm} = -p(1 - P)^m \ln(1 - P) + (1 - p)(1 - P)^{n-m} \ln(1 - P).$$

Setting $d\mathbf{P}(A)/dm = 0$, we get the equality $(1 - P)^{2m-n} = (1 - p)/p$. Thus,

$$m = \frac{n}{2} + \frac{\ln \dfrac{1 - p}{p}}{2 \ln(1 - P)}.$$

The preceding formula is used in solving Problems 6.18 and 6.19.

PROBLEMS

6.1 There are two batches of 10 and 12 items each and one defective item in each batch. An item taken at random from the first batch is transferred to the second, after which one item is taken at random from the second batch. Find the probability of drawing a defective item from the second batch.

6.2 Two domino pieces are chosen at random from a complete set. Find the probability that the second piece will match the first.

6.3 Two urns contain, respectively, m_1 and m_2 white balls and n_1 and n_2 black balls. One ball is drawn at random from each urn and then from the two drawn balls one is taken at random. What is the probability that this ball will be white?

6.4 There are n urns, each containing m white and k black balls. One ball is drawn from the first urn and transferred to the second urn.

Then one ball is taken at random from the second urn and transferred to the third, and so on. Find the probability of drawing a white ball from the last urn.

6.5 There are five guns that, when properly aimed and fired, have respective probabilities of hitting the target as follows: 0.5, 0.6, 0.7, 0.8 and 0.9. One of the guns is chosen at random, aimed and fired. What is the probability that the target is hit?

6.6 For quality control on a production line one item is chosen for inspection from each of three batches. What is the probability that faulty production will be detected if, in one of the batches, 2/3 of the items are faulty and in the other two they are all good?

6.7 A vacuum tube may come from any one of three batches with probabilities p_1, p_2 and p_3, where $p_1 = p_3 = 0.25$ and $p_2 = 0.5$. The probabilities that a vacuum tube will operate properly for a given number of hours are equal to 0.1, 0.2 and 0.4, respectively, for these batches. Find the probability that a randomly chosen vacuum tube will operate for the given number of hours.

6.8 Player A plays two opponents alternately. The probability that he wins from one at the first trial is 0.5 and the probability that he wins from the other at the first trial is 0.6. These probabilities increase by 0.1 each time the opponents repeat the play against A. Assume that A wins the first two games. Find the probability that A will lose the third game if his opponent in the first game is not known and if ties are excluded.

6.9 A particular material used in a production process may come from one of six mutually exclusive categories with probabilities 0.09, 0.16, 0.25, 0.25, 0.16 and 0.09. The probabilities that an item of production will be acceptable if it is made from materials in these categories are, respectively, 0.2, 0.3, 0.4, 0.4, 0.3 and 0.2. Find the probability of producing an acceptable item.

6.10 An insulating plate 100 mm. long covers two strips passing perpendicular to its length. Their boundaries are located, respectively, at the distances of 20, 40 mm. and 65, 90 mm. from the edge of the plate. A hole of 10 mm. diameter is made, so that its center is located equiprobably on the plate. Find the probability of an electric contact with any of the strips if a conductor is applied from above to an arbitrary point located at the same distance from the base of the plate as the center of the hole.

6.11 The probability that k calls are received at a telephone station during an interval of time t is equal to $P_t(k)$. Assuming that the numbers of calls during two adjacent intervals are independent, find the probability $P_{2t}(s)$ that s calls will be received during an interval $2t$.

6.12 Find the probability that 100 light bulbs selected at random from a lot of 1000 will be nondefective if any number of defective bulbs from 0 to 5 per 1000 is equally probable.

6.13 A white ball is dropped into a box containing n balls. What is the probability of drawing the white ball from this box if all the hypotheses about the initial color composition of the balls are equally probable?

6.14 In a box are 15 tennis balls, of which nine are new. For the first game three balls are selected at random and, after play, they are returned to the box. For the second game three balls are also selected at random. Find the probability that all the balls taken for the second game will be new.

6.15 There are three quarters and four nickels in the right pocket of a coat, and six quarters and three nickels in the left pocket. Five coins taken at random from the right pocket are transferred to the left pocket. Find the probability of drawing a quarter at random from the left pocket after this transfer has been made.

6.16 An examination is conducted as follows: Thirty different questions are entered in pairs on 15 cards. A student draws one card at random. If he correctly answers both questions on the drawn card, he passes. If he correctly answers only one question on the drawn card, he draws another card and the examiner specifies which of the two questions on the second card is to be answered. If the student correctly answers the specified question, he passes. In all other circumstances he fails.

If the student knows the answers to 25 of the questions, what is the probability that he will pass the examination?

6.17 Under what conditions does the following equality hold:

$$\mathbf{P}(A) = \mathbf{P}(A \mid B) + \mathbf{P}(A \mid \bar{B})?$$

6.18 One of two urns, each containing 10 balls, has a marked ball. A player has the right to draw, successively, 20 balls from either of the urns, each time returning the ball drawn to the urn. How should one play the game if the probability that the marked ball is in the first urn is 2/3? Find this probability.

6.19 Ten helicopters are assigned to search for a lost airplane; each of the helicopters can be used in one out of two possible regions where the airplane might be with the probabilities 0.8 and 0.2. How should one distribute the helicopters so that the probability of finding the airplane is the largest if each helicopter can find the lost plane within its region of search with the probability 0.2, and each helicopter searches independently? Determine the probability of finding the plane under optimal search conditions.

7. COMPUTATION OF THE PROBABILITIES OF HYPOTHESES AFTER A TRIAL (BAYES' FORMULA)

Basic Formulas

The probability $\mathbf{P}(H_k \mid A)$ of the hypothesis H_k after the event A occurred is given by the formula

$$\mathbf{P}(H_k \mid A) = \frac{\mathbf{P}(H_k)\mathbf{P}(A \mid H_k)}{\mathbf{P}(A)},$$

where

$$\mathbf{P}(A) = \sum_{j=1}^{n} \mathbf{P}(H_j)\mathbf{P}(A \mid H_j),$$

and the hypotheses H_j ($j = 1, \ldots, n$) form a complete set of mutually exclusive events.

Solution for Typical Examples

Example 7.1 A telegraphic communications system transmits the signals dot and dash. Assume that the statistical properties of the obstacles are such that an average of 2/5 of the dots and 1/3 of the dashes are changed. Suppose that the ratio between the transmitted dots and the transmitted dashes is 5:3. What is the probability that a received signal will be the same as the transmitted signal if (a) the received signal is a dot, (b) the received signal is a dash.

SOLUTION. Let A be the event that a dot is received, and B that a dash is received.

One can make two hypotheses: H_1 that the transmitted signal was a dot; and H_2 that the transmitted signal was a dash. By assumption, $\mathbf{P}(H_1):\mathbf{P}(H_2) = 5:3$. Moreover, $\mathbf{P}(H_1) + \mathbf{P}(H_2) = 1$. Therefore $\mathbf{P}(H_1) = 5/8$, $\mathbf{P}(H_2) = 3/8$. One knows that

$$\mathbf{P}(A \mid H_1) = \frac{3}{5}, \qquad \mathbf{P}(A \mid H_2) = \frac{1}{3},$$

$$\mathbf{P}(B \mid H_1) = \frac{2}{5}, \qquad \mathbf{P}(B \mid H_2) = \frac{2}{3}.$$

The probabilities of A and B are determined from the total probability formula:

$$\mathbf{P}(A) = \frac{5}{8}\cdot\frac{3}{5} + \frac{3}{8}\cdot\frac{1}{3} = \frac{1}{2}, \qquad \mathbf{P}(B) = \frac{5}{8}\cdot\frac{2}{5} + \frac{3}{8}\cdot\frac{2}{3} = \frac{1}{2}.$$

The required probabilities are:

(a) $$\mathbf{P}(H_1 \mid A) = \frac{\mathbf{P}(H_1)\mathbf{P}(A \mid H_1)}{\mathbf{P}(A)} = \frac{\frac{5}{8}\cdot\frac{3}{5}}{\frac{1}{2}} = \frac{3}{4};$$

(b) $$\mathbf{P}(H_2 \mid B) = \frac{\mathbf{P}(H_2)\mathbf{P}(B \mid H_2)}{\mathbf{P}(B)} = \frac{\frac{3}{8}\cdot\frac{2}{3}}{\frac{1}{2}} = \frac{1}{2}.$$

Similarly one can solve Problems 7.1 to 7.16.

Example 7.2 There are two lots of items; it is known that all the items of one lot satisfy the technical standards and 1/4 of the items of the other lot are defective. Suppose that an item from a lot selected at random turns out to be good. Find the probability that a second item of the same lot will be defective if the first item is returned to the lot after it has been checked.

SOLUTION. Consider the hypotheses: H_1 that the lot with defective items was selected, and H_2 that the lot with nondefective items was selected. Let A denote the event that the first item is nondefective. By the assumption of the problem $\mathbf{P}(H_1) = \mathbf{P}(H_2) = 1/2$, $\mathbf{P}(A \mid H_1) = 3/4$, $\mathbf{P}(A \mid H_2) = 1$. Thus, using the formula for the total probability, we find that the probability of the event A will be $\mathbf{P}(A) = 1/2[(3/4) + 1] = 7/8$. After the first trial, the probability that the lot will contain defective items is

$$\mathbf{P}(H_1 \mid A) = \frac{\mathbf{P}(H_1)\mathbf{P}(A \mid H_1)}{\mathbf{P}(A)} = \frac{\frac{1}{2}\cdot\frac{3}{4}}{\frac{7}{8}} = \frac{3}{7}.$$

The probability that the lot will contain only good items is given by

$$\mathbf{P}(H_2 \mid A) = \frac{4}{7}.$$

Let B be the event that the item selected in the first trial turns out to be defective. The probability of this event can also be found from the formula for the total probability. If p_1 and p_2 are the probabilities of the hypotheses H_1 and H_2 after a trial, then according to the preceding computations $p_1 = 3/7$, $p_2 = 4/7$. Furthermore, $\mathbf{P}(B \mid H_1) = 1/4$, $\mathbf{P}(B \mid H_2) = 0$. Therefore the required probability is $\mathbf{P}(B) = (3/7)\cdot(1/4) = 3/28$.

One can solve Problems 7.17 and 7.18 similarly.

PROBLEMS

7.1 Consider 10 urns, identical in appearance, of which nine contain two black and two white balls each and one contains five white and one black ball. An urn is picked at random and a ball drawn at random from it is white. What is the probability that the ball is drawn from the urn containing five white balls?

7.2 Assume that k_1 urns contain m white and n black balls each and that k_2 urns contain m white and n black balls each. A ball drawn from a randomly selected urn turns out to be white. What is the probability that the given ball will be drawn from an urn of the first type?

7.3 Assume that 96 per cent of total production satisfies the standard requirements. A simplified inspection scheme accepts a standard production with the probability 0.98 and a nonstandard one with the probability 0.05. Find the probability that an item undergoing this simplified inspection will satisfy the standard requirements.

7.4 From a lot containing five items one item is selected, which turns out to be defective. Any number of defective items is equally probable. What hypothesis about the number of defective items is most probable?

7.5 Find the probability that among 1000 light bulbs none are defective if all the bulbs of a randomly chosen sample of 100 bulbs turn out to be good. Assume that any number of defective light bulbs from 0 to 5 in a lot of 1000 bulbs is equally probable.

7.6 Consider that D plays against an unknown adversary under the following conditions: the game cannot end in a tie; the first move is made by the adversary; in case he loses, the next move is made by D whose gain means winning the game; if D loses, the game is repeated under the same conditions. Between two equally probable adversaries, B and C, B has the probability 0.4 of winning in the first move and 0.3 in the second, C has the probability 0.8 of winning in the first move and 0.6 in the second. D has the probability 0.3 of winning in the first move regardless of the adversary and, respectively, 0.5, 0.7 when playing against B and C in the second move. The game is won by D.

What is the probability that (a) the adversary is B, (b) the adversary is C.

7.7 Consider 18 marksmen, of whom five hit a target with the probability 0.8, seven with the probability 0.7, four with the probability 0.6 and two with the probability 0.5. A randomly selected marksman fires a shot without hitting the target. To what group is it most probable that he belongs?

7.8 The probabilities that three persons hit a target with a dart are equal to 4/5, 3/4 and 2/3. In a simultaneous throw by all three marksmen, there are exactly two hits. Find the probability that the third marksman will fail.

7.9 Three hunters shoot simultaneously at a wild boar, which is killed by one bullet. Find the probability that the boar is killed by the first, second or the third hunter if the probabilities of their hitting the boar are, respectively, 0.2, 0.4 and 0.6.

7.10 A dart thrown at random can hit with equal probability any point of a region S that consists of four parts representing 50 per cent, 30 per cent, 12 per cent and 8 per cent of the entire region. Which part of region S is most likely to be hit?

7.11 In an urn, there are n balls whose colors are white or black with equal probabilities. One draws k balls from the urn, successively, with replacement. What is the probability that the urn contains only white balls if no black balls are drawn?

7.12 The firstborn of a set of twins is a boy, what is the probability that the other is also a boy if among twins the probabilities of two boys or two girls are a and b, respectively, and among twins of different sexes the probabilities of being born first are equal for both sexes?

7.13 Considering that the probability of the birth of twins of the same sex is twice that of twins of different sexes, that the probabilities of twins of different sexes are equal in any succession and that the probabilities of a boy and a girl are, respectively, 0.51 and 0.49, find the probability of a second boy if the firstborn is a boy.

7.14 Two marksmen fire successively at a target. Their probabilities of hitting the target on the first shots are 0.4 and 0.5 and the probabilities of hitting the target in the next shots increase by 0.05 for each of them. What is the probability that the first shot was fired by the first marksman if the target is hit by the fifth shot?

7.15 Consider three independent trials, in which the event A occurs with the probability 0.2. The probability of the occurrence of the event B depends on the number of occurrences of A. If the event A occurs once, this probability is 0.1; if A occurs twice, it is 0.3; if A

occurs three times it is 0.7; if the event A does not occur, the event B is impossible. Find the most probable number of occurrences of A if it is known that B has occurred.

7.16 There are n students in a technical school. Of these, n_k, where $k = 1, 2, 3$, are in their second year. Two students are randomly selected; one of them has been studying for more years than the other. What is the probability that this student has been studying for three years?

7.17 The third item of one of three lots of items is of second grade, the remaining items are of first grade. An item selected from one of the lots turns out to be of first grade. Find the probability that it was taken from the lot containing second grade items. Find the same probability under the assumption that a second item selected from the same lot turns out to be of first grade if the first item is returned to the lot after inspection.

7.18 Consider a lot of eight items of one sample. From the data obtained by checking one-half of the lot, three items turn out to be technically good and one is defective. What is the probability that in checking three successive items one will turn out to be good and two defective if any number of defective items is equally probable in the given lot?

8. EVALUATION OF PROBABILITIES OF OCCURRENCE OF AN EVENT IN REPEATED INDEPENDENT TRIALS

Basic Formulas

The probability $P_{n;m}$ that an event occurs m times in n independent trials, in which the probability of occurrence of the event is p, is given by the binomial distribution formula

$$P_{n;m} = C_n^m p^m q^{n-m},$$

where $q = 1 - p$.

The probability for realization of the event at least m times in n trials can be computed from the formula

$$R_{n;m} = \sum_{k=m}^{n} P_{n;k} \quad \text{or} \quad R_{n;m} = 1 - \sum_{k=0}^{m-1} P_{n;k}.$$

The probability of occurrence of the event at least once in n trials will be

$$R_{n;1} = 1 - q^n.$$

The number of trials that must be carried out in order to claim that a given event occurs at least once, with a probability at least P, is given by the formula

$$n \geqslant \frac{\ln(1 - P)}{\ln(1 - p)},$$

where p is the probability of occurrence of the event in each of the trials.

The most probable value μ of the number m of occurrences of the event A equals the integral part of the number $(n + 1)p$, and if $(n + 1)p$ is an integer, the largest value of the probability is attained for two numbers $\mu_1 = (n + 1)p - 1$ and $\mu_2 = (n + 1)p$.

If the trials are independent, but the probabilities for realization of the event on different trials are different, the probability $P_{n;m}$ that the event occurs m times in n trials equals the coefficient of u^m in the expansion of the generating function

$$G(u) = \prod_{k=1}^{n} (p_k u + q_k) = \sum_{m=0}^{n} P_{n;m} u^m,$$

where $q_k = 1 - p_k$, p_k being the probability that the event occurs in the kth trial.

The coefficients $P_{n;m}$ can be determined by differentiating the function $G(u)$:

$$P_{n;m} = \frac{1}{m!}\left\{\frac{d^m G(u)}{du^m}\right\}_{u=0},$$

which gives, for example,

$$P_{n;0} = q_1 q_2 \cdots q_n.$$

Solution for Typical Examples

Example 8.1 What is more probable in playing against an equal adversary (if the game cannot end in a tie) to win (a) three games out of four or five out of eight, (b) at least three games out of four or at least five out of eight?

Solution. Since the adversaries are equal, the probabilities for them to win or lose a game are equal: i.e., $p = q = \frac{1}{2}$.

(a) The probability of winning three games out of four is

$$P_{4;3} = C_4^1 \frac{1}{2^4} = \frac{1}{4}.$$

The probability of winning five games out of eight is $P_{8;5} = C_8^3(1/2^8) = 7/32$. Since $1/4 > 7/32$, it is more probable to win three games out of four.

(b) The probability of winning at least three games out of four is

$$R_{4;3} = P_{4;3} + P_{4;4} = \frac{1}{4} + \frac{1}{16} = \frac{5}{16},$$

and the probability of winning at least five games out of eight is

$$R_{8;5} = P_{8;5} + P_{8;6} + P_{8;7} + P_{8;8} = \frac{7}{32} + \left(\frac{8\cdot 7}{2} + 8 + 1\right)\frac{1}{2^8} = \frac{93}{256}.$$

Since $93/256 > 5/16$, it is more probable to win at least five games out of eight.

Similarly one can solve Problems 8.1 to 8.31.

Example 8.2 There are six consumers of electric current. The probability that under certain conditions a breakdown will occur that will disconnect one of the consumers is 0.6 for the first consumer, 0.2 for the second, and 0.3 for each of the remaining four. Find the probability that the generator will be

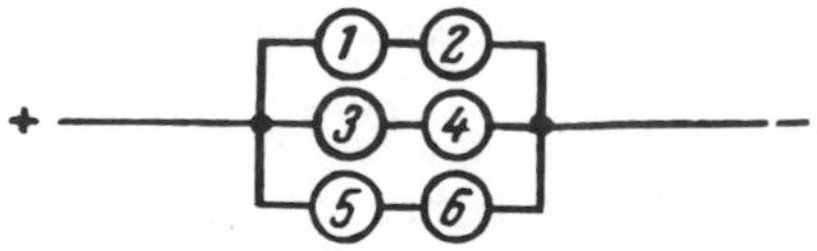

FIGURE 7

completely disconnected if (a) all the consumers are connected in series, (b) all the consumers are connected as shown in the scheme (Figure 7).

SOLUTION. (a) The probability that all six consumers will not be disconnected is equal to the product of the probabilities for each consumer not to be disconnected: that is,

$$q = q_1 q_2 q_3^4 \approx 0.077.$$

The required probability equals the probability that at least one consumer will be disconnected: that is, $p = 1 - q \approx 0.923$.

(b) In this case, the generator is completely disconnected if in each pair of successively connected consumers, there is at least one who is disconnected:

$$p = (1 - q_1 q_2)(1 - q_3^2)^2 \approx 0.177.$$

Problems 8.32 to 8.35 can be solved similarly.

Example 8.3 A lot contains 1 per cent of defective items. What should be the number of items in a random sample so that the probability of finding at least one defective item in it is at least 0.95?

SOLUTION. The required number n is given by the formula $n \geqslant \ln(1 - P)/\ln(1 - p)$. In the present case $P = 0.95$ and $p = 0.01$. Thus, $n \geqslant \ln 0.05/\ln 0.99 \approx 296$.

One can solve Problems 8.36 to 8.40 similarly.

Example 8.4 A wholesaler furnishes products to 10 retail stores. Each of them can send an order for the next day with the probability 0.4, independent of the orders from the other stores. Find the most probable number of orders per day and the probability of this number of orders.

SOLUTION. Here we have $n = 10$, $p = 0.4$, $(n + 1)p = 4.4$. The most probable number μ of orders equals the integral part of the number $(n + 1)p$: that is, $\mu = 4$.

The probability of getting four orders out of 10 is

$$P_{10;4} = C_{10}^4 \cdot 0.4^4 \cdot 0.6^6 = 0.251.$$

Similarly one can solve Problems 8.41 to 8.42.

PROBLEMS

8.1 Find the probability that the license number of the first car encountered on a given day will not contain (a) a 5, (b) two 5's.

All license numbers have four digits, repetitions of digits are permitted and all digits appear with equal probability in all positions.

8.2 There are 10 children in a family. If the probabilities of a boy or a girl are both 0.5, find the probability that this family has (a) five boys, (b) at least three but at most eight boys.

8.3 From a table of random numbers one copies at random 200 two-digit numbers (from 00 to 99). Find the probability that among them, the number 33 appears (a) three times, (b) four times.

8.4 Consider that a library has only books in mathematics and engineering. The probabilities that any reader will select a book in mathematics and engineering are, respectively, 0.7 and 0.3. Find the probability that five successive readers will take books only in engineering or only in mathematics if each of them takes only one book.

8.5 Two light bulbs are connected in series in a circuit. Find the probability that an increase in the voltage above its rated value will break the circuit if, under these assumptions, the probability that a bulb burns out is 0.4 for each of the two bulbs.

8.6 The event B will occur only if the event A occurs at least three times. Find the probability for realization of the event B in (a) five independent trials, (b) seven independent trials, if the probability of occurrence of the event A in one trial is equal to 0.3.

8.7 An electric system containing two stages of type A, one stage of type B and four stages of type C is connected as shown in Figure 8. Find the probability of a break in the circuit such that it cannot be eliminated with the aid of the key K if the elements of type A are out of order with the probability 0.3, the elements of type B with the probability 0.4 and the elements of type C with the probability 0.2.

8.8 The probability that a unit must undergo repairs after m accidents is given by the formula $G(m) = 1 - (1 - 1/\omega)^m$, where ω is the average number of accidents before the unit is submitted for repairs. Prove that the probability that after n cycles the unit will need repairs is given by the formula $W_n = 1 - (1 - p/\omega)^n$, where p is the probability that an accident will occur during one cycle.

8.9 Consider four independent trials in which the event A occurs with the probability 0.3. The event B will occur with the probability 1 if the event A occurs at least twice, it cannot occur if the event A does not occur, and it occurs with a probability 0.6 if the event A occurs once. Find the probability of the occurrence of the event B.

8.10 Consider 200 independent shots fired at a target under identical conditions and leading to 116 hits. Which value, 1/2 or 2/3, for the probability of hitting in one shot is more probable if, before the trial, both hypotheses are equally probable?

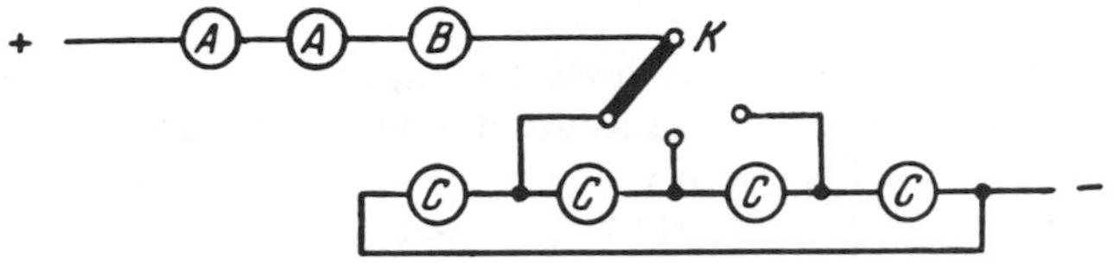

FIGURE 8

8.11 Evaluate the dependence of at least one occurrence of the event A, in 10 independent trials, on the probability p for realization of the event A in each trial for the following values of p: 0.01, 0.05, 0.1, 0.2, 0.3, 0.4, 0.5, 0.6.

8.12 The probability that an event occurs at least once in four independent trials is equal to 0.59. What is the probability of occurrence of the event A in one trial if the probabilities are equal in all trials?

8.13 The probability that an event occurs in each of 18 independent trials is 0.2. Find the probability that this event will occur at least three times?

8.14 The probability of winning with one purchased lottery ticket is 0.02. Evaluate the probabilities of winning a prize with n tickets for $n = 1, 10, 20, 30, 40, 50, 60, 70, 80, 90, 100$ if the tickets belong to different series for each case.

8.15 Given that a lottery ticket wins a prize and that the probabilities that this prize is a bicycle or a washing machine are, respectively, 0.03 and 0.02, find the probability of winning at least one of these items with 10 winning tickets selected from different series.

8.16 A game consists of throwing rings on a peg. A player gets six rings and throws them until the first success. Find the probability that at least one ring remains unused if the probability of a successful throw is 0.1.

8.17 Find the probability of scoring at least 28 points in three shots fired from a pistol at a target, with the maximal score of 10 points per shot, if the probability of scoring 30 points is 0.008. Assume that in one shot the probability of scoring eight points is 0.15, and less than eight points, 0.4.

8.18 Two basketball players each make two attempts at throwing a ball into the basket. The probabilities of making a basket at each throw are, respectively, 0.6 and 0.7. Find the probability that (a) both will have the same numbers of baskets, (b) the first basketball player will have more baskets than the second.

8.19 The probability that a tube will remain in good condition after 1000 hours of operation is 0.2. What is the probability that at least one out of three tubes will remain in good condition after 1000 hours of operation?

8.20 Three technicians produce items of excellent and good qualities on their machines. The first and second technicians make excellent items with the probability 0.9 and the third technician with the probability 0.8. One of the technicians has manufactured eight items, of which two are good. What is the probability that among the next eight items made by this technician there will be two good and six excellent items?

8.21 For victory in a volleyball competition, a team must win three games out of five; the teams are not equally matched. Find the probability that the first team will win each game if for equal chances this team must give odds of (a) two games, (b) one game.

8.22 A competition between two chess players is conducted under the following conditions: draws do not count; the winner is the one who first scores four points under the assumption that the adversary has in this case at most two points; if both players have three points each, the one who scores five points first wins.

For each of the players, find the probability of winning the competition if the probabilities of losing each game are in the ratio 3:2.

8.23 A person uses two matchboxes for smoking. He reaches at random for one box or the other. After some time, he finds out that one box is empty. What is the probability that there will be k matches left in the second box if, initially, each box had n matches (Banach's problem).

8.24 The probability of scoring 10 points is 0.7, and nine points, 0.3. Find the probability of scoring at least 29 points in three shots.

8.25 During each experiment one of two batteries with powers of 120 watts and 200 watts is connected in the circuit for one hour. The probabilities of a favorable outcome of this experiment are 0.06 and 0.08, respectively. One considers that the result of a series of experiments has been attained if one gets at least one favorable outcome in the experiment with the battery of 200 watts or at least two favorable outcomes with the battery of 120 watts. The total energy consumed in all experiments cannot exceed 1200 watts. Which battery is more efficient?

8.26 A device stops if there are at least five defective tubes of type I and at least two defective tubes of type II. Find the probability that the device will stop if five tubes are defective and if the probabilities of a defective tube among the tubes of type I and II are 0.7 and 0.3, respectively.

8.27 The probability of a dangerous overload of a device is 0.4 in each experiment. Find the probability that this device will stop in three independent experiments if the probabilities of a stop in one, two and three experiments are 0.2, 0.5 and 0.8.

8.28 The probability that any of n identical units takes part in an experiment is p ($p < 1/n$). If a given unit participates in the experiments exactly k times, the result of these experiments is considered attained. Find the probability of attaining the desired result in m experiments.

8.29 Under the assumptions of the preceding problem, find the probability of attaining the desired result in $(2k - 1)$ experiments if the experiments are discontinued when the result has been attained.

8.30 The probability that a device will stop in a trial is 0.2. How many devices should be tried so that the probability of at least three stops is 0.9?

8.31 A point A must be connected with 10 telephone subscribers at a point B. Each subscriber keeps the line busy 12 minutes per hour. The calls from any two subscribers are independent. What is the minimal number of channels necessary so that all the subscribers will be served at any instant with the probability 0.99?

8.32 Four radio signals are emitted successively. The probabilities of reception for each of them are independent of the reception of the other signals and equal, respectively, 0.1, 0.2, 0.3 and 0.4. Find the probability that k signals will be received, where $k = 0, 1, 2, 3, 4$.

8.33 Using the assumptions of the preceding problem, find the probability of establishing a two-part radio communication system if the probability of this event is equal to 0.2 for the reception of one signal, 0.6 for two signals and 1 for three and four signals.

8.34 The probabilities that three tubes burn out are respectively, 0.1, 0.2 and 0.3. The probabilities that a device will stop if one, two or three tubes burn out are 0.25, 0.6 and 0.9, respectively. Find the probability that the device will stop.

8.35 A hunter fires a shot at an elk from a distance of 100 m. and hits it with the probability 0.5. If he does not hit it on the first shot, he fires a second shot from a distance of 150 m. If he does not hit the elk in this case, he fires the third shot from a distance of 200 m. If the probability of a hit is inversely proportional to the square of the distance, find the probability of hitting the elk.

8.36 How many numbers should be selected from a table of random numbers to ensure the maximal probability of appearance among them of three numbers ending with a 7?

8.37 The probability of scoring 10 hits in one shot is $p = 0.02$. How many independent shots should be fired so that the probability of scoring 10 hits at least once is at least 0.9?

8.38 During one cycle an automatic machine makes 10 items. How many cycles are necessary so that the probability of making at least one defective item is at least 0.8 if the probability that a part is defective is 0.01?

8.39 Circles of radius 1 cm. have their centers located 60 cm. apart on a line. Several lines of this kind are placed parallel to each other in the same plane; a relative shift of the lines with any amount from 0 to 60 cm. is equally probable. A circle of radius 7 cm. moves in the same plane and perpendicularly to these lines. What should be the number of lines so that the probability of intersection of the moving circle with one of the other circles is at least 0.9?

8.40 From a box containing 20 white and two black balls, n balls are drawn, with replacement one at a time. Find the minimal number of draws so that the probability of getting a black ball at least once exceeds 1/2.

8.41 For a certain basketball player the probability of throwing the ball into the basket in one throw is 0.4. He makes 10 throws. Find the most probable number of successful throws and the corresponding probability.

8.42 Find the most probable number of negative and positive errors and the corresponding probabilities in four measurements if, in each of them, the probability of a positive error equals 2/3, and of a negative one, 1/3.

9. THE MULTINOMIAL DISTRIBUTION. RECURSION FORMULAS. GENERATING FUNCTIONS

Basic Formulas

The probability that in n independent trials, in which the events $A_1, A_2, \ldots, A_m$ occur with the corresponding probabilities $p_1, p_2, \ldots, p_m$, the events A_k, where $k = 1, 2, \ldots, m$, will occur exactly n times ($\sum_{k=1}^{m} n_k = n$), is given by the multinomial distribution formula:

$$P_{n;n_1,n_2,\ldots,n_m} = \frac{n!}{n_1!\,n_2!\cdots n_m!}\,p_1^{n_1}p_2^{n_2}\cdots p_m^{n_m}.$$

The probability $P_{n;n_1,n_2,\ldots,n_m}$ is the coefficient of $u_1^{n_1}u_2^{n_2}\cdots u_m^{n_m}$ in the following generating function:

$$G(u_1, u_2, \ldots, u_m) = (p_1u_1 + p_2u_2 + \cdots + p_mu_m)^n.$$

The generating function for $n + N$ independent trials is the product of the generating functions for n and N trials, respectively. Using this property, one can frequently simplify the calculation of the required probabilities. For the same purpose one applies a proper substitution of the arguments in the generating function. If, for instance, one wishes to find the probability that in n trials the event A_1 will appear l times more than the event A_2, then in the generating function one should set $u_2 = 1/u$, $u_1 = u$, $u_j = 1$, where $j = 3, 4, \ldots, m$. The required probability is the coefficient of u^l in the expansion in a power series for the function

$$G(u) = \left(p_1u + \frac{p_2}{u} + \sum_{j=3}^{m} p_j\right)^n.$$

If $p_k = 1/m$, where $k = 1, 2, \ldots, m$, and one wishes to find the probability that the sum of the numbers of the occurring events is r, one looks for the coefficient of u^r in the expansion in powers of u of the function

$$G(u) = \frac{1}{m^n}\,u^n(1 + u + \cdots + u^{m-1})^n = \frac{u^n}{m^n}\left(\frac{1 - u^m}{1 - u}\right)^n.$$

In the expansion of $G(u)$, it is convenient to use for $(1 - u)^{-n}$ the following expansion:

$$(1 - u)^{-n} = 1 + C_n^{n-1}u + C_{n+1}^{n-1}u^2 + C_{n+2}^{n-1}u^3 + \cdots.$$

Factorials of large numbers can be obtained from logarithm tables (see 2T in the table list) or approximated by Stirling's formula:

$$n! = \sqrt{2\pi n}\,n^n e^{-n}\left(1 + \frac{1}{12n} + \frac{1}{288n^2} + \cdots\right).$$

The probability of occurrence of a given event can sometimes be obtained using relations (recursion formulas) of the form

$$p_k = a_kp_{k-1} + b_kp_{k-2},$$

where a_k and b_k are given constants. The required probability is determined by passage from n to $n + 1$, after an evaluation, based on initial data, of the probabilities for several values of k.

Solution for Typical Examples

Example 9.1 The probabilities that the diameter of any item is less than, greater than or equal to some accepted value are respectively 0.05, 0.10 and

0.85. From the total lot, one selects 100 random samples. Find the probability that among them there will be five items with a smaller diameter and five with a larger diameter than the acceptable diameter.

SOLUTION. Let the event A_1 mean that an item of the first type, an item A_2 of the second type and A_3 of the third type are randomly selected. By assumption, $p_1 = 0.05$, $p_2 = 0.10$, $p_3 = 0.85$. The total number of trials, n, is 100. We seek the probability p that the events A_1 and A_2 will occur five times each. Then $n_1 = n_2 = 5$, $n_3 = 90$. Therefore the required probability

$$p = P_{100;5,5,90} = \frac{100!}{5!\,5!\,90!}\,0.05^5 \cdot 0.1^5 \cdot 0.85^{90}.$$

If we use logarithms, we find

$$\log p = \log 100! - \log 90! - 2\log 5! + 5\log 5 + 90\log 0.85 - 15.$$

Using the logarithm table for factorials and the table for decimal logarithms, we obtain

$$\log p = \bar{3}.7824, \quad \text{or} \quad p \approx 0.006.$$

Similarly one can solve Problems 9.1 to 9.7 and 9.25.

Example 9.2 In each trial the probability of occurrence of an event equals p. What is the probability that the number of occurrences of the event will be even in n trials?

SOLUTION. Let us denote by p_k the probability that in k trials the event will occur an even number of times.

Before the kth trial, one can make two hypotheses: in the $(k - 1)$st trial, the event occurred an even or odd number of times. The probabilities of these hypotheses are p_{k-1} and $1 - p_{k-1}$, respectively. Then

$$p_k = p_{k-1}(1 - p) + (1 - p_{k-1})p,$$

that is,

$$p_k = p + p_{k-1}(1 - 2p).$$

Representing the last expression in the form

$$\left(p_k - \frac{1}{2}\right) = (1 - 2p)\left(p_{k-1} - \frac{1}{2}\right) \qquad (k = 1, 2, \ldots, n)$$

and, respectively, multiplying the left and right sides of n such equalities, we obtain

$$\prod_{k=1}^{n}\left(p_k - \frac{1}{2}\right) = (1 - 2p)^n \prod_{k=1}^{n}\left(p_{k-1} - \frac{1}{2}\right).$$

Simplifying both sides of the last equality by $\prod_{k=1}^{n-1}\left(p_k - \frac{1}{2}\right)$, we find

$$p_n - \frac{1}{2} = (1 - 2p)^n\left(p_0 - \frac{1}{2}\right).$$

Since $p_0 = 1$, the required probability will be

$$p_n = \frac{1}{2}\,[1 + (1 - 2p)^n].$$

Problems 9.8 to 9.13 and 9.26 can be solved similarly.

Example 9.3 Find the probability of purchasing a ticket with a number whose sums of the first three and last three digits are equal if it has six digits and may be any number from 000000 to 999999.

SOLUTION. Let us first consider the first three digits of the number. Since they are arbitrary, one can consider that one performs three trials ($n = 3$), in which any one digit occurs with the probability $p = 1/10$.

In the given case, the number of events, m, is 10, the probability is $p = 1/10$, where $k = 0, 1, \ldots, 9$, and the generating function has the form

$$G_1(u_0, u_1, \ldots, u_9) = \frac{1}{10^3}\left(\sum_{k=0}^{9} u_k\right)^3,$$

where the subscript k of u_k indicates the number k occurring in the trial.

Let us set $u_k = u^k$. Then the coefficient of u^σ in the expansion of the function

$$G_1(u) = \frac{1}{10^3}\left(\sum_{k=0}^{9} u^k\right)^3 = \frac{1}{10^3}\left(\frac{1 - u^{10}}{1 - u}\right)^3$$

gives the probability that the sum of the first three digits of the number on the ticket is σ.

Similarly, the coefficient of $u^{-\sigma}$ in the expansion of

$$G_2(u) = \frac{1}{10^3}\left(\frac{1 - u^{-10}}{1 - u^{-1}}\right)^3$$

gives the probability that the sum of the last three digits of the number is σ.

But in this case, the coefficient of u^0 in the expansion

$$G(u) = G_1(u)G_2(u) = \frac{1}{10^6 u^{27}}\left(\frac{1 - u^{10}}{1 - u}\right)^6$$

is equal to the required probability that the sum of the first three digits and the sum of the last three digits are equal.

We have

$$(1 - u^{10})^6 = 1 - C_6^1 u^{10} + C_6^2 u^{20} - \cdots,$$

$$(1 - u)^{-6} = 1 + C_6^5 u + C_7^5 u^2 + \cdots.$$

Thus the required probability is

$$p = \frac{1}{10^6}(C_{32}^5 - C_6^1 C_{22}^5 + C_6^2 C_{12}^5) = 0.05525.$$

Similarly one can solve Problems 9.14 to 9.24.

PROBLEMS

9.1 Suppose that an urn contains three balls: one black, one red and one white. One draws balls from it five times, one ball at a time, with replacement. Find the probability that the red and white balls will be drawn at least twice each.

9.2 An employee produces a good item with probability 0.90, an item with a defect that can be eliminated with the probability 0.09 and an item with a permanent defect with the probability 0.01. He makes three items. Find the probability that among them there is at least one good item and at least one with a defect that can be eliminated.

9.3 Each of nine balls can be placed with equal probability in one of three initially empty boxes. Find the probability that (a) there will be three balls in each box, (b) there will be four balls in the first box, three in the second box and two in the third box.

9.4 Ten shots are fired at a target consisting of an inner circle and two concentric annuli. The probabilities of hitting these regions in one shot are 0.15, 0.22 and 0.13, respectively. Find the probability that there will be six hits in the circle, three in the first annulus and one in the second annulus.

9.5 A device consists of four units, each made of vacuum tubes. If one tube is out of order, the probabilities that it belongs to a given unit are $p_1 = 0.6111$, $p_2 = p_3 = 0.664$, $p_4 = 0.2561$, respectively, and these do not depend on how many tubes were previously out of order. Find the probability that the device will stop when four tubes are out of order if this event may occur when at least one tube of the first unit or at least one tube in each of the second and third units is out of order.

9.6 Twelve persons get on a train that has six cars; each passenger may select with equal probability each of the cars. Find the probability that (a) there will be two passengers in each car, (b) there will be one car without passengers, one with one passenger, two with two passengers each and the remaining two with three and four passengers, respectively.

9.7 An urn contains l white, m black and n red balls. From it are drawn, with replacement, one at a time, $l_1 + m_1 + n_1$ balls. Find the probability that (a) first, l_1 white balls, then m_1 black balls, and finally, n_1 red balls are drawn, (b) l_1 white, m_1 black, and n_1 red balls are drawn so that balls of identical color appear successively, but the succession of colors may be arbitrary, (c) l_1 white, m_1 black and n_1 red balls are drawn in any succession.

9.8 Find the probability that in n tosses, a coin will show heads an odd number of times.

9.9 Two equally matched adversaries play chess until one of them leads by two games. What is the probability that $2n$ decisive games (that are not draws) will be needed?

9.10 Two persons play until one of them wins all the money from the other. Find the probability that exactly n games will be necessary if all the stakes are equal, each player has at the beginning three stakes, and the probability of winning a game is 1/2 for each of the two players.

9.11 Two persons play until one of them is ruined. The first player has an initial capital of n dollars and the second, m dollars. The probabilities of winning are, respectively, p and q $(p + q = 1)$. In each game, the gain for one player (loss for the other) is one dollar. Find the probabilities of a complete ruin for each of them.

9.12 In a chess competition there are $n + 1$ equally good players. Each man plays each of the others until he loses. The competition continues until one of the players wins n games. What is the probability that m decisive games will be played (draws are not counted)?

9.13 A competition between two equal chess players takes place under the following conditions: the draws are not taken into account; the winner is the one who scores six points if his adversary scores no more than four points; if one wins six games, and the other five, then the competition continues until the difference in points becomes two.

Find the probability that the number of decisive games is (a) at most 10, (b) exactly n.

9.14 The probability that an event occurs in each of n experiments is equal to p. Prove that the generating function for the probabilities of at least $n - m$ occurrences of this event is

$$G(u) = \frac{(p + qu)^n}{1 - u}.$$

9.15 The probability that an event occurs in the kth experiment is equal to p_k $(k = 1, 2, \ldots, n)$. Prove that the generating functions for the probabilities of, respectively, at most m occurrences and at least $n - m$ occurrences of this event, in n independent trials are

$$G_1(u) = \frac{\prod_{k=1}^{n} (q_k + p_k u)}{1 - u} \quad \text{and} \quad G_2(u) = \frac{\prod_{k=1}^{n} (p_k + q_k u)}{1 - u}.$$

9.16 Each of two marksmen fires n shots at his target. Find the probability that they will score the same number of hits if the probability of hitting in each shot is 0.5.

9.17 Each of two identical devices, left and right, has two tubes. After 100 hours of operation, one tube can burn out in only one of them with the probability 1/4 and both tubes can burn out with the probability 1/16. Find the probability that in n pairs of such devices the number of burnt-out tubes in the left devices will exceed at least by m $(m \leqslant 2n)$ the number of burnt-out tubes in the right devices. Find this probability in the case when $n = m = 3$.

9.18 The competition for the title of world champion in 100 square-board checkers consists of 20 games. Find the probability that it will end with the score 12:8 if the probability of winning each game is 0.2 for each of the two players.

9.19 To win the competition for the title of world champion in chess, the challenger must score at least 12.5 points out of a possible 24. In the case of a tie (12:12) the title is kept by the defending champion. The participants are two equal players whose probabilities of winning a game are half as great as the probabilities of a tie. Find (a) the probability that the defending champion will keep his title, and the probability that the challenger will become the world champion, (b) the probability that 20 games will be played in this competition.

9.20 Find the probability that in n throws of a pair of dice, the sum of points marked on the upper faces will be (a) equal to a given number m, (b) not greater than m.

Find these probabilities for $n = 10$ and $m = 20$.

9.21 Find the probability of getting a ticket with a number the sum of whose digits is 21 if all numbers of the ticket from 0 to 999999 are equally probable.

9.22 Any of the n quantities $X_1, X_2, \ldots, X_n$ can take any integral positive value from 1 to m with equal probability. Find the probability that the sum $X_1 + X_2 + \cdots + X_n$ will be (a) equal to a given number N ($nm \geqslant N \geqslant n$), (b) not less than a given number N.

9.23 Two marksmen fire three shots each at their targets. One can score any number of points from seven to 10 with equal probability, whereas for the other, the probability of scoring seven and 10 points is 1/8, and of scoring eight and nine points is 3/8. Find the probability that (a) the first marksman will score 25 points, (b) the second marksman will score 25 points, (c) both marksmen will score the same number of points.

9.24 Two distinguishable coins are tossed simultaneously and repeatedly. Find the probability that at the nth toss (and not before) each will have shown heads as many times as the other.

9.25 Find the probability that a runoff will be necessary in the elections of l persons if n people vote. The probability of being eliminated is the same for each of the k candidates and equal to p, and to be elected a candidate must get the majority of the votes. A runoff takes place only in the case when candidates l and $l + 1$ get an equal number of votes.

9.26 Two equal volleyball teams play one game. The game continues until one of the teams leads by two points; the minimal score necessary is 15. Find the probabilities (a) P_k and Q_k that the game will be won, respectively, by the first team (which serves the ball first) and the second team with the score $15{:}k$ ($k = 0, 1, \ldots, 13$), (b) P_1 and Q_1 that the game will be won by each of the teams if the losing team has at most 13 points, (c) P_k and Q_k that the game will be won with a score of $(16 + k){:}(14 + k)$, where $k = 0, 1, \ldots$, (d) P_{II} and Q_{II} that the game will be won if each team loses at least 14 points, (e) P and Q that the game will be won, respectively, by the first and second teams.

II RANDOM VARIABLES

10. THE PROBABILITY DISTRIBUTION SERIES, THE DISTRIBUTION POLYGON AND THE DISTRIBUTION FUNCTION OF A DISCRETE RANDOM VARIABLE

Basic Formulas

A random variable is said to be discrete if its possible values can be enumerated.

A discrete random variable X can be specified by: (1) a distribution series, (2) a distribution function (integral distribution law).

By a distribution series we mean the set of all possible values x_i of X, and the corresponding probabilities $p_i = \mathbf{P}(X = x_i)$. A distribution series can be represented by a table (see Table 2) or a formula.

The probabilities p_i satisfy the condition

$$\sum_{i=1}^{n} p_i = 1,$$

in which the value of n may be finite or infinite.

The graphic representation of a distribution series is called a distribution polygon. To construct it, one represents the values of the random variable (x_i) on the x-axis, and the probabilities p_i on the y-axis; next, one joins the points A_i with the coordinates (x_i, p_i) by a broken curve (Figure 9).

The distribution function (integral distribution law) of a random variable X is defined as the function $F(x)$, equal to the probability $\mathbf{P}(X < x)$ that the random variable is less than the (arbitrarily chosen) value x. The function $F(x)$ is given by the formula

$$F(x) = \sum_{x_i < x} p_i,$$

in which the summation is extended over all values of i such that $x_i < x$.

TABLE 2

x_i	x_1	x_2	...	x_n
p_i	p_1	p_2	...	p_n

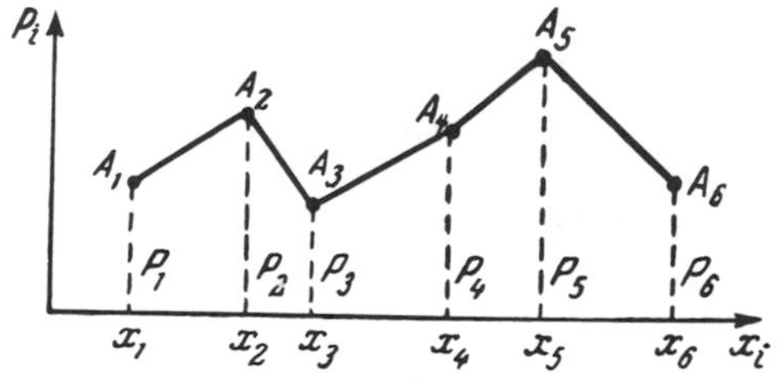

FIGURE 9

SOLUTION FOR TYPICAL EXAMPLES

Example 10.1 From a lot of 100 items, of which 10 are defective, a random sample of size 5 is selected for quality control. Construct the distribution series of the random number X of defective items contained in the sample.

SOLUTION. Since the number of defective items in the sample can be any positive integer from 0 to 5 inclusive, the possible values x_i of the random variable X are

$$x_1 = 0, \quad x_2 = 1, \quad x_3 = 2, \quad x_4 = 3, \quad x_5 = 4, \quad x_6 = 5.$$

The probability $\mathbf{P}(X = k)$ that the sample will contain exactly k ($k = 0, 1, 2, 3, 4, 5$) defective items is

$$\mathbf{P}(X = k) = \frac{C_{10}^{k} C_{90}^{5-k}}{C_{100}^{5}}.$$

The computations with the preceding formula give, with an accuracy of 0.001, the following results:

$$p_1 = \mathbf{P}(X = 0) = 0.583, \qquad p_2 = \mathbf{P}(X = 1) = 0.340,$$
$$p_3 = \mathbf{P}(X = 2) = 0.070,$$
$$p_4 = \mathbf{P}(X = 3) = 0.007, \qquad p_5 = \mathbf{P}(X = 4) = 0,$$
$$p_6 = \mathbf{P}(X = 5) = 0.$$

Using for verification the equality $\sum_{k=1}^{6} p_k = 1$, we can convince ourselves that the computations and the round-off are correct (see Table 3).

TABLE 3

x_i	0	1	2	3	4	5
p_i	0.583	0.340	0.070	0.007	0	0

Similarly one can solve Problems 10.13 and 10.14.

Example 10.2 Items are tested under overload conditions. The probability that each item passes the test is 4/5 and independence prevails. The tests are concluded when an item fails to meet the requirements of the test. Derive the formula for the distribution series of the number of trials.

SOLUTION. The trials end with the kth item ($k = 1, 2, 3, \ldots$) if the first $k - 1$ items pass the test and the kth item fails.

If X is the random number of trials, then

$$\mathbf{P}(X = k) = \left(1 - \frac{1}{5}\right)^{k-1} \cdot \frac{1}{5} = \frac{1}{5}\left(\frac{4}{5}\right)^{k-1}, \qquad (k = 1, 2, 3, \ldots).$$

The formula obtained for the distribution series is equivalent to Table 4.

TABLE 4

x_i	1	2	3	$\ldots$	k	$\ldots$
p_i	$\frac{1}{5}$	$\frac{4}{5^2}$	$\frac{4^2}{5^3}$	$\ldots$	$\frac{4^{k-1}}{5^k}$	$\ldots$

The peculiarity of the current problem is that theoretically the number of trials can be infinite, but the probability of such an event is zero:

$$\lim_{k \to \infty} \mathbf{P}(X = k) = \lim_{k \to \infty} \frac{4^{k-1}}{5^k} = 0.$$

Problems 10.2, 10.4, 10.5, 10.7, 10.10 and 10.12 are solved in a similar manner.

Example 10.3 A car has four traffic lights on its route. Each of them allows it to move ahead or stop with the probability 0.5.

Sketch the distribution polygon of the probabilities of the numbers of lights passed by the car before the first stop has occurred.

SOLUTION. Let X denote the random number of lights passed by the car before the first stop occurs; it can assume the following values:

$$x_1 = 0, \quad x_2 = 1, \quad x_3 = 2, \quad x_4 = 3, \quad x_5 = 4.$$

The probabilities $p_i = \mathbf{P}(X = x_i)$ that the number of traffic lights X passed by the car will equal some given value can be computed with the formula

$$p_i = \mathbf{P}(X = x_i) = \begin{cases} p(1 - p)^{i-1} & \text{for } i = 1, 2, 3, 4, \\ (1 - p)^4 & \text{for } i = 5, \end{cases}$$

in which p is the probability with which the traffic lights can stop the car ($p = 0.5$).

As a result of these computations, we obtain that $p_1 = 0.5$, $p_2 = 0.25$, $p_3 = 0.125$, $p_4 = 0.0625$, $p_5 = 0.0625$. With these results we construct the probability distribution polygon (Figure 10).

Following this example, we can solve Problems 10.3, 10.8 and 10.9.

Example 10.4 A space rocket has a device consisting of four units A_1, A_2, A_3, A_4, each of which fails to operate when at least one elementary particle hits it. The failure of the entire device occurs either if A_1 fails or if A_2, A_3 and A_4 fail simultaneously.

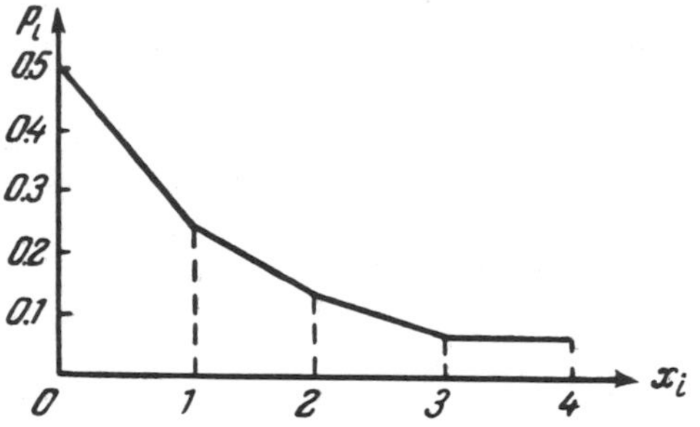

FIGURE 10

Construct the distribution function $F(x)$ of the random number of elementary particles X, for which the entire device will fail if the probability that a particle reaching the device will hit A_1 is $p_1 = 0.4$, and the probabilities for hitting A_2, A_3 and A_4, are, respectively, $p_2 = p_3 = p_4 = 0.2$.

SOLUTION. Let A_1, A_2, A_3, A_4 denote the events that A_1, A_2, A_3, A_4 fail. The required distribution function $F(x)$ equals the probability that the device will continue its operation after $n < x$ hits; i.e.,

$$F(x) = \mathbf{P}(\overline{A_1 \cup A_2A_3A_4}).$$

Using the formula (see Section 5)

$$\overline{A_1 \cup A_2A_3A_4} = \bar{A}_1(\bar{A}_2 \cup \bar{A}_3 \cup \bar{A}_4) = \bar{A}_1\bar{A}_2 \cup \bar{A}_1\bar{A}_3 \cup \bar{A}_1\bar{A}_4$$
$$= (\overline{A_1 \cup A_2}) + (\overline{A_1 \cup A_3}) + (\overline{A_1 \cup A_4})$$

and applying the formula for the addition of probabilities, we obtain

$$\begin{aligned} F(x) &= 1 - \mathbf{P}[(\overline{A_1 \cup A_2}) \cup (\overline{A_1 \cup A_3}) \cup (\overline{A_1 \cup A_4})] \\ &= 1 - \mathbf{P}(\overline{A_1 \cup A_2}) - \mathbf{P}(\overline{A_1 \cup A_3}) - \mathbf{P}(\overline{A_1 \cup A_4}) \\ &\quad + \mathbf{P}[(\overline{A_1 \cup A_2})(\overline{A_1 \cup A_3})] + \mathbf{P}[(\overline{A_1 \cup A_2})(\overline{A_1 \cup A_4})] \\ &\quad + \mathbf{P}[(\overline{A_1 \cup A_3})(\overline{A_1 \cup A_4})] - \mathbf{P}[(\overline{A_1 \cup A_2})(\overline{A_1 \cup A_3})(\overline{A_1 \cup A_4})], \end{aligned}$$

where all the probabilities are defined under the assumption that n particles hit the device. Since $p_1 + p_2 + p_3 + p_4 = 1$, and for each hit of a particle one and only one stage necessarily fails to operate, we have

$$\mathbf{P}(\overline{A_1 \cup A_2}) = (p_3 + p_4)^n; \qquad \mathbf{P}(\overline{A_1 \cup A_3}) = (p_2 + p_4)^n;$$
$$\mathbf{P}(\overline{A_1 \cup A_4}) = (p_2 + p_3)^n;$$
$$\mathbf{P}[(\overline{A_1 \cup A_2})(\overline{A_1 \cup A_3})] = p_4^n; \qquad \mathbf{P}[(\overline{A_1 \cup A_2})(\overline{A_1 \cup A_4})] = p_3^n;$$
$$\mathbf{P}[(\overline{A_1 \cup A_3})(\overline{A_1 \cup A_4})] = p_2^n;$$
$$\mathbf{P}[(A_1 \cup A_2)(A_1 + A_3)(A_1 \cup A_4)] = 0.$$

Thus, taking into account that $p_2 = p_3 = p_4 = 0.2$, we obtain

$$F(x) = 1 + 3p_2^n - 3(2p_2)^n = 1 - 3p_2^n(2^n - 1) = 1 - 3\,\frac{(2^{[x]} - 1)2^{[x]}}{10^{[x]}},$$

where $[x]$ denotes the largest integer less than x, for example, $[5.9] = 5$, $[5] = 4$.

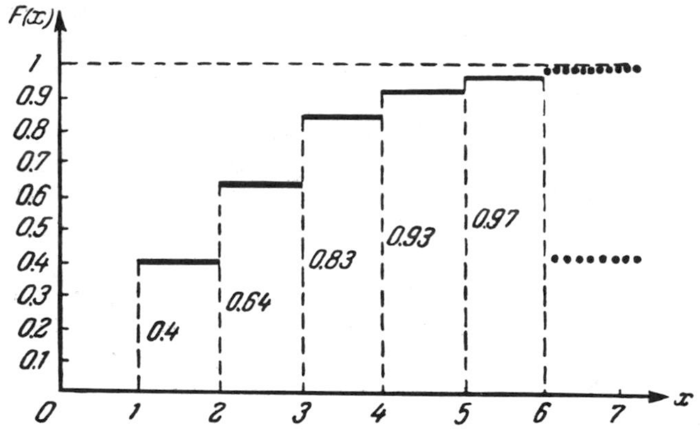

FIGURE 11

Therefore, the graph of the probability distribution function for several initial values of x has the form shown in Figure 11.

Problems 10.6 and 10.11 are solved similarly.

PROBLEMS

10.1 Construct the distribution series and the distribution function for a random number of successful events in one experiment if the experiment consists of throwing a ball into a basket and the probability of a success in one trial is $p = 0.3$.

10.2 An experiment consists of three independent tossings of a coin, in each of which heads shows up with the probability $p = 0.5$. For a random number of heads, construct (a) its distribution series, (b) distribution polygon, (c) distribution function.

10.3 Five devices are subjected to successive reliability tests. Each device is tested only if the preceding one turns out to be reliable. Construct the distribution series of a random number of tests if the probability of passing these tests is 0.9 for each device.

10.4 Some independent experiments are discontinued when the first favorable outcome has occurred. For a random number of experiments find (a) the distribution series, (b) the distribution polygon, (c) the most probable number of experiments, if the probability of a favorable outcome in each trial is 0.5.

10.5 Two basketball players shoot the ball alternately until one of them scores. Construct the distribution series for a random number of shots thrown by each of them if the probability of a success is 0.4 for the first player and 0.6 for the second.

10.6 A target consists of a circle numbered 1 and two annuli numbered 2 and 3. By hitting the circle numbered 1, one scores 10 points; the annulus numbered 2, 5 points; and the annulus numbered 3,

1 point. The corresponding probabilities of hitting the circle numbered 1 and annuli numbered 2 and 3, are 0.5, 0.3 and 0.2. Construct the distribution series for a random sum of scores as a result of three hits.

10.7 An experiment is performed with a series of identical devices that are turned on successively for a period of five seconds each. The lifetime of one device is 16 seconds. The experiment is discontinued when at least one device stops. Find the distribution series for a random number of devices if the probability of stopping is 1/2 for each device.

10.8 There are n patterns for the same item. The probability of producing a nondefective item from each of them is p. (a) Find the distribution series of the number of patterns left after the first nondefective item has been produced. (b) Construct the distribution series for a random number of patterns used.

10.9 A lot of n items is tested for reliability; the probability that each item passes the test is p. Construct the distribution series for a random number of items that pass the test.

10.10 A device consisting of units a, b_1 and b_2 fails to operate if the event $C = A \cup B_1B_2$, where A denotes the failure of the unit a, and B_1 and B_2 denote failure of the units b_1 and b_2, respectively. The failures occur when the device is hit by at least one cosmic particle. Construct the distribution series of a number of random particles hitting the device if the probabilities that a particle hits one of the units are: $\mathbf{P}(A) = 0.5$, $\mathbf{P}(B_1) = \mathbf{P}(B_2) = 0.25$.

10.11 An experiment can be a success with probability p or a failure with probability $(1 - p)$. The probability of a favorable outcome in m successful trials is $\mathbf{P}(m) = 1 - (1 - 1/\omega)^m$. Construct the distribution series of the number of trials necessary for a favorable result.

10.12 The number of trials X is a random integer between 0 and ∞. The probability $\mathbf{P}(X = k) = (n^k e^{-n})/k!$. Each trial can be a success with the probability p and a failure with the probability $(1 - p)$. Construct the distribution series of the number of successful trials.

10.13 The probability of obtaining heads in each of five tosses of a coin is 0.5. Find the distribution series for the ratio of the number X of heads to the number Y of tails.

10.14 Construct the distribution series for the sum of digits of three-digit random numbers.

11. THE DISTRIBUTION FUNCTION AND THE PROBABILITY DENSITY FUNCTION OF A CONTINUOUS RANDOM VARIABLE

Basic Formulas

A random variable is said to be continuous if it can assume any numerical values on a given interval, and for which, for any x on this interval, there exists the limit

$$f(x) = \lim_{\Delta x \to 0} \frac{\mathbf{P}(x < X < x + \Delta x)}{\Delta x},$$

called probability density.

A continuous random variable can be defined either by a distribution function $F(x)$ (the integral distribution law) or by a probability density function $f(x)$ (differential distribution law).

The distribution function $F(x) = \mathbf{P}(X < x)$, where x is an arbitrary real number, gives the probability that a random variable X will be less than x.

The distribution function $F(x)$ has the following basic properties:

(1) $\mathbf{P}(a \leqslant X < b) = F(b) - F(a)$;

(2) $F(x_1) \leqslant F(x_2)$, if $x_1 < x_2$;

(3) $\lim_{x \to +\infty} F(x) = 1$;

(4) $\lim_{x \to -\infty} F(x) = 0$.

The probability density function (differential distribution law) $f(x)$ has the following fundamental properties:

(1) $f(x) \geqslant 0$;

(2) $f(x) = \dfrac{dF(x)}{dx}$;

(3) $\int_{-\infty}^{+\infty} f(x)\, dx = 1$;

(4) $P(a \leqslant X < b) = \int_a^b f(x)\, dx$.

The quantity x_p defined by the equality $F(x_p) = p$ is called a quantile; the quantile $x_{0.5}$ is called the median. If the density has a maximum, the value of x for which $f(x) = \max$ is called the mode.

The notion of probability density $f(x)$ can also be introduced for a discrete random variable by setting

$$f(x) = \sum_{k=1}^{n} p_k \delta(x - x_k),$$

in which x_k denote the possible values of the random variable, p_k are their corresponding probabilities

$$p_k = \mathbf{P}(X = x_k),$$

$\delta(x)$ is the δ-function, that is, a "generalized" function with the properties

$$\delta(x) = \begin{cases} \infty, & x = 0, \\ 0, & x \neq 0, \end{cases}$$

$$\int_{-\infty}^{\infty} \delta(x)\, dx = 1, \qquad \int_{-\infty}^{\infty} \varphi(x)\delta(y - x)\, dx = \varphi(y),$$

where $\varphi(x)$ is any function continuous at the point $x = y$. The function $\delta(x)$ can be represented analytically by

$$\delta(x) = \frac{1}{2\pi} \int_{-\infty}^{\infty} e^{i\omega x}\, d\omega,$$

where the integral is understood in the sense of its principal value.[1]

---

[1] See for example, Gel'fand, I. M., and Shilov, G. E.: Generalized Functions. Vol. 1, Properties and Operations. Translated by E. Saletan. New York, Academic Press, Inc., 1964.

Solution for Typical Examples

Example 11.1 The projection X of the radius-vector of a random point on a circumference of radius a onto the diameter has the distribution function (the arcsine law)

$$F(x) = \begin{cases} 1 \\ \frac{1}{2} + \frac{1}{\pi}\arcsin\frac{x}{a} \\ 0 \end{cases} \text{for} \begin{cases} x \geqslant a; \\ -a < x < a; \\ x \leqslant -a. \end{cases}$$

Determine (a) the probability that X will be on the interval $(-a/2, a/2)$, (b) the quantile $x_{0.75}$, (c) the probability density $f(x)$ of the random variable X, (d) the mode and median of the distribution.

Solution. (a) The probability that X assumes values on the interval $(-a/2, a/2)$ is equal to

$$\mathbf{P}\left(-\frac{a}{2} < X < \frac{a}{2}\right) = F\left(\frac{a}{2}\right) - F\left(-\frac{a}{2}\right) = \frac{2}{\pi}\arcsin\frac{1}{2} = \frac{1}{3}.$$

(b) By assumption $p = 0.75$; solving the equation

$$\frac{1}{2} + \frac{1}{\pi}\arcsin\frac{x_{0.75}}{a} = 0.75,$$

we obtain

$$x_{0.75} = \frac{a\sqrt{2}}{2}.$$

(c) The probability density $f(x)$ of the random variable X is:

(1) for all values of x belonging to the interval $(-a, a)$,

$$f(x) = \frac{dF(x)}{dx} = \frac{d}{dx}\left(\frac{1}{2} + \frac{1}{\pi}\arcsin\frac{x}{a}\right) = \frac{1}{\pi\sqrt{a^2 - x^2}};$$

(2) zero for all the remaining values of x.

(d) We call the value of the argument for which the probability density achieves its maximum the distribution mode. The arcsine law has no mode since the function

$$f(x) = \frac{1}{\pi\sqrt{a^2 - x^2}}$$

has no maxima.

We call the quantity $x_{0.5}$ the distribution median defined by the equality $F(x_{0.5}) = 1/2$.

Solving the equation

$$\frac{1}{2} + \frac{1}{\pi}\arcsin\frac{x_{0.5}}{\pi} = \frac{1}{2},$$

we find that $x_{0.5} = 0$.

Problems 11.1 to 11.8 are solved similarly.

Example 11.2 The probability density of a random variable is

$$f(x) = ax^2e^{-kx} \qquad (k > 0, 0 \leqslant x < \infty).$$

Find (a) the coefficient a, (b) the distribution function of the random variable X, (c) the probability that the random variable belongs to the interval $(0, 1/k)$.

SOLUTION. (a) The coefficient a is given by the equality

$$\int_0^\infty ax^2e^{-kx}\,dx = 1.$$

This implies that

$$a = \frac{1}{\int_0^\infty x^2e^{-kx}\,dx}.$$

Integrating by parts twice, we obtain

$$\int_0^\infty x^2e^{-kx}\,dx = \frac{2}{k^3}.$$

Consequently, $a = k^3/2$ and the probability density has the form

$$f(x) = \frac{k^3}{2}x^2e^{-kx}.$$

(b) The distribution function $F(x)$ of the random variable X is determined by the formula

$$F(x) = \int_0^x \frac{k^3}{2}x^2e^{-kx}\,dx = 1 - \frac{k^2x^2 + 2kx + 2}{2}e^{-kx}.$$

(c) The probability $\mathbf{P}(0 < X < 1/k)$ that the random variable X will assume values on the given interval is computed according to the formula

$$\mathbf{P}\left(0 < X < \frac{1}{k}\right) = F\left(\frac{1}{k}\right) = 1 - \frac{5}{2e} \approx 0.086.$$

Similarly one can solve Problems 11.9, 11.10 and 11.12.

Example 11.3 An electronic device has three parallel lines. The probability that each line fails to operate during the warranty period of the device is 0.1. Using the δ-function, express the probability density for a random number of lines that fail to operate during the warranty period if the failure of one line is independent of whether the other lines operate.

SOLUTION. Let us denote by X the random numbers of lines that fail. The random variable X is discrete, and its distribution series (Table 5) is

TABLE 5

x_k	0	1	2	3
p_k	0.729	0.243	0.027	0.001

Using the notion of probability density for a discrete variable, we obtain

$$f(x) = 0.729\delta(x) + 0.243\delta(x - 1) + 0.027\delta(x - 2) + 0.001\delta(x - 3).$$

Similarly we can solve Problem 11.15.

PROBLEMS

11.1 The distribution function of a uniformly distributed random variable X has the form

$$F(x) = \begin{cases} 0 & \text{for } x < 0, \\ x & \text{for } 0 \leqslant x \leqslant 1, \\ 1 & \text{for } x > 1. \end{cases}$$

Find the probability density of the random variable X.

11.2 Given the distribution function of a random variable

$$F(x) = \frac{1}{\sqrt{2\pi}} \int_{-\infty}^{x} \exp\left\{-\frac{t^2}{2}\right\} dt \quad \text{(normal distribution law)},$$

find the probability density of the random variable X.

11.3 Cramer (1946) gives the distribution function of the yearly incomes of persons who must pay income tax:

$$F(x) = \begin{cases} 1 - \left(\dfrac{x_0}{x}\right)^{\alpha} & \text{for } x \geqslant x_0, \\ 0 & \text{for } x < x_0 \end{cases} \qquad (\alpha > 0).$$

Find the yearly income that can be exceeded by a randomly selected taxpayer with the probability 0.5.

11.4 The distribution function of the random period during which a radio device operates without failures has the form

$$F(t) = 1 - \exp\left\{-\frac{t}{T}\right\} \qquad (t \geqslant 0).$$

Find (a) the probability that the device will operate without failures during a time period T, (b) the probability density $f(t)$.

11.5 The random variable representing the eccentricity of an item is characterized by the Rayleigh distribution

$$F(x) = 1 - \exp\left\{-\frac{x^2}{2\sigma^2}\right\} \qquad (x \geqslant 0).$$

Find (a) the mode of the distribution, (b) the median of the distribution, (c) the probability density $f(x)$.

11.6 The Weibull distribution function

$$F(x) = 1 - \exp\left\{-\frac{x^m}{x_0}\right\} \qquad (x \geqslant 0)$$

characterizes in a series of cases the lifetime of the elements of an electronic instrument.

Find (a) the probability density $f(x)$, (b) the quantile of order p of this distribution, (c) the mode of the distribution.

11.7 The random nonoperating period of a radio device has the probability density

$$f(x) = \frac{M}{x\sigma\sqrt{2\pi}} \exp\left\{-\frac{(\log x - \log x_0)^2}{2\sigma^2}\right\},$$

where $M = \log e = 0.4343\ldots$ (this is the logarithmic normal distribution law).

Find (a) the mode of the distribution for $x_0 = 1$ and $\sigma = \sqrt{5M}$, (b) the distribution function.

11.8 Given the distribution function of a random variable X: $F(x) = a + b \arctan(x/2)$ $(-\infty < x < +\infty)$ (the Cauchy probability law), determine (a) constants a and b, (b) the probability density, (c) $\mathbf{P}(\alpha \leqslant X < \beta)$.

11.9 How large should a be so that $f(x) = ae^{-x^2}$ is the probability density of a random variable X varying between infinite bounds?

11.10 For which value of a is the function

$$f(x) = \frac{a}{1 + x^2} \qquad (-\infty < x < +\infty)$$

equal to the probability density of a random variable X?

Find (a) the distribution function of the random variable X, (b) the probability that the random variable will fall in the interval $(-1, 1)$.

11.11 The scale of a stop watch has divisions of 0.2 seconds each. What is the probability that the error in the time estimate is larger than 0.05 seconds if the estimate is made with an accuracy of one division with a round-off to the nearest integer.

11.12 The azimuthal limb has divisions of 1° each. What is the probability that there will occur an error of $\pm 10'$ in the computation of the azimuth if the angle estimates are rounded off to the nearest degree?

11.13 It is known that the probability of failure for an electronic tube during Δx days is $k\Delta x$ with a precision of higher order of magnitude than Δx, and is independent of the number x of days during which the tube operates prior to the interval Δx. What is the probability of failure for a tube during l days?

11.14 A streetcar line has a length L. The probability that a passenger will get on the streetcar in the vicinity of a point x is proportional to $x(L - x)^2$, and the probability that a passenger who entered at point x will get off at point y is proportional to $(y - x)^h$, $h \geqslant 0$.

Find the probability that (a) the passenger will get on the streetcar before point z, (b) the passenger who got on the streetcar at point x will get off after point z.

11.15 Some devices are subjected to successive accelerated reliability tests that are terminated when the first failure occurs. Using the concept of probability density of a discrete random variable, find the probability density of a random number of devices tested if the probability of failure for each device is 0.5.

12. NUMERICAL CHARACTERISTICS OF DISCRETE RANDOM VARIABLES

Basic Formulas

The most frequently used characteristics of discrete random variables are the moments of these variables.

The moments m_k and the central moments μ_k of the kth order of discrete random variables are defined by the formulas:

$$m_k = \mathbf{M}[X^k] = \sum_{i=1}^{n} x_i^k p_i,$$

$$\mu_k = \mathbf{M}[(X - \bar{x})^k] = \sum_{i=1}^{n} (x_i - \bar{x})^k p_i,$$

in which $\mathbf{M}[X^k]$ is the expectation of X^k, x_i are the possible values of a random variable X, p_i the probabilities of these values and $\bar{x}$ is the expectation of X. Therefore, the first moment is determined by the formula

$$\bar{x} = \mathbf{M}[X] = \sum_{i=1}^{n} x_i p_i,$$

the second central moment or the variance is given by

$$\mathbf{D}[X] = \mathbf{M}[(X - \bar{x})^2] = \sum_{i=1}^{n} (x_i - \bar{x})^2 p_i$$

or by

$$\mathbf{D}[X] = \mathbf{M}[X^2] - (\mathbf{M}[X])^2.$$

The mean-square deviation σ is given by the relation

$$\sigma = +\sqrt{\mathbf{D}[X]}.$$

If the probabilities of different values of X depend on the disjoint or mutually exclusive events A_k, then the conditional expectation of X with the condition that A_k occurs is

$$\mathbf{M}[X \mid A_k] = \sum_{i=1}^{n} x_i \mathbf{P}(X = x_i \mid A_k).$$

If A_k ($k = 1, 2, \ldots, m$) form a complete set of events: that is, $\sum_{k=1}^{n} \mathbf{P}(A_k) = 1$, then the total expectation of X and the conditional expectation are related by the formula

$$\mathbf{M}[X] = \mathbf{M}\{\mathbf{M}[X \mid A_k]\} = \sum_{k=1}^{m} \mathbf{M}[X \mid A_k]\mathbf{P}(A_k).$$

In all the preceding formulas the number of terms in the sums can be infinite; in this case, for the existence of the expectation the sum must converge absolutely.

SOLUTION FOR TYPICAL EXAMPLES

Example 12.1 From a lot containing 100 items of which 10 are defective, a sample of five items is selected at random for quality control. Find the expectation for the number of defective items contained in the random sample.

SOLUTION. The random number of defective items contained in the sample has the following possible values:

$$x_1 = 0, \quad x_2 = 1, \quad x_3 = 2, \quad x_4 = 3, \quad x_5 = 4, \quad x_6 = 5.$$

The probability $p_i = \mathbf{P}(X = x_i)$ that X will assume a given value x_i is (see Example 10.1)

$$p_i = \frac{C_{10}^{i-1} C_{90}^{6-i}}{C_{100}^5} \qquad (i = 1, 2, 3, 4, 5, 6).$$

The required expectation is

$$\bar{x} = \sum_{i=1}^{6} (i-1) \frac{C_{10}^{i-1} C_{90}^{6-i}}{C_{100}^5} = \frac{1}{C_{100}^5} \sum_{j=0}^{5} j C_{10}^j C_{90}^{5-j}.$$

Since $\sum_{j=0}^5 C_{10}^j C_{90}^{5-j}$ is the coefficient of u^5 in the product $(1 + u)^{10}(1 + u)^{90}$, we see $\sum_{j=0}^5 j C_{10}^j C_{90}^{5-j}$ is the coefficient of u^5 in the expression

$$\frac{\partial}{\partial t}\{(1 + tu)^{10}(1 + u)^{10}(1 + u)^{90}\}\big|_{t=1} = 10u(1 + u)^{99}.$$

Consequently, we have

$$\sum_{j=0}^{5} j C_{10}^j C_{90}^{5-j} = 10 C_{99}^4, \quad \text{or} \quad \bar{x} = \frac{10 C_{99}^4}{C_{100}^5} = 0.5.$$

Similarly, one can solve Problems 12.1 and 12.2.

Example 12.2 A discrete random variable X is given by the distribution series $p_k = \mathbf{P}(X = k)$, $k = 1, 2, 3, \ldots$. Express the expectation of X in terms of the generating function $G(u)$ (see Section 9).

SOLUTION. By the definition of the expectation of a random variable:

$$\mathbf{M}[X] = \sum_{k=1}^{\infty} k p_k.$$

On the other hand, the value of the derivative of the generating function computed at $u = 1$ is

$$G'(1) = \frac{dG(u)}{du}\bigg|_{u=1} = \sum_{k=1}^{\infty} k p_k u^{k-1}\bigg|_{u=1} = \sum_{k=1}^{\infty} k p_k.$$

Consequently,

$$\mathbf{M}[X] = G'(1).$$

One can solve Problems 12.3 to 12.6 and 12.24 to 12.26 similarly.

Example 12.3 An experiment can be a success with the probability p and a failure with the probability $1 - p$.

The conditional probability $P(m)$ for achieving the desired result after m successful trials is

$$P(m) = 1 - \left(1 - \frac{1}{\omega}\right)^m \qquad (\omega > 1).$$

Find the expectation of the number of independent trials necessary for achieving the desired result.

SOLUTION. Let $P_n(A)$ denote the probability of achieving the desired result in n trials. If $P_{n,m}$ is the probability of exactly m successes out of a total of n trials, then, according to the formula for the total probability, we have

$$P_n(A) = \sum_{m=0}^{n} P_{n,m}P(m).$$

Since the trials are independent, and the probability of a successful outcome in each of them is p,

$$P_{n,m} = C_n^m p^m(1 - p)^{n-m}.$$

Substituting into the formula for $P_n(A)$ the values of $P_{n,m}$ and $P(m)$, we obtain

$$P_n(A) = \sum_{m=0}^{n} C_n^m p^m(1 - p)^{n-m}\left\{1 - \left(1 - \frac{1}{\omega}\right)^m\right\} = 1 - \left(1 - \frac{p}{\omega}\right)^n.$$

To attain the desired result, exactly n trials are necessary if it will be attained at the nth trial. The probability of the latter circumstance is $P_n(A) - P_{n-1}(A)$. Consequently, $\mathbf{M}[X]$, the expectation of the random number of trials necessary to attain the desired result, is:

$$\mathbf{M}[X] = \sum_{n=1}^{\infty} n[P_n(A) - P_{n-1}(A)] = \sum_{n=1}^{\infty} n\left\{\left(1 - \frac{p}{\omega}\right)^{n-1} - \left(1 - \frac{p}{\omega}\right)^n\right\}$$
$$= \frac{p}{\omega}\sum_{n=1}^{\infty} n\left(1 - \frac{p}{\omega}\right)^{n-1}.$$

To compute the last sum we make use of the equality

$$\sum_{n=1}^{\infty} nx^{n-1} = \frac{d}{dx}\sum_{n=0}^{\infty} x^n = \frac{d}{dx}\left[\frac{1}{1 - x}\right] = \frac{1}{(1 - x)^2},$$

valid for $|x| < 1$. Here, setting $x = 1 - p/\omega$, we obtain

$$\mathbf{M}[X] = \frac{p}{\omega}\frac{1}{\left[1 - \left(1 - \frac{p}{\omega}\right)\right]^2} = \frac{\omega}{p}.$$

Similarly Problems 12.10 to 12.15, 12.21 and 12.31 can be solved.

Example 12.4 A device has n fuses. In the case of overload, one of the fuses burns out and is replaced by a new one. What is the expectation $\mathbf{M}[N]$ of the number of overloads N after which all the initial fuses of the device are replaced by new ones if one assumes that it is equally likely for all fuses (old or new) to burn out?

SOLUTION. Let us denote by $\mathbf{M}[N \mid k]$ the expectation of the number of overloads after which all the initial fuses will be replaced if k fuses have not yet been replaced.

To compute $\mathbf{M}[N \mid k]$ we use the formula for the total expectation. If k fuses ($k \geqslant 1$) remain nonreplaced, then, in order that one of them burns out, a subsequent overload is necessary. The average number of overloads necessary for a remaining fuse to burn out will depend on the result of the subsequent overload.

In the subsequent overload there can occur two events:

A_1 that one of the initial fuses burns out with the probability $\mathbf{P}(A_1) = k/n$;

A_2 that one of the replaced fuses burns out with the probability $\mathbf{P}(A_2) = 1 - k/n$.

If at the subsequent overload A_1 occurs, then the expectation of the number of overloads necessary for the replacement of all k fuses that have not been replaced before this overload is $1 + \mathbf{M}[N \mid k - 1]$. If at the subsequent overload A_2 occurs, then the expectation equals $1 + \mathbf{M}[N \mid k]$. Using the formula for the total expectation, we find

$$\begin{aligned}\mathbf{M}[N \mid k] &= \frac{k}{n}\{1 + \mathbf{M}[N \mid k - 1]\} + \left(1 - \frac{k}{n}\right)\{1 + \mathbf{M}[N \mid k]\}\\ &= 1 + \frac{k}{n}\mathbf{M}[N \mid k - 1] + \frac{n-k}{n}\mathbf{M}[N \mid k]\end{aligned}$$

or after simple transformations,

$$\mathbf{M}[N \mid k] - \mathbf{M}[N \mid k - 1] = \frac{n}{k}.$$

If $k = 1$, that is, only one fuse has not been replaced, the probability of its replacement equals $1/n$. Therefore, according to Example 12.3, we shall have

$$\mathbf{M}[N \mid 1] = n.$$

Thus, we have a chain of equalities

$$\begin{aligned}\mathbf{M}[N \mid n] - \mathbf{M}[N \mid n - 1] &= \frac{n}{n},\\ \mathbf{M}[N \mid n - 1] - \mathbf{M}[N \mid n - 2] &= \frac{n}{n-1},\\ &\cdots\\ \mathbf{M}[N \mid 3] - \mathbf{M}[N \mid 2] &= \frac{n}{3},\\ \mathbf{M}[N \mid 2] - \mathbf{M}[N \mid 1] &= \frac{n}{2},\\ \mathbf{M}[N \mid 1] &= n,\end{aligned}$$

whose sum gives

$$\mathbf{M}[N \mid n] = \frac{n}{n} + \frac{n}{n-1} + \frac{n}{n-2} + \cdots + \frac{n}{3} + \frac{n}{2} + n$$

or

$$\mathbf{M}[N] = \mathbf{M}[N \mid n] = n\left(1 + \frac{1}{2} + \frac{1}{3} + \cdots + \frac{1}{n-1} + \frac{1}{n}\right).$$

Problems 12.16, 12.20, 12.22 and 12.23 can be solved in a similar manner.

Example 12.5 As a result of experiments with two devices A and B, one finds the probability of observing a noise whose level is evaluated in a three-point system (see Table 6).

TABLE 6

Noise level		1	2	3
The probability of observing a noise of a given level	Device A Device B	0.20 0.06	0.06 0.04	0.04 0.10

Using the data from Table 6, select the better device, i.e., the device with lower noise level.

SOLUTION. Let X denote the random noise level. The average noise level for the device A is

$$\mathbf{M}_A[X] = 0.20 \cdot 1 + 0.06 \cdot 2 + 0.04 \cdot 3 = 0.44 \text{ db.}$$

For the device B:

$$\mathbf{M}_B[X] = 0.06 \cdot 1 + 0.04 \cdot 2 + 0.10 \cdot 3 = 0.44 \text{ db.}$$

Thus, compared according to the average number of points, both devices are equivalent.

As an additional criterion for comparison, we use the mean-square deviation of the noise level.

$$\sigma_A = \sqrt{\mathbf{D}_A[X]} = \sqrt{\mathbf{M}_A[X^2] - (\bar{x}_A)^2} = \sqrt{0.80 - 0.44^2} \approx 0.78 \text{ db.},$$

$$\sigma_B = \sqrt{\mathbf{D}_B[X]} = \sqrt{\mathbf{M}_B[X^2] - (\bar{x}_B)^2} = \sqrt{1.12 - 0.44^2} \approx 0.96 \text{ db.}$$

Hence A gives a more stable indication with respect to the means and, consequently, it is better than B.

PROBLEMS

12.1 Find the expectation of the number of devices failing in reliability tests if in each test only one device is tested and the probability of its failure is p.

12.2 Assuming that the mass of a body can take, with equal probability, any integral number of grams on the interval 1 to 10, determine for which of the three sets of weights (a) 1, 2, 2, 5, 10; (b) 1, 2, 3, 4, 10; (c) 1, 1, 2, 5, 10, the average number of necessary weights will be minimum, if one can place weights only on one scale and the selection of weights is made to minimize the number used in the process of weighing.

12.3 A certain device consisting of five elements is tested. The probability that an element numbered i fails to operate is

$$p_i = 0.2 + 0.1(i - 1).$$

Find the expectation and the variance of the number of elements that stop if the failures of the elements are independent.

12.4 Three devices are tested independently. The probabilities of failure for each device are p_1, p_2 and p_3, respectively. Prove that the expectation of the number of devices failing to operate is $p_1 + p_2 + p_3$.

12.5 Determine the expectation of the number of devices failing to operate during a test period if the probability of failure for all devices is p and the number of devices that are tested is n.

12.6 A lottery distributes m_1 prizes worth k_1, m_2; $k_2, \ldots, m_n$; and k_n. The total number of tickets is N. What should be the cost of a ticket so that the expectation of a prize per ticket is equal to half its cost.

12.7 The first player tosses three fair coins and the second, two fair coins. The winner, who gets all five coins, is the one who scores more heads. In the case of a tie the game is repeated until there is a decisive result. What is the expectation of winning for each of the players?

12.8 Three persons A, B and C play a game as follows: two participate in each game; the loser cedes his place to the third person; the first game is played by A and B. The probability of winning each game is 1/2 for each player. They continue to play until one of them wins two games in succession and gets m dollars. What is the expectation of a gain for each of the players (a) after the first game under the assumption that A won it, (b) at the beginning of the game?

12.9 Three persons A, B and C play a game as follows: two players participate in each game; the winner cedes his place to the third person; first A plays with B. The probability of winning each game is 1/2 for each player. They continue to play until one of them wins two consecutive times and gets a sum of money equal to the number of all games played. What is the expectation of winning for A and C at the beginning of the game?

12.10 An automatic line in a state of normal adjustment can produce a defective item with probability p. The readjustment of the line is made immediately after the first defective item has been produced. Find the average number of items produced between two readjustments of the line.

12.11 The probability that a call signal emitted by one radio station is received by another is 0.2 at each emission. The call signals are emitted every five seconds until an answer signal is received. The total passage time for the call and answer signals is 16 seconds. Find the average number of call signals emitted before a two-way connection has been established.

12.12 Find the expectation and the variance of the number of items produced between two readjustments in a production line in normal adjustment if in the state of normal adjustment, the probability of a defective item is p, and the readjustment is made after the kth defective item has been produced.

12.13 The conditional probability that a device stops, computed under the assumption that m elements fail to operate, has the form:

(a) for the device A

$$P(m) = 1 - e^{-\alpha m} \qquad (\alpha > 0;\ m = 0, 1, 2, \ldots);$$

(b) for the device B

$$P(m) = \begin{cases} 0 & \text{for } m = 0, \\ 1 - e^{-\alpha(m-1)} & \text{for } m = 1, 2, 3, \ldots. \end{cases}$$

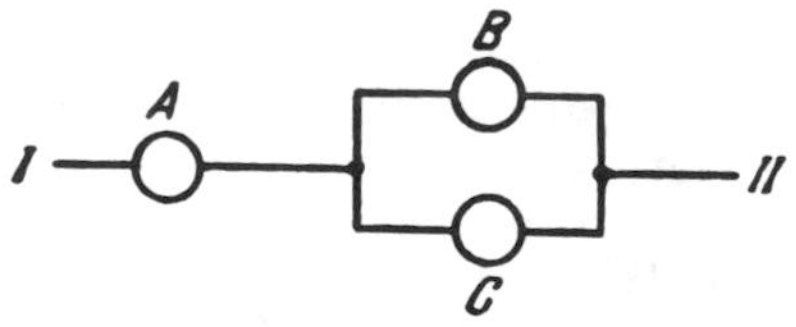

FIGURE 12

Find the expectation of the number of nonoperating elements that lead to stops of the devices A and B.

12.14 A blocking scheme consisting of the relay A connected in series with two relays B and C, which are connected in parallel, must ensure the closing of the circuit between the terminals I and II (Figure 12). As a result of damage, the relay A can stop with the probability 0.18, and the relays B and C with equal probabilities 0.22. Find the average number of times that the scheme is turned on until the first failure occurs.

12.15 A certain device contains the elements A, B and C, which can be affected by cosmic radiation and stop operating if at least one particle hits them. The stoppage of the device occurs in the case of failure of the element A or a simultaneous failure of the elements B and C. Find the expectation of the number of particles that caused the stoppage of the device if the conditional probabilities that a particle reaching the device hits the elements A, B and C are 0.1, 0.2 and 0.2, respectively.

12.16 A certain device has n elements of type A and m elements of type B. If one element of type A ceases to operate, it is not replaced and the device continues to operate until there remains at least one nondefective element of type A. The elements of type B are replaced repeatedly if they fail so that the number of nondefective elements of type B remains constant in the scheme. The failures of each of the nondefective elements of the device are equally probable. Determine the average number of element failures leading to a total stoppage of the device, i.e., to nonoperation of all the n elements of type A.

12.17 Prove that the variance of the number of occurrences of an event in the case of a single experiment does not exceed 1/4.

12.18 Find the conditions under which the third central moment of the binomial distribution is zero.

12.19 The distribution function of a random variable X is given by the equality

$$F(x) = \sum_{m<x} C_n^m p^m (1 - p)^{n-m}.$$

Prove that if $\lim_{n\to\infty} np = a$, then $\lim_{n\to\infty} \mathbf{D}[X] = a$.

12.20 Ten balls are drawn in succession from an urn containing a very large number of white and black balls mixed in equal proportion. The balls drawn before the first black ball occurs are returned to the urn; the first black ball that appears, together with all those that follow, is placed in another urn, which is initially empty. Find the expectation of the number of black and white balls in the second urn.

Solve the same problem under the assertion that the number n of balls drawn is random and obeys Poisson's law with parameter $a = 10$; that is,

$$P(n = k) = \frac{a^k}{k!} e^{-a}.$$

12.21 A game consists of tossing a fair coin until heads shows up. If heads appears at the kth tossing, player A gets k dollars from player B. How many dollars should A pay to B before the game starts so that the expectation of loss for each player is zero (i.e., the game is "fair")?

12.22 A motor transport column can arrive at a service station at any instant of time. If n repairmen are scheduled on duty by method A, the average number of cars serviced equals np. If they are scheduled by method B, the number $n[1 - (1 - p)^2]$ will be serviced if the column arrives during the first two quarters of 24 hours, np if the column arrives during the third quarter of 24 hours, and $0.5\ np$ if the column arrives during the last quarter of 24 hours.

For what values of p should one prefer the scheduling by method B?

12.23 A repairman services n one-type machines, which are in a row at a distance a apart from one another. After finishing the repair on one machine, he moves on to the machine that needs service before all the others. Assuming that malfunctions of all machines are equally probable, compute the average distance this repairman moves.

12.24 A random variable X may assume positive integral values with probabilities decreasing in a geometric progression. Select the first term and the ratio of the progression so that the expectation of X is 10, and under this assumption compute the probability P_{10} that $X \leqslant 10$.

12.25 A random variable X can assume any integral positive value n with a probability proportional to $1/3^n$. Find the expectation of X.

12.26 An experiment is organized so that a random variable X assumes the value $1/n$ with the probability $1/n$, where n is any positive integer. Find $\mathbf{M}[X]$.

12.27 A game consists of repeated independent trials in which the event A can occur with the probability p. If A occurs in $n > 0$ consecutive trials, and does not occur at the $(n + 1)$st trial, the first player gets y^n dollars from the second player. If $n = 0$, the first player pays one dollar to the second. Determine the quantity y under the assumption that the game will be "fair," i.e., the expectation of a gain for both players is 0. Consider the case when $p = 1/13$.

12.28 Balls are drawn from a box containing m white and n black balls until a white ball appears. Find the expectation of the number of balls drawn and its variance, if each ball is returned to the box after each draw.

12.29 Consider two boxes with white and black balls: the first contains M white balls out of a total of N, and the second contains M_1 white balls out of a total of N_1 balls. An experiment consists of a simultaneous random drawing of one ball from each box and transfer to the other box, after which the balls are mixed. Determine the expectation of the number of white balls in the first box after a given number of k trials. Consider the case when $k \to \infty$.

12.30 Communication with a floating research station is maintained by n radio stations. The station that enters in a two-way connection is the one that first receives the call signals from the floating station, and the occurrence of this event is equally probable for each of the radio stations ($p = 1/n$). The floating research station will communicate m times. Determine the probability that radio station No. 1 will be involved k times. Find the expectation and the variance of the number of times radio station No. 1 communicates.

12.31 The independent trials of a device are repeated until a stop occurs. The probability p of a stop is the same for each trial. Find the expectation and the variance of the number of trials before stop.

12.32 Two persons toss a coin in turn until both get the same number of heads. The probability that after $2n$ tossings both will have an equal number of heads is

$$p_n = \frac{(2n - 2)!}{2^{2n-1}n!\,(n - 1)!}.$$

Determine the expectation of the number of tosses.

13. NUMERICAL CHARACTERISTICS OF CONTINUOUS RANDOM VARIABLES

Basic Formulas

The expectation $\bar{x} = \mathbf{M}[X]$ and the variance $\mathbf{D}[X]$ of a random variable X with the probability density $f(x)$ can be computed by the formulas

$$\bar{x} = \mathbf{M}[X] = \int_{-\infty}^{+\infty} xf(x)\,dx,$$

$$\mathbf{D}[X] = \int_{-\infty}^{+\infty} (x - \bar{x})^2 f(x)\,dx.$$

In the first case it is assumed that the integral converges absolutely.

The expectation and the variance of continuous random variables have the same properties as the analogous quantities for discrete random variables. The mean-square or standard deviation σ is defined by the formula

$$\sigma = +\sqrt{\mathbf{D}[X]}.$$

For a symmetric distribution law one may define as a dispersion characteristic of a random variable the mean deviation E, determined by the condition

$$\mathbf{P}(|X - \bar{x}| < E) = \frac{1}{2}.$$

The moment of kth order m_k and the central moment of kth order μ_k can be computed according to the formulas

$$m_k = \int_{-\infty}^{+\infty} x^k f(x)\,dx,$$

$$\mu_k = \int_{-\infty}^{+\infty} (x - \bar{x})^k f(x)\,dx.$$

Solution for Typical Examples

Example 13.1 The probability density for the random rolling amplitudes of a ship has the form (Rayleigh's law)

$$f(x) = \frac{x}{a^2} \exp\left\{-\frac{x^2}{2a^2}\right\} \qquad (x \geqslant 0).$$

Determine (a) the expectation $\mathbf{M}[X]$, (b) the variance $\mathbf{D}[X]$, and the mean-square deviation σ, (c) the central moments of third and fourth order μ_3 and μ_4.

SOLUTION. The computation of the moments reduces to the evaluation of integrals of the form

$$J_n = \int_0^\infty t^n e^{-t^2}\, dt \qquad (n > 0 \text{ is an integer})$$

which for even n are

$$J_{2k} = \frac{1}{2}\Gamma\left(k + \frac{1}{2}\right) = \frac{(2k-1)!!}{2^{k+1}}\sqrt{\pi},$$

where

$$(2k-1)!! = (2k-1)(2k-3)(2k-5)\cdots 7\cdot 5\cdot 3\cdot 1,$$

and for odd n

$$J_{2k+1} = \frac{1}{2}\Gamma(k+1) = \frac{k!}{2}.$$

(a) The expectation of a random rolling amplitude is

$$\bar{x} = \mathbf{M}[X] = \int_0^\infty x f(x)\, dx = \frac{1}{a^2}\int_0^\infty x^2 \exp\left\{-\frac{x^2}{2a^2}\right\} dx.$$

Performing the substitution $x/(a\sqrt{2}) = t$, we obtain

$$\mathbf{M}[X] = 2\sqrt{2}a\int_0^\infty t^2 e^{-t^2}\, dt = 2\sqrt{2}aJ_2 = 2\sqrt{2}\,\frac{\sqrt{\pi}}{4}\, a = a\sqrt{\frac{\pi}{2}}.$$

Thus,

$$\bar{x} = \mathbf{M}[X] = a\sqrt{\frac{\pi}{2}}.$$

(b) Since

$$\sigma_x^2 = \mathbf{D}[X] = \mathbf{M}[X^2] - (x)^2 = 4a^2J_3 - \frac{\pi}{2}a^2 = a^2\left(2 - \frac{\pi}{2}\right).$$

then

$$\sigma_x = a\sqrt{2 - \frac{\pi}{2}}.$$

(c) $$\mu_3 = \mathbf{M}[(X - \bar{x})^3] = m_3 - 3\bar{x}m_2 + 2(\bar{x})^3,$$

where $m_3 = 4\sqrt{2}a^3J_4 = 3a^3\sqrt{\pi/2}$.

Consequently,

$$\mu_3 = a^3(\pi - 3)\sqrt{\frac{\pi}{2}},$$

$$\mu_4 = \mathbf{M}[(X - \bar{x})^4] = m_4 - 4\bar{x}m_3 + 6\bar{x}^2m_2 - 3\bar{x}^4,$$

where $m_4 = 8a^4J_5 = 8a^4$. Hence

$$\mu_4 = a^4\left(8 - \frac{3}{4}\pi^2\right).$$

Similarly one can solve Problems 13.1 to 13.13, 13.22 and 13.23.

Example 13.2 Find the mean deviation of a random variable whose probability density (the Laplace density) has the form

$$f(x) = \frac{1}{2}e^{-|x|}.$$

SOLUTION. Since the probability density is symmetric with respect to zero, it follows that $\bar{x} = 0$. The mean deviation E is computed according to the formula

$$\frac{1}{2} = \mathbf{P}(|X - \bar{x}| < E) = \int_{-E}^{E} \frac{1}{2}e^{-|x|}\,dx = \int_0^E e^{-x}\,dx = 1 - e^{-E}.$$

From this it follows that $E = \ln 2 = 0.6931$.

In a similar way Problems 13.1 and 13.4 can be solved.

PROBLEMS

13.1 The probability density of a random variable X has the form

$$f(x) = \begin{cases} \dfrac{1}{2l} & \text{for} \quad |x - a| \leqslant l, \\ 0 & \text{for} \quad |x - a| > l. \end{cases}$$

Determine (a) $\mathbf{M}[X]$ and (b) $\mathbf{D}[X]$; (c) find the relation between the mean-square and mean deviations of X.

13.2 The distribution function of a random variable X has the form

$$F(x) = \begin{cases} 0 & \text{for} \quad x \leqslant -1, \\ a + b \arcsin x & \text{for} \quad -1 \leqslant x \leqslant 1, \\ 1 & \text{for} \quad 1 \leqslant x. \end{cases}$$

Find the constants a and b. Compute $\mathbf{M}[X]$ and $\mathbf{D}[X]$.

13.3 Determine the expectation and the variance of a random variable X if the probability density is

$$f(x) = \begin{cases} \dfrac{2\rho}{E\sqrt{\pi}} \exp\left\{-\rho^2 \dfrac{x^2}{E^2}\right\} & \text{for} \quad x \geqslant 0, \\ 0 & \text{for} \quad x < 0. \end{cases}$$

13.4 The probability density of a random variable X has the form (the arcsine law)

$$f(x) = \frac{1}{\pi\sqrt{a^2 - x^2}} \qquad (-a \leqslant x \leqslant a).$$

Determine the variance and the mean deviation.

13.5 The probability density of the random rolling amplitudes of a ship is given by the formula (Rayleigh's law)

$$f(a) = \frac{a}{\sigma^2} \exp\left\{-\frac{a^2}{2\sigma^2}\right\},$$

in which σ^2 is the variance of the angle of heel.

Are the amplitudes smaller and greater than the average encountered with the same frequency?

13.6 The velocities of the molecules of a gas have the probability density (Maxwell's law)

$$f(v) = Av^2 e^{-h^2 v^2} \qquad (v \geqslant 0).$$

Find the expectation and the variance of the velocity of the molecules and also the magnitude of A for given h.

13.7 The probability density of a random variable X is given in the form

$$f(x) = \begin{cases} 0 & \text{for } x < 0, \\ \dfrac{x^m}{m!} e^{-x} & \text{for } x \geqslant 0. \end{cases}$$

Find $\mathbf{M}[X]$ and $\mathbf{D}[X]$.

13.8 Find the expectation and the variance of a random variable whose probability density has the form

$$F(x) = \begin{cases} 1 - \dfrac{x_0^3}{x^3} & \text{for } x \geqslant x_0, \qquad (x_0 > 0). \\ 0 & \text{for } x \leqslant x_0 \end{cases}$$

Find $\mathbf{M}[X]$ and $\mathbf{D}[X]$.

13.9 Find the expectation and the variance of a random variable whose probability density has the form (the Laplace density)

$$f(x) = \frac{1}{2} e^{-|x|}.$$

13.10 A random variable X has the probability density (the beta-density)

$$f(x) = \begin{cases} Ax^{\alpha} \exp\left\{-\dfrac{x}{\beta}\right\} & \text{for } x \geqslant 0 \qquad (\alpha > -1;\ \beta > 0), \\ 0 & \text{for } x < 0. \end{cases}$$

Determine the parameter A, the expectation and the variance of the random variable X.

13.11 A random variable X has the probability density (beta-density)

$$f(x) = \begin{cases} Ax^{a-1}(1-x)^{b-1} & \text{for } 0 \leqslant x \leqslant 1 \qquad (a > 0;\ b > 0), \\ 0 & \text{for } x < 0 \text{ and } x > 1. \end{cases}$$

Find the parameter A, the expectation and the variance of the random variable X.

13.12 A random variable X has the probability density

$$f(x) = A(1 + x^2)^{-(n+1)/2},$$

where $n > 1$ is a positive integer. Determine the constant A, the expectation and the variance of the random variable X.

13.13 The probability density of a nonnegative random variable X has the form

$$f(x) = Ax^{n-2} \exp\left\{-\frac{x^2}{2}\right\},$$

in which $n > 1$.

Find A, the expectation and the variance of X.

13.14 Prove that if the conditions

$$\lim_{x \to -\infty} [xF(x)] = 0 \quad \text{and} \quad \lim_{x \to \infty} \{x[1 - F(x)]\} = 0$$

are satisfied, then for the expectation of a random variable the following equality holds true:

$$\mathbf{M}[X] = \int_0^\infty [1 - F(x)]\, dx - \int_{-\infty}^0 F(x)\, dx.$$

13.15 The probability of finding a sunken ship during a search time t is given by the formula

$$p(t) = 1 - e^{-\gamma t} \qquad (\gamma > 0).$$

Determine the average time of search necessary to find the ship.

13.16 Find the expectation $m(t)$ of a mass of radioactive substance after time t if initially the mass of the substance was m_0, and the probability of nuclear disintegration of any atom per unit time is a constant p.

13.17 Find the half-life of a radioactive substance if the probability of nuclear disintegration of any atom per unit time is a constant p. (The half-life period T_n is defined as the instant when the mass of the radioactive substance is one-half its initial value.)

13.18 The processing of the results obtained in a census has shown that the differential distribution law of the ages of persons involved in research can be represented by the formula

$$f(t) = k(t - 22.5)(97.5 - t)^5 \qquad (t \text{ is time in years, } 22.5 \leqslant t \leqslant 97.5).$$

Determine how many times the number of scientific workers under the average age exceeds those above the average.

13.19 Determine for Student's distribution given by the probability density

$$f_n(x) = \frac{1}{\sqrt{n\pi}} \frac{\Gamma\left(\frac{n+1}{2}\right)}{\Gamma\left(\frac{n}{2}\right)} \left(1 + \frac{x^2}{n}\right)^{-(n+1)/2},$$

the moments m_k for $k < n$.

13.20 A random variable X obeys the beta-density; i.e., it has the probability density

$$f(x) = \frac{\Gamma(p+q)}{\Gamma(p)\Gamma(q)} x^{p-1}(1-x)^{q-1} \qquad (0 < x < 1;\ p > 0;\ q > 0).$$

Find the moment of kth order.

13.21 Find the expectation and the variance of a random variable having the probability density $2/\pi \cos^2 x$ on the interval $(-\pi/2, \pi/2)$.

13.22 Express the central moment μ_k in terms of the moments.

13.23 Express the moment m_k in terms of the central moments and the expectation $\bar{x}$.

14. POISSON'S LAW

Basic Formulas

The distribution series of a random variable X has the form

$$\mathbf{P}(X = m) = P_m = P(m, a) = \frac{a^m}{m!} e^{-a},$$

in which $a = \mathbf{M}[X]$, is called the Poisson distribution law.

Poisson's law can approximately replace the binomial distribution in the case when the probability p of occurrence of an event A in each trial is small and the number n of trials is large. In such a case, the approximate equality

$$P_{n,m} = C_n^m p^m (1-p)^{n-m} \approx \frac{a^m}{m!} e^{-a},$$

in which $a = np$, holds true.

SOLUTION FOR TYPICAL EXAMPLES

Example 14.1 A radio device consists of 1000 electronic elements. The probability of nonoperation for one element during one year of operation is 0.001 and is independent of the condition of the other elements. What is the probability that at least two elements will fail to operate during a year?

SOLUTION. Assuming that the random number X of nonoperating elements obeys Poisson's law

$$\mathbf{P}(X = m) = P_m = \frac{a^m}{m!} e^{-a},$$

where $a = np = 1000 \cdot 0.001 = 1$, we obtain the following:

(1) the probability that exactly two elements fail to operate is

$$\mathbf{P}(X = 2) = P_2 = \frac{a^2}{2!} e^{-a} = \frac{1}{2e} = 0.184;$$

(2) the probability that at least two elements fail to operate is

$$\mathbf{P}(X \geqslant 2) = \sum_{m=2}^{\infty} P_m = 1 - P_0 - P_1 = 1 - e^{-a}(1 + a)$$
$$= 1 - \frac{2}{e} \approx 0.264.$$

Similarly one can solve Problems 14.1 to 14.7.

Example 14.2 An explosion of a balloon during a reliability test generates 100 fragments that are uniformly distributed in a cone bounded by angles of 30° and 60° (Figure 13). Find the expectation and the variance of the number of fragments reaching 1 sq. m. of the surface of the sphere located inside the cone if the radius of the sphere is 50 m. and its center coincides with the point of explosion.

SOLUTION. Let a sphere of radius 50 m. intersect the cone formed by fragments and let us determine the expectation of the number of fragments passing through a unit area of the spherical zone formed by the intersection of the cone with the sphere. Let S denote the area of this zone:

$$S = 50^2 \int_0^{2\pi} \int_{\pi/6}^{\pi/3} \sin\theta \, d\theta \, d\varphi = 5000\pi\left(\cos\frac{\pi}{6} - \cos\frac{\pi}{3}\right)$$
$$= 2500\pi(\sqrt{3} - 1) \approx 5725 \text{ sq. m.}$$

Since the total number of fragments is $N = 100$, the expectation for a fragments passing through a unit area of the surface of the spherical zone will be

$$a = \frac{N}{S} = \frac{100}{5725} = 0.01745 \text{ splinters.}$$

The probability that a given fragment will reach a given area $S_0 = 1$ sq. m. is small (it equals $S_0/S = 1.75 \cdot 10^{-4}$), therefore, one may consider that the

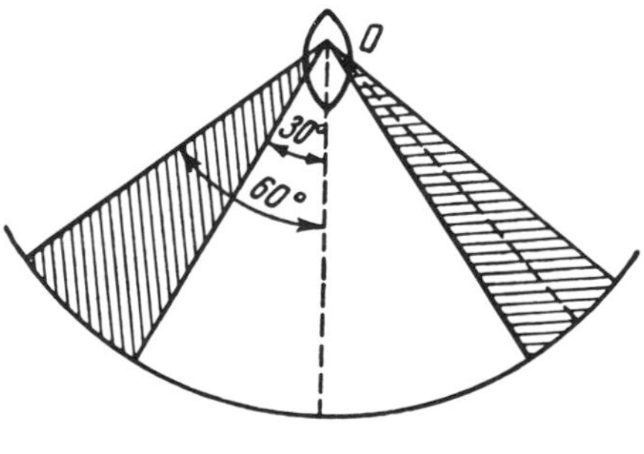

FIGURE 13

random number of fragments reaching 1 sq. m. of the surface of the sphere is distributed according to Poisson's law and, consequently, the following equality is valid:

$$\mathbf{D}[X] = \mathbf{M}[X] = a = 0.01745.$$

In a similar way one can solve Problems 14.10 and 14.12.

PROBLEMS

14.1 The expectation for the number of failures of a radio device during 10,000 hours of operation is 10. Find the probability that the device fails to operate during 100 hours.

14.2 The probability that any telephone subscriber calls the switchboard during one hour is 0.01. The telephone station services 300 subscribers. What is the probability that four subscribers will call the switchboard during one hour?

14.3 A device contains 2000 equally reliable elements with the probability of failure for each of them equal to $p = 0.0005$. What is the probability that the device will fail to operate if failure occurs when at least one element fails to operate?

14.4 A switchboard receives an average of 60 calls during one hour. What is the probability that during 30 seconds in which the operator is away there will be no calls?

14.5 The probability that an item will fail to pass a test is 0.001. Find the probability that from a total of 5000 items more than one item will fail. Compare the results obtained using Poisson's distribution with those obtained with the binomial distribution. In the latter, make use of logarithm tables with seven significant digits.

14.6 During a certain period of time the average number of connections to wrong calls per telephone subscriber is eight. What is the probability that for a preassigned subscriber the number of wrong connections will be greater than four?

14.7 Find the probability that among 200 items tested, more than three will turn out to be defective if the average percentage of defective items is 1 per cent.

14.8 The proofs of a 500-page book contain 500 misprints. Find the probability that there are at least three misprints per page.

14.9 In the observations made by Rutherford and Geiger, a radioactive substance emitted an average of 3.87 α-particles during 7.5 seconds. Find the probability that the substance will emit at least one α-particle per second.

14.10 Determine the asymmetry coefficient of a random variable distributed according to Poisson's law. (The asymmetry coefficient is the quotient $S_k = \mu_3/\sigma^3$.)

14.11 During its flight period, the instrument compartment of a space ship is reached by r elementary particles with the probability

$$P(r, \lambda) = \frac{\lambda^r}{r!} e^{-\lambda}.$$

The conditional probability for each particle to hit a preassigned unit equals p. Find the probability that this unit will be hit by (a) exactly k particles, (b) at least one particle.

14.12 Find the variance for the number of atoms (of a radioactive substance) that decay in a unit time if the mass of the substance is M, the half-life is T_p, the atomic weight is A and the number of atoms in a gram-atomic weight is N_0.[2]

14.13 Determine the probability that a screen of area $S = 0.12$ sq. cm. located at a distance $r = 5$ cm. perpendicular to the flow of α-particles emitted by a radioactive substance is hit during one second by (a) exactly 10 α-particles, (b) not less than two α-particles, if the half-life of the substance is $T_n = 4.4 \cdot 10^9$ years, the mass of the substance is $m = 0.1$ g., and the atomic weight is $A = 238$.[2]

14.14 Prove that the multinomial distribution

$$P_n(k_1, k_2, \ldots, k_m, k_{m+1}) = C_n^{k_1, k_1, \ldots, k_m} p_1^{k_1} p_2^{k_2} \cdots p_m^{k_m} p_{m+1}^{k_{m+1}},$$

in which

$$p_1 + p_2 + \cdots + p_m + p_{m+1} = 1,$$

and

$$k_1 + k_2 + \cdots + k_m + k_{m+1} = n,$$

can be approximated by the multidimensional Poisson law

$$e^{-(\lambda_1 + \lambda_2 + \cdots + \lambda_m)} \frac{\lambda_1^{k_1} \lambda_2^{k_2} \cdots \lambda_m^{k_m}}{k_1!\, k_2! \cdots k_m!},$$

in which $\lambda_i = np_i$, if all the probabilities p_i except for p_{m+1} are small and n is large.

15. THE NORMAL DISTRIBUTION LAW

Basic Formulas

The probability density of a normally distributed random variable has the form

$$f(x) = \frac{1}{\sigma\sqrt{2\pi}} \exp\left\{-\frac{(x - \bar{x})^2}{2\sigma^2}\right\}$$

or

$$f(x) = \frac{\rho}{E\sqrt{\pi}} \exp\left\{-\rho^2 \frac{(x - \bar{x})^2}{E^2}\right\}$$

in which σ is the mean-square deviation, $E = \rho\sqrt{2}\,\sigma$ is the mean deviation (sometimes also called "probable deviation"), and $\rho = 0.476936\ldots$.

---

[2] Ignore scattering and absorption of particles. Avogadro's number, $N_0 = 6.02 \times 10^{23}$, is the number of atoms in a quantity of the substance whose mass in grams equals its atomic weight. The half-life T_p is the time during which a mass of the substance decays to half the original mass.

The probability that a normally distributed random variable X assumes values on the interval (x_1, x_2) can be computed by using one of the following formulas:

$$\text{(1)}\quad \mathbf{P}(x_1 < X < x_2) = \frac{1}{2}\left[\Phi\left(\frac{x_2 - \bar{x}}{\sigma}\right) - \Phi\left(\frac{x_1 - \bar{x}}{\sigma}\right)\right],$$

in which

$$\Phi(x) = \frac{2}{\sqrt{2\pi}} \int_0^x \exp\left\{-\frac{t^2}{2}\right\} dt$$

is the Laplace function (probability integral);

$$\text{(2)}\quad \mathbf{P}(x_1 < X < x_2) = \frac{1}{2}\left[\hat{\Phi}\left(\frac{x_2 - \bar{x}}{E}\right) - \hat{\Phi}\left(\frac{x_1 - \bar{x}}{E}\right)\right],$$

in which

$$\hat{\Phi}(x) = \frac{2\rho}{\sqrt{\pi}} \int_0^x e^{-\rho^2 t^2}\, dt$$

is the normalized Laplace function.

The values of the functions $\Phi(x)$ and $\hat{\Phi}(x)$ are given in 8T and 11T in the table list on pages 471, 472.

Solution for Typical Examples

Example 15.1 The measurement of the distance to a certain object is accompanied by systematic and random errors. The systematic error equals 50 m. in the direction of decreasing distance. The random errors obey the normal distribution law with the mean-square deviation $\sigma = 100$ m. Find (1) the probability of measuring the distance with an error not exceeding 150 m. in absolute value, (2) the probability that the measured distance does not exceed the actual one.

Solution. Let X denote the total error made in measuring the distance. Its systematic component is $\bar{x} = -50$ m. Consequently, the probability density of the total errors has the form

$$f(x) = \frac{1}{100\sqrt{2\pi}} \exp\left\{-\frac{(x + 50)^2}{20{,}000}\right\}.$$

(1) According to the general formula, we have

$$\begin{aligned}\mathbf{P}(|X| < 150) &= \mathbf{P}(-150 < X < 150)\\ &= \frac{1}{2}\left[\Phi\left(\frac{150 + 50}{100}\right) - \Phi\left(\frac{-150 + 50}{100}\right)\right]\\ &= \frac{1}{2}[\Phi(2) - \Phi(-1)].\end{aligned}$$

The probability integral is an odd function and, hence,

$$\Phi(-1) = -\Phi(1).$$

From this we get

$$\mathbf{P}(|X| < 150) = \frac{1}{2}[\Phi(2) + \Phi(1)].$$

From 8T in the table list, we find

$$\Phi(2) = 0.9545, \qquad \Phi(1) = 0.6827;$$

and, finally,

$$\mathbf{P}(|X| < 150) = 0.8186.$$

(2) The probability that the measured distance will not exceed the actual one is

$$\mathbf{P}(-\infty < X < 0) = \frac{1}{2}[\Phi(0.5) + \Phi(\infty)].$$

Since $\Phi(\infty) = \lim_{x\to\infty} \Phi(x) = 1$, and from 8T in the table list, page 471, we find $\Phi(0.5) = 0.3829$, it follows that

$$\mathbf{P}(-\infty < X < 0) = 0.6914.$$

Similarly one can solve Problems 15.1 to 15.4 and 15.10 to 15.14.

Example 15.2 Determine the mean error of an instrument with no systematic errors, and whose random errors are distributed according to the normal law and fall with the probability 0.8 within the bounds ± 20 m.

SOLUTION. From the assumption of the problem, it follows that

$$\mathbf{P}(|X| \leqslant 20) = 0.8.$$

Since the probability density of the random errors is normal, and $\bar{x} = 0$ (the systematic error is absent), we have

$$\mathbf{P}(|X| \leqslant 20) = \frac{1}{2}\left[\hat{\Phi}\left(\frac{20}{E}\right) - \hat{\Phi}\left(-\frac{20}{E}\right)\right] = \hat{\Phi}\left(\frac{20}{E}\right).$$

The unknown value of the mean error is determined as the solution of the transcendental equation

$$\hat{\Phi}\left(\frac{20}{E}\right) = 0.8.$$

Using 11T in the table list on page 472, we find

$$\frac{20}{E} = 1.90,$$

from which it follows that

$$E = \frac{20}{1.90} = 10.5 \text{ m}.$$

In a similar way one can solve Problems 15.8 and 15.18.

PROBLEMS

15.1 A measuring instrument gives a systematic error of 5 m. and a mean error of 50 m. What is the probability that the error of a measurement will not exceed 5 m. in absolute value?

15.2 The systematic error in maintaining the altitude of an airplane is +20 m. and the random error is characterized by a mean deviation of 50 m. For a flight, the plane is assigned a corridor 100 m. high. What are the probabilities that the plane will fly below, inside and above the corridor if the plane is given an altitude corresponding to the midpoint of the corridor?

15.3 The mean error in distance measurements with a radar device is 25 m. Determine (a) the variance of the errors of the measurements, (b) the probability of obtaining errors not exceeding 20 m. in absolute value.

15.4 A measuring instrument has a mean error of 40 m. and no systematic errors. How many measurements should be performed so that in at least one of them, the error will not exceed 7.5 m. in absolute value with a probability greater than 0.9?

15.5 Given two random variables X and Y with equal variance, one being distributed normally and the other uniformly, find the correlation between their mean deviations.

15.6 A normally distributed random variable X has the expectation $\bar{x} = -15$ m. and the mean deviation 10 m. Compute the table for the distribution function for values of the argument increasing by 10 m. and plot the graph.

15.7 An altimeter gives random and systematic errors. The systematic error is +20 m. and the random errors obey the normal distribution law. What should be the mean error of the instrument so that the error in altitude measurement is less than 100 m. with the probability 0.9?

15.8 Find the relation between the arithmetic mean deviation

$$E_1 = \mathbf{M}[|X - \bar{x}|]$$

of a normally distributed random variable and its mean-square deviation.

15.9 For a normally distributed random variable X with $\mathbf{M}[X] = 0$ find (a) $\mathbf{P}(X \geqslant k\sigma)$, (b) $\mathbf{P}(|X| \geqslant k\sigma)$ (for $k = 1, 2, 3$).

15.10 The gunpowder charge of a shotgun is weighed on scales with a mean error of 100 mg. The nominal mass of the gunpowder charge is 2.3 g. Determine the probability of damaging the gun if the maximum admitted mass of the gunpowder charge is 2.5 g.

15.11 Two independent measurements are made with an instrument having a mean error of 20 m. and a systematic error of +10 m. What is the probability that both errors will occur with different signs exceeding 10 m. in absolute value?

15.12 Two parallel lines are drawn in the plane at the distance L. On this plane, a circle of radius R is dropped. The dispersion center is located at distance b outward from one of the parallels. The mean deviation of the center of the circle in the direction perpendicular to this parallel is E.

Determine for one throw (a) the probability that the circle will cover at least one of the parallels, (b) the probability that it will cover both parallels, if $L = 10$ m., $R = 8$ m., $b = 5$ m. and $E = 10$ m.

15.13 A product is considered to be of high quality if the deviation of its dimensions from the standards does not exceed 3.45 mm. in absolute value. The random deviations of its dimensions obey the normal distribution with a mean-square deviation of 3 mm.; systematic errors are absent. Determine the average number of products of high quality from a total of four items produced.

15.14 What should be the width of the tolerance field in order to obtain, with a probability at most 0.0027, an item whose size lies outside the tolerance field if the random deviations of the size from the mid-point of the tolerance field obey the normal distribution with parameters $\bar{x} = 0$ and $\sigma = 5\mu$?

15.15 What should be the distance between two fishing boats sailing on parallel routes so that the probability of sighting a school of fish moving between the boats in the same direction is 0.5, if the width of the strip of search for each boat is a normally distributed random variable with parameters $\bar{x} = 3.7$ km. and $E = 0.74$ km., and for different boats these quantities are independent?

15.16 In many measurements it has been established that 75 per cent of the errors (a) do not exceed $+1.25$ mm., (b) do not exceed 1.25 mm. in absolute value. Replacing the frequencies of occurrences of the errors by their probabilities, determine in both cases the mean deviation of the distribution law of the errors. Assume the distribution is normal with zero expectation.

15.17 The random deviation X of the size of an item from the standard obeys the normal law with the expectation $\bar{x}$ and the mean-square deviation σ. Nondefective items are considered to be those for which $a < X < b$. The items subjected to alteration are those for which $X > b$.

Find (a) the distribution function for the random deviations of the sizes of the items subject to alteration, (b) the distribution function for the random deviations of the sizes of nondefective items.

15.18 A normally distributed random variable X has a zero expectation. Determine the mean deviation E, for which the probability $\mathbf{P}(a < X < b)$ will be largest $(0 < a < b)$.

16. CHARACTERISTIC FUNCTIONS

Basic Formulas

The expectation of the function e^{iuX} (where u is a real variable, and $i = \sqrt{-1}$) is called the characteristic function $E(u)$ of a random variable X:

$$E(u) = \mathbf{M}[e^{iuX}].$$

For a continuous random variable we have

$$E(u) = \int_{-\infty}^{+\infty} e^{iux} f(x)\, dx,$$

where $f(x)$ is the probability density of the random variable X.

For a discrete random variable (and only for a discrete one)

$$E(u) = \sum_{k=1}^{n} p_k e^{iux_k},$$

in which x_k are the particular values of the random variable and $p_k = \mathbf{P}(X = x_k)$ are the probabilities that correspond to them.

If the moment m_k exists, then

$$m_k = \mathbf{M}[X^k] = \frac{1}{i^k} \frac{d^k E(u)}{du^k}\bigg|_{u=0}.$$

The probability density $f(x)$ is determined uniquely by the characteristic function:

$$f(x) = \frac{1}{2\pi} \int_{-\infty}^{+\infty} e^{-iux} E(u)\, du.$$

For discrete random variables the last formula gives the probability density in the form of a sum of δ-functions. There is a one-to-one correspondence between distribution functions and characteristic functions.

Solution for Typical Examples

Example 16.1 A lot of n items contains m defective items. A sample of r items is drawn from the lot for quality control ($m < r < n - m$). Find the characteristic function of the number of defective items contained in the sample.

Solution. The random variable X representing the number of defective items in the sample may assume all the integral values on the interval $(0, m)$. Let us denote

$$p_k = \mathbf{P}(X = k), \quad \text{where} \quad k = 0, 1, 2, \ldots, m.$$

Determining p_k as the ratio between the number of equally probable (unique) mutually exclusive results of the experiment and the total number of results, we find

$$p_k = \frac{C_m^k C_{n-m}^{r-k}}{C_n^r}.$$

Consequently, the characteristic function

$$E(u) = \sum_{k=0}^{m} \frac{C_m^k C_{n-m}^{r-k}}{C_n^r} e^{iku}.$$

Similarly one can solve Problems 16.1 to 16.5.

Example 16.2 Find the characteristic function of a random variable X with the probability density

$$f(x) = \frac{1}{2} e^{-|x|}.$$

Solution. Since the characteristic function is

$$E(u) = \int_{-\infty}^{+\infty} e^{iux} f(x)\, dx,$$

this leads to

$$E(u) = \frac{1}{2}\int_{-\infty}^{+\infty} e^{iux-|x|}\,dx$$
$$= \frac{1}{2}\int_{0}^{\infty} e^{(iu-1)x}\,dx + \frac{1}{2}\int_{-\infty}^{0} e^{(iu+1)x}\,dx$$
$$= \frac{1}{2}\left(\frac{1}{1-iu} - \frac{1}{1+iu}\right) = \frac{1}{1+u^2};$$

that is,

$$E(u) = \frac{1}{1+u^2}.$$

Problems 16.6 to 16.12 can be solved in a similar way.

Example 16.3 A random variable X has the characteristic function

$$E(u) = \frac{1}{1+u^2}.$$

Find the probability density of this random variable.

Solution. The probability density $f(x)$ is related to the characteristic function $E(u)$ by

$$f(x) = \frac{1}{2\pi}\int_{-\infty}^{+\infty} e^{-iux}E(u)\,du.$$

Substituting the value of $E(u)$, we obtain

$$f(x) = \frac{1}{2\pi}\int_{-\infty}^{+\infty} \frac{e^{-iux}}{1+u^2}\,du.$$

We shall consider u as the real part of the complex variable $w = u + iv$.

For $x < 0$ the integral over the real axis is the integral over a closed contour consisting of the real axis and the semicircle "of infinite radius" located in the upper half-plane (Figure 14); that is,

$$f(x) = \frac{1}{2\pi}\int_{-\infty}^{+\infty} \frac{e^{-iux}}{1+u^2}\,du = \frac{1}{2\pi}\oint \frac{e^{-ixw}}{1+w^2}\,dw.$$

By the theorem of residues

$$\oint \frac{e^{-ixw}}{1+w^2}\,dw = 2\pi i\left(\frac{e^{-ixw}}{2w}\right)_{w=i} = \pi e^{x},$$

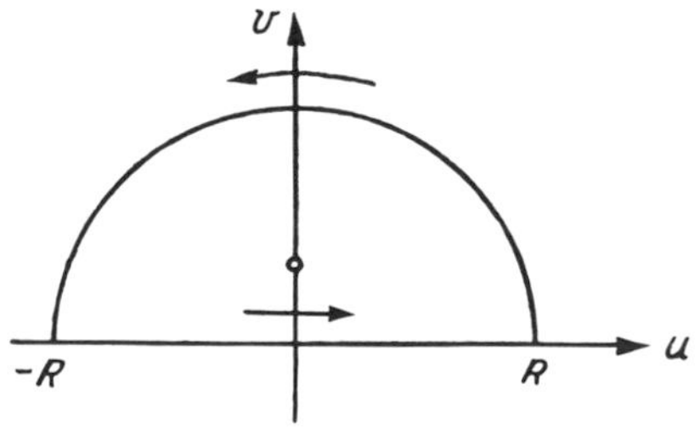

FIGURE 14

or, taking into account that $x < 0$, we have

$$f(x) = \frac{1}{2} e^{-|x|}.$$

Similarly, for $x > 0$

$$f(x) = \frac{1}{2\pi} \int_{-\infty}^{+\infty} \frac{e^{-iux}}{1 + u^2}\, du = \frac{1}{2\pi} \int_{-\infty}^{+\infty} \frac{e^{iu_1x}}{1 + u_1^2}\, du_1 = \frac{1}{2\pi} \oint \frac{e^{ixw}}{1 + w^2}\, dw,$$

where the integration is extended over the same contour (Figure 14).

According to the theorem of residues

$$\oint \frac{e^{ixw}}{1 + w^2}\, dw = 2\pi i \left(\frac{e^{ixw}}{2w}\right)_{w=i} = \pi e^{-x},$$

or, using the fact that $x > 0$, we have

$$f(x) = \frac{1}{2} e^{-|x|}.$$

Therefore, for any value of x

$$f(x) = \frac{1}{2} e^{-|x|}.$$

Similarly one can solve Problems 16.15 and 16.16.

Example 16.4 Find the moments of a random variable X whose characteristic function is $E(u) = 1/(1 + u^2)$.

Solution. The moments exist up to any order since all the derivatives of $E(u)$ are continuous at origin. Consequently,

$$m_k = \frac{1}{i^k} \left.\frac{d^k E(u)}{du^k}\right|_{u=0}.$$

We shall determine the derivatives

$$\left.\frac{d^k E(u)}{du^k}\right|_{u=0}$$

as the coefficients of $u^k/k!$ in the expansion of the function $1/(1 + u^2)$ in a Maclaurin series; that is, we shall use the equality

$$\frac{1}{1 + u^2} = \sum_{k=0}^{\infty} \left.\frac{d^k E(u)}{du^k}\right|_{u=0} \frac{u^k}{k!}.$$

On the other hand, the function $1/(1 + u^2)$ for $|u| < 1$ is the sum of the geometric progression

$$\frac{1}{1 + u^2} = \frac{1}{1 - (iu)^2} = \sum_{m=0}^{\infty} (iu)^{2m} = \sum_{m=0}^{\infty} i^{2m}(2m)! \frac{u^{2m}}{(2m)!}.$$

Thus, the Maclaurin series of the function $1/(1 + u^2)$ contains only even powers of u. It follows from this that

$$\left.\frac{d^kE(u)}{du^k}\right|_{u=0} = \begin{cases} i^k k! & \text{for } k \text{ even}, \\ 0 & \text{for } k \text{ odd}, \end{cases}$$

and the moments

$$m_k = \begin{cases} k! & \text{for } k \text{ even}, \\ 0 & \text{for } k \text{ odd}. \end{cases}$$

In a similar way one can solve Problems 16.3, 16.7, 16.8, 16.10 and 16.14.

PROBLEMS

16.1 Find the characteristic function of the number of occurrences of an event in one trial if its probability of occurrence in one trial is p.

16.2 Find the characteristic function of the number of occurrences of an event A in n independent trials if the probability of occurrence of A varies from one trial to another and equals p_k ($k = 1, 2, \ldots, n$) for the kth trial.

16.3 Determine the characteristic function of a discrete random variable X with a binomial distribution and also the corresponding $\mathbf{M}[X]$ and $\mathbf{D}[X]$.

16.4 Find the characteristic function of a discrete random variable X obeying Pascal's distribution law

$$\mathbf{P}(X = m) = \frac{a^m}{(1 + a)^{m+1}} \qquad (a > 0),$$

and the corresponding $\mathbf{M}[X]$ and $\mathbf{D}[X]$.

16.5 A discrete random variable X obeys Poisson's law

$$\mathbf{P}(X = m) = \frac{a^m}{m!} e^{-a}.$$

Find (a) the characteristic function $E(u)$, and (b) using $E(u)$, find $\mathbf{M}[X]$ and $\mathbf{D}[X]$.

16.6 Find the characteristic function of a normally distributed random variable with expectation $\bar{x}$ and variance σ^2.

16.7 Find the characteristic function and the moments of a random variable with the probability density

$$f(x) = \begin{cases} e^{-x} & \text{for } x \geqslant 0, \\ 0 & \text{for } x < 0. \end{cases}$$

16.8 Find the characteristic function and all the moments of a random variable uniformly distributed over the interval (a, b).

16.9 A random variable X has the probability density

$$f(x) = 2h^2 x e^{-h^2x^2} \qquad (x \geqslant 0).$$

Find its characteristic function.

16.10 A random variable X has the probability density

$$f(x) = \begin{cases} \dfrac{\alpha^\lambda}{\Gamma(\lambda)} x^{\lambda-1} e^{-\alpha x} & \text{for} \quad x > 0, \\ 0 & \text{for} \quad x < 0 \end{cases} \qquad (\alpha, \lambda > 0).$$

Find its characteristic function and moments.

16.11 Find the characteristic function of a random variable X whose probability density (the arcsine law) is

$$f(x) = \frac{1}{\pi\sqrt{a^2 - x^2}} \qquad (|x| \leqslant a).$$

16.12 Find the characteristic function of a random variable X obeying Cauchy's distribution law:

$$f(x) = \frac{a}{\pi} \frac{1}{(x - \bar{x})^2 + a^2}.$$

16.13 Using the expression

$$E(u) = \exp\left\{iu\bar{x} - \frac{u^2\sigma^2}{2}\right\}$$

for the characteristic function of the normal distribution law, determine the characteristic function of the random variable (a) $Y = aX + b$, (b) $\hat{X} = X - \bar{x}$.

16.14 Using the expression

$$E(u) = \exp\left\{-\frac{u^2\sigma^2}{2}\right\}$$

for the characteristic function of a centralized random variable X that obeys a normal distribution law, determine all its central moments.

16.15 The characteristic function of a random variable X is given in the form

$$E(u) = e^{-a|u|} \qquad (a > 0).$$

Determine the probability density of X.

16.16 Given the characteristic functions

$$E_1(u) = \frac{1 + iu}{1 + u^2}, \qquad E_2(u) = \frac{1 - iu}{1 + u^2},$$

determine the corresponding probability densities.

16.17 Given the characteristic function

$$E(u) = \frac{1}{2e^{-iu} - 1},$$

show that it corresponds to a discrete random variable. Find the distribution series of this variable.

17. THE COMPUTATION OF THE TOTAL PROBABILITY AND THE PROBABILITY DENSITY IN TERMS OF CONDITIONAL PROBABILITY

Basic Formulas

The total probability of an event A is given by the formula

$$\mathbf{P}(A) = \int_{-\infty}^{+\infty} f(x)\mathbf{P}(A \mid x)\, dx,$$

in which $f(x)$ is the probability density of the random variable X, on the values of which depends the probability of occurrence of A; $\mathbf{P}(A \mid x)$ is the probability of occurrence of the event A, computed under the assumption that the random variable x assumes the value x.

The conditional probability density $f(x \mid A)$ of a random variable X, i.e., the probability density under the assumption that A occurred, is determined by the formula (the generalized Bayes formula):

$$f(x \mid A) = \frac{f(x)\mathbf{P}(A \mid x)}{\int_{-\infty}^{+\infty} f(x)\mathbf{P}(A \mid x)\, dx},$$

in which $f(x)$ is the probability density, prior to the experiment, of the random variable X.

SOLUTION FOR TYPICAL EXAMPLES

Example 17.1 The probability of an event depends on the random variable X and can be expressed by the following formula:

$$\mathbf{P}(A \mid x) = \begin{cases} 1 - e^{-kx} & \text{for } x \geqslant 0, \\ 0 & \text{for } x < 0 \end{cases} \qquad (k > 0).$$

Find the total probability of the event A if X is a normally distributed random variable with expectation $\bar{x}$ and variance σ^2.

SOLUTION. The total probability of the event A is

$$\mathbf{P}(A) = \int_{-\infty}^{+\infty} f(x)\mathbf{P}(A \mid x)\, dx.$$

Substituting here the given probability density

$$f(x) = \frac{1}{\sigma\sqrt{2\pi}} \exp\left\{-\frac{(x - \bar{x})^2}{2\sigma^2}\right\},$$

we obtain

$$\begin{aligned} \mathbf{P}(A) &= \int_0^{\infty} \frac{1}{\sigma\sqrt{2\pi}} \exp\left\{-\frac{(x - \bar{x})^2}{2\sigma^2}\right\}(1 - e^{-kx})\, dx \\ &= \frac{1}{2}\left[1 + \Phi\left(\frac{\bar{x}}{\sigma}\right)\right] - \frac{1}{\sigma\sqrt{2\pi}} \int_0^{\infty} \exp\left\{-\frac{(x - \bar{x})^2}{2\sigma^2} - kx\right\} dx. \end{aligned}$$

The exponent of e in the last integral can be reduced to the form

$$-\frac{(x-\bar{x})^2}{2\sigma^2} - kx = -\frac{(x-\bar{x}+k\sigma^2)^2}{2\sigma^2} - k\left(\bar{x} - \frac{k\sigma^2}{2}\right),$$

Consequently,

$$\mathbf{P}(A) = \frac{1}{2}\left[1 + \Phi\left(\frac{x}{\sigma}\right)\right] - \frac{1}{\sigma\sqrt{2\pi}} \times \exp\left\{-k\left(\bar{x} - \frac{k\sigma^2}{2}\right)\right\} \int_0^\infty \exp\left\{-\frac{(x-\bar{x}+k\sigma^2)^2}{2\sigma^2}\right\} dx.$$

Since

$$\frac{1}{\sigma\sqrt{2\pi}} \int_0^\infty \exp\left\{-\frac{(x-\bar{x}+k\sigma^2)^2}{2\sigma^2}\right\} dx = \frac{1}{2}\left[1 + \Phi\left(\frac{\bar{x} - k\sigma^2}{\sigma}\right)\right],$$

then

$$\mathbf{P}(A) = \frac{1}{2}\left\{1 + \Phi\left(\frac{\bar{x}}{\sigma}\right) - \left[1 + \Phi\left(\frac{\bar{x} - k\sigma^2}{\sigma}\right)\right] \exp\left[-k\left(\bar{x} - \frac{k\sigma^2}{2}\right)\right]\right\}.$$

Similarly one can solve Problems 17.1 to 17.10.

Example 17.2 The deviation of the size of an item from the midpoint of the tolerance field of width $2d$ equals the sum of two random variables X and Y with probability densities

$$f(x) = \frac{1}{\sigma_x\sqrt{2\pi}} \exp\left\{-\frac{x^2}{2\sigma_x^2}\right\}$$

and

$$\varphi(y) = \frac{1}{\sigma_y\sqrt{2\pi}} \exp\left\{-\frac{y^2}{2\sigma_y^2}\right\}.$$

Determine the (conditional) probability density of the random variable X for the nondefective items if the distribution $\varphi(y)$ does not depend on the value assumed by X.

SOLUTION. Let A denote the event that an item produced turns out to be nondefective. The conditional probability $\mathbf{P}(A \mid x)$ of getting a nondefective item, under the assumption that the random variable X takes the value x, is

$$\mathbf{P}(A \mid x) = \int_{-x-d}^{-x+d} \frac{1}{\sigma_y\sqrt{2\pi}} \exp\left\{-\frac{y^2}{2\sigma_y^2}\right\} dy = \frac{1}{2}\left[\Phi\left(\frac{x+d}{\sigma_y}\right) - \Phi\left(\frac{x-d}{\sigma_y}\right)\right].$$

Let $f(x \mid A)$ be the conditional probability density of X for nondefective items so that

$$f(x \mid A) = \frac{f(x)\mathbf{P}(A \mid x)}{\int_{-\infty}^{+\infty} f(x)\mathbf{P}(A \mid x)\, dx}.$$

Substituting the values of $f(x)$ and $\mathbf{P}(A \mid x)$, we obtain

$$f(x \mid A) = \frac{\dfrac{1}{\sigma_x\sqrt{2\pi}} \exp\left\{-\dfrac{x^2}{2\sigma_x^2}\right\} \dfrac{1}{2}\left[\Phi\left(\dfrac{x+d}{\sigma_y}\right) - \Phi\left(\dfrac{x-d}{\sigma_y}\right)\right]}{\displaystyle\int_{-\infty}^{+\infty} \frac{1}{\sigma_x\sqrt{2\pi}} \exp\left\{-\frac{x^2}{2\sigma_x^2}\right\} \frac{1}{2}\left[\Phi\left(\frac{x+d}{\sigma_y}\right) - \Phi\left(\frac{x-d}{\sigma_y}\right)\right] dx}$$

or

$$f(x \mid A) = \frac{\dfrac{1}{2}\dfrac{1}{\sigma_x\sqrt{2\pi}} \exp\left\{-\dfrac{x^2}{2\sigma_x^2}\right\}\left[\Phi\left(\dfrac{x+d}{\sigma_y}\right) - \Phi\left(\dfrac{x-d}{\sigma_y}\right)\right]}{\Phi\left(\dfrac{d}{\sqrt{\sigma_x^2 + \sigma_y^2}}\right)}.$$

PROBLEMS

17.1 Suppose that a straight line is drawn in the plane and on it are marked points separated by the distance l. Determine the probability that at least one point will coincide with the center of a circle of diameter b and moving in the same plane so that its center describes a straight line intersecting the given line at an angle θ, equally probable over the interval (θ_1, θ_2). The angles θ_1 and θ_2 satisfy the conditions $\sin\theta_1 < b/l$ and $\sin\theta_2 > b/l$.)

17.2 On each of two parallel lines, points are taken independently at a constant interval $l = 100$ m. Determine the probability that at least one point will lie in an infinite strip of width $D = 25$ m., located in the same plane as the two parallels so that the lines that bound it are perpendicular to these parallels.

17.3 Find the probability of hitting a target in one trial if the distance to the target at the instant of the shot is a random variable uniformly distributed over the interval 100 to 200 m. and the conditional probability of hitting the target is $3000/D^2$, where D is expressed in meters.

17.4 On a shore of a bay of width $L = 30$ km. there is an observation station whose distance of observation is a normally distributed random variable with the expectation $\bar{x} = 20$ km. and mean deviation $E = 1$ km. A ship can pass with equal probability through the bay, while moving along the shore at any distance from the station. Find the probability that the observation station will discover the ship.

17.5 On the right pan of a balance a load is placed whose mass obeys the normal distribution law with parameters $\bar{x} = 20$ kg. and $E = 1$ kg. On the left pan another load is placed whose mass is equally probable within the bounds 0 to 50 kg. Determine the probability that the right pan will outweigh the left one. Compare the result with that obtained under the assumption that the load on the right pan is not random, but is exactly 20 kg.

17.6 Consider a number n of independent measurements of a normal random variable X whose expectation coincides with the origin of the reference system and with mean deviation R. Find the probability that the result of at least one measurement will deviate from the random variable Z by at most $\pm r$ if Z is uniformly distributed over the interval $(-l, l)$.

17.7 Given a sequence of random variables $X_1, X_2, \ldots, X_n$ with the same probability density $f(x)$, we call the random variable

$$W_n = X_{\max} - X_{\min},$$

in which $X_{\max}$ is the maximum and $X_{\min}$ the minimum of the obtained values X_j ($j = 1, 2, \ldots, n$), the range.

Find the distribution function of the range

$$F(w) = \mathbf{P}(W_n < w).$$

17.8 What is the probability that two points selected randomly in a circle will lie on one side of a chord parallel to a given direction and whose distance from the center is a uniformly distributed random variable?

17.9 The coordinates X_i of the random points $A_1, A_2, \ldots, A_n$ have the probability densities

$$f_i(x) \qquad (i = 1, 2, \ldots, n).$$

One of these n points coincides with a point A_0 whose deviation of coordinates from a given number has the probability density $f(x)$. Determine the probability that the point A will coincide with A_0.

17.10 A random variable X obeys Poisson's law

$$\mathbf{P}(X = m) = \frac{a^m}{m!} e^{-a},$$

whose parameter is unknown, but prior to the experiment the parameter has the probability density

$$f(a) = ae^{-a} \qquad (a > 0).$$

After the experiment, a random variable X assumes the value m_0. Find the probability density a after the experiment.

III SYSTEMS OF RANDOM VARIABLES

18. DISTRIBUTION LAWS AND NUMERICAL CHARACTERISTICS OF SYSTEMS OF RANDOM VARIABLES

Basic Formulas

The distribution function (integral distribution law) $F(x_1, x_2, \ldots, x_n)$ of a system of n random variables $(X_1, X_2, \ldots, X_n)$ is defined by the formula

$$F(x_1, x_2, \ldots, x_n) = \mathbf{P}(X_1 < x_1, X_2 < x_2, \ldots, X_n < x_n).$$

For a system of continuous random variables, there can exist a probability density (differential distribution law) defined by the formula

$$f(x_1, x_2, \ldots, x_n) = \frac{\partial^n F(x_1, x_2, \ldots, x_n)}{\partial x_1 \, \partial x_2 \cdots \partial x_n}.$$

A system of discrete random variables is characterized by the set of probabilities $\mathbf{P}(X_1 = i_1, X_2 = i_2, \ldots, X_n = i_n)$, which can be reduced to a table with n rows (according to the number of random variables).

The distribution function for continuous random variables can be expressed in the form of a multiple integral

$$F(x_1, x_2, \ldots, x_n) = \int_{-\infty}^{x_1} \int_{-\infty}^{x_2} \cdots \int_{-\infty}^{x_n} f(x_1, x_2, \ldots, x_n) \, dx_1 \, dx_2 \cdots dx_n,$$

and for discrete random variables in the form of the multiple sum

$$F(x_1, x_2, \ldots, x_n) = \sum_{i_1 < x_1} \sum_{i_2 < x_2} \cdots \sum_{i_n < x_n} \mathbf{P}(X_1 = i_1, X_2 = i_2, \ldots, X_n = i_n),$$

in which the summation is extended over all the possible values of each of the random variables for which $i_1 < x_1, i_2 < x_2, \ldots, i_n < x_n$.

For $n = 2$, a system of continuous random variables can be interpreted as a random point in the plane and for $n = 3$ as a random point in space.

The probability that a random point lands in a region S is obtained by integrating the probability density over this region.

The basic numerical characteristics of a system of n random variables are: the expectations,

$$\mathbf{M}[X_i] = \int_{-\infty}^{\infty}\int_{-\infty}^{\infty}\cdots\int_{-\infty}^{\infty} x_i f(x_1, x_2, \ldots, x_n)\, dx_1\, dx_2 \cdots dx_n,$$

the variances,

$$\mathbf{D}[X_i] = k_{ii} = \sigma_i^2 = \int_{-\infty}^{\infty}\int_{-\infty}^{\infty}\cdots\int_{-\infty}^{\infty} (x_i - \bar{x}_i)^2 f(x_1, x_2, \ldots, x_n)\, dx_1\, dx_2 \cdots dx_n$$

and the covariances,

$$\mathbf{M}[(X_i - \bar{x})(X_j - \bar{x}_j)] = k_{j}$$
$$= \int_{-\infty}^{\infty}\int_{-\infty}^{\infty}\cdots\int_{-\infty}^{\infty} (x_i - \bar{x}_i)(x_j - \bar{x}_j) f(x_1, x_2, \ldots, x_n)\, dx_1\, dx_2 \cdots dx_n.$$

The moments for discrete random variables can be computed similarly; i.e., the integration is replaced by summation over all possible values of the random variables.

The second central moments form the covariance matrix

$$\|k_{ij}\| = \begin{Vmatrix} k_{11} & k_{12} & k_{13} & \cdots & k_{1n} \\ k_{21} & k_{22} & k_{23} & \cdots & k_{2n} \\ k_{31} & k_{32} & k_{33} & \cdots & k_{3n} \\ \cdot & \cdot & \cdot & \cdot & \cdot \\ k_{n1} & k_{n2} & k_{n3} & \cdots & k_{nn} \end{Vmatrix},$$

in which $k_{ij} = k_{ji}$. Sometimes it is very convenient to use the formula

$$k_{ij} = \mathbf{M}[X_i X_j] - \mathbf{M}[X_i]\cdot\mathbf{M}[X_j].$$

The random variables $X_1, X_2, \ldots, X_n$ are said to be uncorrelated if the nondiagonal elements of the covariance matrix are zero.

The nondimensional correlation characteristic between the random variables X_i and X_j is the correlation coefficent

$$r_{ij} = \frac{k_{ij}}{\sqrt{\mathbf{D}[X_i]\mathbf{D}[X_j]}}.$$

The correlation coefficients form the normalized covariance matrix

$$\|r_{ij}\| = \begin{Vmatrix} 1 & r_{12} & r_{13} & \cdots & r_{1n} \\ r_{21} & 1 & r_{23} & \cdots & r_{2n} \\ r_{31} & r_{32} & 1 & \cdots & r_{3n} \\ \cdot & \cdot & \cdot & \cdot & \cdot \\ r_{n1} & r_{n2} & r_{n3} & \cdots & 1 \end{Vmatrix},$$

in which $r_{ij} = r_{ji}$.

The continuous random variables $X_1, X_2, \ldots, X_n$ forming a system are called independent if

$$f(x_1, x_2, \ldots, x_n) = f_1(x_1)f_2(x_2)\cdots f_n(x_n),$$

and are called dependent if

$$f(x_1, x_2, \ldots, x_n) \not\equiv f_1(x_1)f_2(x_2)\cdots f_n(x_n),$$

where $f_i(x_i)$ is the probability density of the random variable X_i (see Section 20).

The discrete random variables $X_1, X_2, \ldots, X_n$ are said to be independent if

$$\mathbf{P}(X_1 = i_1, X_2 = i_2, \ldots, X_n = i_n) = \mathbf{P}(X_1 = i_1)\mathbf{P}(X_2 = i_2)\cdots\mathbf{P}(X_n = i_n).$$

Solution for Typical Examples

Example 18.1 As a result of a test, an item can be classified as first grade with the probability p_1, second grade with the probability p_2 or defective with the probability $p_3 = 1 - p_1 - p_2$. A number of n items are tested. Determine the probability density for different numbers of items of first and second grade, their expectations, variances and covariances.

Solution. Let X denote the number of items of first grade and Y, of second grade. Since the tests are independent, the probability that k items will be classified as first grade, s items as second grade and the remaining $n - k - s$ items as defective (taking into account all the possible combinations of the three terms k, s and $n - k - s$ of which the sum is composed) is

$$\mathbf{P}(X = k,\ Y = s) = \frac{n!}{k!\,s!\,(n-k-s)!}\,p_1^k p_2^s p_3^{n-k-s},$$
$$p_1 + p_2 + p_3 = 1.$$

The values of this probability for $k = 0, 1, \ldots, n$, $s = 0, 1, \ldots, n$ and $k + s \leqslant n$ form the required set of probabilities for different numbers of items of first and second grade. The expectation of the number of first grade items is

$$\begin{aligned}\mathbf{M}[X] = \bar{x} &= \sum_{k=0}^{n}\sum_{s=0}^{n-k} k\,\frac{n!}{k!\,s!\,(n-k-s)!}\,p_1^k p_2^s p_3^{n-k-s}\\ &= p_1\frac{\partial}{\partial p_1}\left\{\sum_{k=0}^{n}\sum_{s=0}^{n-k}\frac{n!}{k!\,s!\,(n-k-s)!}\,p_1^k p_2^s p_3^{n-k-s}\right\}\\ &= p_1\frac{\partial}{\partial p_1}(p_1+p_2+p_3)^n = np_1(p_1+p_2+p_3)^{n-1} = np_1.\end{aligned}$$

The variance of the number of first grade items is

$$\begin{aligned}\mathbf{D}[X] &= \sum_{k=0}^{n}\sum_{s=0}^{n-k} k^2\,\frac{n!}{k!\,s!\,(n-k-s)!}\,p_1^k p_2^s p_3^{n-k-s} - \bar{x}^2\\ &= p_1\frac{\partial}{\partial p_1}\left\{\sum_{k=0}^{n}\sum_{s=0}^{n-k} k\,\frac{n!}{k!\,s!\,(n-k-s)!}\,p_1^k p_2^s p_3^{n-k-s}\right\} - \bar{x}^2\\ &= p_1\frac{\partial}{\partial p_1}\{np_1(p_1+p_2+p_3)^{n-1}\} - \bar{x}^2\\ &= np_1 + n(n-1)p_1^2 - n^2p_1^2 = np_1(1-p_1).\end{aligned}$$

Similarly, we find that

$$\mathbf{M}[Y] = np_2, \qquad \mathbf{D}[Y] = np_2(1 - p_2).$$

The covariance between the number of first grade and second grade items is

$$\begin{aligned} k_{xy} &= \sum_{k=0}^{n} \sum_{s=0}^{n-k} ks \frac{n!}{k!\, s!\, (n-k-s)!} p_1^k p_2^s p_3^{n-k-s} - \bar{x}\bar{y} \\ &= p_1 \frac{\partial}{\partial p_1} p_2 \frac{\partial}{\partial p_2} \{(p_1 + p_2 + p_3)^n\} - n^2 p_1 p_2 \\ &= p_1 \frac{\partial}{\partial p_1} \{np_2(p_1 + p_2 + p_3)^{n-1}\} - n^2 p_1 p_2 \\ &= n(n-1)p_1 p_2 - n^2 p_1 p_2 = -np_1 p_2. \end{aligned}$$

Example 18.2 For the probability density of a system of random variables (X, Y):

$$f(x, y) = 0.5 \sin (x + y) \qquad \left(0 \leqslant x \leqslant \frac{\pi}{2},\ 0 \leqslant y \leqslant \frac{\pi}{2}\right),$$

determine (a) the distribution function of the system, (b) the expectations of X and Y, (c) the covariance matrix.

SOLUTION. We first find the distribution function (for $0 \leqslant x \leqslant \pi/2$ and $0 \leqslant y \leqslant \pi/2$):

$$\begin{aligned} F(x, y) = \mathbf{P}(X < x,\ Y < y) &= \int_0^x \int_0^y 0.5 \sin (x + y)\, dx\, dy \\ &= 0.5[\sin x + \sin y - \sin (x + y)]. \end{aligned}$$

The expectation of the random variable X is

$$\begin{aligned} \mathbf{M}[X] &= 0.5 \int_0^{\pi/2} \int_0^{\pi/2} x \sin (x + y)\, dx\, dy \\ &= 0.5 \int_0^{\pi/2} x \left[-\cos\left(x + \frac{\pi}{2}\right) + \cos x\right] dx = \frac{\pi}{4} = 0.785. \end{aligned}$$

The variance of X is

$$\begin{aligned} \mathbf{D}[X] &= 0.5 \int_0^{\pi/2} \int_0^{\pi/2} x^2 \sin (x + y)\, dx\, dy - \frac{\pi^2}{16} \\ &= 0.5 \int_0^{\pi/2} x^2 \left[-\cos\left(x + \frac{\pi}{2}\right) + \cos x\right] dx - \frac{\pi^2}{16} \\ &= \frac{\pi^2}{16} + \frac{\pi}{2} - 2 = 0.188. \end{aligned}$$

From the symmetry of the probability density about X and Y, it follows that

$$\mathbf{M}[Y] = \mathbf{M}[X], \qquad \mathbf{D}[Y] = \mathbf{D}[X].$$

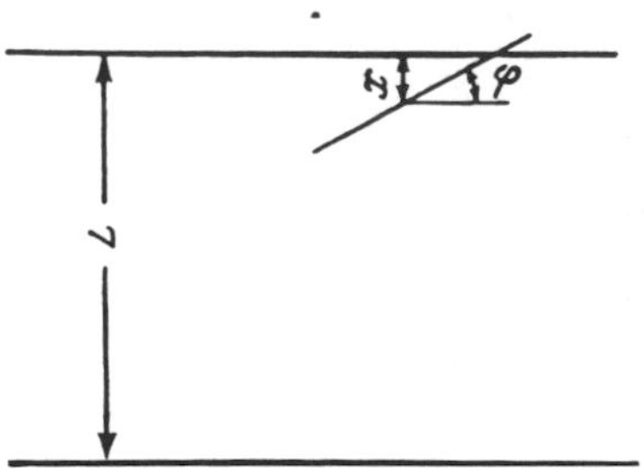

FIGURE 15

The covariance is

$$k_{xy} = 0.5 \int_0^{\pi/2} \int_0^{\pi/2} xy \sin(x + y)\, dx\, dy - \frac{\pi^2}{16}$$
$$= 0.5 \int_0^{\pi/2} x\left[\sin\left(x + \frac{\pi}{2}\right) - \sin x - \frac{\pi}{2} \cos\left(x + \frac{\pi}{2}\right)\right] dx - \frac{\pi^2}{16}$$
$$= \frac{\pi}{2} - 1 - \frac{\pi^2}{16} = -0.046.$$

Therefore, the covariance matrix has the form

$$\|k_{ij}\| = \begin{Vmatrix} 0.188 & -0.046 \\ -0.046 & 0.188 \end{Vmatrix}.$$

In a similar way Problems 18.18 and 18.19 can be solved.

Example 18.3 A needle of length l is dropped on a smooth table ruled with equidistant parallel lines at distance L apart. Determine the probability that the needle will cross one of the lines if $l < L$ (Buffon's problem).

SOLUTION. Introduce a system of random variables (X, Φ), where X is the distance from the midpoint of the needle to the nearest line and Φ is the acute angle made by the needle with this line (Figure 15). Obviously, X can assume all values from 0 to $L/2$, and Φ from 0 to $\pi/2$ with equal probability. Therefore, $f(x, \varphi) = 2/L \cdot 2/\pi = 4/\pi L$ for $0 \leqslant x \leqslant L/2$, $0 \leqslant \varphi \leqslant \pi/2$.

The needle will cross one of the lines for a given φ if $0 \leqslant x \leqslant (l \sin \varphi)/2$. From this, it follows that

$$P = \frac{4}{\pi L} \int_0^{\pi/2} d\varphi \int_0^{l \sin \varphi/2} dx = \frac{2l}{\pi L}.$$

Similarly one can solve Problems 18.20 and 18.21.

PROBLEMS

18.1 The coordinates X, Y of a randomly selected point are uniformly distributed over a rectangle bounded by the abscissas $x = a$, $x = b$ and the ordinates $y = c$, $y = d$ $(b > a,\ d > c)$. Find the probability density and the distribution function of the random variables (X, Y).

18.2 A system of random variables (X, Y) has the probability density

$$f(x, y) = \frac{A}{\pi^2(16 + x^2)(25 + y^2)}.$$

Determine (a) the magnitude of A, (b) the distribution function $F(x, y)$.

18.3 Determine the probability density of a system of three positive random variables (X, Y, Z) if their distribution function is

$$F(x, y, z) = (1 - e^{-ax})(1 - e^{-by})(1 - e^{-cz}) \quad (x \geqslant 0, y \geqslant 0, z \geqslant 0).$$

18.4 Under the assumptions of the preceding problem, find the locus of points with the same probability density

$$f(x, y, z) = f_0, \qquad f_0 \leqslant abc.$$

18.5 From a sample of $n = 6$ items, X turn out to be nondefective and of these Y ($Y \leqslant 3$) are of excellent quality. The system (X, Y) is given by the following two-dimensional probability distribution table (matrix) (Table 7):

(a) Form the distribution function; (b) find the probability of obtaining at least two items of excellent quality; (c) find $\mathbf{M}[X]$, $\mathbf{M}[Y]$ and the covariance matrix.

TABLE 7

$\mathbf{P}(X = i,\ Y = j)$

$j \backslash i$	0	1	2	3	4	5	6
0	0.202	0.174	0.113	0.062	0.049	0.023	0.004
1	0	0.099	0.064	0.040	0.031	0.020	0.006
2	0	0	0.031	0.025	0.018	0.013	0.008
3	0	0	0	0.001	0.002	0.004	0.011

18.6 A system of independent random variables $X_1, X_2, \ldots, X_n$ is given by the probability densities $f_1(x_1), f_2(x_2), \ldots, f_n(x_n)$. Determine the distribution function of this system.

18.7 The probability density of a system of two random variables X_1 and X_2 that can be measured only simultaneously is $f(x_1, x_2)$. The values u and v are observed. Find the probability that u will be the value of the random variable X_1 and v that of X_2.

18.8 Assume that the probability density for a system of three random variables that can be measured only simultaneously is $f(x_1, x_2, x_3)$. The values of u, v, w are observed, but it is not known how these values and the random variables correspond. Determine the probability that u is the realization of X_1 and w that of X_3.

18.9 Find the probability that a randomly selected point is located in the shaded region shown in Figure 16 if the distribution function $F(x, y)$ is known.

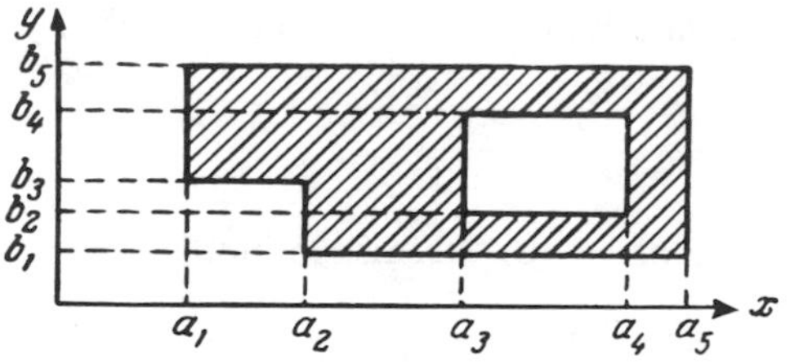

FIGURE 16

18.10 What is the probability that a point with coordinates (X, Y) hits a region specified by the inequalities $(1 \leqslant x \leqslant 2, 1 \leqslant y \leqslant 2)$ if the distribution function $(a > 0)$

$$F(x, y) = \begin{cases} 1 - a^{-x^2} - a^{-2y^2} + a^{-x^2 - 2y^2} & \text{for } x \geqslant 0,\ y \geqslant 0, \\ 0 & \text{for } x < 0 \text{ or } y < 0? \end{cases}$$

18.11 The coordinates of a random point (X, Y) are uniformly distributed over a rectangle bounded by the abscissas 0 and a and ordinates 0 and b. Find the probability that a random point hits a circle of radius R if $a > b$ and the center of the circle coincides with the origin of the coordinates.

18.12 The probability density of a system of random variables is

$$f(x, y) = c(R - \sqrt{x^2 + y^2}) \quad \text{for} \quad x^2 + y^2 \leqslant R^2.$$

Find (a) the constant c, (b) the probability of hitting a circle of radius $a < R$ if the centers of both circles coincide with the origin.

18.13 The random variables X and Y are related by the equality $mX + nY = c$, in which m, n and c are constants $(m \neq 0, n \neq 0)$.

Find (a) the correlation coefficient r_{xy}, (b) the quotient of the mean-square deviations σ_x/σ_y.

18.14 Prove that the absolute value of the correlation coefficient does not exceed one.

18.15 Show that

$$\begin{aligned} k_{xyz} &= \mathbf{M}[(X - \bar{x})(Y - \bar{y})(Z - \bar{z})] \\ &= \mathbf{M}[XYZ] - \bar{x}k_{yz} - \bar{y}k_{xz} - \bar{z}k_{xy} - \bar{x}\bar{y}\bar{z}. \end{aligned}$$

18.16 Suppose that the covariance matrix of a system of random variables (X_1, X_2, X_3) is

$$\|k_{ij}\| = \begin{Vmatrix} 16 & -14 & 12 \\ -14 & 49 & -21 \\ 12 & -21 & 36 \end{Vmatrix}.$$

Form the normalized covariance matrix $\|r_{ij}\|$.

18.17 Some items are classified by their shape as round or oval and by their weight as light or heavy. The probabilities that a randomly selected item will be round and light, oval and light, round and heavy or oval and heavy are α, β, γ and $\delta = 1 - \alpha - \beta - \sigma$, respectively. Find the expectations and variances for the number X of round items and Y of light items, and also the covariance k_{xy} between the number of round items and light items if $\alpha = 0.40$, $\beta = 0.05$. $\gamma = 0.10$.

18.18 Determine the expectations and the covariance matrix of a system of random variables (X, Y) if the probability density is

$$f(x, y) = \frac{2}{\pi(x^2 + y^2 + 1)^3}.$$

18.19 Find the probability density, the expectation and the covariance matrix of a system of random variables (X, Y) defined on the intervals $(0 \leqslant x \leqslant \pi/2)$ and $(0 \leqslant y \leqslant \pi/2)$ if the distribution function of the system is

$$F(x, y) = \sin x \sin y.$$

18.20 Solve Buffon's problem; i.e., find the probability that the needle will cross at least one of the lines, in the case $l > L$ (see Example 18.3).

18.21 A needle of length l is dropped on a plane partitioned into rectangles with sides a and b. Determine the probability that the needle will cross at least one side of a rectangle if $a < l$, $b < l$.

19. THE NORMAL DISTRIBUTION LAW IN THE PLANE AND IN SPACE. THE MULTIDIMENSIONAL NORMAL DISTRIBUTION

Basic Formulas

The probability density of a system of two normal random variables (X, Y) is (for a normal distribution of the coordinates of a point in the plane)

$$f(x, y) = \frac{1}{2\pi\sigma_x\sigma_y\sqrt{1 - r^2}} \times \exp\left\{-\frac{1}{2(1 - r^2)}\left[\frac{(x - \bar{x})^2}{\sigma_x^2} - \frac{2r(x - \bar{x})(y - \bar{y})}{\sigma_x\sigma_y} + \frac{(y - \bar{y})^2}{\sigma_y^2}\right]\right\},$$

where $\bar{x}$, $\bar{y}$ are the expectations of X and Y; σ_x, σ_y are the mean-square deviations and r is the correlation coefficient of X with Y.

The locus of points with equal probability density is an ellipse (dispersion ellipse) defined by the equation

$$\frac{(x - \bar{x})^2}{\sigma_x^2} - \frac{2r(x - \bar{x})(y - \bar{y})}{\sigma_x\sigma_y} + \frac{(y - \bar{y})^2}{\sigma_y^2} = k^2.$$

If $r = 0$, then the symmetry axes of the dispersion ellipse are parallel to the coordinate axes Ox and Oy, the random variables X and Y are uncorrelated and independent and the probability density is

$$f(x, y) = \frac{1}{2\pi\sigma_x\sigma_y} \exp\left\{-\frac{1}{2}\left[\frac{(x - \bar{x})^2}{\sigma_x^2} + \frac{(y - \bar{y})^2}{\sigma_y^2}\right]\right\} = \frac{\rho^2}{\pi E_x E_y} \exp\left\{-\rho^2\left[\frac{(x - \bar{x})^2}{E_x^2} + \frac{(y - \bar{y})^2}{E_y^2}\right]\right\}.$$

where $E_x = \sigma_x \rho\sqrt{2}$, $E_y = \sigma_y \rho\sqrt{2}$ are the mean deviations of X and Y, respectively, and $\rho = 0.4769\ldots$.

The ellipse defined by the equality

$$\frac{(x-\bar{x})^2}{E_x^2} + \frac{(y-\bar{y})^2}{E_y^2} = 1,$$

is called the unit ellipse.

The probability density of a system of n normal random variables (for a multidimensional normal distribution) is

$$f(x_1, x_2, \ldots, x_n) = \frac{1}{(2\pi)^{n/2}\sqrt{\Delta}} \exp\left\{-\frac{1}{2}\sum_{i,j=1}^{n} k_{ij}^{(-1)}(x_i - \bar{x}_i)(x_j - \bar{x}_j)\right\},$$

where

$$\Delta = \begin{vmatrix} k_{11} & k_{12} & \cdots & k_{1n} \\ k_{21} & k_{22} & \cdots & k_{2n} \\ \cdot & \cdot & \cdot & \cdot \\ k_{n1} & k_{n2} & \cdots & k_{nn} \end{vmatrix}$$

is the determinant formed by the elements of the covariance matrix; $k_{ij}^{(-1)}$ are the elements of the inverse matrix,

$$k_{ij}^{(-1)} = \frac{1}{\Delta} A_{ji} = \frac{1}{\Delta} A_{ij};$$

and A_{ij} is the cofactor of the element k_{ij}.

In the case of three independent normal random variables X, Y, Z, we have $k_{xy} = k_{yz} = k_{xz} = 0$ and

$$f(x, y, z) = \frac{1}{(2\pi)^{3/2}\sigma_x\sigma_y\sigma_z} \exp\left\{-\frac{1}{2}\left[\frac{(x-\bar{x})^2}{\sigma_x^2} + \frac{(y-\bar{y})^2}{\sigma_y^2} + \frac{(z-\bar{z})^2}{\sigma_z^2}\right]\right\}$$
$$= \frac{\rho^3}{\pi^{3/2}E_xE_yE_z} \exp\left\{-\rho^2\left[\frac{(x-\bar{x})^2}{E_x^2} + \frac{(y-\bar{y})^2}{E_y^2} + \frac{(z-\bar{z})^2}{E_z^2}\right]\right\},$$

where E_x, E_y, E_z are the mean deviations of X, Y, Z, respectively.

This is a particular case where the symmetry axes of the ellipsoid are parallel to the coordinate axes Ox, Oy and Oz.

Solution for Typical Examples

Example 19.1 Given the covariance matrix of a system of four normal random variables (X_1, X_2, X_3, X_4),

$$\|k_{ij}\| = \begin{Vmatrix} 15 & 3 & 1 & 0 \\ 3 & 16 & 6 & -2 \\ 1 & 6 & 4 & 1 \\ 0 & -2 & 1 & 3 \end{Vmatrix},$$

determine the probability density $f(x_1, x_2, x_3, x_4)$ if $\bar{x}_1 = 10$, $\bar{x}_2 = 0$, $\bar{x}_3 = -10$, $\bar{x}_4 = 1$.

SOLUTION. We first compute the cofactors of the determinant $\Delta = |k_{ij}|$:

$$A_{11} = \begin{vmatrix} 16 & 6 & -2 \\ 6 & 4 & 1 \\ -2 & 1 & 3 \end{vmatrix} = 28; \qquad A_{12} = -\begin{vmatrix} 3 & 6 & -2 \\ 1 & 4 & 1 \\ 0 & 1 & 3 \end{vmatrix} = -13;$$

$$A_{13} = \begin{vmatrix} 3 & 16 & -2 \\ 1 & 6 & 1 \\ 0 & -2 & 3 \end{vmatrix} = 16; \qquad A_{14} = -\begin{vmatrix} 3 & 16 & 6 \\ 1 & 6 & 4 \\ 0 & -2 & 1 \end{vmatrix} = -14;$$

$$A_{22} = \begin{vmatrix} 15 & 1 & 0 \\ 1 & 4 & 1 \\ 0 & 1 & 3 \end{vmatrix} = 162; \qquad A_{23} = -\begin{vmatrix} 15 & 3 & 0 \\ 1 & 6 & 1 \\ 0 & -2 & 3 \end{vmatrix} = -291;$$

$$A_{24} = \begin{vmatrix} 15 & 3 & 1 \\ 1 & 6 & 4 \\ 0 & -2 & 1 \end{vmatrix} = 205; \qquad A_{33} = \begin{vmatrix} 15 & 3 & 0 \\ 3 & 16 & -2 \\ 0 & -2 & 3 \end{vmatrix} = 633;$$

$$A_{34} = -\begin{vmatrix} 15 & 3 & 1 \\ 3 & 16 & 6 \\ 0 & -2 & 1 \end{vmatrix} = -405; \qquad A_{44} = \begin{vmatrix} 15 & 3 & 1 \\ 3 & 16 & 6 \\ 1 & 6 & 4 \end{vmatrix} = 404.$$

Next, we find the value of the determinant:

$$\Delta = \begin{vmatrix} 15 & 3 & 1 & 0 \\ 3 & 16 & 6 & -2 \\ 1 & 6 & 4 & 1 \\ 0 & -2 & 1 & 3 \end{vmatrix} = 15A_{11} + 3A_{12} + A_{13} = 397.$$

In deriving the formula for the probability density, we take into account the fact that for $i \neq j$, the exponent contains equal terms

$$k_{ij}^{(-1)}(x_i - \bar{x}_i)(x_j - \bar{x}_j) = k_{ij}^{(-1)}(x_j - \bar{x}_j)(x_i - \bar{x}_i).$$

The probability density is

$$f(x_1, x_2, x_3, x_4) = \frac{1}{4\pi^2\sqrt{397}} \exp\Big\{-\frac{1}{794}[28(x_1 - 10)^2 - 26(x_1 - 10)x_2 + 32(x_1 - 10)(x_3 + 10) - 28(x_1 - 10) \times (x_4 - 1) + 162x_2^2 - 582x_2(x_3 + 10) + 410x_2(x_4 - 1) + 633(x_3 + 10)^2 - 810(x_3 + 10)(x_4 - 1) + 404(x_4 - 1)^2]\Big\}.$$

Example 19.2 A random point in space is given by three rectangular coordinates forming a system of normal random variables with the probability density

$$f(x, y, z) = \frac{\sqrt{3}}{16\pi^{3/2}} \exp\left\{-\frac{1}{8}[2x^2 + 4y^2 - 2y(z + 5) + (z + 5)^2]\right\}.$$

(a) Find the covariance matrix, (b) determine the locus of points when the probability is 0.01.

SOLUTION. (a) Since

$$f(x, y, z) = f_1(x)f_2(y, z),$$

where

$$f_1(x) = c_1 \exp\left\{-\frac{x^2}{4}\right\}.$$

$$f_2(y, z) = c_2 \exp\left\{-\frac{1}{2}\left[y^2 - \frac{2y(z + 5)}{4} + \frac{(z + 5)^2}{4}\right]\right\},$$

then

$$k_{xy} = k_{xz} = 0.$$

This implies that

$$\mathbf{D}[X] = \sigma_x^2 = 2; \qquad \mathbf{D}[Y] = \sigma_y^2 = \frac{1}{1 - r^2}; \qquad \mathbf{D}[Z] = \sigma_z^2 = \frac{4}{1 - r^2};$$

$$\frac{r}{\sigma_y\sigma_z(1 - r^2)} = \frac{1}{4}; \qquad r = \frac{k_{yz}}{\sigma_y\sigma_z} = \frac{1}{2}; \qquad k_{yz} = \frac{4}{3};$$

$$\|k_{ij}\| = \begin{Vmatrix} 2 & 0 & 0 \\ 0 & \frac{4}{3} & \frac{4}{3} \\ 0 & \frac{4}{3} & \frac{16}{3} \end{Vmatrix}.$$

For verification, we can compute the normalization factor

$$\frac{1}{(2\pi)^{3/2}\sqrt{\Delta}} = \frac{\sqrt{3}}{\pi^{3/2}\cdot 2\sqrt{2}\cdot\sqrt{32}} = \frac{\sqrt{3}}{16\pi^{3/2}}.$$

(b) The required locus of points with constant probability density is the surface of the ellipsoid

$$2x^2 + 4y^2 - 2y(z + 5) + (z + 5)^2 = -8 \ln \frac{16\pi^{3/2}}{100\sqrt{3}}.$$

Example 19.3 Find the probability that a point (X, Y, Z) lands in a region representing a hollow parallelepiped whose outer surface is given by the planes

$$x = a_1, \quad x = b_1, \quad y = c_1, \quad y = d_1, \quad z = m_1, \quad z = n_1,$$

and whose inner surface is given by the planes

$$x = a_2, \quad x = b_2, \quad y = c_2, \quad y = d_2, \quad z = m_2, \quad z = n_2$$
$$(b_i > a_i,\ d_i > c_i,\ n_i > m_i,\ i = 1, 2).$$

The dispersion of points (X, Y, Z) obeys a normal distribution with the principal axes parallel to the coordinate axes, the dispersion center at the point $\bar{x}, \bar{y}, \bar{z}$ and mean deviations E_x, E_y, E_z.

SOLUTION. Since the principal dispersion axes are parallel to the coordinate axes, the event that one of the coordinates, for instance x, will assume values on the interval (a, b) is independent of the values assumed by the other coordinates. Therefore,

$$\mathbf{P}(a \leqslant x \leqslant b,\ c \leqslant y \leqslant d,\ m \leqslant z \leqslant n) = \mathbf{P}(a \leqslant x \leqslant b)\mathbf{P}(c \leqslant y \leqslant d)\mathbf{P}(m \leqslant z \leqslant n),$$

in which

$$\mathbf{P}(a \leqslant x \leqslant b) = \frac{1}{2}\left[\hat{\Phi}\left(\frac{b - \bar{x}}{E_x}\right) - \hat{\Phi}\left(\frac{a - \bar{x}}{E_x}\right)\right].$$

The probabilities of the other inequalities can be determined similarly.

The required probability of reaching the interior of the hollow parallelepiped will be determined as the difference between the probabilities of reaching the parallelepipeds bounded by the outer and inner surfaces; i.e.,

$$P = \frac{1}{8}\left[\hat{\Phi}\left(\frac{b_1 - \bar{x}}{E_x}\right) - \hat{\Phi}\left(\frac{a_1 - \bar{x}}{E_x}\right)\right]\left[\hat{\Phi}\left(\frac{d_1 - \bar{y}}{E_y}\right) - \hat{\Phi}\left(\frac{c_1 - \bar{y}}{E_y}\right)\right]$$
$$\times \left[\hat{\Phi}\left(\frac{n_1 - \bar{z}}{E_z}\right) - \hat{\Phi}\left(\frac{m_1 - \bar{z}}{E_z}\right)\right] - \frac{1}{8}\left[\hat{\Phi}\left(\frac{b_2 - \bar{x}}{E_x}\right) - \hat{\Phi}\left(\frac{a_2 - \bar{x}}{E_x}\right)\right]$$
$$\times \left[\hat{\Phi}\left(\frac{d_2 - \bar{y}}{E_y}\right) - \hat{\Phi}\left(\frac{c_2 - \bar{y}}{E_y}\right)\right]\left[\hat{\Phi}\left(\frac{n_2 - \bar{z}}{E_z}\right) - \hat{\Phi}\left(\frac{m_2 - \bar{z}}{E_z}\right)\right].$$

PROBLEMS

19.1 It is known that X and Y are independent normal random variables with expectations $\bar{x}$ and $\bar{y}$ and mean deviations E_x and E_y, respectively. Express the distribution function of the system (X, Y) in terms of the normalized Laplace functions.

19.2 Given the expectations of two normal random variables $\mathbf{M}[X] = 26$, $\mathbf{M}[Y] = -12$, and their covariance matrix,

$$\|k_{ij}\| = \begin{Vmatrix} 196 & -91 \\ -91 & 169 \end{Vmatrix},$$

determine the probability density of the system (X, Y).

19.3 Given the probability density for the coordinates of a random point in the plane,

$$f(x, y) = c \exp\{-[4(x - 5)^2 + 2(x - 5)(y - 3) + 5(y - 3)^2]\},$$

find (a) constant c, (b) the covariance matrix, (c) the area S_{el} of the unit ellipse.

19.4 Determine at the point $x_1 = 2$, $x_2 = 2$ the probability density of a system of two normal random variables for which $\bar{x}_1 = \bar{x}_2 = 0$ and

$$\|k_{ij}\| = \begin{Vmatrix} 1 & 0 \\ 0 & 2 \end{Vmatrix}.$$

19.5 Given the covariance matrix of a system of three normal random variables (X, Y, Z),

$$\|k_{ij}\| = \begin{Vmatrix} 5 & 2 & -2 \\ 2 & 6 & 3 \\ -2 & 3 & 8 \end{Vmatrix},$$

and expectations $\bar{x} = \bar{y} = \bar{z} = 0$, find the probability density $f(x, y, z)$ and its maximum value.

19.6 A system of n normal random variables has the covariance matrix

$$\begin{Vmatrix} 1 & 1 & 1 & \cdots & 1 & 1 & 1 \\ 1 & 2 & 2 & \cdots & 2 & 2 & 2 \\ 1 & 2 & 3 & \cdots & 3 & 3 & 3 \\ \cdot & \cdot & \cdot & \cdot & \cdot & \cdot & \cdot \\ 1 & 2 & 3 & \cdots & n-2 & n-2 & n-2 \\ 1 & 2 & 3 & \cdots & n-2 & n-1 & n-1 \\ 1 & 2 & 3 & \cdots & n-2 & n-1 & n \end{Vmatrix}.$$

(a) Compute the inverse of this matrix; (b) find the probability $f(x_1, x_2, \ldots, x_n)$ if $\bar{x}_1 = \bar{x}_2 = \cdots \bar{x}_n = 0$.

19.7 The coordinates (X_1, Y_1) and (X_2, Y_2) of two random points in the plane obey the normal distribution law with the expectations of all coordinates zero and the variances of all coordinates equal to 10. The covariances between coordinates with the same symbol are equal: $\mathbf{M}[X_1X_2] = \mathbf{M}[Y_1Y_2] = 2$; the remaining pairs of coordinates are uncorrelated. Find the probability density $f(x_1, y_1, x_2, y_2)$.

19.8 The coordinates (X, Y) of a random point A in the plane obey the normal law

$$f(x, y) = \frac{1}{2\pi ab} \exp\left\{-\frac{1}{2}\left(\frac{x^2}{a^2} + \frac{y^2}{b^2}\right)\right\}.$$

Determine the probability that A will turn out to be inside an ellipse with principal semi-axes ka and kb and coinciding with the coordinate axes Ox and Oy.

19.9 The coordinates (X, Y, Z) of a random point A in space obey the normal distribution law

$$f(x, y, z) = \frac{\rho^3}{\pi^{3/2}E_1E_2E_3} \exp\left\{-\rho^2\left(\frac{x^2}{E_1^2} + \frac{y^2}{E_2^2} + \frac{z^2}{E_3^2}\right)\right\}.$$

Find the probability that A is inside an ellipsoid with the principal semi-axes kE_1, kE_2 and kE_3 coinciding with the coordinate axes Ox, Oy and Oz.

19.10 The determination of the coordinates of a point in the plane is accompanied by a systematic error d in one of its rectangular coordinates and a random error obeying a circular normal distribution with mean deviation E. Find the probability that the deviation of the point from its measured position will not exceed a quantity R.

19.11 A system of random variables (X, Y) obeys a normal distribution with numerical characteristics $\mathbf{M}[X] = \mathbf{M}[Y] = 0$, $E_x = E_y = 10$, $k_{xy} = 0$. Determine the probability that (a) $X < Y$, (b) $X > 0$, $Y < 0$.

19.12 Compute the probability that a random point A with coordinates X, Y and obeying a normal distribution law will lie in a rectangle whose sides are parallel to the principal dispersion axes if the coordinates of its vertices are (a, b), (a, d), (c, b), (c, d) for $a = -5$, $b = 10$, $c = 5$, $d = 20$, and $\bar{x} = 0$, $\bar{y} = 10$, $E_x = 20$, $E_y = 10$.

19.13 A random point is distributed in accordance with a normal circular law with mean deviation $E = 10$ m. Compare the probability of hitting a figure whose area is 314 sq. m. if its shape is (a) a circle, (b) a square, (c) a rectangle whose sides are in the ratio 10:1. The dispersion center coincides with the geometric center of this figure.

19.14 Find the probability that a randomly selected point lies inside the shaded region (Figure 17) bounded by three concentric circles and the rays issuing from their common center if the radius of the exterior circle is R, and the dispersion of the point in the plane obeys a circular normal distribution law with mean deviation E. The dispersion center coincides with the center of the circles.

19.15 Find the probability of hitting a figure bounded by the arcs determined by the radii R_1 and R_2 and the rays issuing from the

FIGURE 17

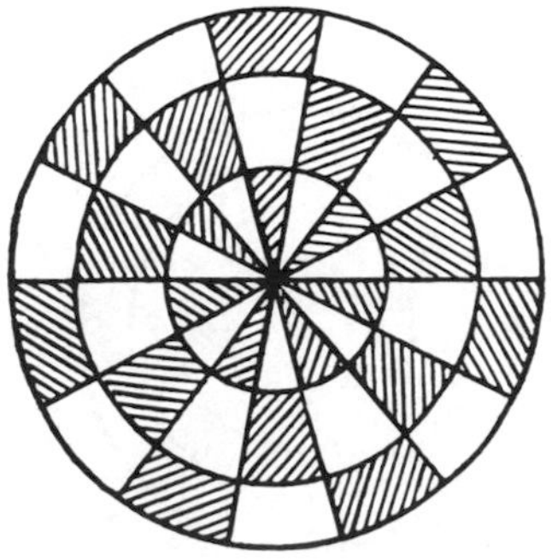

common center O if the dispersion of a random point in the plane obeys a circular normal distribution with mean deviation E, and the angle made by the rays is α. The dispersion center coincides with M $(R_1 < R_2)$.

19.16 The probability of hitting a rectangle with sides $2d$ and $2k$ and parallel to the principal dispersion axes satisfies the following approximate formula:

$$P(\bar{x}, \bar{z}) = \frac{1}{4}\left[\hat{\Phi}\left(\frac{\bar{x}+d}{E_x}\right) - \hat{\Phi}\left(\frac{\bar{x}-d}{E_x}\right)\right]\left[\hat{\Phi}\left(\frac{\bar{z}+k}{E_z}\right) - \hat{\Phi}\left(\frac{\bar{z}-k}{E_z}\right)\right]$$
$$\approx A\,\frac{\rho^2}{\pi\alpha\beta}\exp\left\{-\rho^2\left(\frac{\bar{x}^2}{\alpha^2} + \frac{\bar{z}^2}{\beta^2}\right)\right\},$$

which is recommended when d/E_x and k/E_z do not exceed 1.5. Equating the zero and second moments on the left- and right-hand sides of this equality, find the values of A, α, β.

19.17 Using the approximate formula from the preceding problem, find the probability of hitting a rectangle with sides $2d$ and $2k$ parallel to the principal dispersion axes if the coordinates of the dispersion center are uniformly distributed over the given rectangle and E_x, E_z are known. Compare the result obtained with the probability of a direct hit in the same region when the center of dispersion coincides with the center of the region.

19.18 A target consists of four concentric circles of radii 10, 20, 30 and 40 cm., respectively (Figure 18). By hitting the bull's-eye one scores 5 points, and for each of the three annuli—4, 3, and 2 points. The score is satisfactory if one scores at least 7 points in three shots, and excellent if one scores more than 12 points. What is the probability of a satisfactory score in the case of circular normal dispersion with mean deviation 20 cm.? What is the probability of an excellent score? The dispersion center coincides with the center of the target.

19.19 What is the probability of hitting a right triangle ABC with legs $BC = a$ and $AC = b$ parallel to the principal dispersion axes $(AC \parallel Oy)$, $(BC \parallel Ox)$ if the dispersion center coincides with point A and

$$\frac{a}{E_x} = \frac{b}{E_y} = k\,?$$

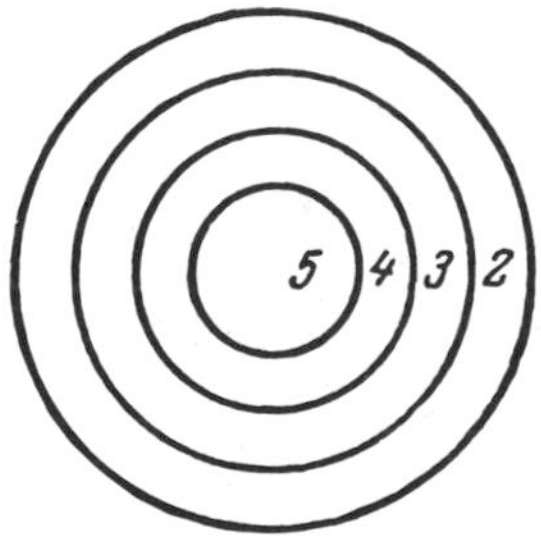

FIGURE 18

19.20 Find the probability that a point with coordinates X, Y, Z will hit a region representing a sphere of radius R, from which a central cube with edge a has been removed (the diagonal of the cube is shorter than the diameter of the sphere). The dispersion center coincides with the common center of the sphere and the cube. The distribution is normal spherical with mean deviation E.

19.21 Find the probability that a point $A(X, Y, Z)$ will lie inside a right cylinder whose base has radius R, and whose height is h if the dispersion in the xy-plane parallel to the base obeys a normal circular distribution with mean deviation E, and the dispersion along the generator is independent of X, Y and obeys (a) a normal distribution with mean deviation B (the dispersion center is located on the axis of the cylinder and divides it in the ratio $m:n$), (b) a uniform distribution over the interval $(-H, H)$ for $H > h$.

19.22 Find the probability that a random point $A(X, Y, Z)$ will lie in a right circular cone whose vertex coincides with the dispersion center, whose height is h and whose base has radius R; the dispersion in the xy-plane, which is parallel to the base, obeys a normal circular law with mean deviation E, and the dispersion along the height is independent of X, Y and obeys a normal distribution with mean deviation a.

19.23 A normal distribution law in the plane is given by the expectations of random variables $\bar{x}_1 = \bar{x}_2 = 10$ and the covariance matrix

$$\|k_{ij}\| = \begin{Vmatrix} 36 & -18 \\ -18 & 25 \end{Vmatrix}.$$

Find the locus of points with probability density 10^{-5}.

19.24 A normal distribution law in space is given by the expectations $\bar{x}_1 = 2$, $\bar{x}_2 = 0$, $\bar{x}_3 = -2$ and the covariance matrix

$$\|k_{ij}\| = \begin{Vmatrix} 4 & 0 & -1 \\ 0 & 4 & 0 \\ -1 & 0 & 4 \end{Vmatrix}.$$

Find the locus of points whose probability density is 10^{-5}.

19.25 For the multidimensional normal distribution given in Problem 19.6, find the locus of points with probability density 10^{-5}. Find the value of n, for which this problem has no solutions.

20. DISTRIBUTION LAWS OF SUBSYSTEMS OF CONTINUOUS RANDOM VARIABLES AND CONDITIONAL DISTRIBUTION LAWS

Basic Formulas

If $F(x, y)$ is the distribution function of a system of two random variables, then the distribution function of the random variable X is

$$F_x(x) = F(x, \infty) = \int_{-\infty}^{x} \int_{-\infty}^{\infty} f(x, y)\, dy\, dx.$$

Similarly, the distribution function of Y is

$$F_y(y) = F(\infty, y) = \int_{-\infty}^{y} \int_{-\infty}^{\infty} f(x, y)\, dx\, dy.$$

The probability densities of the random variables contained in the system are

$$f_x(x) = \int_{-\infty}^{\infty} f(x, y)\, dy,$$

$$f_y(y) = \int_{-\infty}^{\infty} f(x, y)\, dx.$$

If $F(x_1, x_2, \ldots, x_n)$ is the distribution function of a system of n random variables, then the distribution function of some of these variables (subsystems of random variables), for example, $X_1, X_2, \ldots, X_k$ is

$$F_{1,2,\ldots,k}(x_1, x_2, \ldots, x_k) = F(x_1, x_2, \ldots, x_k, \infty, \ldots, \infty),$$

and the corresponding probability density is

$$f_{1,2,\ldots,k}(x_1, x_2, \ldots, x_k) = \int_{-\infty}^{\infty} \cdots \int_{-\infty}^{\infty} f(x_1, x_2, \ldots, x_n)\, dx_{k+1} \cdots dx_n.$$

The probability density of one of two random variables, computed under the assumption that the other random variable assumes a certain value (the conditional probability density), is

$$f(x \mid y) = \frac{f(x, y)}{f_y(y)}, \qquad f(y \mid x) = \frac{f(x, y)}{f_x(x)}.$$

The probability density of the subsystem of the random variables ($X_1, X_2, \ldots, X_k$), computed under the assumption that the remaining random variables $X_{k+1}, X_{k+2}, \ldots, X_n$ assume certain values, is

$$f(x_1, x_2, \ldots, x_k \mid x_{k+1}, x_{k+2}, \ldots, x_n) = \frac{f(x_1, x_2, \ldots, x_n)}{f_{k+1,\ldots,n}(x_{k+1}, \ldots, x_n)}.$$

The probability density of a system can be expressed in terms of the conditional densities by the formula

$$f(x_1, x_2, x_3, \ldots, x_n) = f_1(x_1) f_2(x_2 \mid x_1) f_3(x_3 \mid x_1, x_2) \ldots f_n(x_n \mid x_1, x_2, \ldots, x_n).$$

Solution for Typical Examples

Example 20.1 The position of a random point $A(X, Y)$ is equally probable at any point of an ellipse with the principal semi-axes a and b coinciding with the coordinate axes Ox and Oy, respectively.

(a) Determine the probability density of each of the two rectangular coordinates and their mutual conditional probability densities, (b) analyze the dependence and the correlation of the random variable forming the system.

SOLUTION. (a) Since

$$f(x, y) = \begin{cases} \dfrac{1}{\pi ab} & \text{for } \dfrac{x^2}{a^2} + \dfrac{y^2}{b^2} \leqslant 1, \\ 0 & \text{for } \dfrac{x^2}{a^2} + \dfrac{y^2}{b^2} > 1, \end{cases}$$

for a given x on the interval $(-a, a)$, the probability density $f(x, y)$ differs from zero only if $-b\sqrt{1 - x^2/a^2} \leqslant y \leqslant b\sqrt{1 - x^2/a^2}$; this implies that

$$f_x(x) = \int_{-b\sqrt{1-(x^2/a^2)}}^{b\sqrt{1-(x^2/a^2)}} \frac{dy}{\pi ab} = \frac{2\sqrt{1 - \dfrac{x^2}{a^2}}}{\pi a}.$$

For $|x| > a, f_x(x) = 0$. From this we obtain

$$f(y \mid x) = \begin{cases} \dfrac{1}{2b\sqrt{1 - \dfrac{x^2}{a^2}}} & \text{for } |x| < a,\ |y| \leqslant b\sqrt{1 - \dfrac{x^2}{a^2}}, \\ 0 & \text{for } |x| > a \text{ or } |y| > b\sqrt{1 - \dfrac{x^2}{a^2}}. \end{cases}$$

Similarly,

$$f_y(y) = \frac{2}{\pi b}\sqrt{1 - \frac{y^2}{b^2}}, \qquad y \leqslant b,$$

$$f(x \mid y) = \frac{1}{2a\sqrt{1 - \dfrac{y^2}{b^2}}} \quad \text{for } |y| < b,\ |x| \leqslant a\sqrt{1 - \frac{y^2}{b^2}}$$

and

$$f_y(y) = f(x \mid y) = 0 \quad \text{for } |y| > b \text{ or } |x| > a\sqrt{1 - \frac{y^2}{b^2}}.$$

(b) The covariance between X and Y is

$$k_{xy} = \int_{-\infty}^{\infty}\int_{-\infty}^{\infty} xyf(x, y)\, dx\, dy,$$

where the function integrated is different from zero inside the ellipse

$$\frac{x^2}{a^2} + \frac{y^2}{b^2} = 1.$$

Making the change of variables

$$x = ar\cos\varphi, \qquad y = br\sin\varphi,$$

we obtain

$$k_{xy} = \int_0^{2\pi}\int_0^1 arbr\cos\varphi\sin\varphi \frac{1}{\pi ab} abr\, dr\, d\varphi = 0.$$

Thus, the random variables X and Y are uncorrelated ($k_{xy} = 0$) but dependent, since

$$f_x(x)f_y(y) \neq f(x, y).$$

Example 20.2 The coordinates of a random point in the plane obey the normal distribution law

$$f(x, y) = \frac{1}{2\pi\sigma_1\sigma_2\sqrt{1 - r^2}} \times \exp\left\{-\frac{1}{2(1 - r^2)}\left[\frac{(x - \bar{x})^2}{\sigma_1^2} - \frac{2r(x - \bar{x})(y - \bar{y})}{\sigma_1\sigma_2} + \frac{(y - \bar{y})^2}{\sigma_2^2}\right]\right\}.$$

Determine (a) the probability density of the coordinates X and Y, (b) the conditional densities $f(y \mid x)$ and $f(x \mid y)$, (c) the conditional expectations, (d) the conditional variances.

SOLUTION. (a) For the probability density of the coordinate X, we find

$$f_x(x) = \int_{-\infty}^{\infty} f(x, y)\, dy.$$

Making the change of variables

$$\frac{x - \bar{x}}{\sigma_1} = u, \qquad \frac{y - \bar{y}}{\sigma_2} = v$$

and considering the fact that

$$\frac{1}{1 - r^2}[u^2 - 2ruv + v^2] = u^2 + \frac{(v - ru)^2}{1 - r^2},$$

we obtain

$$f_x(x) = \exp\left\{-\frac{u^2}{2}\right\}\frac{1}{\sigma_1 2\pi\sqrt{1 - r^2}}\int_{-\infty}^{\infty}\exp\left\{-\frac{(v - ru)^2}{2(1 - r^2)}\right\}dv$$
$$= \frac{1}{\sigma_1\sqrt{2\pi}}\exp\left\{-\frac{u^2}{2}\right\},$$

or

$$f_x(x) = \frac{1}{\sigma_1\sqrt{2\pi}}\exp\left\{-\frac{(x - \bar{x})^2}{2\sigma_1^2}\right\}.$$

Similarly we find that

$$f_y(y) = \frac{1}{\sigma_2\sqrt{2\pi}}\exp\left\{-\frac{(y - \bar{y})^2}{2\sigma_2^2}\right\}.$$

(b) Dividing $f(x, y)$ by $f_x(x)$, we obtain

$$f(y \mid x) = \frac{1}{\sigma_2\sqrt{2\pi}\,\sqrt{1 - r^2}}\exp\left\{-\frac{\left[y - \bar{y} - r\frac{\sigma_2}{\sigma_1}(x - \bar{x})\right]^2}{2(1 - r^2)\sigma_2^2}\right\}$$

and similarly

$$f(x \mid y) = \frac{1}{\sigma_1\sqrt{2\pi}\sqrt{1 - r^2}} \exp\left\{ -\frac{\left[x - \bar{x} - r\frac{\sigma_1}{\sigma_2}(y - \bar{y})\right]^2}{2(1 - r^2)\sigma_1^2} \right\}.$$

(c) From the expressions for conditional probability densities, it follows that the conditional expectation of the random variable Y for a fixed value $X = x$ is

$$\bar{y}_x = \mathbf{M}[Y \mid x] = \bar{y} + r\frac{\sigma_2}{\sigma_1}(x - \bar{x}).$$

Similarly

$$\bar{x}_y = \mathbf{M}[X \mid y] = \bar{x} + r\frac{\sigma_1}{\sigma_2}(y - \bar{y}).$$

These equations expressing the linear dependence of the conditional expectation of one of the random variables on a fixed value of the other variable are called the regression equations.

(d) From the expressions for conditional distribution densities, it follows that the conditional variances are

$$\mathbf{D}[Y \mid x] = \sigma_{y|x}^2 = \sigma_1^2(1 - r^2),$$

$$\mathbf{D}[X \mid y] = \sigma_{x|y}^2 = \sigma_2^2(1 - r^2).$$

Example 20.3 Determine the probability density of the length of a radius-vector if the coordinates of its end A obey the normal circular distribution law

$$f(x, y) = \frac{1}{2\pi a^2} \exp\left\{ -\frac{x^2 + y^2}{2a^2} \right\}.$$

SOLUTION. We pass now from the rectangular coordinates of A to the polar coordinates (r, φ). The probability that the radius-vector assumes values on the interval $(r, r + dr)$ is approximately $f_r(r)\,dr$, and can be interpreted as the probability for a random point A to lie in an infinitely narrow annulus shown in Figure 19.

Consequently,

$$f_r(r)\,dr = \iint\limits_{r^2 \leqslant x^2 + y^2 \leqslant (r + dr)^2} f(x, y)\,dx\,dy.$$

FIGURE 19

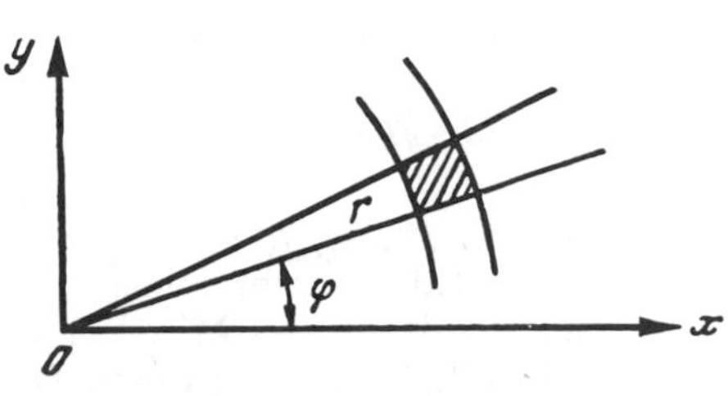

Integrating with respect to the variables r, φ and considering the expression for $f(x, y)$, we obtain

$$f_r(r) = \int_0^{2\pi} \frac{r}{2\pi a^2} \exp\left\{-\frac{r^2}{2a^2}\right\} d\varphi = \frac{r}{a^2} \exp\left\{-\frac{r^2}{2a^2}\right\}$$

(Rayleigh's distribution).

PROBLEMS

20.1 A system of random variables (X, Y, Z) is uniformly distributed inside a rectangular parallelepiped determined by the planes $x = a_1$, $x = a_2$, $y = b_1$, $y = b_2$, $z = c_1$, $z = c_2$. Find the probability densities of the system (X, Y, Z), of the subsystem (Y, Z) and of the random variable Z. Verify the dependence of the random variables forming the system.

20.2 The position of a random point (X, Y) is equally probable anywhere on a circle of radius R and whose center is at the origin. Determine the probability density and the distribution function of each of the rectangular coordinates. Are random variables X and Y dependent?

20.3 Under the assumption made in the preceding problem, find the probability density $f(y \mid x)$ for $|x| < R$, $|x| = R$ and $|x| > R$.

20.4 Under the assumptions of Problem 20.2, compute the covariance matrix of the system of variables X and Y. Are these variables correlated?

20.5 A system of random variables X, Y obeys a uniform distribution law over a square with side a. The diagonals of the square coincide with the coordinate axes.

(a) Find the probability density of the system (X, Y), (b) determine the probability density for each of the rectangular coordinates, (c) find the conditional probability densities, (d) compute the covariance matrix of the system of random variables (X, Y), (e) verify their dependence and correlation.

20.6 The random variables (X, Y, Z) are uniformly distributed inside a sphere of radius R. Determine for points lying inside this sphere, the probability density of the coordinate Z and the conditional probability density $f(x, y \mid z)$.

20.7 Given the differential distribution law for a system of nonnegative random variables

$$f(x, y) = kxye^{-(x^2+y^2)} \qquad (x \geqslant 0,\ y \geqslant 0),$$

determine k, $f_x(x)$, $f_y(y)$, $f(x \mid y)$, $f(y \mid x)$, and the first and second moments of the distribution.

20.8 Given $f_y(y)$, $\mathbf{M}[X \mid y]$ and $\mathbf{D}[X \mid y]$ for a system of random variables (X, Y), find $\mathbf{M}[X]$ and $\mathbf{D}[X]$.

20.9 A system of two random variables (X, Y) obeys the normal distribution law

$$f(x, y) = k \exp\left\{-\frac{1}{0.72\sigma^2}[(x-5)^2 + 0.8(x-5) \times (y+2) + 0.25(y+2)^2]\right\}.$$

Determine (a) the conditional expectations and variances, (b) the probability density of each of the random variables forming the system, (c) the conditional probability densities $f(y \mid x)$ and $f(x \mid y)$.

20.10 The probability density of a system of two random variables (X, Y) is given in the form

$$f(x, y) = Ae^{-ax^2 + bxy - cy^2} \qquad (a > 0,\ c > 0).$$

Find the distribution law $f_x(x)$ and $f_y(y)$. Under what conditions are X and Y independent random variables?

20.11 Given the probability density of a system of two random variables,

$$f(x, y) = ke^{-4x^2 - 6xy - 9y^2},$$

find the constant k, the covariance between X and Y, and the conditional distributions $f(x \mid y)$ and $f(y \mid x)$.

20.12 The position of a reference point in the plane is distributed according to a normal law with $\bar{x} = 125$ m., $\bar{y} = -30$ m., $\sigma_x = 40$ m., $\sigma_y = 30$ m. and $r_{xy} = 0.6$. The coordinate X defines the deviation of the reference point with respect to the "distance," i.e., with respect to a direction parallel to the observation line. The coordinate Y defines the deviation of the reference point with respect to a lateral "direction" perpendicular to the observation line. The deviations are estimated from the origin of coordinates.

Determine (a) the probability density of the deviations of the reference point with respect to the distance, (b) the probability density of the deviations of the reference point with respect to the lateral direction, (c) the conditional probability density of the deviations of the reference point with respect to distance in absence of lateral deviations, (d) the conditional probability density of the deviations of the reference point with respect to lateral direction for a deviation with respect to the distance equal to $+25$ m.

20.13 Under the assumptions of the preceding problem, find the regression equations of Y on X and X on Y.

20.14 Determine the probability density of the length of the radius-vector for a random point and its expectation if the coordinates (X, Y, Z) of this point obey the normal distribution law

$$f(x, y, z) = \frac{1}{(2\pi)^{3/2}a^3} \exp\left\{-\frac{1}{2a^2}(x^2 + y^2 + z^2)\right\}.$$

20.15 The coordinates of a random point A in the xy-plane obey the normal distribution law

$$f(x, y) = \frac{1}{2\pi ab} \exp\left\{-\frac{1}{2}\left(\frac{x^2}{a^2} + \frac{y^2}{b^2}\right)\right\}.$$

Find the probability densities $f_r(r)$ and $f_\varphi(\varphi)$ for the polar coordinates of this point.

20.16 Under the assumptions of the preceding problem, find the conditional probability densities $f(r \mid \varphi)$ and $f(\varphi \mid r)$.

20.17 A random point in space obeys the normal distribution law

$$f(x, y, z) = \frac{1}{(2\pi)^{3/2}abc} \exp\left\{-\frac{1}{2}\left(\frac{x^2}{a^2} + \frac{y^2}{b^2} + \frac{z^2}{c^2}\right)\right\}.$$

Find (a) the probability density of the spherical coordinates of this point (R, Θ, Φ) if $x = r\cos\theta\cos\varphi$, $y = r\cos\theta\sin\varphi$, $z = r\sin\theta$, (b) the probability densities of the subsystems (R, Θ) and (Θ, Φ), (c) the conditional probability densities $f(r \mid \theta, \varphi)$ and $f(\varphi \mid r, \theta)$.

20.18 For the system of random variables X_1, Y_1, X_2, Y_2 of Problem 19.7, find the probability densities of the subsystems $f_{x_1,x_2}(x_1, x_2)$ and $f_{x_1,y_1}(x_1, y_1)$.

20.19 Under the assumptions of the preceding problem, determine the probability density $f(x_2, y_2 \mid x_1, y_1)$, the conditional expectations and the conditional variances

$$\mathbf{M}[X_2 \mid x_1, y_1], \quad \mathbf{M}[Y_2 \mid x_1, y_1], \quad \mathbf{D}[X_2 \mid x_1, y_1], \quad \mathbf{D}[Y_2 \mid x_1, y_1]$$

$$\text{for} \quad x_1 = 0, \; y_1 = 10.$$

IV NUMERICAL CHARACTERISTICS AND DISTRIBUTION LAWS OF FUNCTIONS OF RANDOM VARIABLES

21. NUMERICAL CHARACTERISTICS OF FUNCTIONS OF RANDOM VARIABLES

Basic Formulas

The expectation and variance of a random variable Y that is a given function, $Y = \varphi(X)$, of a random variable X whose probability density $f(x)$ is known, is given by the formulas

$$\bar{y} = \mathbf{M}[Y] = \int_{-\infty}^{\infty} \varphi(x)f(x)\,dx,$$

$$\mathbf{D}[Y] = \int_{-\infty}^{\infty} [\varphi(x)]^2 f(x)\,dx - \bar{y}^2.$$

In a similar way, one may find the moments and central moments of any order:

$$m_k[Y] = \int_{-\infty}^{\infty} [\varphi(x)]^k f(x)\,dx,$$

$$\mu_k[Y] = \int_{-\infty}^{\infty} [\varphi(x) - \bar{y}]^k f(x)\,dx, \qquad k = 1, 2, \ldots.$$

The foregoing formulas extend to any number of random arguments; if $Y = \varphi(X_1, X_2, \ldots, X_n)$, then

$$m_k[Y] = \int_{-\infty}^{\infty} \underset{(n)}{\cdots} \int_{-\infty}^{\infty} [\varphi(x_1, x_2, \ldots, x_n)]^k f(x_1, x_2, \ldots, x_n)\,dx_1 \cdots dx_n,$$

$$\mu_k[Y] = \int_{-\infty}^{\infty} \underset{(n)}{\cdots} \int_{-\infty}^{\infty} [\varphi(x_1, x_2, \ldots, x_n) - \bar{y}]^k f(x_1, x_2, \ldots, x_n)\,dx_1 \cdots dx_n,$$

where $f(x_1, x_2, \ldots, x_n)$ is the probability density of the system of random variables $(X_1, X_2, \ldots, X_n)$.

For discrete random variables, the integrals in the preceding formulas are replaced by sums, and the densities by probabilities of the corresponding sets of values of $X_1, X_2, \ldots, X_n$.

If the function $\varphi(X_1, X_2, \ldots, X_n)$ is linear; that is,

$$Y = \sum_{j=1}^{n} a_j X_j + b,$$

then

$$\mathbf{M}[Y] = \sum_{j=1}^{n} a_j \mathbf{M}[X_j] + b,$$

$$\mathbf{D}[Y] = \sum_{j=1}^{n} a_j^2 \mathbf{D}[X_j] + \sum_{i=1}^{n} \sum_{j=1}^{n} a_i a_j k_{ij},$$

where k_{ij} is the covariance between the random variables X_i and X_j.

Knowledge of the distribution law of the random arguments for the determination of the moments of the function is unnecessary in some cases. Let $Z = XY$, then $\mathbf{M}[Z] = \mathbf{M}[X]\mathbf{M}[Y] + k_{xy}$. Furthermore, if X and Y are uncorrelated, i.e., the covariance k_{xy} vanishes, then

$$\mathbf{D}[Z] = \mathbf{D}[X]\mathbf{D}[Y] + \bar{x}^2\mathbf{D}[Y] + \bar{y}^2\mathbf{D}[X],$$

$$\mathbf{M}[Z] = \mathbf{M}[X]\mathbf{M}[Y].$$

The last formula can be generalized for any number of independent random variables:

$$\mathbf{M}\left[\prod_{j=1}^{n} X_j\right] = \prod_{j=1}^{n} \mathbf{M}[X_j].$$

If the moments of the linear function

$$Y = \sum_{j=1}^{n} a_j X_j + b$$

of independent random variables exist, they can be determined by the formula

$$m_k[Y] = \frac{1}{i^k} \frac{d^k}{dt^k} e^{ibt} \prod_{j=1}^{n} E_{x_j}(a_j t)\Big|_{t=0}, \qquad k = 1, 2, \ldots,$$

where $E_{x_j}(t) = \int_{-\infty}^{\infty} f_j(x) e^{itx}\, dx$ is the characteristic function of the random variable X_j.

The asymmetry coefficient and the excess of the random variable Y in this case are given by the formulas

$$\mathrm{Sk}[Y] = \frac{\psi'''(t)}{[\psi''(t)]^{3/2}}\bigg|_{t=0}, \qquad \mathrm{Ex}[Y] = \frac{\psi''''(t)}{[\psi''(t)]^2}\bigg|_{t=0},$$

where $\psi(t) = \ln\left\{e^{ibt} \prod_{j=1}^{n} E_{x_j}(a_j t)\right\}$.

SOLUTION FOR TYPICAL EXAMPLES

Example 21.1 A random variable X obeys a binomial distribution law. Find the expectation and variance of the random variable $Y = e^{aX}$.

SOLUTION. The random variable X can assume values 0, 1, 2, ..., n. The probability that it will assume the value m is determined by the formula $P_{n,m} = C_n^m p^m q^{n-m}$.

Therefore,

$$\mathbf{M}[Y] = \sum_{m=0}^{n} y_m P_{n,m} = \sum_{m=0}^{n} e^{am} C_n^m p^m q^{n-m} = (q + pe^a)^n,$$

$$\mathbf{D}[Y] = \sum_{m=0}^{n} y_m^2 P_{n,m} - \bar{y}^2 = \sum_{m=0}^{n} C_n^m (pe^{2a})^m q^{n-m} - \bar{y}^2$$
$$= (q + pe^{2a})^n - (q + pe^a)^{2n}.$$

Example 21.2 The screen of a navigational radar station represents a circle of radius a. As a result of noise, a spot may appear with its center at any point of the circle. Find the expectation and variance of the distance between the center of the spot and the center of the circle.

SOLUTION. The random distance R from the center of the circle to the spot can be expressed in terms of rectangular coordinates X and Y as

$$R = \sqrt{X^2 + Y^2}.$$

The probability density of the system of random variables (X, Y) is known and is given by the formula

$$f(x, y) = \begin{cases} \dfrac{1}{\pi a^2} & \text{for } x^2 + y^2 \leqslant a^2, \\ 0 & \text{for } x^2 + y^2 > a^2. \end{cases}$$

Therefore,

$$\mathbf{M}[R] = \frac{1}{\pi a^2} \iint_{x^2 + y^2 \leqslant a^2} \sqrt{x^2 + y^2}\, dx\, dy = \frac{1}{\pi a^2} \int_0^{2\pi} d\varphi \int_0^a r^2\, dr = \frac{2}{3} a,$$

$$\mathbf{D}[R] = \frac{1}{\pi a^2} \iint_{x^2 + y^2 \leqslant a^2} (x^2 + y^2)\, dx\, dy - \bar{r}^2 = \frac{1}{\pi a^2} \int_0^{2\pi} d\varphi \int_0^a r^3\, dr - \frac{4}{9} a^2 = \frac{a^2}{18}.$$

In a manner similar to that used in Examples 21.1 and 21.2, one can solve Problems 21.1 to 21.14, 21.20 to 21.24, 21.26, 21.27, 21.29 and 21.30.

Example 21.3 A sample of n items is drawn without replacement from a lot of N items, of which $T = Np$ are defective. Find the expectation and variance of the number of defective items in the sample.

SOLUTION. Let X denote the random number of defective items in the sample.

The random variable X can be represented as $X = \sum_{j=1}^{n} X_j$, where the random variable X_j equals 1 if the jth item selected turns out to be defective, and zero otherwise. The probability is p that the value is 1 and, consequently,

$\bar{x}_j = \mathbf{M}[X_j] = 0\cdot(1 - p) + 1\cdot p = p$ (as in Example 6.1, one can show that the probability of obtaining a defective item does not depend on j).

Then

$$\mathbf{M}[X] = \mathbf{M}\left[\sum_{j=1}^{n} X_j\right] = \sum_{j=1}^{n} \mathbf{M}[X_j] = np.$$

If sampling is done without replacement, the random variables X_j are dependent and, hence,

$$\mathbf{D}[X] = \mathbf{D}\left[\sum_{i=1}^{n} X_i\right] = \sum_{i=1}^{n} \mathbf{D}[X_i] + \sum_{i=1}^{n}\sum_{\substack{j=1\\ i\neq j}}^{n} k_{ij},$$

where

$$\mathbf{D}[X_i] = (1 - \bar{x}_i)^2 p + (0 - \bar{x}_i)^2(1 - p) = (1 - p)^2 p + p^2(1 - p) = pq,$$

$$k_{ij} = \mathbf{M}[(X_i - \bar{x}_i)(X_j - \bar{x}_j)] = \mathbf{M}[X_i X_j] - \{\mathbf{M}[X_i]\}^2$$

$$= \mathbf{P}(X_i = 1)\mathbf{P}(X_j = 1 \mid X_i = 1) - p^2 = p\,\frac{Np - 1}{N - 1} - p^2 = -\frac{pq}{N - 1}.$$

Finally,

$$\mathbf{D}[X] = npq\left(1 - \frac{n - 1}{N - 1}\right).$$

Similarly one can solve Problems 21.15 to 21.17, 21.25 and 21.28.

Example 21.4 Find the expectation for the square of the distance between two points selected at random on the boundary of a rectangle.

SOLUTION. By selecting two random points on the boundary of a rectangle, the following unique mutually exclusive events (hypotheses) may occur (see Figure 20): H_1 that the points lie on the same side a, H_2 that the points lie on the same side b, H_3 that the points lie on adjacent sides, H_4 that the points lie on opposite sides a, H_5 that the points lie on opposite sides b.

For the probabilities of these hypotheses, we have

$$\mathbf{P}(H_1) = 2\left(\frac{a}{2p}\cdot\frac{a}{2p}\right) = \frac{a^2}{2p^2},$$

$$\mathbf{P}(H_2) = 2\left(\frac{b}{2p}\cdot\frac{b}{2p}\right) = \frac{b^2}{2p^2},$$

$$\mathbf{P}(H_3) = 8\left(\frac{a}{2p}\cdot\frac{b}{2p}\right) = 2\,\frac{ab}{p^2},$$

$$\mathbf{P}(H_4) = 2\left(\frac{a}{2p}\cdot\frac{a}{2p}\right) = \frac{a^2}{2p^2},$$

$$\mathbf{P}(H_5) = 2\left(\frac{b}{2p}\cdot\frac{b}{2p}\right) = \frac{b^2}{2p^2},$$

where $2p$ is the perimeter of the rectangle.

FIGURE 20

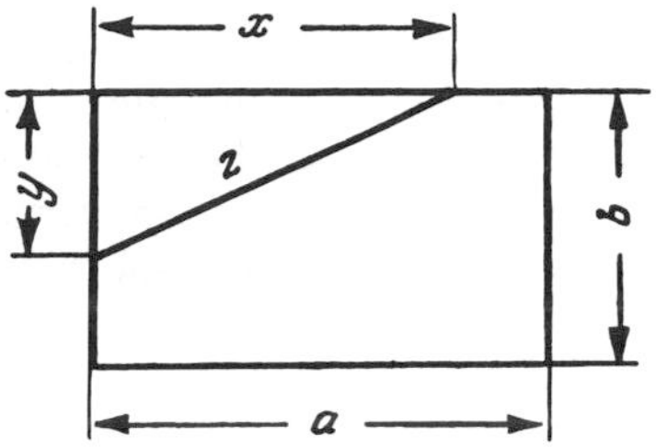

Determine the conditional expectation (i.e., the expectation with the assumption that the hypothesis H_i occurs) for the square of the distance between two points:

$$\mathbf{M}[Z^2 \mid H_1] = \int_0^a \int_0^a f(x, y)(x - y)^2 \, dx \, dy = \frac{1}{a^2} \int_0^a \int_0^a (x - y)^2 \, dx \, dy = \frac{a^2}{6},$$

$$\mathbf{M}[Z^2 \mid H_2] = \frac{1}{b^2} \int_0^b \int_0^b (x - y)^2 \, dx \, dy = \frac{b^2}{6},$$

$$\mathbf{M}[Z^2 \mid H_3] = \frac{1}{ab} \int_0^a \int_0^b (x^2 + y^2) \, dx \, dy = \frac{1}{3}(a^2 + b^2),$$

$$\mathbf{M}[Z^2 \mid H_4] = \mathbf{M}[b^2 + (X - Y)^2] = b^2 + \mathbf{M}[(X - Y)^2] = b^2 + \frac{a^2}{6},$$

$$\mathbf{M}[Z^2 \mid H_5] = \mathbf{M}[a^2 + (X - Y)^2] = a^2 + \mathbf{M}[(X - Y)^2] = a^2 + \frac{b^2}{6}.$$

We find that the total expectation of the random variable Z^2 is:

$$\mathbf{M}[Z^2] = \sum_{j=1}^{5} \mathbf{P}(H_j)\mathbf{M}[Z^2 \mid H_j] = \frac{1}{6p^2}(a^4 + 4a^3b + 6a^2b^2 + 4ab^3 + b^4)$$

$$= \frac{(a + b)^4}{6p^2} = \frac{p^2}{6}.$$

Problems 21.18 and 21.19 can be solved similarly.

PROBLEMS

21.1 Find the expectation of the length of a chord joining a given point on a circle of radius a with an arbitrary point on the circle.

21.2 Find the expectation of the length of a chord drawn in a circle of radius a, perpendicular to a chosen diameter and crossing it at an arbitrary point.

21.3 Some steel balls are sorted according to their size so that the group with rated size 10 mm. contains balls that pass through a circular slot of 10.1 mm. and do not pass through a slot of diameter 9.9 mm. The balls are made of steel with specific weight 7.8 g./cc. Find the expectation and variance of the weight of a ball belonging to a given group if the distribution of the radius in the tolerance range is uniform.

21.4 A fixed point O is located at altitude h above the endpoint A of a horizontal segment AK of length l. A point B is randomly selected on AK. Find the expectation of the angle between segments OA and OB.

21.5 The legs of a compass, each 10 cm. long, make a random angle φ, whose values are uniformly distributed over the interval [0, 180°]. Find the expectation of the distance between the ends of the legs.

21.6 A random variable X obeys a normal distribution law. Find the expectation of the random variable Y if

$$Y = \exp\left\{\frac{\bar{x}^2 - 2\bar{x}X}{2\sigma_x^2}\right\}.$$

21.7 The vertex C of the right angle of an isosceles right triangle is connected by a segment with an arbitrary point M on the base; the length of the base is $2m$. Find the expectation of the length of segment CM.

21.8 A point is selected at random on a circumference of radius a. Find the expectation of the area of a square whose side equals the abscissa of this point.

21.9 An urn contains white and black balls. The probability of drawing a white ball is p and drawing a black one, q. A number n of balls are drawn one by one with replacement. What is the expectation of the number of instances in which a white ball follows a black one?

21.10 A system of random variables X, Y obeys the normal distribution law

$$f(x, y) = \frac{\rho^2}{\pi E^2} \exp\left\{-\rho^2 \frac{x^2 + y^2}{E^2}\right\}.$$

Find the expectation of the random variable

$$R = \sqrt{X^2 + Y^2}.$$

21.11 Two points X and Y are randomly selected in a semicircle of radius a. These points and one end of the bounding diameter form a triangle. What is the expectation of the area of this triangle?

21.12 Three points A, B and C are placed at random on a circumference of unit radius. Find the expectation of the area of the triangle ABC.

21.13 The number of cosmic particles reaching a given area in time t obeys Poisson's law

$$P_m = \frac{(\lambda t)^m}{m!} e^{-\lambda t}.$$

The energy of a particle is a random variable characterized by a mean value $\bar{w}$. Find the average energy gained by the area per unit time.

21.14 An electronic system contains n elements. The probability of failure (damage) of the kth element is p_k ($k = 1, 2, \ldots, n$). Find the expectation of the number of damaged elements.

21.15 A system consisting of n identical units stops operating if at least one unit fails, an event that occurs with equal probability for all the units. The probability that the system will stop during a given cycle is p. A new cycle starts after the preceding one has been completed or, if the preceding cycle has not been completed, after the damaged unit has been repaired. Find the expectation of the number of units subject to repairs at least once during m cycles.

21.16 There are n units operating independently of each other, and carrying out a series of consecutive cycles. The probability of failure for any unit during one cycle is p. A new cycle starts after the preceding one is completed (separately for each unit), or after repairs if the preceding cycle is not completed. Find the probability of the number of units subject to repairs at least once if each unit operates for m cycles.

21.17 In an electronic device the number of elements failing to operate during some time interval obeys Poisson's law with parameter a. The duration t_m of repairs depends on the number m of damaged elements and is given by $t_m = T(1 - e^{-\alpha m})$. Find the expectation of the duration of repairs and the loss caused by delay if the loss is proportional to the square of the duration of repairs: $S_m = kt_m^2$.

21.18 A system has n units operating independently. If at least one unit fails, the system will stop. The probability of occurrence of this event is p and the failures of all units are equally probable. A new cycle starts after the completion of the preceding one or after the damaged unit has been repaired if the preceding cycle has not been completed.

The system must run $2m$ cycles and, moreover, after the first m cycles ($m < n/2$), all the units subject to repairs at least once are discarded and a number m of cycles are repeated with the remaining units under the previous conditions. Find the expectation of the number of units repaired at least once after two series of m cycles each.

21.19 A marksman fires two series of m shots each at n targets. The shots are fired successively at each target and the detailed results of each series of shots are not recorded. The bullet can strike, with probability p, only the target aimed at by the marksman. A target is considered hit if at least one bullet reaches it. The second series is fired after the targets hit in the first series are noted. The rules are the same as in the first series, except that shots are not fired at those targets hit in the first series. Find the expectation of the number of targets hit during the whole experiment for $n = m = 8$ and $n \geqslant 2m$.

21.20 Two points are selected at random on adjacent sides of a rectangle with sides a and b. Find the expectation of the distance between these two points.

21.21 Find the expectation of the distance between two randomly selected points on opposite sides of a rectangle with sides a, b.

21.22 Obtain the formulas for the expectation and variance of the number of occurrences of an event in n independent trials if the probability for its realization varies from one trial to another and equals p_k ($k = 1, 2, \ldots, n$) at the kth trial.

21.23 Ten weights are placed on a scale. The precision of manufacture of each weight is characterized by a mean error of 0.1 g. The precision in the process of weighing is characterized by a mean error of 0.02 g. Find the mean error in the determination of the mass of a body.

21.24 Two points are taken at random on a segment of length l. Find the expectation and variance of the distance between them.

21.25 The probability density of a system of random variables (X, Y) is specified by the formula

$$f(x, y) = \frac{1}{300\pi\sqrt{0.75}} \exp\left\{-\frac{1}{1.5}\left[\frac{(x-5)^2}{100} + \frac{y(x-5)}{150} + \frac{y^2}{225}\right]\right\}.$$

Find the expectation and variance of the random variable $Z = aX + bY$.

21.26 A random variable X obeys the normal distribution law

$$f(x) = \frac{\rho}{E_x\sqrt{\pi}} \exp\left\{-\rho^2 \frac{x^2}{E_x^2}\right\}.$$

Evaluate the expectation and variance of the random variable $Y = |X|$.

21.27 A random variable X obeys Poisson's law. Find the expectation and variance of the random variable $Y = \cos bX$.

21.28 The distance from a lighthouse is given as the arithmetic mean of three measurements. The relation between errors depends on the rate of measurements and is characterized by the following values of the correlation coefficients:

(a) for a rate of 3 sec., $r_{12} = r_{23} = 0.9$, $r_{13} = 0.7$;
(b) for a rate of 5 sec., $r_{12} = r_{23} = 0.7$, $r_{13} = 0.4$;
(c) for a rate of 12 sec., $r_{ij} = 0$, $j \neq i$.

Determine the value of the variance for the arithmetic mean in measurements with different rates if the errors of each measurement are characterized by a variance of 30 sq. m.

21.29 A random variable X obeys a distribution law with a probability density

$$f(x) = \begin{cases} \dfrac{x^2}{a^3}\sqrt{\dfrac{2}{\pi}} \exp\left\{-\dfrac{x^2}{2a^2}\right\} & \text{for} \quad x \geqslant 0, \\ 0 & \text{for} \quad x \leqslant 0. \end{cases}$$

The probability density of a random variable Y is given by the formula

$$f(y) = \begin{cases} \dfrac{2\rho^2}{E^2}\, y \exp\left\{-\dfrac{\rho^2 y^2}{E^2}\right\} & \text{for} \quad y \geqslant 0, \\ 0 & \text{for} \quad y \leqslant 0. \end{cases}$$

Determine the expectation and variance of the random variable $Z = X - Y$ if the random variables X and Y are independent.

21.30 Given a random point in the plane with coordinates (X, Y), and $\bar{x} = 10$, $\bar{y} = -10$, $\sigma_x = 100$, $\sigma_y = 20$, $k_{xy} = 0$, find the expectation and variance of the distance Z from the origin to the projection of this point on OZ, which makes with OX an angle $\alpha = 30°$.

21.31 Determine the correlation coefficient for the random variables X and Y if X is a centralized random variable, and $Y = X^n$, where n is a positive integer.

21.32 Find the expectation and variance of a random variable $Z = X(Y - \bar{y})$ if the probability density of the system (X, Y) is given by the formula

$$f(x, y) = \frac{x^2}{\sigma^3 \Delta\sqrt{2\pi}} \exp\left\{-\frac{x^2}{2\sigma^2\Delta^2}[(y - \bar{y})^2 + \Delta^2]\right\}.$$

21.33 A wheel is spun and then slows down because of friction. When the wheel stops, a fixed radius a makes a random angle φ with the horizontal diameter; φ is distributed uniformly over the interval 0 to 360°. Find the expectation and variance of the distance from the end of radius a to the horizontal diameter.

21.34 As a result of a central force, a mass point describes an elliptic trajectory. The major semi-axis a and the eccentricity of the ellipse e are known. Assuming that it is equally probable to sight the moving point at any instant, find the expectation and variance of the distance at the instant of observation if the observer is located at the center of attraction, at one of the foci of the ellipse, and the distance R to the point is given by the formula $R = \alpha(1 - e^2)/(1 - \cos u)$, where u is the angle made by the radius-vector R with the major axis a. (In the case of a motion in a central field, the sector velocity $R^2\, du/dt = \text{const.}$)

22. THE DISTRIBUTION LAWS OF FUNCTIONS OF RANDOM VARIABLES

Basic Formulas

The probability density $f_y(y)$ of a random variable Y, where $Y = \varphi(X)$ is a monotonic function (i.e., the inverse function $X = \psi(Y)$ is single-valued), is defined by the formula

$$f_y(y) = f_x[\psi(y)]|\psi'(y)|.$$

If the inverse $X = \psi(Y)$ is not single-valued, i.e., to one value of Y there correspond several values of X—$\psi_1(y), \psi_2(y), \psi_3(y), \ldots, \psi_k(y)$ (Figure 21)—then the probability density of Y is given by the formula

$$f_y(y) = \sum_{j=1}^{k} f_x[\psi_j(y)]|\psi_j'(y)|.$$

For a function of several random arguments, it is proper to start from the formula for the distribution function $F_y(y)$. For example, $Y = \varphi(X_1, X_2)$, and let $f_x(x_1, x_2)$ be the probability density of the system of random variables (X_1, X_2). If D_y is a region in the plane X_1OX_2 for which $Y < y$, then the distribution function is

$$F_y(y) = \iint_{D_y} f_x(x_1, x_2)\, dx_1\, dx_2,$$

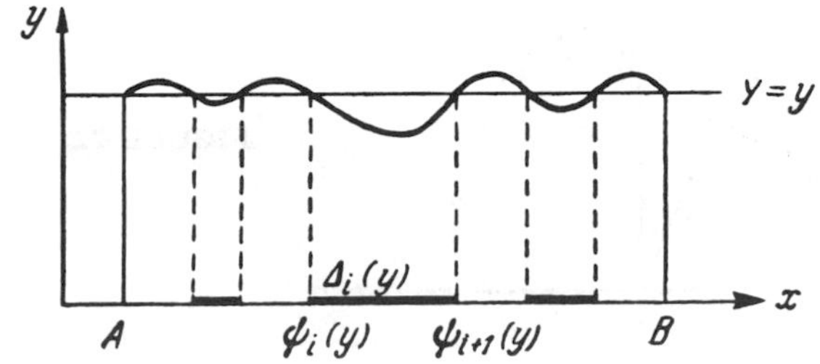

FIGURE 21

and the probability density of the random variable Y is $f_y(y) = dF_y(y)/dy$. In the general case, if the Jacobian determinant for the transformation of the random variables $(X_1, X_2, \ldots, X_n)$ to the random variables $(Y_1, Y_2, \ldots, Y_n)$ is

$$D = \frac{\partial(x_1, x_2, \ldots, x_n)}{\partial(y_1, y_2, \ldots, y_n)} = \begin{vmatrix} \frac{\partial x_1}{\partial y_1} & \frac{\partial x_1}{\partial y_2} & \cdots & \frac{\partial x_1}{\partial y_n} \\ \frac{\partial x_2}{\partial y_1} & \frac{\partial x_2}{\partial y_2} & \cdots & \frac{\partial x_2}{\partial y_n} \\ \cdot & \cdot & \cdot & \cdot \\ \frac{\partial x_n}{\partial y_1} & \frac{\partial x_n}{\partial y_2} & \cdots & \frac{\partial x_n}{\partial y_n} \end{vmatrix}$$

and if this is a one-to-one transformation, then

$$f_y(y_1, y_2, \ldots, y_n) = |D| f_x(x_1, x_2, \ldots, x_n),$$

in which $x_1, \ldots, x_n$ are expressed in terms of $y_1, \ldots, y_n$.

Solution for Typical Examples

Example 22.1 A straight line is drawn at random through a point $(0, l)$ (Figure 22). Find the probability density of the random variable $\eta = l \cos \varphi$.

Solution. The angle φ is a random variable, uniformly distributed over the interval $(0, \pi)$ (Figure 22).

Since here the inverse $\psi(\eta)$ is single-valued (when angle φ varies from 0 to π, the function decreases monotonically), to determine the probability density for η, we apply the formula

$$f_\eta(\eta) = f_\varphi[\psi(\eta)]|\psi'(\eta)|,$$

where

$$\psi(\eta) = \arccos \frac{\eta}{l},$$

$$|\psi'(\eta)| = \frac{1}{l\sqrt{1 - \left(\frac{\eta}{l}\right)^2}}, \qquad f_\varphi(\varphi) = \begin{cases} \frac{1}{\pi} & \text{for } 0 \leqslant \varphi \leqslant \pi \\ 0 & \text{otherwise}. \end{cases}$$

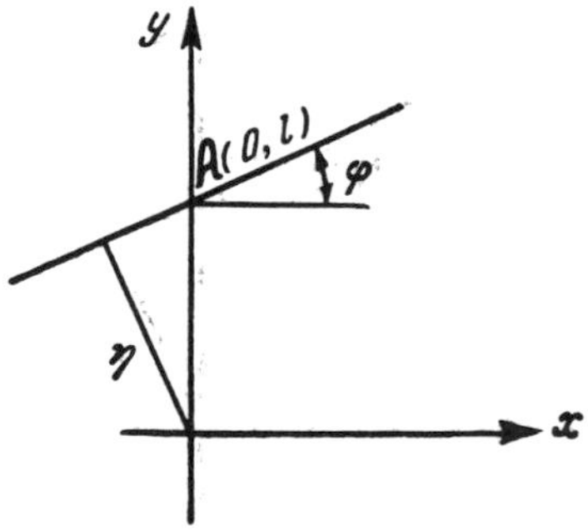

FIGURE 22

Finally, we have

$$f_\eta(\eta) = \begin{cases} \dfrac{1}{\pi\sqrt{l^2 - \eta^2}} & \text{for} \quad |\eta| \leqslant l, \\ 0 & \text{for} \quad |\eta| > l. \end{cases}$$

Similarly one can solve Problems 22.2, 22.5 to 22.7, 22.9 to 22.13 and 22.19.

Example 22.2 A random variable Y is given by the formula

$$Y = \begin{cases} +\sqrt{X} & \text{for} \quad X \geqslant 0, \\ +\sqrt{-X} & \text{for} \quad X < 0. \end{cases}$$

Find the probability density of Y if X is a normal random variable with parameters $\bar{x} = 0$, $\mathbf{D}[X] = 1$.

SOLUTION. In this example, the inverse is two-valued (Figure 23); since to one value of Y there correspond two values of X,

$$X_1 = -Y^2 \equiv \psi_1(Y)$$

and

$$X_2 = Y^2 \equiv \psi_2(Y),$$

by the general formula, we have

$$f_y(y) = f_x(-y^2)\left|-\frac{dy^2}{dy}\right| + f_x(y^2)\left|\frac{dy^2}{dy}\right|$$

$$= \begin{cases} \dfrac{4y}{\sqrt{2\pi}} \exp\left\{-\dfrac{y^4}{2}\right\} & \text{for} \quad 0 \leqslant y < \infty, \\ 0 & \text{for} \quad y < 0. \end{cases}$$

Problems 22.3, 22.4 and 22.8 can be solved in a similar manner.

Example 22.3 The position of a random point with coordinates (X, Y) is equally probable inside a square with side 1 and whose center coincides with the origin. Determine the probability density of the random variable $Z = XY$.

SOLUTION. We shall consider separately two cases: (a) $0 < z < 1/4$, and (b) $-1/4 < z < 0$. For these cases, we shall construct in the plane two hyperbolas with equations $z = xy$.

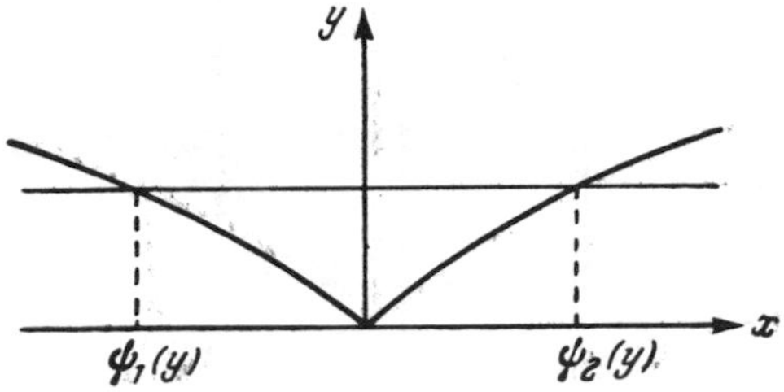

FIGURE 23

In Figure 24A and B, a region is shaded inside which the condition $Z < z$ is satisfied.

The distribution function of the random variable Z is defined for $0 < z < 1/4$ as

$$F_z(z) = \mathbf{P}(Z < z) = 1 - \mathbf{P}(Z > z) = 1 - 2S_{D'_z}$$
$$= 1 - 2\int_{2z}^{1/2} dy \int_{z/y}^{1/2} dx = \frac{1}{2} + 2z - 2z \ln 4z,$$

where $S_{D'_z}$ is the area of the region D'_z;

for $-1/4 < z < 0$,

$$F_z(z) = 2S_{D_z} = 2\int_{-2z}^{1/2} dy \int_{-1/2}^{z/y} dx = \frac{1}{2} + 2z - 2z \ln(-4z).$$

Differentiating these expressions with respect to z, we obtain the probability density:

for $0 < z < 1/4$,

$$f_z(z) = \frac{d}{dz} F_z(z) = -2 \ln 4z;$$

for $-1/4 < z < 0$,

$$f_z(z) = \frac{d}{dz} F_z(z) = -2 \ln(-4z).$$

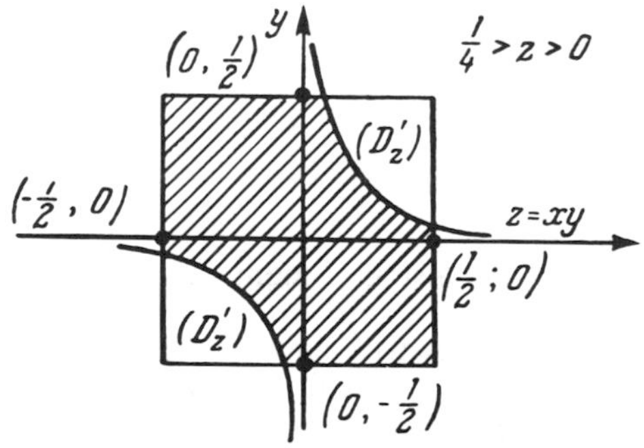

FIGURE 24A

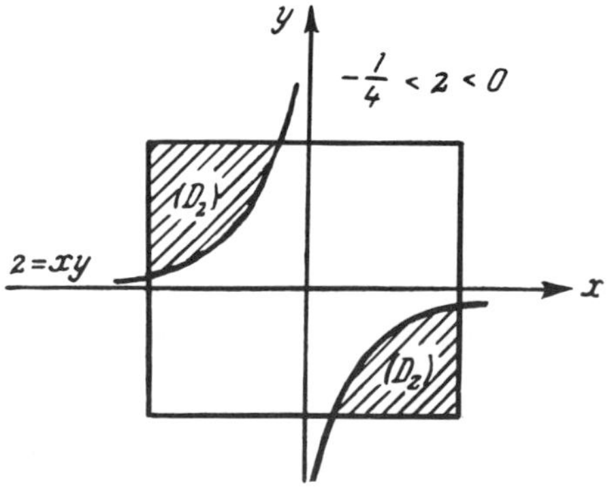

FIGURE 24B

Finally, the probability density for the random variable $Z = XY$ can be written as follows:

$$f_z(z) = \begin{cases} -2 \ln 4|z| & \text{for} \quad |z| < \frac{1}{4}, \\ 0 & \text{for} \quad |z| \geqslant \frac{1}{4}. \end{cases}$$

Problems 22.16 to 22.19 and 22.21 are solved similarly.

Example 22.4 A system of random variables (X, Y) is normally distributed with the probability density

$$f(x, y) = \frac{1}{2\pi\sigma^2} \exp\left\{-\frac{x^2 + y^2}{2\sigma^2}\right\}.$$

Find the probability density of the system (R, Φ) if

$$X = R \cos \Phi,$$

$$Y = R \sin \Phi.$$

SOLUTION. To determine the probability density of the system (R, Φ), apply the formula

$$f(r, \varphi) = f[x(r, \varphi), y(r, \varphi)] \left| \frac{\partial(x, y)}{\partial(r, \varphi)} \right|,$$

where $\partial(x, y)/\partial(r, \varphi)$ is the Jacobian determinant of the transformation from the given system to the system (R, Φ):

$$\frac{\partial(x, y)}{\partial(r, \varphi)} = \begin{vmatrix} \frac{\partial x}{\partial r} & \frac{\partial x}{\partial \varphi} \\ \frac{\partial y}{\partial r} & \frac{\partial y}{\partial \varphi} \end{vmatrix} = r.$$

Therefore,

$$f(r, \varphi) = \frac{r}{2\pi\sigma^2} \exp\left\{-\frac{r^2 \cos^2 \varphi + r^2 \sin^2 \varphi}{2\sigma^2}\right\} = \frac{r}{2\pi\sigma^2} \exp\left\{-\frac{r^2}{2\sigma^2}\right\}.$$

The random variables R and Φ are independent so that

$$f(r, \varphi) = f_r(r) f_\varphi(\varphi),$$

where $f_r(r) = (r/\sigma^2)e^{-r^2/2\sigma^2}$ is Rayleigh's law, and $f_\varphi(\varphi)$ is the uniform distribution law.

Similarly one can solve Problems 22.22, 22.23 and 22.25 to 22.27.

PROBLEMS

22.1 The distribution function of a random variable X is $F_x(x)$. Find the distribution function of the random variable $Y = aX + b$.

22.2 Given the probability density $f(x)$ of a random variable X $(0 < x < \infty)$, find the probability density of the random variable $Y = \ln X$.

22.3 Find the probability density of the random variable $Z = aX^2$ if X is a normal random variable, $\bar{x} = 0$, $\mathbf{D}[X] = \sigma_2$ and $a > 0$.

22.4 Evaluate the probability density of the random variable $Y = |X|$ if X is a normal random variable for which $\bar{x} = 0$, and the mean deviation E is given.

22.5 A random variable X is uniformly distributed over the interval (0, 1) and related to Y by the equation $\tan \pi Y/2 = e^X$. Find the probability density of the random variable Y.

22.6 Find the probability density of the volume of a cube, whose edge X is a random variable uniformly distributed in the interval $(0, a)$.

22.7 A straight line is drawn at random through the point $(0, l)$. Find the probability density of the x-intercept of this line with the Ox-axis.

22.8 A random variable X is uniformly distributed over the interval $(-T/2; T/2)$. Find the probability density of the random variable $Y = a \sin (2\pi/T)X$.

22.9 A random variable X obeys Cauchy's distribution law,

$$f_x(x) = \frac{1}{\pi(1 + x^2)}.$$

Find the probability density of the random variable Y if (a) $Y = 1 - X^3$, (b) $Y = aX^2$, (c) $Y = \arctan X$.

22.10 Determine the probability density of the random variable $Y = X^n$, where n is a positive integer, if the probability density for X is

$$f_x(x) = \frac{1}{\pi} \frac{a}{a^2 + x^2}.$$

22.11 A random variable X is distributed over the interval $(0, \infty)$ with the probability density $f_x(x) = e^{-x}$. Evaluate the probability density of the random variable Y if (a) $Y^2 = X$ and the signs of Y are equally probable, (b) $Y = +\sqrt{X}$.

22.12 A random variable X obeys Pearson's distribution law:

$$f_x(x) = \begin{cases} \dfrac{\Gamma(k + 1.5)}{\sqrt{\pi}\,\Gamma(k + 1)} (1 - x^2)^k & \text{for } |x| \leqslant 1, \\ 0 & \text{for } |x| > 1. \end{cases}$$

Find the probability density of the random variable $Y = \arcsin X$.

22.13 A random variable X is uniformly distributed in the interval (0, 1). Evaluate the probability density of the random variable Y if:

(a) $$X = \frac{1}{2}\left[1 + \frac{2}{\sqrt{2\pi}} \int_0^Y \exp\left\{-\frac{t^2}{2}\right\} dt\right],$$

(b) $$X = \frac{1}{2}\left[1 + \frac{2}{\sqrt{2\pi}} \int_0^{(Y - \bar{y})/\sigma_y} \exp\left\{-\frac{t^2}{2}\right\} dt\right].$$

22.14 The random variables X and Y are connected by the functional dependence $Y = F_x(X)$. The random variable X is uniformly distributed over the interval (a, b) and $F_x(x)$ is its distribution function. Find the probability density of random variable Y.

22.15 A random variable X is uniformly distributed over the interval (0, 1). Assume that there is a function $f_t(t) \geqslant 0$ satisfying the condition $\int_{-\infty}^{\infty} f_t(t)\,dt = 1$. The random variables X and Y are related by the equation $X = \int_{-\infty}^{Y} f_t(t)\,dt$. Prove that $f_t(t)$ is the probability density of random variable Y.

22.16 A system of random variables (X, Y) obeys the normal distribution law

$$f(x, y) = \frac{1}{\sigma_x \sigma_y \sqrt{2\pi}} \exp\left\{-\frac{1}{2}\left(\frac{x^2}{\sigma_x^2} + \frac{y^2}{\sigma_y^2}\right)\right\}.$$

What distribution law does the random variable $Z = X - Y$ obey?

22.17 Find the probability density of the random variable $Z = XY$ if:

(a) the probability density $f(x, y)$ of the system of random variables (X, Y) is given,

(b) X and Y are independent random variables with probability densities

$$f_x(x) = \frac{1}{\sqrt{2\pi}} \exp\left\{-\frac{x^2}{2}\right\} \qquad (-\infty < x < \infty),$$

$$f_y(y) = \begin{cases} y \exp\left\{-\frac{y^2}{2}\right\} & \text{for } 0 \leqslant y \leqslant \infty, \\ 0 & \text{for } y \leqslant 0, \end{cases}$$

(c) X and Y are independent normal random variables with $\bar{x} = \bar{y} = 0$ and variances σ_x^2 and σ_y^2, respectively.

(d) X and Y are independent random variables with probability densities

$$f_x(x) = \begin{cases} \dfrac{1}{\pi\sqrt{1 - x^2}} & \text{for } |x| \leqslant 1, \\ 0 & \text{for } |x| > 1, \end{cases}$$

$$f_y(y) = \begin{cases} y \exp\left\{-\frac{y^2}{2}\right\} & \text{for } 0 \leqslant y \leqslant \infty, \\ 0 & \text{for } y \leqslant 0. \end{cases}$$

22.18 Find the probability density of the random variable $Z = X/Y$ if:

(a) the probability density $f(x, y)$ of the system of random variables (X, Y) is given,

(b) X and Y are independent random variables obeying Rayleigh's distribution law

$$f_x(x) = \begin{cases} \dfrac{x}{a^2} \exp\left\{-\dfrac{x^2}{2a^2}\right\} & \text{for } x \geqslant 0, \\ 0 & \text{for } x \leqslant 0, \end{cases}$$

$$f_y(y) = \begin{cases} \dfrac{y}{a^2} \exp\left\{-\dfrac{y^2}{2a^2}\right\} & \text{for } y \geqslant 0, \\ 0 & \text{for } y \leqslant 0, \end{cases}$$

(c) X and Y are independent random variables with probability densities

$$f_x(x) = \frac{1}{\sqrt{\frac{2\pi}{n}}} \exp\left\{-\frac{nx^2}{2}\right\} \qquad (-\infty < x < \infty),$$

$$f_y(y) = \begin{cases} 2\left(\frac{n}{2}\right)^{n/2} \dfrac{y^{n-1} \exp\left\{-\frac{ny^2}{2}\right\}}{\Gamma\left(\frac{n}{2}\right)} & \text{for} \quad y \geq 0, \\ 0 & \text{for} \quad y < 0, \end{cases}$$

(d) the system of random variables (X, Y) obeys the normal distribution law

$$f(x, y) = \frac{1}{2\pi\sigma_x\sigma_y\sqrt{1 - r^2}} \exp\left\{-\frac{1}{2(1 - r^2)}\left(\frac{x^2}{\sigma_x^2} - 2r\frac{xy}{\sigma_x\sigma_y} + \frac{y^2}{\sigma_y^2}\right)\right\}.$$

22.19 Find the probability density for the modulus of the radius-vector $R = \sqrt{X^2 + Y^2}$ if:

(a) the probability density $f(x, y)$ for the system of random variables (X, Y) is given,

(b) the random variables X and Y are independent and obey the same normal distribution law with zero expectation and mean deviation E,

(c) the probability density for the system of random variables (X, Y) is given by the formula

$$f(x, y) = \begin{cases} \dfrac{1}{\pi a^2} & \text{for} \quad x^2 + y^2 \leq R^2, \\ 0 & \text{for} \quad x^2 + y^2 > R^2, \end{cases}$$

(d) X and Y are independent normal random variables with the probability density

$$f(x, y) = \frac{1}{2\pi\sigma^2} \exp\left\{-\frac{(x - h)^2 + y^2}{2\sigma^2}\right\},$$

(e) the random variables X and Y are independent and obey a normal distribution law with $\bar{x} = \bar{y} = 0$, and variances σ_x^2 and σ_y^2, respectively.

22.20 A system of random variables (X, Y) has the probability density

$$f(x, y) = \frac{1}{2\pi\sigma_x\sigma_y\sqrt{1 - r^2}}$$

$$\times \exp\left\{-\frac{1}{2(1 - r^2)}\left[\left(\frac{x - \bar{x}}{\sigma_x}\right)^2 + \left(\frac{y - \bar{y}}{\sigma_y}\right)^2 - 2r\frac{(x - \bar{x})(y - \bar{y})}{\sigma_x\sigma_y}\right]\right\}.$$

Find the linear transformation leading from random variables X, Y to the independent random variables U, V. Evaluate the mean-square deviations of the new random variables.

22.21 Both roots of the quadratic equation $x^2 + \alpha x + \beta = 0$ can take all values from -1 to $+1$ with equal probabilities. Evaluate the probability density for the coefficients α and β.

22.22 The rectangular coordinates (X, Y) of a random point are dependent random variables, and $\bar{x}$, $\bar{y}$, σ_x, σ_y, r_{xy} are given. Find the probability density of the polar coordinates (T, φ) of this point if

$$\frac{X - \bar{x}}{\sigma_x} = T \cos \Phi, \qquad \frac{Y - \bar{y}}{\sigma_y} = T \sin \Phi.$$

What distribution laws do T and Φ obey if $r_{xy} = 0$?

22.23 Let $S = S_0 + V_0 t + (At^2/2)$, where S_0, V_0 and A are normal random variables whose expectations and covariance matrix are known. Evaluate the probability density $f(s \mid t)$.

22.24 Find the probability density of the nonnegative square root of the arithmetic mean for squares of normal centralized random variables, $Y = \sqrt{1/n \sum_{j=1}^{n} X_j^2}$, if the variance $\mathbf{D}[X_j] = \sigma^2$ $(j = 1, 2, \ldots, n)$.

22.25 The rectangular coordinates of a random point $(X_1, X_2, \ldots, X_n)$ have the probability density

$$f(x_1, x_2, \ldots, x_n) = \frac{1}{(2\pi)^{n/2}\sigma^n} \exp\left\{-\frac{1}{2\sigma^2} \sum_{j=1}^{n} x_j^2\right\}.$$

Find the probability density for n-dimensional spherical coordinates of this point $R, \Phi_1, \Phi_2, \ldots, \Phi_n$ if:

$$\begin{aligned}
X_1 &= R \sin \Phi_1, \\
X_2 &= R \sin \Phi_2 \cos \Phi_1, \\
X_3 &= R \sin \Phi_3 \cos \Phi_1 \cos \Phi_2, \\
&\cdots\cdots\cdots\cdots \\
X_{n-1} &= R \sin \Phi_{n-1} \cos \Phi_1 \cos \Phi_2 \cdots \cos \Phi_{n-2}, \\
X_n &= R \cos \Phi_1 \cos \Phi_2 \cdots \cos \Phi_{n-1}.
\end{aligned}$$

22.26 Two systems of random variables $(X_1, X_2, \ldots, X_n)$ and $(Y_1, Y_2, \ldots, Y_n)$ are related by linear equations

$$X_k = \sum_{j=1}^{n} a_{kj} Y_j \qquad (k = 1, 2, \ldots, n),$$

where $|a_{ij}| \neq 0$. Evaluate the probability density $f_y(y_1, y_2, \ldots, y_n)$ if the probability density $f_x(x_1, x_2, \ldots, x_n)$ is given.

22.27 Find the distribution law of the system of random variables (R, Θ), where $R = \sqrt{X^2 + Y^2 + Z^2}$ is the radius-vector of a random point in space, and $\Theta = \arcsin Y/R$ is the latitude, if the probability density of the rectangular coordinates (X, Y, Z) is $f(x, y, z)$.

23. THE CHARACTERISTIC FUNCTIONS OF SYSTEMS AND FUNCTIONS OF RANDOM VARIABLES

Basic Formulas

We define the characteristic function of a system of random variables $(X_1, X_2, \ldots, X_n)$ as the expectation of the function $\exp\{i \sum_{k=1}^{n} u_k X_k\}$, where u_k $(k = 1, 2, \ldots, n)$ are real quantities, and $i = \sqrt{-1}$:

$$E_{x_1,x_2,\ldots,x_n}(u_1, u_2, \ldots, u_n) = \mathbf{M}\left[\exp\left\{i \sum_{k=1}^{n} u_k X_k\right\}\right].$$

For continuous random variables,

$$E_{x_1,x_2,\ldots,x_n}(u_1, u_2, \ldots, u_n) = \int_{-\infty}^{\infty}\int_{-\infty}^{\infty}\cdots\int_{-\infty}^{\infty} \exp\left\{i \sum_{k=1}^{n} u_k x_k\right\} f(x_1, x_2, \ldots, x_n)\, dx_1\, dx_2 \cdots dx_n.$$

The characteristic function of a system of independent random variables equals the product of the characteristic functions of the random variables contained in the system:

$$E_{x_1,x_2,\ldots,x_n}(u_1, u_2, \ldots, u_n) = \prod_{j=1}^{n} E_{x_j}(u_j).$$

For a multidimensional normal distribution with expectations $\bar{x}_1, \bar{x}_2, \ldots, \bar{x}_n$ and covariance matrix

$$\|k_{rs}\| = \begin{Vmatrix} k_{11} & k_{12} & \cdots & k_{1n} \\ k_{21} & k_{22} & \cdots & k_{2n} \\ \cdot & \cdot & \cdot & \cdot \\ k_{n1} & k_{n2} & \cdots & k_{nn} \end{Vmatrix}$$

we have

$$E(u_1, u_2, \ldots, u_n) = \exp\left\{i \sum_{r=1}^{n} u_r \bar{x}_r - \frac{1}{2} \sum_{r=1}^{n} \sum_{s=1}^{n} k_{rs} u_r u_s\right\}.$$

If the appropriate moments of a system of random variables exist,

$$\mathbf{M}[X_1^{r_1} X_2^{r_2} \cdots X_n^{r_n}] = \exp_i\left\{-\sum_{k=1}^{n} r_k\right\} \frac{\partial^{r_1+r_2+\cdots+r_n} E_{x_1,x_2,\ldots,x_n}(u_1, u_2, \ldots, u_n)}{\partial u_1^{r_1}\, \partial u_2^{r_2} \cdots \partial u_n^{r_n}}\Bigg|_{u_1=u_2=\cdots=u_n=0}.$$

If the random variable $Y = \varphi(X)$, then

$$E_y(u) = \mathbf{M}[e^{iuY}] = \int_{-\infty}^{\infty} e^{iu\varphi(x)} f(x)\, dx.$$

The characteristic function of a system of random variables, $(Y_1, Y_2, \ldots, Y_n)$, of which each is a function of other random variables

$$Y_1 = \varphi_1(X_1, X_2, \ldots, X_m),$$
$$Y_2 = \varphi_2(X_1, X_2, \ldots, X_m),$$
$$\cdots\cdots\cdots$$
$$Y_n = \varphi_n(X_1, X_2, \ldots, X_m),$$

equals

$$E_{y_1,\ldots,y_n}(u_1, \ldots, u_n)$$
$$= \int_{-\infty}^{\infty} \cdots \int_{-\infty}^{\infty} \exp\left\{i \sum_{k=1}^{n} u_k \varphi_k(x_1, \ldots, x_m)\right\} f(x_1, \ldots, x_m)\, dx_1 \cdots dx_m.$$

The characteristic function of a subsystem of random variables can be obtained from the characteristic functions of the system by replacing the variables u_k, corresponding to random variables not in the subsystem, by zeros.

Solution for Typical Examples

Example 23.1 A particle starts from the origin and moves in a certain direction for a distance l_1. Then it changes its direction many times, making a random walk: for a distance l_2, then for a distance l_3, and so forth. The trajectory of the wandering particle consists thus of segments of lengths $l_1, l_2, \ldots, l_n$, the direction of each being determined by the angle α_k made with the Ox-axis. These angles are uniformly distributed in the interval $(0, 2\pi)$, and they are independent. Find the characteristic function of the coordinate X of the end-point of the trajectory and the corresponding probability density.

Solution. The coordinate X is determined as the sum of the projections of segments l_k on the Ox-axis:

$$X = \sum_{k=1}^{n} l_k \cos \alpha_k.$$

Since α_k are independent,

$$f(\alpha_1, \alpha_2, \ldots, \alpha_n) = \prod_{k=1}^{n} f(\alpha_k),$$

and

$$f(\alpha_k) = \begin{cases} \dfrac{1}{2\pi} & \text{for } 0 \leqslant \alpha_k \leqslant 2\pi, \\ 0 & \text{for } \alpha_k < 0,\ \alpha_k > 2\pi. \end{cases}$$

Therefore,

$$E_x(u) = \int_{-\infty}^{\infty} \cdots \int_{-\infty}^{\infty} \exp\left\{iu \sum_{k=1}^{n} l_k \cos \alpha_k\right\} f(\alpha_1, \ldots, \alpha_n)\, d\alpha_1 \cdots d\alpha_n$$
$$= \prod_{k=1}^{n} \int_0^{2\pi} \exp\{iul_k \cos \alpha_k\} \frac{d\alpha_k}{2\pi} = \prod_{k=1}^{n} J_0(l_k u),$$

where J_0 is the Bessel function of the first kind of zero order.

From this,

$$f(x) = \frac{1}{2\pi} \int_{-\infty}^{\infty} e^{-iux} \prod_{k=1}^{n} J_0(l_k u)\, du,$$

or

$$f(x) = \frac{1}{2\pi} \int_{-\infty}^{\infty} \cos ux \prod_{k=1}^{n} J_0(l_k u)\, du.$$

Example 23.2 Given the covariance matrix $\|k_{rs}\|$ of a system of six normal random variables $X_1, X_2, \ldots, X_6$ with zero expectations, evaluate the expectation of the product $X_3^2X_2X_4$ by applying the method of characteristic functions.

SOLUTION. The expectation $\mathbf{M}[X_3^2X_2X_4]$ is determined by the distribution of the subsystem (X_2, X_3, X_4). The characteristic function corresponding to this subsystem has the form

$$E_{x_2,x_3,x_4}(u_2, u_3, u_4) = \exp\left\{-\frac{1}{2}\sum_{r=2}^{4}\sum_{s=2}^{4} k_{rs}u_ru_s\right\}.$$

The required expectation can be obtained by differentiating the characteristic function four times:

$$\mathbf{M}[X_3^2X_2X_4] = \left.\frac{\partial^4 E_{x_2,x_3,x_4}(u_2, u_3, u_4)}{\partial u_3^2\,\partial u_2\,\partial u_4}\right|_{u_2=u_3=u_4=0}.$$

The first method. If we expand the characteristic function in a power series according to its exponent, then we find that in calculating the desired mixed partial derivative for $u_2 = u_3 = u_4 = 0$, only one term of the expansion is different from zero:

$$\mathbf{M}[X_3^2X_2X_4] = \frac{1}{8}\left.\frac{\partial^4(\sum_{r=2}^{4}\sum_{s=2}^{4} k_{rs}u_ru_s)}{\partial u_3^2\,\partial u_2\,\partial u_4}\right|_{u_2=u_3=u_4=0}.$$

The mixed derivative of the square of the polynomial for $u_2 = u_3 = u_4 = 0$ will have terms different from zero if before differentiation they were proportional to $u_3^2u_2u_4$; that is,

$$\mathbf{M}[X_3^2X_2X_4] = \frac{1}{4}\frac{\partial^4(2k_{33}k_{24}u_3^2u_2u_4 + 4k_{23}k_{34}u_3^2u_2u_4)}{\partial u_3^2\,\partial u_2\,\partial u_4}$$
$$= k_{33}k_{24} + 2k_{23}k_{34}.$$

The second method. For convenience, we introduce the notation

$$\tau_r = \sum_{s=1}^{n} k_{rs}u_s.$$

Then

$$\frac{\partial E}{\partial u_3} = \frac{\partial}{\partial u_3}\exp\left\{-\frac{1}{2}\sum_{r=1}^{6}\sum_{s=1}^{6} k_{rs}u_ru_s\right\} = -E\tau_3,$$

$$\frac{\partial^2 E}{\partial u_3^2} = E\tau_3^2 - E\frac{\partial \tau_3}{\partial u_3} = E\tau_3^2 - Ek_{33},$$

$$\frac{\partial^3 E}{\partial u_3^2\,\partial u_2} = -E\tau_3^2\tau_2 + 2E\tau_3k_{23} + E\tau_2k_{33},$$

$$\frac{\partial^4 E}{\partial u_3^2\,\partial u_2\,\partial u_4} = E\tau_3^2\tau_2\tau_4 - 2E\tau_3\tau_2k_{34} - E\tau_3^2k_{24} - 2E\tau_3\tau_4k_{23}$$
$$+ 2Ek_{34}k_{23} - E\tau_2\tau_4k_{33} + Ek_{24}k_{33}.$$

For $u_1 = u_2 = \cdots = u_6 = 0$, we have

$$E = 1, \qquad \tau_r = 0,$$

which implies that

$$\mathbf{M}[X_3^2X_2X_4] = k_{33}k_{24} + 2k_{23}k_{34}.$$

Similarly one can solve Problems 23.11 to 23.14.

PROBLEMS

23.1 Prove that the characteristic function of the sum of independent random variables is the product of the characteristic functions of its terms.

23.2 Given $E_{x_1,\ldots,x_n}(u_1, u_2, \ldots, u_n)$, the characteristic function of the system $(X_1, X_2, \ldots, X_n)$, find the characteristic function of the sum $Z = X_1 + X_2 + \cdots + X_n$.

23.3 Find the characteristic function of the linear function $Y = \sum_{k=1}^{n} a_k X_k + c$ of the random variables $X_1, X_2, \ldots, X_n$ whose characteristic functions are given.

23.4 Find the characteristic function for the square of the deviation of a normal random variable from its expectation, $Y = (X - \bar{x})^2$, and the moments of Y.

23.5 Find the characteristic function of the random variable $Y = aF(X) + b$, where X is a random variable and $F(x)$ is its distribution function.

23.6 Find the characteristic function of the random variable $Y = \ln F(X)$, where X is a random variable and $F(x)$ its distribution function. Evaluate the moments of Y.

23.7 Find the characteristic function of the projection of a segment a on the Oy-axis if the angle made by this segment with the Oy-axis is uniformly distributed in the interval $(0, 2\pi)$. Evaluate the probability density of the projection.

23.8 Find the characteristic function of a system of two random variables obeying the normal distribution law

$$f(x, y) = \frac{1}{2\pi\sigma_1\sigma_2\sqrt{1 - r^2}}$$
$$\times \exp\left\{-\frac{1}{2(1 - r^2)}\left[\frac{(x - \bar{x})^2}{\sigma_1^2} - \frac{2r(x - \bar{x})(y - \bar{y})}{\sigma_1\sigma_2} + \frac{(y - \bar{y})^2}{\sigma_2^2}\right]\right\}.$$

23.9 Find the characteristic function of a system of n random variables $(X_1, X_2, \ldots, X_n)$, obeying a normal distribution law, if the expectations of the random variables forming the system are all equal to a and their covariance matrix is

$$\|k_{rs}\| = \left\|\begin{matrix} \sigma^2 & \alpha\sigma^2 & 0 & 0 & 0 & \cdots & 0 & 0 \\ \alpha\sigma^2 & \sigma^2 & \alpha\sigma^2 & 0 & 0 & \cdots & 0 & 0 \\ 0 & \alpha\sigma^2 & \sigma^2 & \alpha\sigma^2 & 0 & \cdots & 0 & 0 \\ \cdot & \cdot & \cdot & \cdot & \cdot & \cdot & \cdot & \cdot \\ 0 & 0 & 0 & 0 & 0 & \cdots & \sigma^2 & \alpha\sigma^2 \\ 0 & 0 & 0 & 0 & 0 & \cdots & \alpha\sigma^2 & \sigma^2 \end{matrix}\right\|$$

$$(r, s = 1, 2, \ldots, n).$$

23.10 Find the characteristic function of

$$Y = \sum_{m=1}^{n} X_m,$$

in which $(X_1, X_2, \ldots, X_n)$ is a system of normal random variables and

$$\bar{x}_m = m,$$
$$k_{m,l} = n - |m - l|.$$

23.11 Using the method of characteristic functions, find $\mathbf{M}[(X_1^2 - \sigma^2)(X_2^2 - \sigma^2)]$ if X_1, X_2 are normal random variables for which $\bar{x}_1 = \bar{x}_2 = 0$,

$$\mathbf{M}[X_1^2] = \mathbf{M}[X_2^2] = \sigma^2, \quad \text{and} \quad \mathbf{M}[X_1X_2] = k_{12}.$$

23.12 Applying the method of characteristic functions, evaluate (a) $\mathbf{M}[X_1^2X_2^2X_3^2]$, (b) $\mathbf{M}[(X_1^2 - \sigma^2)(X_2^2 - \sigma^2)(X_3^2 - \sigma^2)]$, if X_1, X_2, X_3 are normal random variables for which $\bar{x}_1 = \bar{x}_2 = \bar{x}_3 = 0$, $\mathbf{M}[X_1^2] = \mathbf{M}[X_2^2] = \mathbf{M}[X_3^3] = \sigma^2$, and k_{12}, k_{13}, k_{23} are the covariances between the corresponding random variables.

23.13 Applying the method of characteristic functions, evaluate $\mathbf{M}[X_1X_2X_3]$ if X_1, X_2, X_3 are normal centralized random variables.

23.14 Using the method of characteristic functions, express $\mathbf{M}[X_1X_2X_3X_4]$ in terms of the elements of the covariance matrix k_{ml} of the system of random variables (X_1, X_2, X_3, X_4), whose expectations are zero.

23.15 Prove that the central moment of even order of a system of n normal random variables is given by the formula

$$\mu_{r_1,r_2,\ldots,r_n} = \mathbf{M}[(X_1 - \bar{x}_1)^{r_1}(X_2 - \bar{x}_2)^{r_2}\cdots(X_n - \bar{x}_n)^{r_n}]$$
$$= \frac{r_1!\,r_2!\cdots r_n!}{2^s s!}\sum k_{m_1l_1}\cdots k_{m_sl_s},$$

where $r_1 + r_2 + \cdots + r_n = 2s$, and the summation is extended over all possible permutations of $2s$ indices $m_1, m_2, \ldots, m_n$ and $l_1, l_2, \ldots, l_n$, of which r_1 indices equal 1, r_2 indices equal 2, ..., r_n indices equal n.

23.16 Given a system of dependent normal random variables $(X_1, X_2, \ldots, X_n)$, prove that the random variable $Y = \sum_{j=1}^{n} a_jX_j + b$ also obeys a normal distribution law.

23.17 The output of a factory consists of identical units, each of which in the rth quarter of the year ($r = 1, 2, 3, 4$) is with probability p_r of first quality and with probability $q_r = 1 - p_r$ of second quality. An item of first quality costs S_1 dollars and an item of second quality, S_2 dollars. Evaluate the characteristic function of the system of random variables (X, Y), where X is the cost of items produced during the first three quarters of the year, and Y the cost during the last three quarters of the year. Evaluate the covariance of X and Y. The number of items produced in the rth quarter is N_r.

24. CONVOLUTION OF DISTRIBUTION LAWS

Basic Formulas

The operation of finding the distribution law of a sum of mutually independent random variables in terms of the distribution laws of its summands is called convolution (composition) of distribution laws. If X and Y are independent discrete random variables, the distribution series of the random variable $Z = X + Y$ is given by the formula

$$\mathbf{P}(Z = z_i) = \sum_j \mathbf{P}(X = x_j)\mathbf{P}(Y = z_i - x_j) = \sum_k \mathbf{P}(Y = y_k)\mathbf{P}(X = z_i - y_k),$$

where the summation is extended over all possible values of the random variables.

If X and Y are continuous random variables, the probability density for the random variable $Z = X + Y$ is

$$f_z(z) = \int_{-\infty}^{\infty} f_x(x)f_y(z - x)\,dx = \int_{-\infty}^{\infty} f_y(y)f_x(z - y)\,dy,$$

and the distribution function $F_z(z)$ is determined by the formula

$$F_z(z) = \iint_{x+y<z} f_x(x)f_y(y)\,dx\,dy.$$

The probability density $f_y(y)$ of a sum of independent random variables $X_1, X_2, \ldots, X_n$ $(Y = X_1 + X_2 + \cdots + X_n)$ can be found either by using the characteristic functions in accordance with the formula

$$f_y(y) = \frac{1}{2\pi}\int_{-\infty}^{\infty} e^{-iyt}\left\{\prod_{j=1}^{n} E_{x_j}(t)\right\} dt,$$

where

$$E_{x_j}(t) = \int_{-\infty}^{\infty} e^{ixt} f_{x_j}(x)\,dx,$$

or by successive applications of the convolution formula for two random variables.

Solution for Typical Examples

Example 24.1 Find the probability density of the sum of two independent random variables $Z = X + Y$, where X is uniformly distributed over the interval (0, 1), and Y has Simpson's distribution (Figure 25):

$$f_y(y) = \begin{cases} y & \text{for } 0 \leqslant y \leqslant 1, \\ 2 - y & \text{for } 1 \leqslant y \leqslant 2, \\ 0 & \text{otherwise.} \end{cases}$$

FIGURE 25

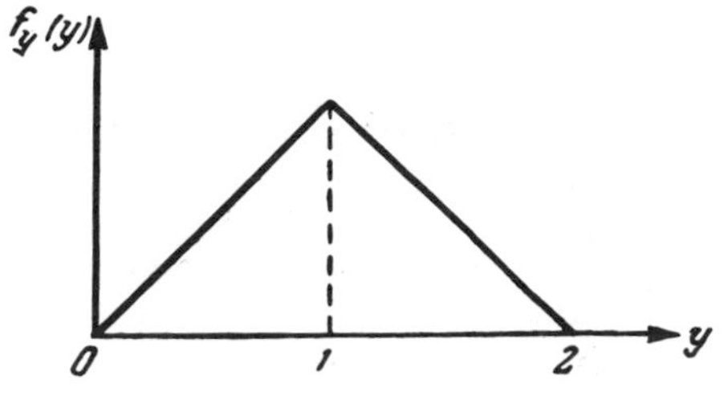

SOLUTION. Since the functions $f_x(x)$ and $f_y(y)$ are different from zero only for particular values of their arguments, it is more convenient first to find the distribution function of the random variable Z. We have

$$F_z(z) = \mathbf{P}(Z < z) = \iint\limits_{D_z} f_x(x) f_y(y)\, dx\, dy,$$

where D_z is the region inside which $x + y < z$, and none of the functions $f_x(x)$ and $f_y(y)$ vanishes (Figure 26).

The shape of the integration domain depends on which of the three intervals (0, 1), (1, 2) or (2, 3) contains z. Computing the integrals for these cases, we obtain

$$F_z(z) = \begin{cases} 0 & \text{for } z < 0, \\ \int_0^z f_y(y)\, dy \int_0^{z-y} f_x(x)\, dx = \dfrac{z^3}{6} & \text{for } 0 \leqslant z \leqslant 1, \\ \int_0^{z-1} dx \int_0^1 y\, dy + \int_0^{z-1} dx \int_1^{z-x} (2 - y)\, dy + \int_{z-1}^1 dx \int_0^{z-x} y\, dy & \\ \qquad = z - 1 + \dfrac{(2 - z)^3}{6} - \dfrac{(z - 1)^3}{6} & \text{for } 1 \leqslant z \leqslant 2, \\ 1 - \int_{z-1}^2 (2 - y)\, dy \int_{z-y}^1 dx = 1 - \dfrac{1}{6}(3 - z)^3 & \text{for } 2 \leqslant z \leqslant 3, \\ 1 & \text{for } z > 3. \end{cases}$$

By differentiation with respect to z, we find the probability density:

$$f_z(z) = \begin{cases} \dfrac{z^2}{2} & \text{for } 0 \leqslant z \leqslant 1, \\ -z^2 + 3z - \dfrac{3}{2} & \text{for } 1 \leqslant z \leqslant 2, \\ \dfrac{1}{2}(z^2 - 6z + 9) & \text{for } 2 \leqslant z \leqslant 3, \\ 0 & \text{for } z < 0 \text{ or } z > 3. \end{cases}$$

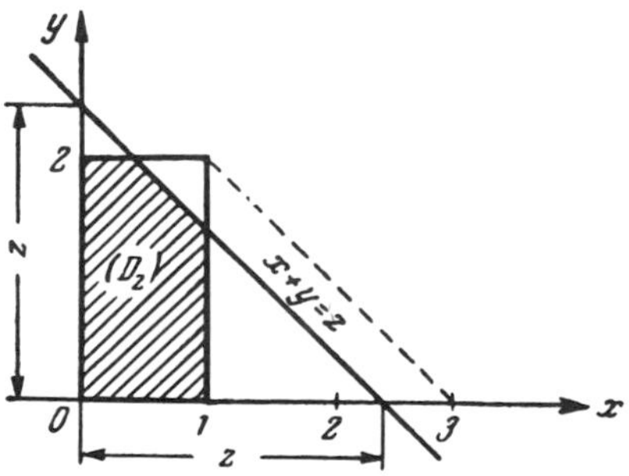

FIGURE 26

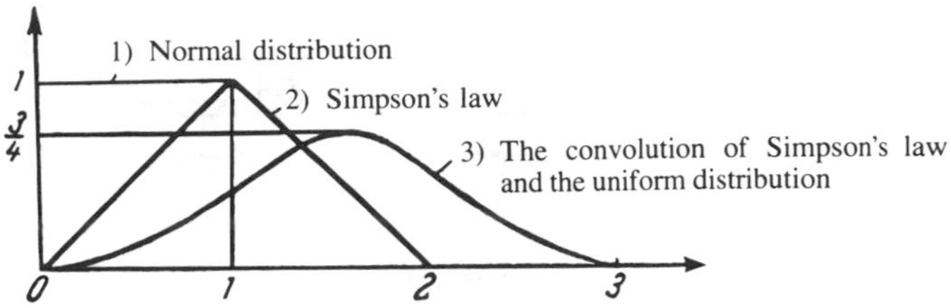

FIGURE 27

The functions $f_x(x)$, $f_y(y)$ and $f_z(z)$ are represented in Figure 27. Problems 24.1, 24.2, 24.4 and 24.8 can be solved similarly.

Example 24.2 A point C is chosen at random on a segment A_1A_2 of length $2L$. The possible deviation of the midpoint of segment $F_1F_2 = 2B$ from the midpoint of A_1A_2 has a normal distribution with mean deviation E. Find the probability that the distance from C to the midpoint of segment F_1F_2 does not exceed a given quantity $(d + B)$.

SOLUTION. Let X denote the random deviation of the point C from the midpoint of A_1A_2, and let Y be the deviation of the midpoint of F_1F_2 from the midpoint of A_1A_2 (Figure 28). Then the deviation of the point C from the midpoint of segment F_1F_2 is $Z = Y - X$. Since the function $f_y(y)$ does not vanish on the real axis,

$$f_z(z) = \int_{-\infty}^{\infty} f_x(x)f_y(z + x)\, dx = \frac{1}{2L}\int_{-L}^{L} \frac{\rho}{E\sqrt{\pi}} \exp\left\{-\rho^2 \frac{(z + x)^2}{E^2}\right\} dx$$

$$= \frac{1}{2L}\frac{\rho}{\sqrt{\pi}}\int_{(z-L)/E}^{(L+z)/E} e^{-\rho^2 t^2}\, dt = \frac{1}{4L}\left[\hat{\Phi}\left(\frac{L + z}{E}\right) + \hat{\Phi}\left(\frac{L - z}{E}\right)\right].$$

The distance from C to the midpoint of F_1F_2 will not exceed the quantity $d + B$ if $|z| < d + B$. Therefore, the probability of this event is given by the formula

$$P = \mathbf{P}(|z| < d + B) = \int_{-(d+B)}^{d+B} f_z(z)\, dz$$

$$= \frac{1}{4L}\int_{-(d+B)}^{d+B}\left[\hat{\Phi}\left(\frac{L + z}{E}\right) + \hat{\Phi}\left(\frac{L - z}{E}\right)\right] dz$$

$$= \frac{E}{4L}\left[\int_{(L-d-B)/E}^{(L+d+B)/E} \hat{\Phi}(t)\, dt - \int_{(L+d+B)/E}^{(L-d-B)/E} \hat{\Phi}(t)\, dt\right] = \frac{E}{2L}\int_{(L-d-B)/E}^{(L+d+B)/E} \hat{\Phi}(t)\, dt$$

$$= \frac{1}{2L}\left\{(L + B + d)\hat{\Phi}\left(\frac{L + B + d}{E}\right) - (L - B - d)\hat{\Phi}\left(\frac{L - B - d}{E}\right)\right.$$

$$\left. + \frac{E}{\rho\sqrt{\pi}}\left[\exp\left\{-\frac{\rho^2}{E^2}(L + B + d)^2\right\} - \exp\left\{-\frac{\rho^2}{E^2}(L - d - B)^2\right\}\right]\right\}.$$

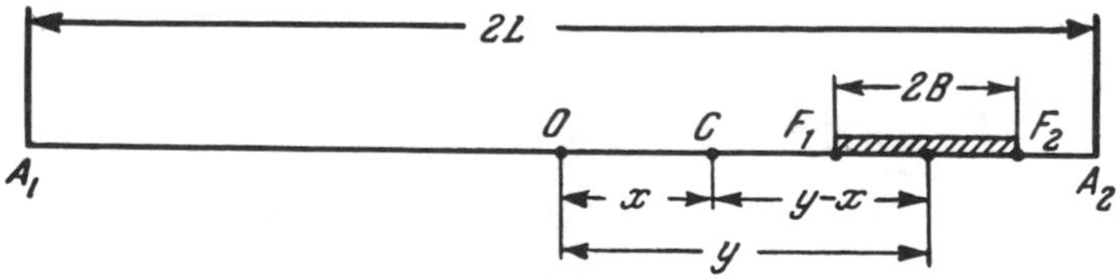

FIGURE 28

In a similar manner Problems 24.3, 24.5 to 24.7, 24.13 to 24.15 can be solved.

Example 24.3 Two groups of identical items, of n_1 and n_2 items each, are mixed together. The number of defective items in each group (X and Y, respectively) has the binomial distribution:

$$\mathbf{P}(X = m) = C_{n_1}^{m} p^m q^{n_1 - m},$$

$$\mathbf{P}(Y = m) = C_{n_2}^{m} p^m q^{n_2 - m}.$$

Find the distribution series of the random variable $Z = X + Y$.

SOLUTION. For the probability $\mathbf{P}(Z = z)$ to be different from zero, Z must be integral-valued and lie on the interval $(0, n_1 + n_2)$. Applying the general formula and taking into account that $0 \leqslant x \leqslant z$, we obtain

$$\mathbf{P}(Z = z) = \sum_{x=0}^{z} C_{n_1}^{x} p^x q^{n_1 - x} C_{n_2}^{z-x} p^{z-x} q^{n_2 - z + x}$$

$$= p^z q^{n_1 + n_2 - z} \sum_{x=0}^{z} C_{n_1}^{x} C_{n_2}^{z-x} = C_{n_1 + n_2}^{z} p^z q^{n_1 + n_2 - z}$$

$$(z = 0, 1, 2, \ldots, n_1 + n_2).$$

(The equality $\sum_{x=0}^{z} C_{n_1}^{x} C_{n_2}^{z-x} = C_{n_1+n_2}^{z}$ can be proved, for example, by induction. First, one proves it for $n_1 = 1$ and for any n_2.)

This problem can also be solved by using characteristic functions. For the random variables X and Y, we have

$$E_x(t) = \mathbf{M}[e^{iXt}] = (pe^{it} + q)^{n_1},$$

$$E_y(t) = \mathbf{M}[e^{iYt}] = (pe^{it} + q)^{n_2}.$$

Since X and Y are by hypothesis independent, we have

$$E_z(t) = E_x(t)E_y(t) = (pe^{it} + q)^{n_1 + n_2}.$$

From this, it follows that the random variable Z also has a binomial distribution.

Similarly one can solve Problems 24.12 and 24.16 to 24.21.

Example 24.4 Let $X_1, X_2, \ldots, X_n$ be independent random variables, each of which obeys Poisson's law

$$\mathbf{P}(X_j = m) = \frac{a^m}{m!} e^{-a}$$

with the same parameter a.

Find the distribution series of the random variable $Y = \sum_{j=1}^{n} X_j$ and prove that the centralized and normalized random variable $(Y - \bar{y})/\sigma_y$ for $n \to \infty$ has a normal distribution.

SOLUTION. We find the characteristic function of the random variable X_j:

$$E_{x_j}(t) = \mathbf{M}[e^{itX_j}] = \sum_{k=0}^{\infty} e^{ikt} \frac{a^k}{k!} e^{-a} = e^{-a} \sum_{k=0}^{\infty} \frac{(ae^{it})^k}{k!} = e^{-a} e^{ae^{it}} = e^{a(e^{it}-1)}.$$

Since the random variables X_j are independent, the characteristic function of Y is given by the formula

$$E_y(t) = \prod_{j=1}^{n} E_{x_j}(t) = e^{na(e^{it}-1)}.$$

Consequently, the random variable Y has Poisson's distribution law with parameter na. Use the notation $Z = (Y - \bar{y})/\sigma_y$. The random variable Z is obtained as a result of normalizing and centralizing the random variable Y. It is known that for Poisson's law, the expectation and variance are numerically equal quantities, both equal to the parameter of this law. Thus,

$$Z = \frac{Y - na}{\sqrt{na}}.$$

Evaluate the characteristic function of Z:

$$\begin{aligned}
E_z(t) &= \mathbf{M}[e^{itZ}] = \mathbf{M}\left[\exp\left\{\frac{it(Y - na)}{\sqrt{na}}\right\}\right] = e^{-i\sqrt{na}} E_y\left(\frac{t}{\sqrt{na}}\right) \\
&= e^{-i\sqrt{na}} \exp\left\{na\left[\exp\left(\frac{it}{\sqrt{na}}\right) - 1\right]\right\} \\
&= e^{-na} e^{-it\sqrt{na}} \exp\left\{na \exp\left[\frac{it}{\sqrt{na}}\right]\right\} \\
&= e^{-na} e^{-it\sqrt{na}} \exp\left\{na\left(1 + \frac{it}{\sqrt{na}} - \frac{t^2}{2na} + \cdots\right)\right\} \\
&= \exp\left\{na\left(-\frac{t^2}{2na} + \frac{i^3 t^3}{3!\,(na)^{3/2}} - \cdots\right)\right\}.
\end{aligned}$$

Consequently,

$$\lim_{n \to \infty} E_z(t) = \exp\left\{-\frac{t^2}{2}\right\}.$$

The limit of $E_z(t)$ is the characteristic function of the random variable with a normal distribution, with expectation zero and variance one.

One can solve Problems 24.6, 24.10, 24.19 and 24.20 similarly.

PROBLEMS

24.1 Find the probability density of the sum of two independent variables, each of which is uniformly distributed over the interval (a, b).

24.2 Find the convolution of two uniform distributions with parameters a and b $(b > a)$, if the dispersion centers for both distributions coincide and the parameter of a uniform distribution law is defined as being half the length of the interval of the possible values of a random variable.

24.3 The random variable X obeys a normal distribution law with parameters $\bar{x}$ and σ_x, Y obeys a uniform distribution law with parameter $(b - a)/2$ and $\bar{y} = (a + b)/2$. Find the probability density of the random variable $Z = X - Y$ if X and Y are independent.

24.4 Find the probability density of the sum of three independent random variables, each of which is uniformly distributed over the interval (a, b).

24.5 Find the convolution of a normal law (with expectation $\bar{x}$ and mean deviation E) and a uniform distribution law, given in the interval $(\bar{x} - l, \bar{x} + l)$. Find the relative error caused by replacing the resulting law by a normal law with the same variance and expectation. (Perform the computations for $\bar{x} = 0$, $l = E$, $l = 2E$, $l = 3E$ and $l = 4E$ at point $z = 0$.)

24.6 Find the probability density of the random variable $Z = X + Y$ if the random variables X and Y are independent and obey Cauchy's law:

$$f_x(x) = \frac{1}{\pi}\frac{h}{1 + h^2(x - a)^2}, \qquad f_y(y) = \frac{1}{\pi}\frac{k}{1 + k^2(y - b)^2}.$$

24.7 Find the probability density of the sum of two random variables X and Y, obeying the hyperbolic secant law:

$$f_x(x) = \frac{1}{\pi}\frac{1}{\operatorname{sech} x}, \qquad f_y(y) = \frac{1}{\pi}\frac{1}{\operatorname{sech} y}.$$

24.8 Let X and Y be independent random variables with probability densities given by the formulas

$$f_x(x) = \frac{1}{2}\exp\left\{-\frac{x}{2}\right\} \qquad (0 \leqslant x \leqslant \infty),$$

$$f_y(y) = \frac{1}{3}\exp\left\{-\frac{y}{3}\right\} \qquad (0 \leqslant y \leqslant \infty).$$

Find the probability density of the random variable $Z = X + Y$.

24.9 Find the probability density of the distance between the points $A_1(X_1, Y_1)$ and $A_2(X_2, Y_2)$ if the systems (X_1, Y_1) and (X_2, Y_2) are independent and uniformly distributed. The unit dispersion ellipses of the points A_1 and A_2 have major semi-axes (a_1, b_1) and (a_2, b_2). The angle made by a_1 and a_2 is α. The centers of the unit ellipses coincide.

24.10 Let X_j $(j = 1, 2, \ldots, n)$ be normally distributed independent random variables with $\bar{x}_j = 0$ and $\mathbf{D}[X_j] = 1$. Prove that for the random variable $Y = \sum_{j=1}^{n} X_j^2$, the probability density is determined by the formula

$$f_y(y) = \frac{y^{(n/2)-1}}{2^{n/2}\Gamma\left(\frac{n}{2}\right)} \exp\left\{-\frac{y}{2}\right\} \qquad (0 \leqslant y \leqslant \infty).$$

24.11 An instrument gives a systematic error a and a random error obeying a normal distribution law with mean deviation E. Prove that for $E \geqslant d$, the probability $p(a)$ of an error within a given tolerance range $\pm d$ is approximately given by the formula

$$p(a) \approx \frac{2\rho d}{E_x\sqrt{\pi}} \exp\left\{-\rho^2\left(\frac{a}{E_x}\right)^2\right\}.$$

where

$$E_x = \sqrt{E^2 + \frac{2}{3}\rho^2 d^2}.$$

24.12 Two persons fire independent shots, each at his target, until the first hit is scored. Find the expectation and variance for the total number of failures and the distribution function for the number of failures if the probability of hitting a target at each shot is p_1 for the first marksman and p_2 for the second.

24.13 What should be the reserve shear strength of a sample so that the probability that it will support a load is at least 98 per cent? The errors in the determination of the given load and of the maximal load obey a normal distribution with mean deviations: $E_{q_1} = 10\%\bar{q}_1$, $E_{q_2} = 5\%\bar{q}_2$, where $\bar{q}_1$ and $\bar{q}_2$ are the expectations for the given and maximal loads and $\bar{q}_1 = 20$ kg.

24.14 A navigational transmitter is installed on each shore of a sound of width L. The transmitters serve the ships passing through the sound. The maximal ranges of each of the transmitters are independent random variables with expectation $\bar{x}$ and mean deviation E. Assuming that any distance between the course of a ship and the shores is equally probable, and that $2\bar{x} < L$, find (a) the probability that a ship will be served by two transmitters, (b) the probability that a ship will be served by at least one transmitter.

24.15 Observer A moves from infinity toward observer B. The maximal distances for sighting each other are independent random variables with expectations $\bar{x}_A$ and $\bar{x}_B$, respectively, and mean deviations E_A, E_B. Find the probability that A will sight B first.

24.16 Find the convolution of m exponential distributions with the same parameter λ.

24.17 Let X and Y be independent random variables, assuming integral nonnegative values i and j with probabilities $\mathbf{P}(X = i) = (1 - a)a^i$ and $\mathbf{P}(Y = j) = (1 - b)b^j$, where a and b are positive integers less than one. Find the distribution function of the random variable $Z = X + Y$.

24.18 Let X and Y be independent random variables; X assumes three possible values 0, 1, 3 with probabilities 1/2, 3/8, 1/8, and Y assumes two possible values 0 and 1 with probabilities 1/3, 2/3. Find the distribution series of the random variable $Z = X + Y$.

24.19 Let X, Y be independent random variables, each of which obeys Poisson's distribution:

$$\mathbf{P}(X = m) = \frac{a^m}{m!} e^{-a},$$

$$\mathbf{P}(Y = m) = \frac{a^m}{m!} e^{-a}.$$

Find the distribution series of the random variable $Z = X + Y$.

24.20 Let X_j $(j = 1, 2, \ldots, n)$ be independent random variables, each of which takes only two values: 1 with probability p, and zero with probability $q = 1 - p$. Find the distribution series of the random variable $Y = \sum_{j=1}^{n} X_j$.

24.21 Let X and Y be independent discrete random variables assuming positive integral values k from 1 to ∞ with probability $(1/2)^k$. Find the distribution function of the random variable $Z = X + Y$.

25. THE LINEARIZATION OF FUNCTIONS OF RANDOM VARIABLES

Basic Formulas

Any continuous differentiable function, whose derivative is finite at a given point and for sufficiently small variations about the point, can be replaced approximately by a linear function, by using a Taylor series and retaining only the linear terms. If the probability is small that the arguments of the function will assume values outside the region where the function can be considered linear, this function can be expanded in the vicinity of the point corresponding to the expectations of its arguments. The approximate values of the expectation and variance in this case are given by:

(a) for the function of one random argument $Y = \varphi(X)$,

$$\bar{y} \approx \varphi(\bar{x}), \qquad \mathbf{D}[Y] \approx [\varphi'(\bar{x})]^2 \mathbf{D}[X];$$

(b) for a function of many arguments $Y = \varphi(X_1, X_2, \ldots, X_n)$,

$$\bar{y} \approx \varphi(\bar{x}_1, \bar{x}_2, \ldots, \bar{x}_n),$$

$$\mathbf{D}[Y] \approx \sum_{i=1}^{n} \left(\frac{\partial \bar{\varphi}}{\partial x_i}\right)^2 \mathbf{D}[X_i] + \sum_{\substack{i=1 \\ i \neq j}}^{n} \sum_{j=1}^{n} \left(\frac{\partial \bar{\varphi}}{\partial x_i}\right)\left(\frac{\partial \bar{\varphi}}{\partial x_j}\right) k_{ij},$$

where k_{ij} denotes the covariance for the random variables X_i and X_j, and $\partial\bar{\varphi}/\partial x_i$ are the derivatives computed for values of the arguments equal to the expectations.

If the random arguments are mutually uncorrelated, then

$$\mathbf{D}[Y] \approx \sum_{i=1}^{n} \left(\frac{\partial \bar{\varphi}}{\partial x_i}\right)^2 \mathbf{D}[X_i].$$

For more accuracy in the results of linearization, in the expansion of the function one must retain, beside the first two terms, some higher-order terms as well. If one retains the first three terms of the series, then the expectation and variance are determined by the formulas:

(a) for a function of one argument $Y = \varphi(X)$,

$$\bar{y} \approx \varphi(\bar{x}) + \frac{1}{2}\varphi''(\bar{x})\mathbf{D}[X],$$

$$\mathbf{D}[Y] \approx [\varphi'(\bar{x})]^2\mathbf{D}[X] + \frac{1}{4}[\varphi''(\bar{x})]^2\{\mu_4[X] - \mathbf{D}^2[X]\} + \varphi'(\bar{x})\varphi''(\bar{x})\mu_3[X];$$

(b) for a function of several random arguments $Y = \varphi(X_1, X_2, \ldots, X_n)$ the expectation is given by the formula

$$\bar{y} \approx \varphi(\bar{x}_1, \bar{x}_2, \ldots, \bar{x}_n) + \frac{1}{2}\sum_{i=1}^{n} \frac{\partial^2 \bar{\varphi}}{\partial x_i^2}\mathbf{D}[X_i] + \sum_{i<j} \frac{\partial^2 \bar{\varphi}}{\partial x_i\, \partial x_j} k_{ij}$$

in the general case and by the formula

$$\bar{y} \approx \varphi(\bar{x}_1, \bar{x}_2, \ldots, \bar{x}_n) + \frac{1}{2}\sum_{i=1}^{n} \frac{\partial^2 \bar{\varphi}}{\partial x_i^2}\mathbf{D}[X_i]$$

in the case when the random arguments are mutually uncorrelated. If the random arguments are mutually independent, then the variance is given by the formula

$$\mathbf{D}[Y] \approx \sum_{i=1}^{n} \left(\frac{\partial \bar{\varphi}}{\partial x_i}\right)^2 \mathbf{D}[X_i] + \frac{1}{4}\sum_{i=1}^{n} \left(\frac{\partial^2 \bar{\varphi}}{\partial x_i^2}\right)^2 \{\mu_4[X_i] - \mathbf{D}^2[X_i]\}$$
$$+ \sum_{i<j} \left(\frac{\partial^2 \bar{\varphi}}{\partial x_i\, \partial x_j}\right)^2 \mathbf{D}[X_i]\mathbf{D}[X_j] + \sum_{i=1}^{n} \left(\frac{\partial \bar{\varphi}}{\partial x_i}\right)\left(\frac{\partial^2 \bar{\varphi}}{\partial x_i^2}\right)\mu_3[X_i].$$

SOLUTION FOR TYPICAL EXAMPLES

Example 25.1 The expectation of the number of defective devices is given by the formula

$$T = N\left[1 - \left(1 - \frac{P}{\Omega N}\right)^m\right].$$

where P is the probability that the trial of one device is considered successful, Ω is the average number of successful trials until the first failure occurs, N is the number of devices tested and m is the number of trials (successes and failures) for each device.

Using the linearization method, find the dependence of the expectation and variance of the random variable T on m if N, P and Ω are independent random variables, whose expectations and variances are:

$$\mathbf{M}[N] = 5, \quad \mathbf{M}[P] = 0.8, \quad \mathbf{M}[\Omega] = 4,$$

$$\mathbf{D}[N] = 1, \quad \mathbf{D}[P] = 0.1, \quad \mathbf{D}[P] = 0.2.$$

SOLUTION. Applying the general formulas of the linearization method, we obtain

$$\mathbf{M}[T] \approx \bar{n}\left[1 - \left(1 - \frac{p}{\bar{\omega}\bar{n}}\right)^m\right] = 5(1 - 0.96^m),$$

$$\mathbf{D}[T] \approx \left(\frac{\partial \bar{T}}{\partial n}\right)^2 \mathbf{D}[N] + \left(\frac{\partial \bar{T}}{\partial p}\right)^2 \mathbf{D}[P] + \left(\frac{\partial \bar{T}}{\partial \omega}\right)^2 \mathbf{D}[\Omega],$$

where

$$\frac{\partial \bar{T}}{\partial n} = \left.\frac{\partial T}{\partial n}\right|_{N = \bar{n},\, P = \bar{p},\, \Omega = \bar{\omega}} = 1 - \left(1 - \frac{\bar{p}}{\bar{\omega}\bar{n}}\right)^m - \frac{m\bar{p}}{\bar{\omega}\bar{n}}\left(1 - \frac{\bar{p}}{\bar{\omega}\bar{n}}\right)^{m-1} = 1 - 0.96^m - 0.04m0.96^{m-1},$$

$$\frac{\partial \bar{T}}{\partial p} = \frac{m}{\bar{\omega}}\left(1 - \frac{\bar{p}}{\bar{\omega}\bar{n}}\right)^{m-1} = 0.25m0.96^{m-1},$$

$$\frac{\partial \bar{T}}{\partial \omega} = -0.05m0.96^{m-1},$$

$$\mathbf{D}[T] \approx 0.00835m^2 0.96^{2(m-1)} - 0.08m(1 - 0.96^m)0.96^{m-1} + (1 - 0.96^m)^2.$$

The approximate values of the expectations and variance of T for different values of m are given in Table 8.

TABLE 8

m	2	10	30	100
$\mathbf{D}[T]$	0.025	0.327	0.684	0.854
$\mathbf{M}[T]$	0.390	1.675	3.530	4.915

Similarly one can solve Problems 25.1 to 25.11, 25.14, 25.17 and 25.19 to 25.22.

Example 25.2 The maximal altitude of a satellite is given by the formula

$$Y = y_0 + (R + y_0)\left[\frac{1 + l}{2(1 - \lambda)} - 1\right],$$

where

$$\lambda = \frac{V^2}{2gR}\left(1 + \frac{y_0}{R}\right), \qquad l = \sqrt{1 - 4\lambda(1 - \lambda)\cos^2\Theta},$$

y_0 is the altitude of the active part of the trajectory, g the acceleration of gravity on the surface of the earth and R the radius of the earth.

The function Y can be linearized in the domain of practically possible values of the random arguments. The initial velocity V and the launching angle Θ are normal random variables with probability density

$$f(v, \theta) = \frac{1}{2\pi\sigma_v\sigma_\theta\sqrt{1 - r^2}} \times \exp\left\{-\frac{1}{2(1 - r^2)}\left[\left(\frac{v - \bar{v}}{\sigma_v}\right)^2 + \left(\frac{\theta - \bar{\theta}}{\sigma_\theta}\right)^2 - 2r\frac{(v - \bar{v})(\theta - \bar{\theta})}{\sigma_v\sigma_\theta}\right]\right\}.$$

Find the approximate value of the variance for the maximal altitude of the satellite.

SOLUTION. Since the given function is linearizable in the domain of the practically possible values of the random arguments,

$$\mathbf{D}[Y] \approx \left(\frac{\partial \bar{Y}}{\partial v}\right)^2 \mathbf{D}[V] + \left(\frac{\partial \bar{Y}}{\partial \theta}\right)^2 \mathbf{D}[\Theta] + 2\left(\frac{\partial \bar{Y}}{\partial v}\right)\left(\frac{\partial \bar{Y}}{\partial \theta}\right)k_{v\theta},$$

where $k_{v\theta} = r\sigma_v\sigma_\theta$,

$$\frac{\partial \bar{Y}}{\partial v} = \frac{\lambda(R + y_0)[2(1 - \lambda)(2\lambda - 1)\cos^2\bar{\theta} + l(1 + l)]}{vl(1 - \lambda)^2},$$

$$\frac{\partial \bar{Y}}{\partial \theta} = \frac{\lambda(R + y_0)\sin 2\bar{\theta}}{l},$$

and λ and l are computed for $V = \bar{v}$, $\Theta = \bar{\theta}$.

One can solve Problems 25.13 and 25.23 in a similar way.

Example 25.3 Let X and Y be independent random variables with probability density

$$f_x(x) = \frac{2}{\pi\sqrt{1 - x^2}} \qquad (0 \leqslant x \leqslant 1),$$

$$f_y(y) = \frac{2}{\pi\sqrt{1 - y^2}} \qquad (0 \leqslant y \leqslant 1).$$

Using the linearization method, find the expectation and variance of the random variable $Z = \arctan X/Y$. Correct the results obtained by using the first three terms of the Taylor series.

SOLUTION. Using the general formulas of linearization, we have

$$\mathbf{M}[Z] \approx \arctan\frac{\bar{x}}{\bar{y}}, \qquad \mathbf{D}[Z] \approx \left(\frac{\partial \bar{Z}}{\partial x}\right)^2 \mathbf{D}[X] + \left(\frac{\partial \bar{Z}}{\partial y}\right)^2 \mathbf{D}[Y],$$

where

$$\bar{x} = \bar{y} = \frac{2}{\pi}\int_0^1 \frac{x\,dx}{\sqrt{1-x^2}} = \frac{2}{\pi},$$

$$\mathbf{D}[X] = \mathbf{D}[Y] = \frac{2}{\pi}\int_0^1 \frac{x^2\,dx}{\sqrt{1-x^2}} - \bar{x}^2 = \frac{1}{2} - \frac{4}{\pi^2},$$

$$\frac{\partial \bar{Z}}{\partial x} = \left.\frac{\partial Z}{\partial x}\right|_{x=\bar{x},\, y=\bar{y}} = \frac{\bar{y}}{\bar{x}^2 + \bar{y}^2} = \frac{\pi}{4},$$

$$\frac{\partial \bar{Z}}{\partial y} = -\frac{\bar{x}}{\bar{x}^2 + \bar{y}^2} = -\frac{\pi}{4}.$$

Thus, the linearization method gives

$$\mathbf{M}[Z] \approx \arctan\frac{\bar{x}}{\bar{y}} = \frac{\pi}{4},$$

$$\mathbf{D}[Z] \approx 2\,\frac{\pi^2}{16}\,\frac{\pi^2 - 8}{2\pi^2} = \frac{\pi^2 - 8}{16},$$

Considering the next term of Taylor's series, we obtain

$$\mathbf{M}[Z] \approx \arctan\frac{\bar{x}}{\bar{y}} + \frac{1}{2}\left\{\left(\frac{\partial^2 \bar{Z}}{\partial x^2}\right)\mathbf{D}[X] + \left(\frac{\partial^2 \bar{Z}}{\partial y^2}\right)\mathbf{D}[Y]\right\},$$

$$\begin{aligned}\mathbf{D}[Z] \approx{}& \left(\frac{\partial \bar{Z}}{\partial x}\right)^2\mathbf{D}[X] + \left(\frac{\partial \bar{Z}}{\partial y}\right)^2\mathbf{D}[Y] \\ &+ \frac{1}{4}\left\{\left(\frac{\partial^2 \bar{Z}}{\partial x^2}\right)^2(\mu_4[X] - \mathbf{D}^2[X]) + \left(\frac{\partial^2 \bar{Z}}{\partial y^2}\right)^2(\mu_4[Y] - \mathbf{D}^2[Y])\right\} \\ &+ \frac{\partial^2 \bar{Z}}{\partial x\,\partial y}\mathbf{D}[X]\mathbf{D}[Y] + \frac{\partial \bar{Z}}{\partial x}\frac{\partial^2 \bar{Z}}{\partial x^2}\mu_3[X] + \frac{\partial \bar{Z}}{\partial y}\frac{\partial^2 \bar{Z}}{\partial y^2}\mu_3[Y],\end{aligned}$$

where

$$\frac{\partial^2 \bar{Z}}{\partial x^2} = -\frac{2\bar{x}\bar{y}}{(\bar{x}^2 + \bar{y}^2)^2} = -\frac{\pi^2}{8},$$

$$\frac{\partial^2 \bar{Z}}{\partial y^2} = \frac{2\bar{x}\bar{y}}{(\bar{x}^2 + \bar{y}^2)^2} = \frac{\pi^2}{8},$$

$$\frac{\partial^2 \bar{Z}}{\partial x\,\partial y} = \frac{\bar{x}^2 - \bar{y}^2}{(\bar{x}^2 + \bar{y}^2)^2} = 0,$$

$$\frac{\partial \bar{Z}}{\partial x}\frac{\partial^2 \bar{Z}}{\partial x^2} = \frac{\pi}{4}\left(-\frac{\pi^2}{8}\right) = -\frac{\pi^3}{32},$$

$$\frac{\partial \bar{Z}}{\partial y}\frac{\partial^2 \bar{Z}}{\partial y^2} = \left(-\frac{\pi}{4}\right)\frac{\pi^2}{8} = -\frac{\pi^3}{32},$$

$$\mu_3[X] = \mu_3[Y] = m_3 - 3m_1m_2 + 2m_1^3$$
$$= \frac{2}{\pi}\int_0^1 \frac{x^3\,dx}{\sqrt{1-x^2}} - 3\bar{x}\int_0^1 \frac{2}{\pi}\frac{x^2\,dx}{\sqrt{1-x^2}} + 2\bar{x}^3$$
$$= \frac{4}{3\pi} - \frac{3}{\pi} + \frac{16}{\pi^3} = \frac{16}{\pi^3} - \frac{5}{3\pi},$$

$$\mu_4[X] = \mu_4[Y] = m_4 - 4m_1m_3 + 6m_1^2m_2 - 3m_1^4$$
$$= \frac{3}{8} - 4\frac{2}{\pi}\cdot\frac{4}{3\pi} + 6\frac{4}{\pi^2}\cdot\frac{1}{2} - 3\frac{16}{\pi^4} = \frac{4}{3\pi^2} + \frac{3}{8} - \frac{48}{\pi^4}.$$

Therefore, taking into account the quadratic terms of the Taylor series, we obtain

$$\mathbf{M}[Z] \approx \frac{\pi}{4},$$

$$\mathbf{D}[Z] \approx \frac{\pi^2 - 8}{16} + \frac{\pi^4}{8\cdot 128} + \frac{7\pi^2}{48} - \frac{3}{2}.$$

Similarly one can solve Problems 25.12, 25.15, 25.16 and 25.18.

PROBLEMS

25.1 The amount of heat Q in calories produced in a conductor with resistance R by a current I in time T is given by the formula

$$Q = 0.24I^2RT.$$

The errors in the measurements of I, R and T are independent random variables with expectations $\bar{\imath} = 10$ amps, $\bar{r} = 30\ \Omega$, $t = 10$ min. and mean deviations $E_I = 0.1$ amp., $E_R = 0.2\ \Omega$, $E_T = 0.5$ sec. Find the approximate value of the mean deviation of the random variable Q.

25.2 The fundamental frequency of a string is given by the formula

$$\Omega = \frac{1}{2}\sqrt{\frac{P}{ML}},$$

where P is the tension, M the mass of the string and L the length of the string.

Given the expectations $\bar{p}$, $\bar{m}$, $\bar{l}$ and mean-square deviations σ_p, σ_m, σ_l, find the variance of the fundamental frequency caused by the variances of the tension, mass and length of the string if the corresponding correlation coefficients are r_{pl}, r_{pm}, r_{ml}.

25.3 The resistance of a section of an electric circuit is given by the formula

$$Z = \sqrt{R^2 + \left(\Omega L - \frac{1}{\Omega C}\right)^2},$$

where R denotes the ohmic resistance, L the inductance of the conductor, C its capacity and Ω the frequency of the current.

Evaluate the mean error in the magnitude of the resistance as a result of errors in independent measurements of R, L, C and Ω, if one knows $\bar{r}$, $\bar{l}$, $\bar{c}$, $\bar{\omega}$ and the mean deviations E_R, E_L, E_C, E_Ω.

25.4 If the elements of a circuit are connected in parallel, the intensity of the current in the circuit is given by the formula

$$I = \frac{E}{R + \frac{W}{n}},$$

where E is the electromotive force across the system, W is its internal resistance, n is the number of elements and R is the resistance of an external section of the circuit.

Using the linearization method, find the expectation and variance of the intensity of the current if the random variables E, R and W are independent and $\bar{e}$, $\bar{r}$, $\bar{w}$, σ_E, σ_R, σ_W are given.

25.5 Applying the linearization method, find the mean deviations E_x and E_y, which characterize the variance of coordinates of a mass point moving in a vacuum, if

$$X = VT\cos\Theta, \qquad Y = VT\sin\Theta - \frac{gT^2}{2},$$

where V is the initial velocity of the point ($\bar{v} = 800$ m./sec., $E_V = 0.1$ per cent of $\bar{v}$), T is the time of the flight ($\bar{t} = 40$ sec., $E_T = 0.1$ sec.), Θ is the launching angle ($\bar{\theta} = 45°$, $E_0 = 4'$) and g is the acceleration of gravity.

The random variables V, T and θ are independent and normal.

25.6 Find the approximate value of the mean value of the error in estimating the projection V_1 of the velocity of a ship on a given direction. Errors are due to measuring the velocity V and the angle q of the course. Here $V_1 = -V\cos q$, $E_V = 1$ m./sec., $E_q = 1°$, and the most probable values of V and q are 10 m./sec. and 60°, respectively (V and q are independent normal random variables).

25.7 Is the linearization method applicable under the assumptions made in the preceding problem if the error in the computation formulas must not exceed 0.2 m./sec.?

25.8 Find the approximate value of the mean-square deviations for rectangular coordinates of a random point

$$X = H\cot\varepsilon\cos\beta,$$

$$Y = H\cot\varepsilon\sin\beta,$$

$$Z = H,$$

if the random variables H, ε and β are independent and their expectations and mean-square deviations are equal, respectively, to: $\bar{h} = 6200$ m., $\bar{\varepsilon} = 45°$, $\bar{\beta} = 30°$, $\sigma_H = 25$ m., $\sigma_\beta = \sigma_\varepsilon = 0.001$ radians.

25.9 The passage from spherical to Cartesian coordinates is given by the formulas:

$$X = R \sin \Theta \cos \Phi,$$

$$Y = R \sin \Theta \sin \Phi,$$

$$Z = R \cos \Theta.$$

The errors in the determination of Θ, R and Φ are independent with mean-square deviations $\sigma_R = 10$ m., $\sigma_\Theta = \sigma_\Phi = 0.001$ radians. Find approximate values for mean-square deviations of the rectangular coordinates if $\bar{\theta} = \bar{\varphi} = 45°$, $\bar{r} = 10{,}000$ m.

25.10 The approximate expression for the velocity of a rocket at the end of the operation of its engine is given by Tsiolkovskiy's formula

$$V = U \ln \frac{q + \Omega}{q},$$

where U is the effective velocity of gas flow, q the weight of the rocket without fuel and Ω the weight of the fuel.

The variance of the weight of the fuel is characterized by the deviation E_Ω. Find the approximate value of the mean deviation of the velocity caused by the variance of the weight of the fuel if the expectation $\mathbf{M}[\Omega] = \bar{\omega}$.

25.11 The altitude of a mountain peak H expressed in terms of the distance D on the slope and the inclination angle ε is

$$H = D \sin \varepsilon.$$

Find the approximate value of the mean error in estimating the altitude if $E_D = 80$ m., $E_\varepsilon = 0.001°$, and the most probable values are $\bar{d} = 12{,}300$ m. and $\bar{\varepsilon} = 31°.2$, respectively. (The random variables D and ε are independent and normal.)

25.12 Let $Z = \sin XY$, where X and Y are independent random variables. Find the approximate value of σ_z if $\bar{x} = \bar{y} = 0$, $\sigma_x = \sigma_y = 0.001$.

25.13 The altitude of a mountain peak is given by the formula $H = D \sin \varepsilon$. The probability density of the errors in estimating the distance D on the slope and the inclination angle ε is given by

$$f(d, \varepsilon) = \frac{1}{2\pi\sigma_d\sigma_\varepsilon\sqrt{0.91}} \times \exp\left\{-\frac{1}{1.82}\left[\left(\frac{d - \bar{d}}{\sigma_d}\right)^2 + \left(\frac{\varepsilon - \bar{\varepsilon}}{\sigma_\varepsilon}\right)^2 + 0.6\,\frac{(d - \bar{d})(\varepsilon - \bar{\varepsilon})}{\sigma_d\sigma_\varepsilon}\right]\right\},$$

where $\sigma_d = 40$ m., $\sigma_\varepsilon = 0.001$ radians, $\bar{d} = 10{,}000$ m. and $\bar{\varepsilon} = 30°$. Find the approximate value for the mean deviation of the errors made in estimating the altitude.

25.14 The distance D_1 (Figure 29) is determined by a radar station whose errors have the mean deviation $E_p = 20$ m. Distance D_2 can be

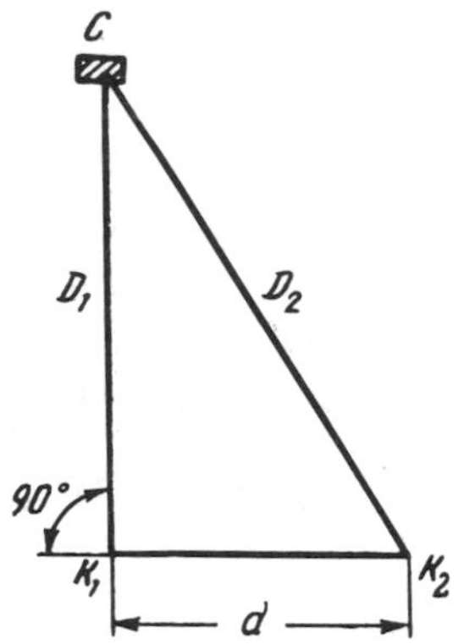

FIGURE 29

determined either with a range finder, which gives errors with mean deviation $E_D = 40$ m., or by the formula

$$D_2 = \sqrt{D_1^2 + d^2}.$$

Find which method of determination of distance K_2C is more accurate if the errors in estimating the distance between K_1 and K_2 have mean deviation $E_d = 50$ m.

25.15 Retaining the first three terms of the expansion of the function $Y = \varphi(X)$ in a Taylor series, find the expectation and variance of the random variable Y if X obeys a normal distribution law.

25.16 The area of a triangle is given by the formula

$$S = \frac{ab}{2} \sin \gamma.$$

Retaining all the terms of the Taylor series of the function $S = \varphi(\gamma)$ up to γ^3 inclusive, find the expectation of the area of the triangle and the variance of its area caused by the dispersion of the angle if the random variable γ is normally distributed, and $\bar{\gamma}$ and $\mathbf{D}[\gamma]$ are given.

25.17 In the triangle ABC (Figure 30) the side a and the opposite angle α are random variables, which can be considered uncorrelated and

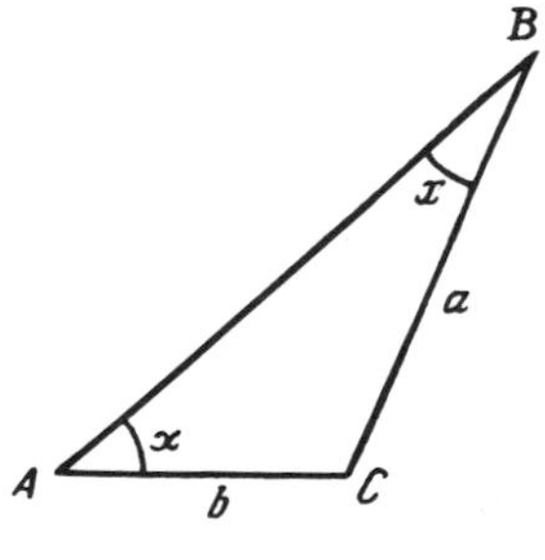

FIGURE 30

normal. Find an approximate value for the exectation X of the angle and its mean deviation if the base b is known and the expectations and mean deviations of the random variables a and α are known.

25.18 A random variable X obeys the normal distribution law

$$f_x(x) = \frac{1}{10\sqrt{2\pi}} \exp\left\{-\frac{(x+5)^2}{200}\right\}.$$

Find an approximate value for the expectation and variance of the random variable $Y = 1/X$. Retain two and then three terms of the Taylor series.

25.19 The radius of a sphere can be considered a normal random variable with expectation $\bar{r}$ and variance σ_r^2 ($\bar{r} \gg \sigma_r$). Find the expectation and variance of the volume of the sphere by using the exact formulas. Compare the results obtained with those of the linearization method.

25.20 To determine the volume of a cone one measures (a) the diameter of the base and the height, (b) the diameter of the base and the length of the generator. In which of these two cases is the error in the determination of volume smaller if the expectation for the height is $\bar{h} = 8$ dm., for the diameter of the base $d = 12$ dm., for the length of the generator $\bar{l} = 10$ dm. and $\sigma_k = \sigma_d = \sigma_l = 0.1$ dm.?

25.21 In a weighing process one uses a bar whose average diameter is 2 mm. What is the mean error if the mean deviation of the diameter of the roll is 0.04 mm. and the density of the metal of which the roll is made is 11.2 g./cc.? Fifty bars are used in the process of weighing.

25.22 The acceleration g of gravity is computed by the formula $g = 4\pi^2 L/T^2$, where L is the length of a physical pendulum, and T, its period. Find the mean error in g if a measurement of the length of the pendulum with mean error $E_L = 5$ mm. yields $L = 5$ m. and the measured period of oscillation is 4.5 sec. The period of oscillation of the pendulum is estimated for the duration of $n = 10$ complete displacements, measured with a mean error $E_t = 0.1$ sec., and the mean error in determining the instant when the pendulum passes through a position of equilibrium is $E_t = 0.5$ per cent T.

25.23 Using the linearization method, find an approximate value for the variance of the random variable $Z = \sqrt{kX^2 + Y^2}$ if $X = \sin V$, $Y = \cos V$, the random variable V is uniformly distributed over the interval $(0, \pi/2)$ and k is a known constant.

26. THE CONVOLUTION OF TWO-DIMENSIONAL AND THREE-DIMENSIONAL NORMAL DISTRIBUTION LAWS BY USE OF THE NOTION OF DEVIATION VECTORS

Basic Formulas

Any two-dimensional (three-dimensional) normal distribution law can be considered as the convolution of two (three) degenerate normal distribution laws, describing the distribution of independent oblique coordinates of a random point in the plane (space) if the coordinate axes are chosen as conjugate directions of the unit distribution ellipse (ellipsoid).[1]

A degenerate normal distribution law is uniquely characterized by a vector passing through the distribution center of this law, in the direction of one of the conjugate diameters of the unit ellipse, and equal in magnitude to this diameter. A vector defined in this way is called a deviation vector.

The convolution of normal distributions in the plane (space) is equivalent to the convolution of deviation vectors. The convolution of normal distributions lying in one plane and given by deviation vectors $\mathbf{a}_i$ ($i = 1, 2, \ldots, k$) is formed according to the following rules:

(1) the coordinates $\bar{x}$, $\bar{y}$ of the center of the compound distribution are given by the formulas,

$$\bar{x} = \sum_{i=1}^{k} \bar{x}_i, \qquad \bar{y} = \sum_{i=1}^{k} \bar{y}_i,$$

where $\bar{x}_i$, $\bar{y}_i$ are the coordinates of the origin of the deviation vector $\mathbf{a}_i$;

(2) the elements k_{ij} of the covariance matrix of the compound distribution are given by the formulas,

$$k_{11} = \frac{1}{2\rho^2}\sum_{i=1}^{k} a_{ix}^2 = \frac{A}{2\rho^2}, \qquad k_{22} = \frac{1}{2\rho^2}\sum_{i=1}^{k} a_{iy}^2 = \frac{C}{2\rho^2},$$

$$k_{12} = \frac{1}{2\rho^2}\sum_{i=1}^{k} a_{ix}a_{iy} = \frac{B}{2\rho^2},$$

where a_{ix} and a_{iy} are the projections of the deviation vector $\mathbf{a}_i$ on the axis of an arbitrarily selected unique rectangular system of coordinates;

(3) the principal directions (ξ, η) of the compound distribution, their corresponding variances (σ_ξ^2, σ_η^2) and the angle α made by the axis $O\xi$ with Ox are determined by the formulas,

$$\begin{aligned}\sigma_\xi^2 &= k_{11}\cos^2\alpha + k_{12}\sin 2\alpha + k_{22}\sin^2\alpha \\ &= \frac{1}{4\rho^2}[A + C + \sqrt{(A - C)^2 + 4B^2}] \\ &= \frac{1}{4\rho^2}[A + C + (A - C)\sec 2\alpha],\end{aligned}$$

$$\begin{aligned}\sigma_\eta^2 &= k_{11}\sin^2\alpha + k_{12}\sin 2\alpha + k_{22}\cos^2\alpha \\ &= \frac{1}{4\rho^2}[A + C - \sqrt{(A - C)^2 + 4B^2}] \\ &= \frac{1}{4\rho^2}[A + C - (A - C)\sec 2\alpha],\end{aligned}$$

[1] If one chooses as conjugate directions the principal diameters of the ellipse (ellipsoid), the degenerate distribution laws characterize the distributions of independent rectangular coordinates of a random point.

FIGURE 31

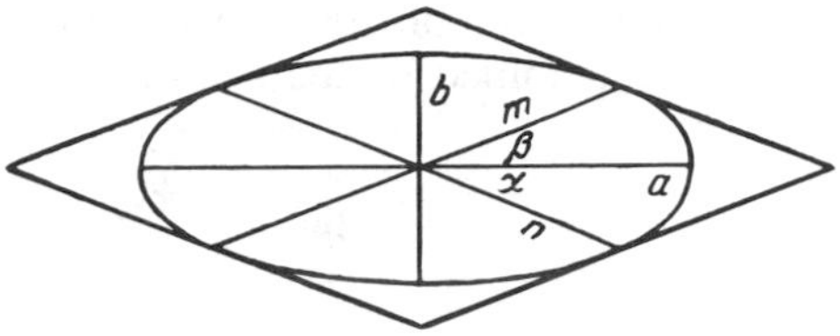

where α is any of the roots of the equation

$$\tan 2\alpha = \frac{2B}{A - C}.$$

The principal semi-axes of the unit ellipse are

$$a = \sigma_\xi \rho \sqrt{2}, \qquad b = \sigma_\eta \rho \sqrt{2}.$$

If a and b are the principal semi-axes of the unit ellipse, if m and n are two conjugate semi-axes of the same ellipse, if α and β are the angles made by n and m with the semi-axis a, and if $\beta + \alpha$ is the angle between the conjugate semi-axes, then in accordance with Apollonius' theorem (Figure 31):

$$m^2 + n^2 = a^2 + b^2,$$

$$mn \sin(\alpha + \beta) = ab,$$

where

$$\tan\alpha \tan\beta = \frac{b^2}{a^2},$$

$$n^2 = \frac{a^2 b^2}{b^2 + (a^2 - b^2)\sin^2\alpha}.$$

The convolution of deviation vectors in space is formed following the same rules. It is convenient to perform the necessary computations by using Table 9.

TABLE 9

No. П/П (i)	a_i	a_{ix}	a_{iy}	a_{iz}	a_{ix}^2	a_{iy}^2	a_{iz}^2	$a_{iz}a_{iy}$	$a_{ix}a_{iz}$	$a_{iy}a_{ix}$	$\kappa_i = a_{ix} + a_{iy} + a_{iz}$	κ_i^2
1 2 . . . k												
Σ					A_1	A_2	A_3	B_1	B_2	B_3		$A_1 + A_2 + A_3 + 2 \times (B_1 + B_2 + B_3)$

The elements of the covariance matrix $\|k_{ij}\|$ of the compound distribution law are determined by the formulas:

$$k_{11} = \sigma_x^2 = \frac{A_1}{2\rho^2}, \qquad k_{22} = \sigma_y^2 = \frac{A_2}{2\rho^2}, \qquad k_{33} = \sigma_z^2 = \frac{A_3}{2\rho^2},$$

$$k_{23} = \frac{B_1}{2\rho^2}, \qquad k_{31} = \frac{B_2}{2\rho^2}, \qquad k_{12} = \frac{B_3}{2\rho^2}.$$

The last two columns of Table 9 serve for checking the accuracy of computations; the following equality must be satisfied:

$$\sum_{i=1}^{k} \kappa_i^2 = A_1 + A_2 + A_3 + 2(B_1 + B_2 + B_3).$$

The variances ξ, η, ζ with respect to the principal directions of the compound distribution ellipsoid (σ_ξ^2, σ_η^2, σ_ζ^2) are given by the formulas

$$\sigma_\xi^2 = \frac{a^2}{2\rho^2}, \qquad \sigma_\eta^2 = \frac{b^2}{2\rho^2}, \qquad \sigma_\zeta^2 = \frac{c^2}{2\rho^2},$$

where a, b, c are the principal semi-axes of the unit ellipsoid of the compound distribution, and are related to the roots (u_1, u_2, u_3) of the equation $u^3 + pu + q = 0$ as follows:

$$a^2 = u_1 + \frac{l}{3},$$

$$b^2 = u_2 + \frac{l}{3},$$

$$c^2 = u_3 + \frac{l}{3},$$

$$p = -\frac{1}{3} l^2 + m, \qquad q = -\frac{2}{27} l^3 + \frac{1}{3} lm + n,$$

$$l = A_1 + A_2 + A_3,$$

$$m = A_1A_2 + A_2A_3 + A_3A_1 - B_1^2 - B_2^2 - B_3^2,$$

$$n = A_1B_1^2 + A_2B_2^2 + A_3B_3^2 - A_1A_2A_3 - 2B_1B_2B_3.$$

The roots of the cubic equation can be found either from special tables or the formulas:

$$u_1 = 2\sqrt{-\frac{1}{3}p}\cos\frac{\varphi}{3}, \qquad u_2 = 2\sqrt{-\frac{1}{3}p}\cos\frac{\varphi - 2\pi}{3},$$

$$u_3 = 2\sqrt{-\frac{1}{3}p}\cos\frac{\varphi + 2\pi}{3},$$

where

$$\cos\varphi = \frac{-9q}{2\sqrt{-3p^3}}.$$

The direction cosines of axes ξ, η, ζ in the coordinate system $Oxyz$ are the solutions of a system of three equations ($i = 1, 2, 3$):

$$\left.\begin{aligned}(A_1 - \lambda_i)\alpha_{i1} + B_3\alpha_{i2} + B_2\alpha_{i3} &= 0\\ B_3\alpha_{i1} + (A_2 - \lambda_i)\alpha_{i2} + B_1\alpha_{i3} &= 0\\ \alpha_{i1}^2 + \alpha_{i2}^2 + \alpha_{i3}^2 &= 1\end{aligned}\right\},$$

where

$$\lambda_1 = a^2, \qquad \lambda_2 = b^2, \qquad \lambda_3 = c^2,$$

and α_{ij} denotes the cosine of the angle made by the ith coordinate axis of the system $O\xi\eta\zeta$ with the jth axis of the system $Oxyz$.

Solution for Typical Examples

Example 26.1 The position of a point A is defined from a point of observation O by distance $OA = D$, and the angular deviation from a reference line OB.

The mean error in estimating the distance is $100k$ per cent of the distance; the mean error in estimating the angular deviation is ε radians. The error made in representing the point A on a chart obeys a normal circular distribution with mean deviation r; the error in the position of the point O also obeys a normal circular distribution law with mean deviation R. Find the compound distribution characterizing the error in position resulting from the representation of point A on the chart. How will the probability that point A lies in a rectangle of size 100×100 sq. m. change if D decreases from 20 to 10 km. ($r = 20$ m., $R = 40$ m., $\varepsilon = 0.003$, $k = 0.005$)?

Solution. Independent deviation vectors kD, r and R act along the direction of OA, and perpendicular to it there act the independent deviation vectors εD, r and R.[2] The distribution of the errors made in the position of A on the chart is defined by a unit ellipse with semi-axes

$$\sqrt{k^2D^2 + r^2 + R^2} \quad \text{and} \quad \sqrt{\varepsilon^2D^2 + r^2 + R^2},$$

and, consequently,

$$P = \hat{\Phi}\left(\frac{50}{\sqrt{k^2D^2 + r^2 + R^2}}\right)\hat{\Phi}\left(\frac{50}{\sqrt{\varepsilon^2D^2 + r^2 + R^2}}\right).$$

For distance $OA = 20{,}000$ m.,

$$P_1 = \hat{\Phi}\left(\frac{50}{109.5}\right)\hat{\Phi}\left(\frac{50}{74.8}\right) = 0.083.$$

If the distance becomes 10,000 m.,

$$P_2 = \hat{\Phi}\left(\frac{50}{67.1}\right)\hat{\Phi}\left(\frac{50}{53.8}\right) = 0.181.$$

[2] Since the angle ε is small, the deviation along the arc εD can be replaced by a deviation of magnitude εD along the tangent and one can consider this deviation perpendicular to the radius D.

Example 26.2 The position of a point K in the plane is defined by measuring the distance from it to two points M and N. The coordinates of the point obey a normal distribution law given by principal semi-axes $a = 60$ m. and $b = 40$ m., and angle $\alpha_1 = 47°52'$ between the semi-axis a and the direction of NK.

How will the distribution of coordinates of point K change if the mean error for distance MK decreases to one-half?

SOLUTION. The deviation errors of the coordinates of K arising from errors in the measurements of MK and NK are the conjugate semi-axes m and n of a unit ellipse, directed along the normals to MK and NK, respectively (see Figure 31). Therefore, $\alpha = 90° - \alpha_1 = 42°8'$.

$$n^2 = \frac{a^2b^2}{b^2 + (a^2 - b^2)\sin^2\alpha} = 2304 \text{ sq. m.}, \qquad n = 48.0 \text{ m.},$$

$$\tan\beta = \frac{b^2}{a^2\tan\alpha} = 0.4913, \qquad \beta = 26°\,10', \qquad \alpha + \beta = 68°\,18',$$

$$m = \frac{ab}{n\sin(\alpha + \beta)} = 53.8 \text{ m.},$$

The principal semi-axes of the unit ellipse of the new distribution can be determined if one considers the fact that the conjugate semi-axes of this ellipse are the segments $n/2 = 24.0$ m. and $m = 53.8$ m., the angle between them being, as before, equal to $\alpha + \beta = 68°\,18'$. Using Apollonius' theorem here, we obtain

$$a_1^2 + b_1^2 = \left(\frac{n}{2}\right)^2 + m^2, \qquad a_1b_1 = \frac{n}{2}m\sin(\alpha + \beta);$$

that is,

$$a_1^2 + b_1^2 = 3470 \text{ sq. m.}, \qquad a_1b_1 = 483.9 \text{ sq. m.},$$

$$a_1 = 58.3 \text{ m.}, \qquad b_1 = 8.3 \text{ m.}$$

Example 26.3 Find the covariance matrix of a three-dimensional distribution, representing the convolution of four degenerate normal distributions with the following deviation vectors (Table 10):

TABLE 10

i	a_i	$\cos(a_i, x)$	$\cos(a_i, y)$	$\cos(a_i, z)$
1	20	0.5	$-0.5\sqrt{3}$	0
2	30	$\sqrt{0.4}$	$\sqrt{0.3}$	$\sqrt{0.3}$
3	50	-0.8	0	-0.6
4	25	0	1	0

Find the principal semi-axes of the unit compound ellipsoid and the direction cosines of the angles between the major semi-axes and the axes of coordinates.

SOLUTION. (1) The computation of the elements of the covariance matrix is given in Table 11.

TABLE 11

i	a_i	a_{ix}	a_{iy}	a_{iz}	a_{ix}^2	a_{iy}^2	a_{iz}^2	$a_{iy}a_{iz}$	$a_{iz}a_{ix}$	$a_{ix}a_{iy}$	κ_i	κ_i^2
1	20	10	−17.32	0	100	300	0	0	0	−173.2	−7.32	54
2	30	18.97	16.43	16.43	360	270	270	270	311.7	311.7	51.83	2686
3	50	−40	0	−30	1600	0	900	0	1200	0	−70	4900
4	25	0	25	0	0	625	0	0	0	0	25	625
					A_1 =2060	A_2 =1195	A_3 =1170	B_1 =270	B_2 =1512	B_3 =138.5		8265

Sum: 2060 + 1195 + 1170 + 2(270 + 1512 + 138.5) = 8265

Check:

$$k_{11} = \frac{A_1}{2\rho^2} = 4528, \qquad k_{22} = \frac{A_2}{2\rho^2} = 2627, \qquad k_{33} = \frac{A_3}{2\rho^2} = 2572,$$

$$k_{23} = \frac{B_1}{2\rho^2} = 593.5, \qquad k_{31} = \frac{B_2}{2\rho^2} = 3323, \qquad k_{12} = \frac{B_3}{2\rho^2} = 304.4.$$

(2) The computation of the principal semi-axes of the unit compound ellipsoid proceeds as follows.

According to the preceding formulas, we find

$$l = 4425; \qquad m = 3892 \cdot 10^3; \qquad n = -8871 \cdot 10^4;$$

$$p = -2635 \cdot 10^3; \qquad q = -767 \cdot 10^6;$$

$$\cos\varphi = 0.4658; \qquad \varphi = 62^\circ\, 14'\, 20'';$$

$$\cos\frac{\varphi}{3} = 0.9352; \qquad \cos\frac{\varphi - 2\pi}{3} = -0.1608;$$

$$\cos\frac{\varphi + 2\pi}{3} = -0.7744;$$

$$u_1 = 1752; \qquad u_2 = -301.3; \qquad u_3 = -1451;$$

$$a^2 = 3227; \qquad b^2 = 1174; \qquad c^2 = 24;$$

$$a = 56.8; \qquad b = 34.3; \qquad c = 4.9.$$

(3) The computation of the cosines of the angles made by the principal semi-axes a with the axes of coordinates proceeds as follows.

We form the system of equations

$$-1167\alpha_{11} + 138\alpha_{12} + 1512\alpha_{13} = 0,$$

$$138\alpha_{11} - 2032\alpha_{12} + 270\alpha_{13} = 0,$$

$$\alpha_{11}^2 + \alpha_{12}^2 + \alpha_{13}^2 = 1.$$

From the first two equations we find

$$\alpha_{12} = 0.1684\alpha_{11}, \qquad \alpha_{13} = 0.7563\alpha_{11};$$

and from the third equation,

$$\alpha_{11} = \pm 0.7905.$$

Thus,

$$\cos(a, x) = \alpha_{11} = \pm 0.7905; \qquad \cos(a, y) = \alpha_{12} = \pm 0.1331;$$

$$\cos(a, z) = \alpha_{13} = \pm 0.5978.$$

Similarly one can solve Problem 26.9.

PROBLEMS

26.1 Find the convolution of two deviation vectors $\mathbf{c}_1$ and $\mathbf{c}_2$ if the angle between them is $\gamma = 30°$, $c_1 = 30$ m., $c_2 = 40$ m. and the distribution centers coincide.

26.2 Solve the preceding problem for $\gamma = 0°$ and $\gamma = 90°$

26.3 Find the compound distribution that is the convolution of deviation vectors $\mathbf{a}_i$, lying in the same plane if their magnitudes are a_i, and the angles α_i between $\mathbf{a}_i$ and the positive direction of the axis of abscissas are given in Table 12.

TABLE 12

i	a_i, m.	α_i	i	a_i, m.	α_i
1	0.9	30°37′	5	0.4	158°48′
2	0.5	59°36′	6	0.5	189° 3′
3	0.7	92°12′	7	0.2	273°18′
4	0.8	127°17′	8	0.3	316°54′

26.4 Find the unit ellipse of the compound variance law of the points in a plane obtained from the composition of the following deviation vectors lying in this plane (Table 13).

TABLE 13

i	a_i, m.	α_i, deg.	i	a_i, m.	α_i, deg.
1	10	297	5	40	117
2	30	117	6	60	27
3	30	117	7	70	297
4	40	297	8	80	207

26.5 Find the convolution of the deviation vector $\mathbf{\Delta}$ ($\Delta = 18$ m.), making an angle $\beta = 75°$ with the direction of Ox, and a normal distribution given by a unit ellipse, one of the semi-axes of which coincides with Ox and has length $a = 30$ m. and the other of which has length $b = 20$ m.

26.6 Find the convolution of two normal distributions in the plane if (a) the principal semi-axes of the unit ellipses are $a_1 = b_1 = 50$ m., $a_2 = b_2 = 25$ m., (b) the principal semi-axes of the unit ellipses are $a_1 = 50$ m., $b_1 = 25$ m., $a_2 = 50$ m., $b_2 = 25$ m. if the angle between a_1 and a_2 is 30°.

26.7 The coordinates of a random point in the plane obey a normal distribution law given by a unit ellipse with principal semi-axes $a = 24$ m., $b = 7$ m. Find the probability of hitting a diamond with side $2l = 60$ m. and acute angle $\gamma = 34°.3$. The center of the diamond coincides with the center of the distribution and the adjacent sides of the diamond are parallel to two conjugate semi-axes.

26.8 Find two deviation vectors equivalent to a normal distribution in the plane, characterized by a unit ellipse with principal semi-axes 80 m. and 60 m. if one of the deviation vectors makes an angle of 30° with the major semi-axis.

26.9 The coordinates of a ship are determined by a radar station, which estimates the distance from a reference point on the shore and the direction of the sighting. The errors in measurement are given by a unit ellipse with principal semi-axes $E_x = 80$ m. in the direction of the axis Ox, and $E_z = 30$ m. in the direction of Oz. The unit ellipse of the errors made in estimating the coordinates of the reference point and caused by inaccurate knowledge of its position has major semi-axes $E_1 = 100$ m., $E_2 = 40$ m. and E_1 makes an angle of 20° with the axis Ox.

Find (a) the probability density for the compound errors made in determining the position of the ship in the system of coordinates xOz, (b) the principal semi-axes and the orientation with respect to the axis Ox of the unit ellipse of the compound errors in the coordinates of the ship.

26.10 The errors in determining the position of a ship at sea are due to three deviation errors, whose magnitudes and directions with respect to the meridian are given in Table 14.

TABLE 14

Δ_i, m.	$\Delta_1 = 400$	$\Delta_2 = 800$	$\Delta_3 = 200$
θ_i°	$\theta_1 = 65$	$\theta_2 = 35$	$\theta_3 = 335$

Find the unit ellipse of the errors in determining the position of the ship.

26.11 Find the distribution law for the coordinates of a point C by sighting it from two points A and B if the base $\tilde{B}$, the angles β_1 and

β_2 and the mean angular errors in sighting from both points $E_{\beta_1} = E_{\beta_2} = E_\beta$ are given. The positions of A and B are known with certainty (Figure 32).

26.12 Under the assumptions made in the preceding problem, compute the major semi-axes of the unit ellipse and its orientation with respect to direction AB for $\tilde{B} = 15$ km., $\beta_1 = 60°$, $\beta_2 = 75°$, $E_{\beta_1} = E_{\beta_2} = 0.0005$.

26.13 Under the assumptions made in Problems 26.11 and 26.12, find the compound distribution law for the errors of coordinates of point C with respect to A if beside the errors in sighting E_{β_1} and E_{β_2}, there is given as well the distribution law for the errors in the position of point B with respect to A with the major semi-axes along the base $E_1 = 30$ m. and perpendicular to the base $E_2 = 15$ m.

26.14 To determine the actual course of a ship and its velocity, one makes two estimates of the position of the ship (at the points A_1 and A_2) with respect to some reference points located on the shore and during an interval of time $\tau = 20$ sec. The distribution of the errors in the position of the ship is circular with the radius of the unit circle $r = 30$ m. Find the mean error in estimating the magnitude of the velocity and the course of the ship if the distance A_1A_2 is measured as $D = 1000$ m.

26.15 The coordinates of a ship at time $t = 0$ are known with an error obeying a normal circular distribution with the radius of the unit circle of 100 m. The mean error in the magnitude of the velocity is 2 m./sec., representing 10 per cent of its velocity, and the mean error in estimating its course is 0.08 radians. Calculate the unit ellipse of the errors made in the position of the ship at time $t = 1$ min.

26.16 The position of a meteorological balloon at the instant of observation is known with an error obeying a normal spherical distribution law with the radius of the unit sphere equal to 50 m.; the velocity of the balloon is known with mean error 2 m./sec. The errors in finding the velocity vector in a plane perpendicular to its course are given by a normal distribution law with radius of the unit circle equal to 3 m./sec. Find the unit ellipsoid of the errors in the position of the balloon 20 seconds after the coordinates and the velocity vector have been determined.

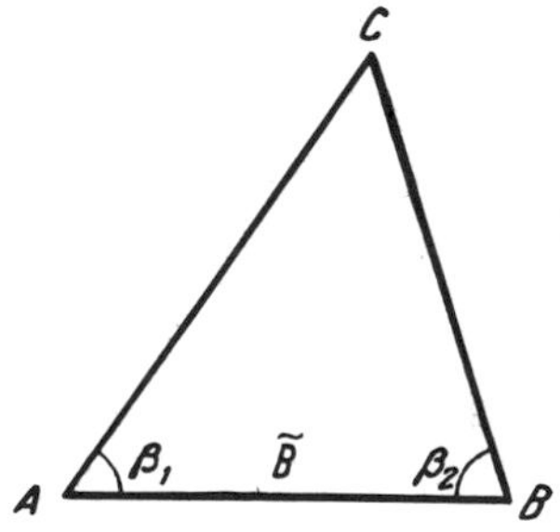

FIGURE 32

26.17 Find the probability density for the sum of two random normal vectors in the space $Oxyz$, and a random vector in the plane Oxz, for which the first moments are

$$\bar{x}_1 = 20, \quad \bar{y}_1 = -10, \quad \bar{z}_1 = -15,$$
$$\bar{x}_2 = 10, \quad \bar{y}_2 = 25, \quad \bar{z}_2 = -40,$$
$$\bar{x}_3 = 15, \quad \bar{z}_3 = -20,$$

respectively, and the covariance matrices for the projections of the vectors on axes of coordinates are

$$\|k_{ij}^{(1)}\| = \begin{Vmatrix} 12 & 2 & 0 \\ -2 & 8 & 1 \\ 0 & 1 & 14 \end{Vmatrix}, \qquad \|k_{ij}^{(2)}\| = \begin{Vmatrix} 10 & 2 & 1 \\ 2 & 8 & -1 \\ 1 & -1 & 17 \end{Vmatrix},$$

$$\|k_{ij}^{(3)}\| = \begin{Vmatrix} 3 & 0 & -1 \\ 0 & 0 & 0 \\ -1 & 0 & 5 \end{Vmatrix}.$$

The random vectors are mutually independent.

26.18 Find the convolution of the deviation vector $\mathbf{x}$, parallel to the axis Ox, $\bar{x} = 25$, $E_x = 40$ of a normal distribution in the plane xOy with the unit ellipse

$$\frac{(x+5)^2}{400} + \frac{(y+10)^2}{900} = 1$$

and the normal distribution in space with the unit ellipsoid

$$\frac{(x-10)^2}{100} + \frac{(y-10)^2}{225} + \frac{z^2}{64} = 1,$$

if x, y, z are the rectangular coordinates of a point in space.

26.19 Construct the covariance matrix of a system of three random variables (the coordinates of a point in space) that corresponds to the resultant of the following deviation vectors (Table 15):

Table 15

i	a_i	$\cos(\mathbf{a}_i, x)$	$\cos(\mathbf{a}_i, y)$	$\cos(\mathbf{a}_i, z)$
1	40	0.6	-0.8	0
2	60	$\frac{1}{3}\sqrt{3}$	$\frac{1}{3}\sqrt{3}$	$-\frac{1}{3}\sqrt{3}$
3	80	-0.5	0.5	$0.5\sqrt{2}$

26.20 Under the conditions of the preceding problem, determine the major semi-axes of a unit joint distribution ellipsoid and the

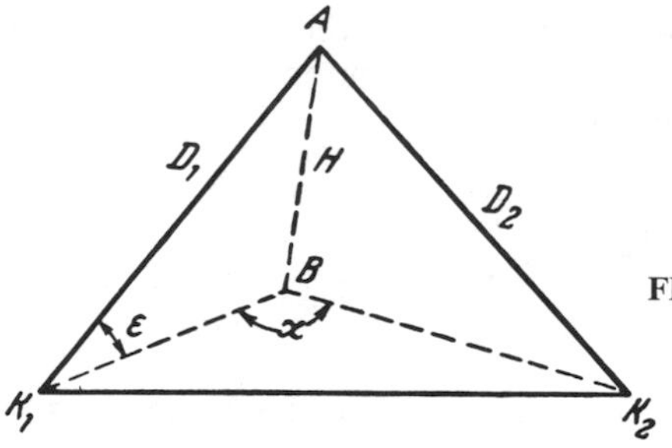

FIGURE 33

direction cosines of the angles between the greatest of the major semi-axes a and the coordinate axes.

26.21 The position of a point K_2 relative to a point K_1 is determined on the basis of measured distances D_1 and D_2 from a point A and of the angle in the horizontal plane $\angle K_1BK_2 = \alpha$ (see Figure 33). Find the covariance matrix of the errors in the determination of the position of the point K_2 relative to the point K_1 if we know that the mean errors made in the determination of the distance are equal to E_D and those made in the determination of the angle are equal to E_α. The measuring errors are mutually independent and they obey normal distribution laws. Assume that the altitude H of the point A over the horizontal plane K_1BK_2 is known exactly.

26.22 Solve Problem 26.21 with the hypothesis that we know (exactly), not the altitude H, but the angle $\varepsilon = \angle AK_1B$.

V ENTROPY AND INFORMATION

27. THE ENTROPY OF RANDOM EVENTS AND VARIABLES

Basic Formulas

Let $A_1, A_2, \dots, A_n$ be a complete set of mutually exclusive events. Then the entropy of this set of events is defined as[1]

$$H = -\sum_{j=1}^{n} \mathbf{P}(A_j) \log_a \mathbf{P}(A_j)$$

and represents the average quantity of information received by knowing which of the events $A_1, A_2, \dots, A_n$ occurred in a certain trial. Thus, the entropy is a measure of uncertainty arising after performing trials involving a complete set of mutually exclusive events $A_1, A_2, \dots, A_n$.

A similar formula defines the entropy $H[X]$ of a discrete variable X, assuming values $x_1, x_2, \dots, x_n$ with probabilities $p_1, p_2, \dots, p_n$:

$$H[X] = -\sum_{j=1}^{n} p_j \log_a p_j.$$

The same formulas hold for $n = \infty$.

The measure of uncertainty of a random variable X assuming a continuous series of values and having a given probability density $f(x)$ is the differential entropy $H[X]$, defined by the formula

$$H[X] = -\int_{-\infty}^{\infty} f(x) \log_a f(x)\, dx,$$

where $f(x) \log_a f(x) \equiv 0$ for those values of x for which $f(x) = 0$.

The conditional entropy of a random variable X with respect to a random variable Y is defined by

$$H[X \mid y_j] = -\sum_{i=1}^{n} \mathbf{P}(X = x_i \mid Y = y_j) \log_a \mathbf{P}(X = x_i \mid Y = y_j)$$

for discrete X and Y, and for continuous X and Y by the conditional differential entropy

$$H[X \mid y] = -\int_{-\infty}^{\infty} f(x \mid y) \log_a f(x \mid y)\, dx.$$

[1] $\mathbf{P}(A_j)$ is the probability of event A_j; $\mathbf{P}(A_j) \log_a \mathbf{P}(A_j) = 0$ if $\mathbf{P}(A_j) = 0$.

We call the expectation of the conditional entropy the conditional mean entropy $H_y[X]$. For discrete random variables

$$H_y[X] = \mathbf{M}[H[X \mid y]]$$
$$= -\sum_{i=1}^{n}\sum_{j=1}^{m} \mathbf{P}(Y = y_j)\mathbf{P}(X = x_i \mid Y = y_j) \log_a \mathbf{P}(X = x_i \mid Y = y_j),$$

and for continuous random variables

$$H_y[X] = \mathbf{M}[H[X \mid y]] = -\int_{-\infty}^{\infty}\int_{-\infty}^{\infty} f_y(y)f(x \mid y) \log_a f(x \mid y)\, dx\, dy.$$

Similar formulas hold for systems of random variables. For example,

$$H[X_1, X_2, \ldots, X_n] = -\int_{-\infty}^{\infty} \cdots \int_{-\infty}^{\infty} f(x_1, x_2, \ldots, x_n) \times \log_a f(x_1, x_2, \ldots, x_n)\, dx_1 \cdots dx_n$$

represents the entropy of a system of n random variables,

$$H_z[X, Y] = -\int_{-\infty}^{\infty}\int_{-\infty}^{\infty}\int_{-\infty}^{\infty} f_z(z)f(x, y \mid z) \log_a f(x, y \mid z)\, dx\, dy\, dz$$

the conditional mean entropy of the subsystem (X, Y) with respect to Z and

$$H_{x,y}[Z] = -\int_{-\infty}^{\infty}\int_{-\infty}^{\infty}\int_{-\infty}^{\infty} f(x, y)f(z \mid x, y) \log_a f(z \mid x, y)\, dx\, dy\, dz$$

the conditional mean entropy of the random variable Z with respect to random variables X and Y. We also have the inequalities

$$H[X, Y] = H[X] + H_x[Y] \leqslant H[X] + H[Y]$$

and

$$H[X_1, X_2, \ldots, X_n] \leqslant \sum_{k=1}^{n} H[X_k],$$

in which equality corresponds to the case of independent random variables.

For $a = 2$, the unit of measure for entropy represents the entropy of a complete set of two mutually exclusive equally possible events. For $a \neq 2$ the value of the entropy computed for $a = 2$ must be multiplied by $\log_a 2$. The unit of measure for entropy is called binary for $a = 2$, decimal for $a = 10$, and so on.

Solution for Typical Examples

Example 27.1 A number of shots are fired at two targets: two shots at the first target and three at the second one. The probabilities of hitting a target in one shot are equal to 1/2 and 1/3, respectively. Which of the two targets yields a more certain outcome?

Solution. The outcome is determined by the number of hits scored, which obeys the binomial distribution law $\mathbf{P}(X = m) = C_n^m p^m(1 - p)^{n-m}$.

We form the distribution series of the first target for $n = 2$ and $p = 1/2$ (Table 16), and of the second target for $n = 3$, $p = 1/3$ (Table 17).

TABLE 16

m	0	1	2
$\mathbf{P}(X = m)$	1/4	1/2	1/4

TABLE 17

m	0	1	2	3
$\mathbf{P}(X = m)$	1/27	2/9	4/9	8/27

The entropy of the number of hits is a measure of the uncertainty of the outcome. For the first target, we have

$$H_1 = -\frac{1}{4}\log\frac{1}{4} - \frac{1}{2}\log\frac{1}{2} - \frac{1}{4}\log\frac{1}{4} = 0.452 \text{ decimal units},$$

and for the second one

$$H_2 = -\frac{1}{27}\log\frac{1}{27} - \frac{2}{9}\log\frac{2}{9} - \frac{4}{9}\log\frac{4}{9} - \frac{8}{27}\log\frac{8}{27} = 0.511 \text{ decimal units}.$$

The outcome in the case of the first target has a greater certainty.

Similarly one can solve Problems 27.1 to 27.11.

Example 27.2 Among all distribution laws of a continuous random variable X with the same known variation D, find the distribution with the maximal differential entropy.

SOLUTION. According to a theorem in calculus of variations, to find a function $y = y(x)$ that realizes an extremum of the integral

$$I = \int_a^b \Phi(x, y)\, dx$$

under constraints

$$\int_a^b \psi_s(x, y)\, dx = c_s \qquad (s = 1, 2, \ldots, m),$$

it is necessary to solve the Euler equation

$$\frac{\partial \Phi_1}{\partial y} = 0,$$

where $\Phi_1 = \Phi + \sum_{s=1}^{m} \lambda_s \psi_s$, and constants λ_s are found from the given constraints. In our example, we are looking for the maximum of the integral

$$-\int_{-\infty}^{\infty} f \ln f \, dx$$

under the constraints

$$\int_{-\infty}^{\infty} f \, dx = 1$$

and

$$\int_{-\infty}^{\infty} (x - \bar{x})^2 f \, dx = D.$$

From this, it follows that

$$\Phi_1(x, f) = -f \ln f + \lambda_1 f + \lambda_2 (x - \bar{x})^2 f.$$

Consequently, the equation for $f(x)$ has the form

$$-\ln f - 1 + \lambda_1 + \lambda_2 (x - \bar{x})^2 = 0,$$

and, therefore,

$$f(x) = ce^{-\lambda_2 (x - \bar{x})^2},$$

where

$$c = e^{-\lambda_1 + 1}.$$

From the constraints, we find that

$$c = \frac{1}{\sqrt{2\pi D}}, \qquad \lambda_2 = \frac{1}{2D}.$$

The solution obtained corresponds to maximal entropy.

Therefore, for a given variation D, the maximal entropy has the normal distribution law

$$f(x) = \frac{1}{\sqrt{2\pi D}} \exp\left\{-\frac{1}{2D}(x - \bar{x})^2\right\}.$$

Problems 27.12 to 27.15 can be solved in a similar manner.

Example 27.3 Prove that the maximal entropy of a discrete random variable is $\log_a n$ (n being the number of possible values of the random variable) and is attained for $p_1 = p_2 = \cdots = p_n = 1/n$.

SOLUTION. We shall make use of the inequality $\ln x \geqslant 1 - 1/x$ ($x > 0$) (equality occurs only for $x = 1$). Applying this inequality, we obtain

$$-H + \log_a n = \sum_{k=1}^{n} p_k \log_a (np_k) \geqslant \frac{1}{\ln a} \sum_{k=1}^{n} p_k \left(1 - \frac{1}{np_k}\right)$$
$$= \frac{\sum_{k=1}^{n} p_k - 1}{\ln a} = 0.$$

It follows that

$$H = -\sum_{k=1} p_k \log_a p_k \leqslant \log_a n.$$

To the case $np_k = 1$, there corresponds maximal entropy: $\log_a n$.

One can solve Problem 27.16 similarly.

PROBLEMS

27.1 Two urns contain 15 balls each. The first urn contains five red, seven white and three black balls; the second urn contains four red, four white and seven black balls. One ball is drawn from each urn. Find the urn for which the outcome of the experiment is more certain.

27.2 The probability of occurrence of an event is p, and of non-occurrence, $q = 1 - p$. For which value of p does the result of the trial have the maximal uncertainty?

27.3 For which of the following two experiments does the outcome have the greatest uncertainty: (a) a random point is taken inside an equilateral triangle and "success" means the point lands inside the inscribed circle; (b) a random point is taken inside a circle and "success" means the point lands inside a given equilateral triangle inscribed in the circle.

27.4 By joining the midpoints of adjacent sides of a regular n-polygon, one constructs another regular n-polygon inscribed in the first. A point taken inside the first polygon may turn out to be inside or outside the inscribed polygon.

Find (a) the entropy of the experiment, (b) the value of n for which the entropy is maximal.

27.5 The probability for realization of an event A at one trial is p. The trials are repeated until A occurs for the first time. Find the entropy of the number of trials and clarify the character of variation of the entropy with the change of p.

27.6 Determine the entropy of a random variable obeying a binomial distribution law (a) in the general case, (b) for $n = 2$, $p = q = 0.5$.

27.7 Determine the entropy of a continuous random variable obeying (a) a uniform probability distribution over the interval (c, d), (b) a normal distribution law with variance σ_x^2, (c) an exponential distribution of the form

$$f(x) = \begin{cases} ce^{-cx} & \text{for } x \geqslant 0 \quad (c > 0), \\ 0 & \text{for } x < 0. \end{cases}$$

27.8 Find the entropy of a random variable X with a distribution function

$$F(x) = \begin{cases} 0 & \text{for } x \leqslant 0, \\ x^2 & \text{for } 0 < x \leqslant 1, \\ 1 & \text{for } x > 1. \end{cases}$$

27.9 Estimate the conditional differential entropy $H[X \mid y]$ and the conditional mean differential entropy $H_y[X]$ of a random variable X with respect to Y, and also $H[Y \mid x]$ and $H_x[Y]$ of the random variable Y with respect to X, for the system of normal random variables (X, Y).

27.10 Find the entropy of a system of n random variables obeying a normal distribution law.

27.11 Given the entropies $H[X]$ and $H[Y]$ of two random variables X and Y, and the conditional mean entropy $H_y[X]$ of the random variable X with respect to Y, find the conditional mean entropy $H_x[Y]$ of Y with respect to X.

27.12 Among all distribution laws of a continuous random variable X whose probability density vanishes outside the interval $a < x < b$, determine the distribution law with maximal differential entropy.

27.13 Among all distribution laws of a continuous random variable X whose probability density vanishes for $x < 0$, for a known expectation $\mathbf{M}[X]$ find the distribution law with maximal differential entropy.

27.14 Find the probability density for which the differential entropy of a random variable is maximal if its second moment is m_2.

27.15 Among all the distribution laws for continuous systems of random variables with a known covariance matrix, find the distribution law for which the entropy of the system is maximal.

27.16 A message is encoded by using two groups of symbols. The first group has k symbols with probabilities of occurrence p_{11}, $p_{12}, \ldots, p_{1k}$ $(\sum_{i=1}^{k} p_{1i} = \alpha)$, the second group has n symbols with probabilities of occurrence $p_{21}, p_{22}, \ldots, p_{2n}$ $(\sum_{j=1}^{n} p_{2j} = 1 - \alpha)$. For a fixed value of α, find the probabilities p_{1i} and p_{2j}, corresponding to the maximal entropy.

27.17 Experiment A consists of selecting an integer from 1 to 1050 at random, experiment B of communicating the values of the remainders upon dividing the selected number by 5 and 7. Find the entropy of experiment A and the conditional mean entropy of A with respect to experiment B.

27.18 Between two systems of random variables $(X_1, X_2, \ldots, X_n)$ and $(Y_1, Y_2, \ldots, Y_n)$, there exists a one-to-one correspondence $Y_k = \varphi_k(X_1, X_2, \ldots, X_n)$, $X_k = \psi_k(Y_1, Y_2, \ldots, Y_n)$, where $k = 1, 2, \ldots, n$. Find the entropy $H[Y_1, Y_2, \ldots, Y_n]$ if the probability density $f_x(x_1, x_2, \ldots, x_n)$ is known.

27.19 Two systems of random variables $(X_1, X_2, \ldots, X_n)$ and $(Y_1, Y_2, \ldots, Y_n)$ are related by linear expressions

$$Y_k = \sum_{j=1}^{n} a_{kj} X_j \qquad (k = 1, 2, \ldots, n).$$

Evaluate the difference of the entropies,

$$H[Y_1, Y_2, \ldots, Y_n] - H[X_1, X_2, \ldots, X_n],$$

(a) in the general case, (b) for $n = 3$ and the transformation matrix

$$\|a_{kj}\| = \begin{Vmatrix} 3 & 2 & -1 \\ 1 & 4 & -2 \\ 0 & -3 & 5 \end{Vmatrix}.$$

28. THE QUANTITY OF INFORMATION

Basic Formulas

The quantity of information obtained in the observation of a complete set of mutually exclusive events is measured by its entropy H; the quantity of information that can be obtained by observing the value of a discrete random variable X is measured by its entropy $H[X]$.

The quantity of information about a random variable X that can be obtained by observing another random variable Y is measured by the difference between the entropy of X and its conditional mean entropy with respect to Y:

$$I_y[X] = H[X] - H_y[X].$$

For discrete random variables,

$$\begin{aligned} I_y[X] &= \mathbf{M}\left[\log_a \frac{\mathbf{P}(X = x \mid Y = y)}{\mathbf{P}(X = x)}\right] \\ &= \mathbf{M}\left[\log_a \frac{\mathbf{P}(X = x, Y = y)}{\mathbf{P}(X = x)\mathbf{P}(Y = y)}\right] \\ &= \sum_{i=1}^{n} \sum_{j=1}^{m} \mathbf{P}(X = x_i, Y = y_j) \log_a \frac{\mathbf{P}(X = x_i, Y = y_j)}{\mathbf{P}(X = x_i)\mathbf{P}(Y = y_j)}. \end{aligned}$$

If, after receiving a message about the discrete random variable Y, the value of the random variable X is completely defined, then $H_y[X] = 0$ and $I_y[X] = H[X]$.

If X and Y are independent, then $H_y[X] = H[X]$ and $I_y[X] = 0$.

For continuous random variables,

$$\begin{aligned} I_y[X] &= \mathbf{M}\left[\log_a \frac{f(X \mid Y)}{f_x(X)}\right] = \mathbf{M}\left[\log_a \frac{f(X, Y)}{f_x(X)f_y(Y)}\right] \\ &= \int_{-\infty}^{\infty}\int_{-\infty}^{\infty} f(x, y) \log_a \frac{f(x, y)}{f_x(x)f_y(y)}\, dx\, dy. \end{aligned}$$

From the symmetry of the formulas defining the quantity of information with respect to X and Y, it follows that

$$I_y[X] = I_x[Y].$$

SOLUTION FOR TYPICAL EXAMPLES

Example 28.1 Using the method of Shannon-Fano,[2] encode an alphabet that consists of four symbols A, B, C and D if the probabilities of occurrence

[2] In the case of encoding by the method of Shannon-Fano, a collection of symbols (alphabet), originally ordered according to the decreasing probabilities of occurrence of the symbols, is divided into two groups so that the sums of the probabilities of the symbols appearing in each group are approximately equal. Each of the groups is then subdivided into two subgroups by using the same principle. The process continues until only one symbol remains in each group. Each symbol is denoted by a binary number whose digits (zeros and ones) show to which group a given symbol belongs in a particular division.

of each symbol in a message are:

$$\mathbf{P}(A) = 0.28, \qquad \mathbf{P}(B) = 0.14, \qquad \mathbf{P}(C) = 0.48, \qquad \mathbf{P}(D) = 0.10.$$

Find the efficiency of the code, i.e., the quantity of information per symbol.

SOLUTION. We order the symbols of the alphabet according to the decreasing probabilities of C, A, B, D and then divide them successively into groups.

In the first division, the first group contains C and the second A, B and D, since $\mathbf{P}(C) = 0.48$ and $\mathbf{P}(A + B + D) = 0.52$. We assign the coded symbol 1 to the first group, and to the second, 0. Similarly, from the second group we obtain the subgroups A and $B + D$ with probabilities 0.28 and 0.24, and with the codes 01 and 00. Finally, the group $B + D$ is divided into B and D with probabilities 0.14 and 0.10, and codes 001 and 000.

It is convenient to represent the coding process by Table 18.

TABLE 18

Symbols of the Alphabet	*Probability of the Symbols*	*Subgroups and their Codes*	*Codes of the Symbols of the Alphabet*
C	0.48	}1	1
A	0.28	}0 }1	01
B	0.14	}0 }0 }1	001
D	0.10	}0 }0 }0	000

A complete set of mutually exclusive events corresponds to the occurrence of one symbol of the alphabet, and the total quantity of information in this particular example is the entropy of the alphabet. Therefore the quantity of information per coded symbol (efficiency of the code) equals the ratio of the entropy of the alphabet to the expected length of the coded versions of symbols:

$$\frac{-0.48 \log_2 0.48 - 0.28 \log_2 0.28 - 0.14 \log_2 0.14 - 0.10 \log_2 0.10}{1\cdot 0.48 + 2\cdot 0.28 + 3\cdot 0.14 + 3\cdot 0.10} = \frac{1.751}{1.760} = 0.995.$$

Similarly one can solve Problems 28.9 and 28.11 to 28.13.

Example 28.2 The probabilities for a signal to be received or not received are α and $\bar{\alpha} = 1 - \alpha$, respectively. As a result of noise, a signal entering the receiver can be recorded at its output with probability β and not recorded with probability $\bar{\beta} = 1 - \beta$. In the absence of the signal at the input, it can be recorded at the output (because of noise) with probability γ and not recorded with probability $\bar{\gamma} = 1 - \gamma$. What quantity of information about the presence of the signal at the input do we obtain by observing it at the output.

SOLUTION. Let X denote the random number of input signals, and Y the random number of output signals.

Then,

$$\mathbf{P}(X = 1) = \alpha, \qquad \mathbf{P}(X = 0) = \bar{\alpha},$$

$$\mathbf{P}(Y = 1 \mid X = 1) = \beta, \qquad \mathbf{P}(Y = 0 \mid X = 1) = \bar{\beta},$$

$$\mathbf{P}(Y = 1 \mid X = 0) = \gamma, \qquad \mathbf{P}(Y = 0 \mid X = 0) = \bar{\gamma},$$

$$\mathbf{P}(Y = 1) = \alpha\beta + \bar{\alpha}\gamma, \qquad \mathbf{P}(Y = 0) = \alpha\bar{\beta} + \bar{\alpha}\bar{\gamma}.$$

This implies that

$$I_y[X] = \alpha\beta \log_a \frac{\beta}{\alpha\beta + \bar{\alpha}\gamma} + \bar{\alpha}\gamma \log_a \frac{\gamma}{\alpha\beta + \bar{\alpha}\gamma} + \alpha\bar{\beta} \log_a \frac{\bar{\beta}}{\alpha\bar{\beta} + \bar{\alpha}\bar{\gamma}} + \bar{\alpha}\bar{\gamma} \log_a \frac{\bar{\gamma}}{\alpha\bar{\beta} + \bar{\alpha}\bar{\gamma}}.$$

One can also use the formula

$$I_y[X] = I_x[Y] = H[Y] - H_x[Y],$$

where the unconditional entropy is

$$H[Y] = -(\alpha\beta + \bar{\alpha}\gamma) \log_a (\alpha\beta + \bar{\alpha}\gamma) - (\alpha\bar{\beta} + \bar{\alpha}\bar{\gamma}) \log_a (\alpha\bar{\beta} + \bar{\alpha}\bar{\gamma}),$$

and the conditional mean entropy is

$$H_x[Y] = -\alpha(\beta \log_a \beta + \bar{\beta} \log_a \bar{\beta}) - \bar{\alpha}(\gamma \log_a \gamma + \bar{\gamma} \log_a \bar{\gamma}).$$

Example 28.3 There are 12 coins of equal value; however, one coin is counterfeit, differing from the others by its weight. How many weighings, using a balance but no weights, are necessary in order to identify the counterfeit coin and to determine whether it is lighter or heavier than the rest.

SOLUTION. Any of the 12 coins may turn out to be the counterfeit one and thus may be lighter or heavier than a genuine coin. Consequently, there are 24 possible outcomes that, for equal probabilities of these outcomes, give as the entropy for the whole experiment used to identify the counterfeit coin the value $\log_2 24 = 3 + \log_2 3 = 3 + 0.477/0.301 = 4.58$.

Each weighing process has three outcomes, which under the assumption of equal probabilities give an entropy equal to $\log_2 3 = 1.58$.

Therefore, the minimal number of weighings cannot be smaller than $\log_2 24/\log_2 3 = 4.58/1.58 = 2.90$; i.e., it is at least three. In fact, it will be shown that for an optimal planning of the experiment, exactly three weighings will be necessary.

In order that the number of weighings is the minimum, each weighing must furnish the maximal quantity of information and, for this purpose, the outcome of a weighing must have maximal entropy.

Suppose that in the first weighing there are i coins on each of the two pans. As mentioned previously, in this case three outcomes are possible:

(1) the pans remain in equilibrium;
(2) the right pan outweighs the left;
(3) the left pan outweighs the right.

For the first outcome, the counterfeit coin is among the $12 - 2i$ coins put aside and, consequently, the probability of this outcome is

$$P_1 = \frac{12 - 2i}{12}.$$

For the second and third outcomes, the counterfeit coin is on one of the pans. Thus, the probabilities of these outcomes are

$$P_2 = P_3 = \frac{i}{12}.$$

In order that a weighing give the maximal information, the probability distribution of the outcomes must have maximal entropy, which means that all probabilities must be equal. From this, it follows that

$$\frac{12 - 2i}{12} = \frac{i}{12}, \qquad i = 4;$$

i.e., in the first weighing process, four coins should be placed on each pan.

Next, we consider separately the following two cases: (a) in the first weighing, the pans remain in equilibrium, (b) one of the pans outweighs the other.

In case (a) we have eight genuine coins, and four suspect coins that are not used in the first weighing. For the second weighing, we can place i suspect coins on the right pan ($i \leqslant 4$) and $j \leqslant i$ suspect and $i - j$ genuine coins on the left pan. In this case $i + j \leqslant 4$, since the number of suspect coins is 4. All possible values for i and j and the corresponding probabilities of the outcomes in the second weighing in case (a) are included in Table 19.

TABLE 19

Experiment number	i	j	P_1	P_2	P_3	H_{ij}
1	1	1	0.5	0.25	0.25	0.452
2	1	0	0.75	0.125	0.125	0.320
3	2	2	0	0.5	0.5	0.301
4	2	1	0.25	0.375	0.375	0.470
5	2	0	0.5	0.25	0.25	0.452
6	3	1	0	0.5	0.5	0.301
7	3	0	0.25	0.375	0.375	0.470
8	4	0	0	0.5	0.5	0.301

In this table, the entropy of the experiment is also given. It is

$$H_{ij} = -P_1 \log P_1 - P_2 \log P_2 - P_3 \log P_3.$$

The maximal entropy is given by experiments 4 and 7. Thus, there are two equivalent versions of the second weighing: it is necessary either to place two suspect coins on one pan and, on the other, one suspect and one genuine coin (experiment 4); or to place three suspect coins on one pan and three genuine coins on the other (experiment 7).

In both versions, the third weighing solves the problem; that is, it identifies the counterfeit coin and determines whether it is lighter or heavier than the rest.

In case (b), in which one of the pans outweighs the other in the first weighing, the coins are divided into the following three groups: four suspect coins, which are placed on the right pan; four suspect coins on the left pan (4 "right" and 4 "left"); and four genuine coins, which are not used in the first weighing.

If, in the second weighing, one places i_1 "right" and i_2 "left" coins on the right pan, j_1 "right," j_2 "left" and $i_1 + i_2 - j_1 - j_2$ genuine coins on the left pan and then compares the entropy of all the possible versions, there will be 13 equivalent versions with maximal (equal) entropy. Any of these versions, for example, $i_1 = 3$, $i_2 = 2$, $j_1 = 1$, $j_2 = 0$ or $i_1 = 1$, $i_2 = 2$, $j_1 = 0$, $j_2 = 2$, gives maximal information and permits us to identify the counterfeit coin, in the third weighing and to find out whether it is lighter or heavier than the rest.

Problems 28.2 and 28.5 can be solved in a similar manner.

PROBLEMS

28.1 A rectangle is divided into 32 squares by four vertical and eight horizontal lines. A point can be inside any one of these squares with equal probability.

Find the quantity of information in the messages that (a) the point is in square 27, (b) the point lies in the third vertical and the first horizontal line, (c) the point lies in the sixth horizontal line.

28.2 There are N coins of equal value, of which one is counterfeit, that is, lighter than the rest.

How many weighings, on a balance without weights, are necessary to identify the counterfeit coin? What is the maximal N for which five weighings are sufficient?

28.3 The symbols of the Morse Code can appear in a message with probabilities 0.51 for a dot, 0.31 for a dash, 0.12 for a space between letters and 0.06 for a space between words. Find the average quantity of information in a message of 500 symbols if there is no relation between successive symbols.

28.4 A composite system can be in one of N equally probable states A_j. The state of the system can be determined by performing some control experiments, the result of each showing the group of states in which the system can be.

In one of the experiments a signal is observed in the states $A_1, A_2, \ldots, A_k$, and not observed in states $A_{k+1}, A_{k+2}, \ldots, A_N$. In another experiment, the signal is observed if the system is in one of the states $A_1, A_2, \ldots, A_l$ $(l \leqslant k)$ or $A_{k+1}, A_{k+2}, \ldots, A_{k+r}$ $(r \leqslant N - k)$ and not observed in the rest. What is the quantity of information in the first and second experiments?

28.5 A defective television set can be in one out of five different states, to which there are corresponding different types of failures. To identify the type of failure, one performs several tests out of a total of seven possible tests, which for different states of the television set make a control light bulb turn on or off. In the following table, these states are denoted by ones and zeros.

Test No.	*State No.*				
	1	2	3	4	5
1	0	0	0	0	1
2	0	0	0	1	1
3	0	1	1	0	0
4	1	0	1	0	0
5	1	0	1	0	1
6	1	1	1	0	0
7	1	1	1	1	0

Find a sequence consisting of the *minimal* number of tests that permit determination of the type of failure.

28.6 Some messages use the symbols of the alphabet A_1, A_2, A_3, A_4 with probabilities $\mathbf{P}(A_1) = 0.45$, $\mathbf{P}(A_2) = 0.10$, $\mathbf{P}(A_3) = 0.15$, $\mathbf{P}(A_4) = 0.30$.

To transmit a message through a communication channel, one can use two codes, 1 and 2. In the first code, the symbols a, b, c and d and in the second code, the symbols a, d, b and c correspond to the symbols of the alphabet.

Determine the efficiency of the codes, i.e., the average quantity of information transmitted per time unit, if the transmission times of the symbols of the code through the communication channel for conventional time units are

$$t_a = 8, \qquad t_b = 6, \qquad t_c = 5, \qquad t_d = 3.$$

28.7 Under the assumptions made in the preceding problem, along with codes 1 and 2, consider other possible codes and find the most efficient one.

28.8 For the transmission of some messages, one uses a code of three symbols whose probabilities of occurrence are 0.8, 0.1 and 0.1. There is no correlation among the symbols of the code. Determine the redundancy of the code; that is, the difference between 1 and the ratio of the entropy of the given code to the maximal entropy of a code containing the same number of symbols.

28.9 A message consists of a sequence of two letters A and B, whose probabilities of occurrence do not depend on the preceding letter and are $\mathbf{P}(A) = 0.8$, $\mathbf{P}(B) = 0.2$.

Perform the coding by using the method of Shannon-Fano for (a) separate letters, (b) blocks consisting of two-letter combinations, (c) blocks of three-letter combinations.

Compare the codes according to their efficiency.

28.10 Compare the codes of the preceding problem according to their redundancy by calculating the mean probabilities of occurrence of the symbol a_j by the formula

$$\mathsf{P}(a_j) = \frac{\sum_{i=1}^{m} \mathsf{P}(A_i)Z_{ij}}{\sum_{i=1}^{m} \mathsf{P}(A_i)R_i},$$

where Z_{ij} is the number of symbols a_j in the ith coded combination, and R_i is the number of all symbols in the ith combination.

28.11 A message consists of a sequence of letters A, B and C, whose probabilities of occurrence do not depend on the preceding combination of letters and are $\mathsf{P}(A) = 0.7$, $\mathsf{P}(B) = 0.2$ and $\mathsf{P}(C) = 0.1$.

(a) Perform the coding by the method of Shannon-Fano for separate letters and two-letter combinations; (b) compare the efficiencies of the codes; (c) compare the redundancies of the codes.

28.12 The probabilities of occurrence of separate letters of the Russian alphabet are given in Table 20, where the symbol "—" denotes the space between words.

Perform the coding of the alphabet by the method of Shannon-Fano, if the probability of occurrence of a letter is independent of the occurrences of the preceding letters.

TABLE 20

Letter	—	о	е, ё	а	и	т	н	с
Probability	0.175	0.090	0.072	0.062	0.062	0.053	0.053	0.043
Letter	р	в	л	к	м	д	п	у
Probability	0.040	0.038	0.035	0.028	0.026	0.025	0.023	0.021
Letter	я	ы	з	ь, ъ	б	г	ч	й
Probability	0.018	0.016	0.016	0.014	0.014	0.013	0.012	0.010
Letter	х	ж	ю	ш	ц	щ	э	ф
Probability	0.009	0.007	0.006	0.006	0.004	0.003	0.003	0.002

28.13 An alphabet consists of n symbols A_j ($j = 1, 2, \ldots, n$) whose occurrences in a message are independent and have probability

$$\mathsf{P}(A_j) = 2^{-k_j},$$

where k_j are positive integers, and

$$\sum_{j=1}^{n} \mathsf{P}(A_j) = 1.$$

Show that if one codes this alphabet by the method of Shannon-Fano, each coded symbol contains a maximal quantity of information equal to one binary unit (one bit).

28.14 Two signals, A_1 and A_2, are transmitted through a communication channel with the probabilities $\mathbf{P}(A_1) = \mathbf{P}(A_2) = 0.5$. At the output of the channel the signals are transformed into symbols a_1 and a_2 and, as a result of noise to which A_1 and A_2 are subjected equally, errors appear in transmission so that an average of one signal out of 100 is distorted (a_1 becomes a_2 or a_2 becomes a_1).

Estimate the average quantity of information per symbol. Compare it with the quantity of information in the absence of noise.

28.15 Signals $A_1, A_2, \ldots, A_n$ are transmitted with equal probabilities through a communication channel. In the absence of noise, the symbol a_j corresponds to the signal A_j $(j = 1, 2, \ldots, m)$. In the presence of noise, each symbol is correctly received with probability p and is distorted to another symbol with probability $q = 1 - p$. Evaluate the average quantity of information per symbol in the cases of absence and of presence of noise.

28.16 Signals $A_1, A_2, \ldots, A_m$ are transmitted through a communication channel with equal probabilities. In the absence of noise, the symbol a_j corresponds to the signal A_j $(j = 1, 2, \ldots, m)$. Because of the presence of noise, signal A_j can be received correctly with probability p_{jj} or as symbol a_i with probability p_{ij} $(i, j = 1, 2, \ldots, m, \sum_{i=1}^{m} p_{ij} = 1)$. Estimate the average quantity of information per symbol that is transmitted through the channel whose noise is characterized by the matrix $\|p_{ij}\|$.

VI THE LIMIT THEOREMS

29. THE LAW OF LARGE NUMBERS

Basic Formulas

If a random variable X has a finite variance, then for any $\varepsilon > 0$, Chebyshev's inequality holds:

$$\mathbf{P}(|X - \bar{x}| \geqslant \varepsilon) \leqslant \frac{\mathbf{D}[X]}{\varepsilon^2}.$$

If $X_1, X_2, \ldots, X_n \ldots$ is a sequence of random variables, pairwise independent, whose variances are bounded by the same constant $\mathbf{D}[X_k] \leqslant C$, $k = 1, 2, \ldots$, then for any constant $\varepsilon > 0$,

$$\lim_{n\to\infty} \mathbf{P}\left(\left|\frac{1}{n}\sum_{k=1}^{n} X_k - \frac{1}{n}\sum_{k=1}^{n} \bar{x}_k\right| < \varepsilon\right) = 1$$

(Chebyshev's theorem).

If the random variables $X_1, X_2, \ldots, X_n \ldots$ all have the same distribution and have finite expectations $\bar{x}$, then for any constant $\varepsilon > 0$,

$$\lim_{n\to\infty} \mathbf{P}\left(\left|\frac{1}{n}\sum_{k=1}^{n} X_k - \bar{x}\right| < \varepsilon\right) = 1$$

(Khinchin's theorem).

For a sequence of dependent random variables $X_1, X_2, \ldots, X_n, \ldots$, satisfying the condition

$$\lim_{n\to\infty} \frac{1}{n^2}\mathbf{D}\left[\sum_{k=1}^{n} X_k\right] = 0,$$

for any constant $\varepsilon > 0$, we have

$$\lim_{n\to\infty} \mathbf{P}\left\{\left|\frac{1}{n}\sum_{k=1}^{n} X_k - \frac{1}{n}\sum_{k=1}^{n} \bar{x}_k\right| < \varepsilon\right\} = 1$$

(Markov's theorem).

In order that the law of large numbers be applicable to any sequence of dependent random variables $X_1, X_2, \ldots, X_n, \ldots$, i.e., for any constant $\varepsilon > 0$, for the relation

$$\lim_{n\to\infty} \mathbf{P}\left\{\left|\frac{1}{n}\sum_{k=1}^{n} X_k - \frac{1}{n}\sum_{k=1}^{n} \bar{x}_k\right| < \varepsilon\right\} = 1$$

to be fulfilled it is necessary and sufficient that the following equality hold true:

$$\lim_{n\to\infty} \mathbf{M} \frac{\left\{\frac{1}{n}\sum_{k=1}^{n}(X_k - \bar{x}_k)\right\}^2}{1 + \left\{\frac{1}{n}\sum_{k=1}^{n}(X_k - \bar{x}_k)\right\}^2} = 0.$$

Solution for Typical Examples

Example 29.1 Prove that if $\varphi(x)$ is a monotonic increasing positive function and $\mathbf{M}[\varphi(X)] = m$ exists, then

$$\mathbf{P}(X > t) \leqslant \frac{m}{\varphi(t)}.$$

Solution. Taking into account the properties of $\varphi(x)$, we obtain a chain of inequalities

$$\mathbf{P}(X > t) = \int_{x>t} f(x)\,dx \leqslant \frac{1}{\varphi(t)}\int_{x>t}\varphi(x)f(x)\,dx$$
$$\leqslant \frac{1}{\varphi(t)}\int_{-\infty}^{+\infty}\varphi(x)f(x)\,dx = \frac{m}{\varphi(t)},$$

since $m = \mathbf{M}[\varphi(X)] = \int_{-\infty}^{+\infty}\varphi(x)f(x)\,dx$. This implies that $\mathbf{P}(X > t) \leqslant m/\varphi(t)$, which we wish to prove.

Similarly one can solve Problems 29.2 to 29.5.

Example 29.2 Given a sequence of independent random variables $X_1, X_2, \ldots, X_n, \ldots$ with the same distribution function

$$F(x) = \frac{1}{2} + \frac{1}{\pi}\arctan\frac{x}{a},$$

determined whether Khinchin's theorem can be applied to this sequence.

Solution. For the applicability of Khinchin's theorem, it is necessary that the expectation of the random variable X exist; i.e., $\int_{-\infty}^{+\infty} x\,dF(x)/dx\,dx$ converge absolutely. However,

$$\int_{-\infty}^{+\infty}|x|\frac{dF(x)}{dx}\,dx = \frac{2a}{\pi}\int_0^{\infty}\frac{x\,dx}{x^2+a^2} = \lim_{A\to\infty}\frac{2a}{\pi}\int_0^{A}\frac{x\,dx}{x^2+a^2}$$
$$= \frac{a}{\pi}\lim_{A\to\infty}\ln\left(1+\frac{A^2}{a^2}\right) = \infty,$$

i.e., the integral does not converge, the expectation does not exist and Khinchin's theorem is not applicable.

Example 29.3 Can the integral $J = \int_a^{\infty}(\sin x)/x\,dx$ $(a > 0)$, after the change of variables $y = a/x$, be calculated by a Monte-Carlo method according to the formula

$$J_n = \frac{1}{n}\sum_{k=1}^{n}\frac{1}{y_k}\sin\frac{a}{y_k},$$

where y_k are random numbers on the interval [0, 1]?

SOLUTION. Performing the previously mentioned change of variables, we obtain

$$J = \int_0^1 \frac{1}{y} \sin \frac{a}{y}\, dy.$$

The quantity J_n can be considered an approximate value of J only if the limit equality $\lim_{n\to\infty} \mathbf{P}(|J_n - J| < \varepsilon) = 1$ holds true.

The random numbers y_k have equal distributions and, thus, the functions $(1/y_k)\sin(a/y_k)$ also have equal distributions. To apply Khinchin's theorem, one should make sure that the expectation $\mathbf{M}[(1/Y)\sin(a/Y)]$ exists, where Y is a random variable uniformly distributed over the interval [0, 1]; i.e., one should prove that $\int_0^1 (1/y)\sin(a/y)$ converges absolutely.

However, if we denote by s the minimal integer satisfying the inequality $s \geqslant a/\pi$, then

$$\int_0^1 \left|\frac{1}{y}\sin\frac{a}{y}\right| dy = \int_a^\infty \frac{|\sin x|}{x}\, dx \geqslant \sum_{k=s}^{\infty} \int_{k\pi}^{\pi(k+1)} \frac{|\sin x|}{x}\, dx$$
$$= \sum_{k=s}^{\infty} \int_0^\pi \frac{\sin y}{y + k\pi}\, dy.$$

Since

$$\sum_{k=s}^{\infty} \int_0^\pi \frac{\sin y}{y + k\pi}\, dy > \sum_{k=s}^{\infty} \frac{1}{\pi(k+1)} \int_0^\pi \sin y\, dy = \frac{2}{\pi} \sum_{k=s}^{\infty} \frac{1}{k+1} = \infty,$$

the integral diverges too,

$$\int_0^1 \left|\frac{1}{y}\sin\frac{a}{y}\right| dy.$$

The latter means that $\mathbf{M}[(1/Y)\sin(a/Y)]$ does not exist and, consequently, the Monte-Carlo method is not applicable in this particular case.

Example 29.4 Can the quantity

$$S_n^2 = \frac{1}{n} \sum_{k=1}^{n} (X_k - a)^2$$

be taken as an approximate value of the variation of errors given by a device if $X_1, X_2, \ldots, X_n, \ldots$ are independent measurements of a constant quantity a and if they all have the same distribution functions?

SOLUTION. Let us denote the true value of the variance by σ^2. The quantity S_n^2 can be considered as an approximate value for σ^2 if

$$\lim_{n\to\infty} \mathbf{P}\{|S_n^2 - \sigma^2| < \varepsilon\} = 1.$$

Since $X_1, X_2, \ldots, X_n, \ldots$ are independent random variables with equal distributions, the variables $Y_k = (X_k - a)^2$ are independent and have equal distributions.

We have

$$\mathbf{M}[Y_k] = \mathbf{M}[(X_k - a)^2] = \mathbf{M}[X_k^2] - 2a\mathbf{M}[X_k] + a^2 = \sigma^2 + \bar{x}^2 - 2a\bar{x} + a^2$$
$$= \sigma^2 + (\bar{x} - a)^2,$$

where $\bar{x} = \mathbf{M}[X_k]$. To satisfy the equality $\mathbf{M}[Y_k] = \sigma^2$, it is necessary that $\bar{x} = a$, which means absence of systematic errors in measurements.

Thus, if the measuring device does not give systematic errors, the conditions for applicability of the law of large numbers are satisfied and, consequently,

$$\lim_{n\to\infty} \mathbf{P}\left\{\left|\frac{1}{n}\sum_{k=1}^{n}(X_k - a)^2 - \sigma^2\right| < \varepsilon\right\} = 1.$$

PROBLEMS

29.1 Use Chebyshev's inequality to estimate the probability that a normal random variable will deviate from its expectation by more than (a) four mean deviations, (b) three mean-square deviations.

29.2 Prove that for any random variable X and any $\varepsilon > 0$, the following inequality holds:

$$\mathbf{P}(\varepsilon X > t^2 + \ln J) < e^{-t^2},$$

where $J = \mathbf{M}[e^{\varepsilon X}]$.

29.3 Prove that if $\mathbf{M}[e^{aX}]$ exists,

$$\mathbf{P}(X \geqslant \varepsilon) \leqslant e^{-a\varepsilon}\mathbf{M}[e^{aX}] \qquad (a > 0).$$

29.4 A random variable X obeys the exponential distribution law

$$f(x) = \frac{x^m}{m!}e^{-x} \qquad (x \geqslant 0).$$

Prove that the following inequality holds true:

$$\mathbf{P}(0 < X < 2(m+1)) > \frac{m}{m+1}.$$

29.5 The probability of occurrence of an event A in one experiment is 1/2. Can one assert that with probability greater than 0.97 the number of occurrences of A in 1000 independent trials will be within the limits of 400 to 600?

29.6 Is the law of large numbers valid for the arithmetic mean of n pairwise independent random variables X_k specified by the distribution series in Table 21?

TABLE 21

x_i	$\sqrt{2}$	0	$-\sqrt{2}$
p_i	$\frac{1}{4}$	$\frac{1}{2}$	$\frac{1}{4}$

29.7 Let X_k be a random variable that can assume with equal probability one of two values, k^s or $-k^s$. For which value of s does the law of large numbers apply to the arithmetic mean of the sequence of independent random variables $X_1, X_2, \ldots, X_k, \ldots$?

29.8 Prove that the law of large numbers is applicable to the arithmetic mean of a sequence of independent random variables X_k specified by the distribution series included in Table 22.

TABLE 22

x_i	$\sqrt{\ln k}$	$-\sqrt{\ln k}$
p_i	$\frac{1}{2}$	$\frac{1}{2}$

29.9 Are the sufficient conditions satisfied for the applicability of the law of large numbers to a sequence of mutually independent random variables X_k with distributions specified by the formulas:

(a) $\mathbf{P}(X_k = \pm 2^k) = \frac{1}{2}$;

(b) $\mathbf{P}(X_k = \pm 2^k) = 2^{-(2k+1)}, \qquad \mathbf{P}(X_k = 0) = 1 - 2^{-2k}$;

(c) $\mathbf{P}(X_k = \pm k) = \frac{1}{2\sqrt{k}}, \qquad \mathbf{P}(X_k = 0) = 1 - \frac{1}{\sqrt{k}}$?

29.10 The random variables $X_1, X_2, \ldots, X_n, \ldots$ have equal expectations and finite variations. Is the law of large numbers applicable to this sequence if all the covariances $k_{ij} = \mathbf{M}[(X_i - \bar{x}) \times (X_j - \bar{x})$ are negative?

29.11 Prove that the law of large numbers is applicable to a sequence of random variables in which each random variable can depend only on random variables with adjacent numbers, and all the random variables contained in the sequence have finite variances and expectations.

29.12 A sequence of independent and equally distributed random variables $X_1, X_2, \ldots, X_i, \ldots$ is specified by the distribution series

$$\mathbf{P}(X_i = k) = \frac{1}{k^3\zeta(3)} \qquad (k = 1, 2, 3, \ldots),$$

where $\zeta(3) = \sum_{k=1}^{\infty} 1/k^3 = 1.20256$ is the value of the Riemann function for argument 3. Is the law of large numbers applicable to this sequence?

29.13 Given a sequence of random variables $X_1, X_2, \ldots, X_n, \ldots$, for which $\mathbf{D}[X_n] \leqslant c$ and $r_{ij} \to 0$ for $|i - j| \to \infty$ (r_{ij} is the correlation coefficient between X_i and X_j), prove that the law of large numbers can be applied to this sequence (Bernstein's theorem).

29.14 A sequence of independent and equally distributed random variables $X_1, X_2, \ldots, X_i, \ldots$ is specified by the distribution series

$$\mathbf{P}(X_i = (-1)^{k-1}k) = \frac{6}{k^2\pi^2} \qquad (k = 1, 2, \ldots);$$

determine whether the law of large numbers applies to this sequence.

30. THE DE MOIVRE-LAPLACE AND LYAPUNOV THEOREMS

Basic Formulas

According to the ιe Moivre-Laplace theorem, for a series of n independent trials in each of which an event A occurs with the same probability $p(0 < p < 1)$, there obtains the relation:

$$\lim_{n\to\infty} \mathbf{P}\left(a \leqslant \frac{m - np}{\sqrt{npq}} < b\right) = \frac{1}{\sqrt{2\pi}} \int_a^b e^{-(t^2/2)}\, dt = \frac{1}{2}\,[\Phi(b) - \Phi(a)],$$

where m is the number of occurrences of event A in n trials and

$$\Phi(x) = \frac{2}{\sqrt{2\pi}} \int_0^x e^{-(t^2/2)}\, dt$$

is the Laplace function (probability integral) whose values are included in 8T in the table list on page 471.

According to Lyapunov's theorem, for a sequence of mutually independent random variables $X_1, X_2, \ldots, X_k, \ldots$ satisfying for some $\delta > 0$ the condition

$$\lim_{n\to\infty} \frac{1}{B_n^{2+\delta}} \sum_{k=1}^{n} \mathbf{M}\{|X_k - a_k|^{2+\delta}\} = 0,$$

the following equality holds:

$$\lim_{n\to\infty} \mathbf{P}\left(a < \frac{1}{B_n} \sum_{k=1}^{n} (X_k - \bar{x}_k) < b\right) = \frac{1}{\sqrt{2\pi}} \int_a^b e^{-(t^2/2)}\, dt = \frac{1}{2}\,[\Phi(b) - \Phi(a)],$$

where $\bar{x}_k = \mathbf{M}[X_k]$ is the expectation of X_k, $\sigma_k^2 = \mathbf{D}[X_k]$ is the variance of X_k, $B_n^2 = \sum_{k=1}^{n} \sigma_k^2$.

To prove that Lyapunov's theorem is applicable to equally distributed random variables, it is sufficient to show that the variances of the terms are finite and different from zero.

SOLUTION FOR TYPICAL EXAMPLES

Example 30.1 The probability that an item will fail during reliability tests is $p = 0.05$. What is the probability that during tests with 100 items, the number failing will be (a) at least five, (b) less than five, (c) between five and ten.

SOLUTION. By the de Moivre-Laplace theorem,

$$\mathbf{P}(m_1 \leqslant m < m_2) \approx \frac{1}{2}\left[\Phi\left(\frac{m_2 - np}{\sqrt{npq}}\right) - \Phi\left(\frac{m_1 - np}{\sqrt{npq}}\right)\right],$$

if n is sufficiently large. By assumption, $n = 100$, $p = 0.05$, $q = 1 - p = 0.95$.

(a) The probability that at least five items fail is

$$\mathbf{P}(m \geqslant 5) = \mathbf{P}(5 \leqslant m < 100) \approx \frac{1}{2}\left[\Phi\left(\frac{100 - 5}{\sqrt{4.75}}\right) - \Phi\left(\frac{5 - 5}{\sqrt{4.75}}\right)\right]$$

$$= \frac{1}{2}\,[\Phi(43.6) - \Phi(0)] = 0.5.$$

(b) The probability that less than five items fail is

$$\mathbf{P}(m < 5) = \mathbf{P}(0 \leqslant m < 5) \approx \frac{1}{2}\left(\Phi\left(\frac{5-5}{\sqrt{4.75}}\right) - \Phi\left(\frac{0-5}{\sqrt{4.75}}\right)\right]$$

$$= \frac{1}{2}[\Phi(0) + \Phi(2.29)] = \frac{1}{2}\cdot 0.9780 = 0.489.$$

(c) The probability that five to ten items fail is

$$\mathbf{P}(5 \leqslant m \leqslant 10) \approx \frac{1}{2}\left[\Phi\left(\frac{10-5}{\sqrt{4.75}}\right) - \Phi\left(\frac{5-5}{\sqrt{4.75}}\right)\right]$$

$$= \frac{1}{2}[\Phi(2.29) - \Phi(0)] = \frac{1}{2}\cdot 0.9780 = 0.489.$$

Problems 30.1 to 30.4 can be solved similarly.

Example 30.2 How many independent trials should be performed so that at least five occurrences of an event A will be observed with probability 0.8, if the probability of A in one trial is $\mathbf{P}(A) = 0.05$.

SOLUTION. From the de Moivre-Laplace theorem, we see that

$$\mathbf{P}(m \geqslant 5) \approx \frac{1}{2}\left[\Phi\left(\frac{n - 0.05n}{\sqrt{0.0475n}}\right) - \Phi\left(\frac{5 - 0.05n}{\sqrt{0.0475n}}\right)\right]$$

$$= \frac{1}{2}\left[\Phi(4.36\sqrt{n}) - \Phi\left(\frac{5 - 0.05n}{\sqrt{0.0475n}}\right)\right].$$

For $n = 1$, we have $\Phi(4.36\sqrt{n}) \approx 1$; therefore, substituting $\mathbf{P}(m \geqslant 5) = 0.8$, we obtain

$$\frac{1}{2}\left[1 - \Phi\left(\frac{5 - 0.05n}{\sqrt{0.0475n}}\right)\right] \approx 0.8$$

or

$$\Phi\left(\frac{5 - 0.05n}{\sqrt{0.0475n}}\right) = -0.6.$$

From 8T in the table list on page 471, we find the argument $x = -0.8416$ corresponding to the value of the function $\Phi(x) = -0.6$. Solving the equation

$$\frac{5 - 0.05n}{\sqrt{0.0475n}} = -0.8416,$$

we find the unique root $n = 144$. Thus, in order that A occur at least five times with probability 0.8, 144 trials are necessary.

Following this example, one can solve Problems 30.5 to 30.7.

Example 30.3 How many trials should be performed to calculate the integral

$$J = \int_0^{\pi/2} \cos x \, dx$$

by a Monte-Carlo method so that with probability 0.9 the relative error in calculating the value of the integral is less than 5?

SOLUTION. The integral $(2/\pi)J = (2/\pi)\int_0^{\pi/2} \cos x\, dx$ can be looked upon as the expectation of the function $\cos x$ of the random variable X, uniformly distributed over the interval $(0, \pi/2)$. Then the approximate value of the integral is

$$J_n = \frac{\pi}{2n} \sum_{k=1}^{n} \cos X_k,$$

where X_k are random numbers on the interval $(0, \pi/2)$.

Let us form the random variable

$$T_n = \frac{J_n - J}{\sqrt{\mathbf{D}[J_n]}},$$

which according to Lyapunov's theorem has the distribution function

$$f(t) = \frac{1}{\sqrt{2\pi}} e^{-(t^2/2)}$$

because the variables $\cos X_k$ are independent and equally distributed, with a finite variance different from zero, and $J = \mathbf{M}[J_n]$. We have

$$\mathbf{D}[J_n] = \frac{\pi^2}{4n}\mathbf{D}[\cos X] = \frac{\pi^2 - 8}{8n}.$$

Applying Lyapunov's theorem, for $b = -a = \varepsilon$ we get

$$\mathbf{P}\left\{\sqrt{\frac{8n}{\pi^2 - 8}}\,|J_n - J| < \varepsilon\right\} \approx \Phi(\varepsilon) = 0.9;$$

consequently, it follows that $\varepsilon = 1.645$.

In order that the relative error $(J_n - J)/J$ be less than 0.05, since $J = 1$, it is necessary to perform n trials so that

$$\sqrt{\frac{8n}{\pi^2 - 8}}\cdot 0.05 > 1.645,$$

thus, we obtain $n > 252$.

Similarly one can solve Problems 30.10 to 30.12.

PROBLEMS

30.1 The probability of occurrence of an event in one trial is 0.3. What is the probability that the relative frequency of this event in 100 trials will lie within the range 0.2 to 0.4?

30.2 There are 100 machines of equal power operating independently so that each is turned on during 0.8 of the entire operating time. What is the probability that at an arbitrary instant of time, 70 to 86 machines will be turned on?

30.3 The probability that a condenser fails during a time T is 0.2. Find the probability that among 100 condensers during time T (a) at least 20 condensers, (b) fewer than 28 condensers, (c) 14 to 26 condensers will fail.

30.4 Using the de Moivre-Laplace theorem, show that for a sufficiently large number of trials,

$$\mathbf{P}\left(p - \varepsilon \leqslant \frac{m}{n} \leqslant p + \varepsilon\right) \approx \hat{\Phi}\left(\frac{\varepsilon\sqrt{n}}{\rho\sqrt{2pq}}\right) = \Phi\left(\frac{\varepsilon\sqrt{n}}{\sqrt{pq}}\right),$$

where m/n is the frequency of occurrence of the event whose probability of occurrence is p.

30.5 The probability of an event is evaluated by a Monte-Carlo method. Find the number of independent trials that insure with probability at least 0.99 that the value of the required probability will be determined with an error not exceeding 0.01. Apply Chebyshev's and Laplace's theorems.

30.6 The probability that an item selected at random is defective in each test is 0.1. A lot is rejected if it contains at least 10 defective items. How many items should be tested so that with probability 0.6 a lot containing 10 per cent defective items will be rejected?

30.7 How many trials are necessary so that with probability 0.9 the frequency of a given event will differ from the probability of occurrence of this event by at most 0.1 if the probability of the event is 0.4?

30.8 The probability of occurrence of a certain event in one trial is 0.6. What is the probability that this event will appear in most of 60 trials?

30.9 The probability of event A is 1/3 and 45,000 independent trials are performed. What is the mean deviation E of the number of occurrences of event A from the expectation of this number?

30.10 The calculation of the integral $J = \int_0^1 x^2\,dx$ is made by a Monte-Carlo method based on 1000 independent trials. Evaluate the probability that the absolute error in the estimate of J will not exceed 0.01.

30.11 How many trials should be performed to calculate the integral

$$J = \int_0^{\pi/2} \sin x\,dx$$

by a Monte-Carlo method so that with probability $P \geqslant 0.99$ the absolute error of the computed value will not exceed 0.1 per cent of J?

30.12 The probability $\mathbf{P}(C) = \mathbf{P}(A + B)$, where $\mathbf{P}(B \mid \bar{A})$ is known, is estimated by a Monte-Carlo method in two ways: (1) the approximate value of $\mathbf{P}(C)$ is found as the frequency of occurrence of the event C in a series of n independent trials, (2) the frequency m/n of occurrence of the event A in a series of n independent trials is found and the approximate value of $\mathbf{P}(C)$ is evaluated by the formula

$$\mathbf{P}(C) \approx P_n(C) = \frac{m}{n} + \left(1 - \frac{m}{n}\right)\mathbf{P}(B \mid \bar{A}).$$

(a) Prove that both ways lead to the same result, (b) find the necessary number of trials in each case so that the error in the estimate of $\mathbf{P}(C)$ does not exceed 0.01 with probability 0.95, if $\mathbf{P}(B \mid \bar{A}) = 0.3$ and the value of $\mathbf{P}(A)$ is of order 0.4.

30.13 There are 100 urns containing five red and 95 black balls each. The experiment is such that after a ball is drawn it is returned to the same urn and the outcome of the trial is not communicated to the observer. How many trials should be performed so that (a) the probability is 0.8 that at least one red ball is drawn from each urn, (b) the probability is 0.8 that at least one red ball is drawn from at least 50 urns?

30.14 Compute the characteristic function $E_{Y_n}(u)$ of the random variable

$$Y_n = \frac{\sum_{i=1}^{n} (X_i - \bar{x}_i)}{\sqrt{\sum_{i=1}^{n} \mathbf{D}[X_i]}}$$

and find its limit for $n \to \infty$ if the random variables $X_1, X_2, \ldots, X_n, \ldots$ are independent and have equal probability densities or distribution series of the form:

$$\text{(a)} \quad f(x_i) = \begin{cases} \dfrac{1}{2h} & \text{for} \quad |x_i| \leqslant h, \\ 0 & \text{for} \quad |x_i| > h; \end{cases}$$

$$\text{(b)} \quad \mathbf{P}(X_i = m) = \frac{a^m}{m!} e^{-a};$$

$$\text{(c)} \quad f(x_i) = \begin{cases} 0 & \text{for} \quad x_i < 0, \\ \dfrac{\beta^\alpha}{\Gamma(\alpha)} x_i^{\alpha-1} e^{-\beta x_i} & \text{for} \quad x_i \geqslant 0. \end{cases}$$

30.15 Find the limit for $n \to \infty$ of the characteristic function $E_{Y_n}(u)$ of the random variable

$$Y_n = \frac{\sum_{i=1}^{n} (X_i - \bar{x}_i)}{\sqrt{\sum_{i=1}^{n} \mathbf{D}[X_i]}},$$

if the random variables $X_1, X_2, \ldots, X_n, \ldots$ are independent, have equal distribution laws, expectations and variances, and the moments of higher order are bounded.

VII THE CORRELATION THEORY OF RANDOM FUNCTIONS

31. GENERAL PROPERTIES OF CORRELATION FUNCTIONS AND DISTRIBUTION LAWS OF RANDOM FUNCTIONS

Basic Formulas

A random function of a real variable t is a function $X(t)$ that for each t is a random variable. If the variable t can assume any values on some (finite or infinite) interval, then the random function is called a stochastic process; if the variable t can assume only discrete values, $X(t)$ is called a random sequence.

The (nonrandom) function $\bar{x}(t)$, which for each t is the expectation $\mathbf{M}[X(t)]$ of the random variable $X(t)$, is called the expectation of the random function $X(t)$.

The correlation (autocorrelation) function $K_x(t_1, t_2)$ of the random function $X(t)$ is defined by the formula,

$$K_x(t_1, t_2) = \mathbf{M}\{[X^*(t_1) - \bar{x}^*(t_1)][X(t_2) - \bar{x}(t_2)]\} = K_x^*(t_2, t_1),$$

where * denotes the complex conjugate.[1]

For stationary random functions we have

$$K_x(t_1, t_2) = K_x(t_2 - t_1), \qquad \bar{x}(t) = \text{const.}$$

The variance of the ordinate of a random function is related to $K_x(t_1, t_2)$ by the formula $\mathbf{D}[X(t)] = \sigma^2_{x(t)} = K_x(t_1, t_2)$. The normalized correlation function is defined by the formula

$$k_x(t_1, t_2) = \frac{K_x(t_1, t_2)}{\sigma_{x(t_1)}\sigma_{x(t_2)}}.$$

The total character of a random function is given by the collection of distribution laws

$$f(x_1 \mid t_1), f(x_1, x_2 \mid t_1, t_2), f(x_1, x_2, x_3 \mid t_1, t_2, t_3), \ldots,$$

where $f(x_1, \ldots, x_n \mid t_1, \ldots, t_n)$ is the density of the joint distribution of the values of the random function at times $(t_1, t_2, t_3, \ldots, t_n)$. The expectation $\bar{x}(t)$

[1] When not otherwise specified, $X(t)$ is real.

and correlation function $K_x(t_1, t_2)$ are expressed in terms of the functions $f(x_1 \mid t_1)$ and $f(x_1, x_2 \mid t_1, t_2)$ by the formulas (for continuous random functions)[2]

$$\bar{x}(t) = \int_{-\infty}^{\infty} xf(x \mid t)\,dx,$$

$$K_x(t_1, t_2) = \int_{-\infty}^{\infty}\int_{-\infty}^{\infty} x_1x_2f(x_1, x_2 \mid t_1, t_2)\,dx_1\,dx_2 - \bar{x}(t_1)\bar{x}(t_2).$$

For a normal stochastic process, the joint distribution at n times is completely defined by the functions $\bar{x}(t)$ and $K_x(t_1, t_2)$ by the formulas for the distribution of a system of normal random variables with expectations

$$\bar{x}(t_1), \bar{x}(t_2), \bar{x}(t_3), \ldots, \bar{x}(t_n),$$

and whose elements of the covariance matrix are $k_{jl} = K_x(t_j, t_l)$, $l, j = 1, 2, \ldots, n$.

The mutual correlation function $R_{xy}(t_1, t_2)$ of two random functions $X(t)$ and $Y(t)$ is specified by the formula

$$R_{xy}(t_1, t_2) = \mathbf{M}\{[X^*(t_1) - \bar{x}^*(t_1)][Y(t_2) - \bar{y}(t_2)]\} = R_{xy}^*(t_2, t_1).$$

For stationary processes,

$$R_{xy}(t_1, t_2) = R_{xy}(t_2 - t_1).$$

The notion of correlation function extends to random functions of several variables. If, for example, the random function $X(\xi, \eta)$ is a function of two nonrandom variables, then

$$K_x(\xi_1, \xi_2; \eta_1, \eta_2) = \mathbf{M}\{[X^*(\xi_1, \eta_1) - \bar{x}^*(\xi_1, \eta_1)][X(\xi_2, \eta_2) - \bar{x}(\xi_2, \eta_2)]\}.$$

Solution for Typical Examples

The problems of this section are of two types. Those of the first type ask for the correlation function of a random function and for the general properties of the correlation function. In solving these problems one should start directly from the definition of the correlation function. The problems of the second type ask for the probability that the ordinates of a random function assume certain values. To solve these problems, it is necessary to use the corresponding normal distribution law specified by its expectation and correlation function.

Example 31.1 Find the correlation function $K_x(t_1, t_2)$ if

$$X(t) = \sum_{j=1}^{k} [A_j \cos \omega_j t + B_j \sin \omega_j t],$$

where ω_j are known numbers, the real random variables A_j and B_j are mutually uncorrelated and have zero expectations and variances defined by the equalities

$$\mathbf{D}[A_j] = \mathbf{D}[B_j] = \sigma_j^2 \qquad (j = 1, 2, \ldots, k).$$

[2] $X(t)$ is considered real.

SOLUTION. Since $\bar{x}(t) = \sum_{j=1}^{k} (\bar{a}_j \cos \omega_j t + \bar{b}_j \sin \omega_j t) = 0$, by the definition of the correlation function

$$K_x(t_1, t_2) = \mathbf{M}\left\{\sum_{j=1}^{k} \sum_{l=1}^{k} [A_j \cos \omega_j t_1 + B_j \sin \omega_j t_1][A_l \cos \omega_l t_2 + B_l \sin \omega_l t_2]\right\}.$$

If we open the parentheses and apply the expectation theorem, we notice that all the terms containing factors of the form $\mathbf{M}[A_j A_l]$, $\mathbf{M}[B_j B_l]$ for $j \neq l$ and $\mathbf{M}[A_j B_l]$ for any j and l are zero, and $\mathbf{M}[A_j^2] = \mathbf{M}[B_j^2] = \sigma_j^2$. Therefore $K_x(t_1, t_2) = \sum_{j=1}^{k} \sigma_j^2 \cos \omega_j (t_2 - t_1)$.

Similarly one can solve Problems 31.3 to 31.6 and 31.10.

Example 31.2 Let $X(t)$ be a normal stationary random function with zero expectation. Prove that if

$$Z(t) = \frac{1}{2}\left[1 + \frac{X(t)X(t + \tau)}{|X(t)X(t + \tau)|}\right],$$

then

$$\bar{z}(t) = \frac{1}{\pi} \arccos [-k_x(\tau)],$$

where $k_x(\tau)$ is the normalized correlation function of $X(t)$.

SOLUTION. Using the fact that $X(t)$ is normal, we see that the distribution law of second order can be represented as

$$f(x_1, x_2 \mid t, t + \tau) = \frac{1}{2\pi\sigma_x^2\sqrt{1 - k_x^2(\tau)}} \exp\left\{-\frac{x_1^2 + x_2^2 - 2k_x(\tau)x_1x_2}{2\sigma_x^2[1 - k_x^2(\tau)]}\right\}.$$

The required expectation can be represented in the form

$$\bar{z}(t) = \int_{-\infty}^{\infty} \int_{-\infty}^{\infty} \frac{1}{2}\left[1 + \frac{x_1x_2}{|x_1x_2|}\right] f(x_1, x_2 \mid t, t + \tau)\, dx_1\, dx_2.$$

Since $(1/2)[1 + (x_1x_2/|x_1x_2|)]$ is identically equal to zero if the signs of ordinates x_1 and x_2 are different, and equal to one otherwise, we see that

$$\bar{z}(t) = \int_{-\infty}^{0} \int_{-\infty}^{0} f(x_1, x_2 \mid t, t + \tau)\, dx_1\, dx_2 + \int_{0}^{\infty} \int_{0}^{\infty} f(x_1, x_2 \mid t, t + \tau)\, dx_1\, dx_2$$
$$= 2\int_{0}^{\infty} \int_{0}^{\infty} f(x_1, x_2 \mid t, t + \tau)\, dx_1\, dx_2,$$

which by integration leads to the result mentioned in the Example. (For integration it is convenient to introduce new variables r, φ, setting $x_1 = r \cos \varphi$, $x_2 = r \sin \varphi$.)

PROBLEMS

31.1 Prove that

(a) $|K_x(t_1, t_2)| \leqslant \sigma_{x(t_1)}\sigma_{x(t_2)}$;

(b) $|K_x(t_1, t_2)| \leqslant \frac{1}{2}[\sigma_{x(t_1)}^2 + \sigma_{x(t_2)}^2]$.

31.2 Prove that $|R_{xy}(t_1, t_2)| \leqslant \sigma_{x(t_1)}\sigma_{y(t_2)}$.

31.3 Prove that the correlation function does not change if any nonrandom function is added to a random function.

31.4 Find the variance of a random function $X(t)$ whose ordinates vary stepwise by quantities Δ_j at random times. The number of steps during a time interval τ obeys a Poisson distribution with a constant $\lambda\tau$, and the magnitudes of the steps Δ_j are mutually independent, with equal variances σ^2 and zero expectations, and $X(0)$ is a nonrandom variable.

31.5 Find the correlation function of a random function $X(t)$ assuming two values, $+1$ and -1; the number of changes of sign of the function obeys a Poisson distribution with a constant temporal density λ, and $\bar{x}(t)$ can be assumed zero.

31.6 A random function $X(t)$ consists of segments of horizontal lines of unit length whose ordinates can assume either sign with equal probability, and their absolute values obey the distribution law

$$f(|x|) = \frac{|x|^\lambda}{\Gamma(\lambda + 1)} e^{-|x|} \quad \text{(gamma-distribution)}.$$

Evaluate $K_x(\tau)$.

31.7 The correlation function of the heel angle of of a ship $\Theta(t)$ has the form

$$K_\theta(\tau) = ae^{-\alpha|\tau|} \cos \beta\tau.$$

Find the probability that at time $t_2 = t_1 + \tau$ the heel angle $\Theta(t_2)$ will be greater than 15°, if $\Theta(t)$ is a normal random function, $\bar{\theta} = 0$, $\Theta(t_1) = 5°$, $\tau = 2$ sec., $a = 30$ deg.^{2}, $\alpha = 0.02$ sec.$^{-1}$ and $\beta = 0.75$ sec.$^{-1}$.

31.8 It is possible to use a sonic depth finder on a rolling ship whose heel angle $\Theta(t)$ satisfies $|\Theta(t)| \leqslant \theta_0$. The time for the first measurement is selected so that this condition is satisfied. Find the probability that the second measurement can be performed after τ_0 sec. if $\Theta(t)$ is a normal function, $\bar{\theta} = 0$, the variance $D[\Theta(t)] = \sigma_\theta^2$ and the normalized correlation function $k(\tau) = K_\theta(\tau)/\sigma_\theta^2$ are known.

31.9 The correlation function of the heel angle $\Theta(t)$ of a ship is $K_\theta(\tau) = ae^{-\alpha|\tau|}[\cos \beta\tau + (\alpha/\beta) \sin \beta|\tau|]$, where $a = 36$ deg.^{2}, $\alpha = 0.25$ sec.$^{-1}$ and $\beta = 1.57$ sec.$^{-1}$. At time t the heel angle is 2°, $\Theta(t) \geqslant 0$. Find the probability that at time $(t + 2)$ second the heel angle will have an absolute value less than 10° if $\Theta(t)$ is a normal random function and $\bar{\theta}(t) = 0$.

31.10 Find the expectation and variance of the random function $Y(t) = a(t)X(t) + b(t)$, where $a(t)$ and $b(t)$ are numerical (nonrandom) functions and $K_x(t_1, t_2)$ and $\bar{x}(t)$ are known.

31.11 Find the distribution law of first order for the values of the random function

$$X(t) = A(t) \cos [\omega t + \Theta(t)],$$

if the distribution laws of first order for the random functions $A(t)$ and $\Theta(t)$ have the form

$$f_a(a \mid t) = \frac{a}{\sigma^2} e^{-(a^2/2\sigma^2)} \quad (a \geqslant 0), \qquad f_\theta(\theta \mid t) = \frac{1}{2\pi} \quad (0 \leqslant \theta \leqslant 2\pi),$$

where ω is a constant, and at the same time $A(t)$ and $\Theta(t)$ are mutually independent.

31.12 Random points are distributed over the real axis so that the probability P_n of occurrence of n points on a prescribed interval τ is given by Poisson's law $P_n = (\lambda\tau)^n/n!\, e^{-\lambda\tau}$, where λ is a positive constant. Find the distribution law of first order for a random function $X(m)$ defining the distance between the mth and the $(m + n + 1)$st random points.

31.13 Find the distribution law for the values of a random function of two variables $U(x, y)$, if

$$U(x, y) = \zeta(x + \xi_0, y + \eta_0) - \zeta(x, y),$$

and the correlation function $K_\zeta(\xi, \eta)$ defined by

$$K_\zeta(\xi, \eta) = \mathbf{M}\{[\zeta(x, y) - \bar{\zeta}(x, y)] \times [\zeta(x + \xi, y + \eta) - \bar{\zeta}(x + \xi, y + \eta)]\},$$

is given in the form

$$K_\zeta(\xi, \eta) = ae^{-\alpha_1|\xi| - \alpha_2|\eta|} \cos \beta_1\xi \cos \beta_2\eta,$$

where $\zeta(\xi, \eta)$ is a normal random function, $a = 100$, $\alpha_1 = 0.2$, $\alpha_2 = 0.1$, $\beta_1 = 0.5$, $\beta_2 = 1.0$, $\xi_0 = 1$ and $\eta_0 = 2$.

32. LINEAR OPERATIONS WITH RANDOM FUNCTIONS

Basic Formulas

An operator is a mapping of functions into functions.[3] The operator L_0 is called linear and homogeneous if it fulfills the conditions

$$\mathrm{L}_0 A\varphi(t) = A\mathrm{L}_0\varphi(t), \qquad \mathrm{L}_0[\varphi_1(t) + \varphi_2(t)] = \mathrm{L}_0\varphi_1(t) + \mathrm{L}_0\varphi_2(t),$$

where A is any constant, and $\varphi(t)$, $\varphi_1(t)$ and $\varphi_2(t)$ are any functions.

A linear nonhomogeneous operator L is any operator related to a linear homogeneous operator L_0 by the expression

$$\mathrm{L}\varphi(t) = \mathrm{L}_0\varphi(t) + F(t),$$

where $F(t)$ is some fixed function.

---

[3] For a more rigorous definition of the notion of "operator," see Taylor, A. E.: Introduction to Functional Analysis. New York, John Wiley & Sons, Inc., 1958; and Heider, L. J., and Simpson, J. E.: Theoretical Analysis. Philadelphia, W. B. Saunders Company, 1967.

If $Y(t) = \mathrm{L}_0 X(t)$ and the operator L_0 is linear and homogeneous, then

$$\bar{y}(t) = \mathrm{L}_0 \bar{x}(t), \qquad K_y(t_1, t_2) = \mathrm{L}^*_{0t_1} \mathrm{L}_{0t_2} K_x(t_1, t_2),$$

where L^* is the operator L in which all coefficients have been replaced by their complex conjugates; the indices t_1 and t_2 in the notation of the operator L_0 show that in the first case the operator acts on variable t_1, and in the second, on the variable t_2. (The possibility of applying the operator to the given random function should be verified in each concrete case.)

If L is a nonhomogeneous operator corresponding to the homogeneous operator L_0 and to the function $F(t)$, and if $Z(t) = \mathrm{L}X(t)$, then

$$\bar{z}(t) = \mathrm{L}\bar{x}(t) = \mathrm{L}_0\bar{x}(t) + F(t), \qquad K_z(t_1, t_2) = \mathrm{L}^*_{0t_1} \mathrm{L}_{0t_2} K_x(t_1, t_2);$$

i.e., the correlation function does not depend on $F(t)$, the function engendering the nonhomogeneity of the operator L.

A random function is differentiable (once) if its correlation function has a second mixed partial derivative for equal values of the variables, which, in the case of stationary functions, is equivalent to the existence of a second derivative of $K(\tau)$ for $\tau = 0$.

It is considerably more difficult to find the expectation and correlation function for the result of the application of a nonlinear operator to a random function whose probability properties are known. An exceptional case is represented by a normal stochastic process for some types of nonlinear operators. For example, if $X(t)$ is a normal random function (we consider $X(t)$ real) and $Y(t) = X^2(t)$, then

$$\begin{aligned} \bar{y}(t) &= \mathbf{M}[X^2(t)] = \mathbf{D}[X(t)] + \bar{x}^2(t), \\ K_y(t_1, t_2) &= \mathbf{M}[X^2(t_1)X^2(t_2)] - \{\mathbf{D}[X(t_1)] + \bar{x}^2(t_1)\}\{\mathbf{D}[X(t_2)] + \bar{x}^2(t_2)\} \\ &= 2K_x^2(t_1, t_2), \end{aligned}$$

since the expectation of the product of four normal variables $X(t_1)$, $X(t_1)$, $X(t_2)$ and $X(t_2)$ can be obtained by a differentiation of the characteristic function of a system of random variables (see Section 23, page 124).

In the same way one can obtain the expectation and correlation function of an essentially nonlinear expression

$$Y(t) = \operatorname{sgn} X(t),$$

if $X(t)$ is normal (see Example 32.2).

Solution for Typical Examples

The problems in this section can be solved by using the general formula for the correlation function of the result obtained by applying a linear operator to a random function; however, in some problems it is more convenient to start directly from the definition of the correlation function. The second way cannot be avoided if, in addition to linear operators, a given expression also contains nonlinear operators. The following are considered examples of applications of both methods.

Example 32.1 Find the standard deviation of the angle Ψ of rotation of a direction gyroscope after 10 minutes of rotation as a result of the random moment $M(t)$, appearing on the axis of the inner suspension ring, if the law of variation of $\Psi(t)$ can be represented by the equation $\dot{\Psi}(t) = M(t)/H$, where the kinetic moment $H = 21.10^5$ g. cm.^{2}/sec.^{2}, and

$$K_m(\tau) = n^2 e^{-\alpha|\tau|}\left(\cos\beta\tau - \frac{\alpha}{\beta}\sin\beta|\tau|\right),$$

$$n = 1.36\cdot 10^4 \text{ g.}\cdot\text{cm.}^2/\text{sec.}^2, \qquad \beta = 0.7 \text{ sec.}^{-1}, \qquad \alpha = 0.1 \text{ sec.}^{-1}.$$

SOLUTION. Since by integration we have $\Psi(t) = 1/H \int_0^t M(t_1)\,dt_1$ (the initial conditions by the nature of the problem are zero), i.e., $\Psi(t)$ and $M(t)$ are linearly related, for the correlation function $K_\psi(t_1, t_2)$ we obtain

$$K_\psi(t_1, t_2) = \frac{1}{H^2}\int_0^{t_1}\int_0^{t_2} K_m(t'' - t')\,dt''\,dt',$$

and for the variance

$$\mathbf{D}[\Psi(t)] = \sigma^2_{\psi(t)} = \frac{1}{H^2}\int_0^t\int_0^t K_m(t'' - t')\,dt''\,dt' = \frac{2}{H^2}\int_0^t (t - \tau)K_m(\tau)\,d\tau.$$

Since

$$e^{-\alpha|\tau|}\left(\cos\beta\tau - \frac{\alpha}{\beta}\sin\beta|\tau|\right) = \frac{-1}{\alpha^2 + \beta^2}\frac{d^2}{d\tau^2}\left\{e^{-\alpha|\tau|}\left(\cos\beta\tau + \frac{\alpha}{\beta}\sin\beta|\tau|\right)\right\},$$

the last integral can be calculated by integration by parts, leading to

$$\mathbf{D}[\Psi(t)] = \frac{2n^2}{H^2(\alpha^2 + \beta^2)}\left\{1 - e^{-\alpha t}\left(\cos\beta t + \frac{\alpha}{\beta}\sin\beta t\right)\right\}$$

$$\approx \frac{2n^2}{H^2(\alpha^2 + \beta^2)}, \qquad \sigma_\psi = 45'.$$

Example 32.2 Find the variance of the angle $\Psi(t)$ of rotation of a direction gyroscope after $T = 10$ minutes of rotation if Ψ is defined by the equation

$$\frac{d\Psi}{dt} = \frac{b}{H}\operatorname{sgn}\dot{\Theta}(t),$$

where Θ is a normal stationary random function with a correlation function

$$K_\theta(\tau) = ae^{-\alpha|\tau|}\left(\cos\beta\tau + \frac{\alpha}{\beta}\sin\beta|\tau|\right),$$

where $\bar{\theta} = 0$ and b, H are constants.

SOLUTION. Here, besides the linear operations of integration and differentiation, the given expression contains the nonlinear operation signum. Thus, using the temporary notation $\dot{\Theta}(t) = X(t)$, we set $Y(t) = \operatorname{sgn} X(t)$. Using the definition of $K_y(\tau)$ as the second central mixed moment of the random variables $Y_1 = \operatorname{sgn} X(t)$ and $Y_2 = \operatorname{sgn} X(t + \tau)$, we obtain

$$K_y(\tau) = 2\int_0^\infty\int_0^\infty f(x_1, x_2)\,dx_1\,dx_2 - 2\int_{-\infty}^0\int_0^\infty f(x_1, x_2)\,dx_1\,dx_2,$$

where the distribution law $f(x_1, x_2)$ is normal.

Substituting the value of this distribution law and changing from rectangular coordinates x_1, x_2 to polar coordinates, one easily calculates both integrals and obtains

$$K_y(\tau) = \frac{2}{\pi} \arcsin k_x(\tau),$$

where the normalized correlation function $k_x(\tau)$ is given by the formula

$$k_x(\tau) = k_{\dot\theta}(\tau) = e^{-\alpha|\tau|}\left(\cos \beta\tau - \frac{\alpha}{\beta} \sin \beta|\tau|\right).$$

The required variance

$$\mathbf{D}[\Psi(t)] = \frac{2b^2}{H^2} \int_0^t (t - \tau)K_y(\tau)\, d\tau$$
$$= \frac{4b^2}{\pi H^2} \int_0^t (t - \tau) \arcsin \left[e^{-\alpha|\tau|}\left(\cos \beta\tau - \frac{\alpha}{\beta} \sin \beta|\tau|\right)\right] d\tau.$$

The problem can be solved by another method, too. If we use the formula $\operatorname{sgn} X = 1/\pi_i \int_{-\infty}^{\infty} e^{iuX}\, du/u$ and set it in the initial differential equation, then after we integrate with respect to time and estimate the expectation of $\Psi^2(t)$, we obtain

$$\mathbf{D}[\Psi(t)] = -\frac{b^2}{4\pi^2 H^2} \int_0^t \int_0^t \int_{-\infty}^{\infty} \int_{-\infty}^{\infty} E(u_1, u_2) \frac{du_1\, du_2}{u_1 u_2}\, dt_1\, dt_2,$$

where $E(u_1, u_2)$ is the characteristic function for the system of normal variables $X(t_1)$ and $X(t_2)$.

If we substitute in the last integral the expression for $E(u_1, u_2)$ and integrate it three times, we find for $\mathbf{D}[\Psi(t)]$ the same expression as just obtained.

Example 32.3 Find the expectation and correlation function of the random function

$$Y(t) = a(t)X(t) + b(t) \frac{dX(t)}{dt},$$

where $a(t)$ and $b(t)$ are given (numerical) functions, $X(t)$ is a differentiable random function and $\bar{x}(t)$, $K_x(t_1, t_2)$ are known.

SOLUTION. The function $Y(t)$ is the result of application of the linear operator $[a(t) + b(t)\, d/dt]$ to the random function $X(t)$. Therefore, the required result can be obtained by applying the general formulas. However, the solution can be found more easily by direct computation of $\bar{y}(t)$ and $K_y(t_1, t_2)$. We have

$$\bar{y}(t) = \mathbf{M}\left[a(t)X(t) + b(t)\frac{dX(t)}{dt}\right] = a(t)\bar{x}(t) + b(t)\frac{d\bar{x}(t)}{dt},$$

$$K_y(t_1, t_2) = \mathbf{M}\left[\left\{a^*(t_1)[X^*(t_1) - \bar{x}^*(t_1)] + b^*(t_1)\left[\frac{dX^*(t_1)}{dt_1} - \frac{d\bar{x}^*(t_1)}{dt_1}\right]\right\}\right.$$
$$\left.\times \left\{a(t_2)[X(t_2) - \bar{x}(t_2)] + b(t_2)\left[\frac{dX(t_2)}{dt_2} - \frac{d\bar{x}(t_2)}{dt_2}\right]\right\}\right]$$
$$= a^*(t_1)a(t_2)K_x(t_1, t_2) + a^*(t_1)b(t_2)\frac{\partial}{\partial t_2} K_x(t_1, t_2)$$
$$+ b^*(t_1)a(t_2)\frac{\partial}{\partial t_1} K_x(t_1, t_2) + b^*(t_1)b(t_2)\frac{\partial^2}{\partial t_1\, \partial t_2} K_x(t_1, t_2).$$

PROBLEMS

32.1 Find the correlation function of the derivative of a random function $X(t)$ if

$$K_x(\tau) = ae^{-\alpha|\tau|}(1 + \alpha|\tau|).$$

32.2 Find the correlation function and variance of the random function

$$Y(t) = \frac{dX(t)}{dt}$$

if $K_x(\tau) = ae^{-\alpha|\tau|}[\cos \beta\tau + (\alpha/\beta) \sin \beta|\tau|]$.

32.3 Let $X(t)$ be a stationary random function with a known correlation function. Find the mutual correlation function of $X(t)$ and $dX(t)/dt$.

32.4 How many derivatives does a random function $X(t)$, with a correlation function $K_x(\tau) = \sigma^2 e^{-\alpha^2\tau^2}$, have?

32.5 How many times can onc differentiate a random function $X(t)$ if $K_x(\tau) = \sigma^2 e^{-\alpha|\tau|}[1 + \alpha|\tau| + (1/3)\alpha^2\tau^2]$?

32.6 Up to what order do the derivatives of a random function $X(t)$ exist if the correlation function has the form

$$K_x(\tau) = \sigma_x^2 e^{-\alpha|\tau|}\left(1 + \alpha|\tau| - 2\alpha^2\tau^2 + \frac{1}{3}\alpha^3|\tau^3|\right)?$$

32.7 A random function $X(t)$ has a correlation function

$$K_x(\tau) = \sigma_x^2 e^{-\alpha|\tau|}(1 + \alpha|\tau|).$$

Find the mutual correlation function of

$$\dot{X}(t + t_0) \quad \text{and} \quad X(t).$$

32.8 The correlation function of a random function $X(t)$ has the form

$$K_x(\tau) = \sigma_x^2 e^{-\alpha|\tau|}(1 + \alpha|\tau|),$$

find the variances for the functions

$$Y(t) = X(t + \tau) \quad \text{and} \quad Z(t) = \dot{X}(t + \tau).$$

32.9 Given the correlation function $K_x(\tau)$ of the stationary random function $X(t)$,

$$K_x(\tau) = \sigma_x^2 e^{-\alpha^2\tau^2},$$

find the correlation function of

$$Y(t) = a\frac{dX(t)}{dt}.$$

32.10 Find the probability P that the derivative V of a normal stationary function $X(t)$ will have a value greater than $b = \sqrt{5}$ m./sec. if

$$K_x(\tau) = ae^{-\alpha|\tau|}\left(\cos \beta\tau + \frac{\alpha}{\beta} \sin \beta|\tau|\right),$$

where $a = 4$ sq. m., $\alpha = 1$ sec.$^{-1}$, $\beta = 2$ sec.$^{-1}$.

32.11 Given the expectations, correlation functions and mutual correlation function between two random functions $X(t)$ and $Y(t)$, find the expectations and the correlation function of the random function

$$Z(t) = X(t) + Y(t).$$

32.12 Express in terms of the distribution laws of a system of n random functions $X_j(t)$ ($j = 1, 2, \ldots, n$) the expectation and the correlation function of

$$X(t) = \sum_{j=1}^{n} X_j(t).$$

32.13 The correlation function $K_x(\tau)$ of a stationary random function $X(t)$ is known. Find the correlation function of $Y(t)$ if

$$Y(t) = X(t) + \frac{dX(t)}{dt} + \frac{d^2X(t)}{dt^2}.$$

32.14 A random function $X(t)$ has the correlation function

$$K_x(\tau) = \sigma_x^2 e^{-\alpha|\tau|}\left(1 + \alpha|\tau| + \frac{1}{3}\alpha^2\tau^2\right),$$

find the correlation function of

$$Z(t) = X(t) + \frac{d^2X(t)}{dt^2}.$$

32.15 Given the correlation function $K_x(\tau)$ of a random function $X(t)$, find the variance of

$$Y(t) = \int_0^t X(\xi)\, d\xi.$$

32.16 A stationary random function $Y(t)$ is related to another function $X(t)$ by

$$Y(t) = \frac{dX(t)}{dt}.$$

Find the correlation function of $X(t)$ if $X(t) = 0$ for $t = 0$ and $K_y(\tau)$ is known.

32.17 Find the correlation function of $X(t)$ and $Y(t) = \int_0^t X(\xi)\, d\xi$ if $K_x(t_1, t_2)$ is known.

32.18 Find the variance of $Y(t)$ for $t = 20$ sec. if

$$Y(t) = \int_0^t X(t_1)\, dt_1,$$

$$K_x(\tau) = ae^{-\alpha|\tau|}(1 + \alpha|\tau|), \quad a = 10 \text{ cm.}^2/\text{sec.}^2, \quad \alpha = 0.5 \text{ sec.}^{-1}.$$

32.19 Find the correlation function and the expectation of

$$Y(t) = a_0X(t) + a_1\frac{dX(t)}{dt} + b_1\int_0^t e^{-\lambda t_1}X(t_1)\, dt_1 + c$$

if $\bar{x}(t)$ and $K_x(t_1, t_2)$ are known and the constants a_0, a_1 and b_1 are real.

32.20 Find the mutual correlation function of $R_{yz}(t_1, t_2)$ if

$$Y(t) = aX(t) + b\frac{dX(t)}{dt},$$

$$Z(t) = c\frac{dX(t)}{dt} + d\frac{d^2X(t)}{dt^2},$$

where a, b, c and d are real constants.

32.21 The speed of an airplane is estimated with the aid of a gyroscopic integrator that gives an error

$$\Delta V(t) = g\int_0^t \sin\theta(t_1)\,dt_1.$$

Here $\theta(t)$ is the error in the stabilization of the axis of the integrator, the correlation function is

$$K_\theta(\tau) = 4\cdot 10^{-8}e^{-0.08|\tau|}\text{ rad.}^2 = ae^{-\alpha|\tau|}$$

and g is the acceleration of gravity. Find the mean-square error in the estimate of the velocity after 10 hours of flight (τ is given in seconds).

32.22 A random function Θ is real, normal and stationary and $\bar{\theta} = 0$. Find the correlation function

$$X(t) = a\ddot{\Theta}(t) + b\Theta + c\dot{\Theta}^2(t),$$

where a, b and c are real constants.

32.23 The perturbation moment acting on the rotor of a gyroscope installed on a ship is expressed in terms of the heel angle $\Theta(t)$ and the angle of trim difference $\Psi(t)$ by the relation

$$M(t) = a\Theta^2(t) + b\Psi^2(t) + c\Theta(t)\dot{\Psi}(t).$$

Find the correlation function $M(t)$ if $K_\theta(\tau)$ and $K_\psi(\tau)$ are known, $R_{\theta\psi}(\tau) \equiv 0$, and $\Theta(t)$ and $\Psi(t)$ are normal.

32.24 Given that $K_x(\tau) = e^{-\alpha^2\tau^2}$, find the correlation function $K_y(\tau)$ if

$$Y(t) = X(t) + \frac{dX(t)}{dt}.$$

32.25 Given

$$K_x(\tau) = ae^{-\alpha|\tau|}\left(1 + \alpha|\tau| + \frac{1}{3}\alpha^2\tau^2\right),$$

find the mutual correlation function between $X(t)$ and $d^2X(t)/dt^2$.

32.26 Given the correlation function $K_x(\tau)$, find $K_x(t_1, t_2)$ if $Y(t) = a(t)X(t) + b(t)\,d^2X(t)/dt^2$, where $a(t)$ and $b(t)$ are numerical (nonrandom) functions.

32.27 Let

$$Y(t) = \int_0^t X(\xi)\,d\xi.$$

Is there a function $X(\xi)$ different from zero for which $Y(t)$ is a stationary random function?

32.28 Is the function $Z(t) = X(t) + Y$ stationary in the broad sense if $X(t)$ is a stationary random function and Y is (a) a random variable uncorrelated with $X(t)$, (b) $Y = X(t_0)$?

32.29 Find the variance of the error $Y(t)$ of a nonperturbed gyro-inertial system after one hour of its operation if $Y(t)$ is defined by the equation

$$\frac{d^2 Y(t)}{dt^2} + \nu^2 Y(t) = X(t),$$

where $\nu = 1.24 \cdot 10^{-3}$ sec.$^{-1}$ is the frequency of Shuler and $X(t)$ is the accelerometer error, which can be considered a stationary normal function of time

$$\bar{x} = 0, \qquad K_x(\tau) = \sigma_x^2 e^{-\alpha|\tau|},$$
$$\sigma_x = 0.01 \text{ m./sec.}^2, \qquad \alpha = 0.1 \text{ sec.}^{-1}.$$

32.30 The angular deviations α and β of a free gyroscope used as a vertical indicator on a rolling ship are defined approximately by the system of equations

$$I_1\ddot{\beta} - H\dot{\alpha} = -k_1 \operatorname{sgn}[\dot{\Psi}(t)],$$
$$I_2\ddot{\alpha} + H\dot{\beta} = k_2 \operatorname{sgn}[\dot{\Theta}(t)],$$

where the moments of inertia I_1, I_2, the kinetic moment of the rotor H and the coefficients of dry friction k_1 and k_2 are constants, and the heel angle $\Theta(t)$ and the angle of trim difference $\Psi(t)$ can be assumed to be two stationary normal functions of time with known correlation functions.

Find $\mathbf{D}[\alpha(t)]$ and $\mathbf{D}[\beta(t)]$ if t is large.

Hint. Introduce a new function $\gamma(t) = (1/\sqrt{q})\alpha(t) + (1/\sqrt{p})\beta(t)$, $q = H/I_2$, $p = H/I_1$ and replace sgn $[\dot{\Psi}(t)]$ and sgn $[\dot{\Theta}(t)]$ by integrals as shown in Example 32.2.

32.31 Find the variance of the function $Z(t)$ defined by the equation $\dot{Z}(t) + a^2[1 + Y(t)]Z(t) = X(t)$, $Z(0) = 0$, where $X(t)$ and $Y(t)$ are independent stationary normal functions with zero expectations and whose correlation functions are known.

33. PROBLEMS ON PASSAGES

Basic Formulas

A passage (time) at a given level a for a random function $X(t)$ is a time t when some graph of this function crosses the horizontal line $X = a$ (from below).

The probability that a passage (time) lies in an infinitely small time interval dt around point t is $p(a \mid t)\,dt$; the temporal probability density $p(a \mid t)$ is expressed in terms of the differential distribution law $f(x, v \mid t)$ of the ordinate of random function $X(t)$ and its derivative $V(t) = \dot{X}(t)$ computed at time t by:

$$p(a \mid t) = \int_0^\infty f(a, v \mid t)v\,dv.$$

The temporal probability density for the intercept of the random function (going down) at the level a is

$$p_1(a \mid t) = -\int_{-\infty}^{0} f(a, v \mid t)v\,dv.$$

For normal functions,

$$\begin{aligned} p(a \mid t) &= \frac{1}{2\pi}\frac{\sigma_v}{\sigma_x}\exp\left\{-\frac{(a-\bar{x})^2}{2\sigma_x^2}\right\} \\ &= \frac{1}{2\pi}\exp\left\{-\frac{(a-\bar{x})^2}{2\sigma_x^2}\right\}\sqrt{\frac{1}{K_x(t,t)}\left.\frac{\partial^2 K_x(t_1,t_2)}{\partial t_1\,\partial t_2}\right|_{t_1=t_2=t}}. \end{aligned}$$

For normal stationary functions,

$$p(a \mid t) = p_1(a \mid t) = p(a) = \frac{1}{2\pi}\exp\left\{-\frac{(a-\bar{x})^2}{2\sigma_x^2}\right\}\frac{\sigma_v}{\sigma_x}.$$

The average number of passages $\bar{n}_a$ of a stationary random function per unit time is $p(a)$.

The average number of passages of a stationary function during a time interval T is $\bar{N}_a = Tp(a)$.

The average duration $\bar{\tau}_a$ of a passage of a stationary function is

$$\bar{\tau}_a = \frac{\int_a^{\infty} f(x)\,dx}{p(a)},$$

where $f(x)$ is the probability density for the ordinates of this random function

For a stationary normal process

$$\bar{\tau}_a = \pi\frac{\sigma_x}{\sigma_v}\exp\left\{\frac{(a-\bar{x})^2}{2\sigma_x^2}\right\}\left[1 - \Phi\left(\frac{a-\bar{x}}{\sigma_x}\right)\right].$$

Similar formulas hold for nonstationary processes:

$$\bar{N}_a = \int_0^T\int_0^{\infty} vf(a, v \mid t)\,dv\,dt, \qquad \bar{\tau}_a = \frac{\int_0^T\int_a^{\infty} f(x \mid t)\,dx\,dt}{\int_0^T\int_0^{\infty} vf(a, v \mid t)\,dv\,dt}.$$

The problem of finding the average number of maxima of a random function (the passage of the first derivative through zero from above) and some other problems can be reduced to problems on passages. For a small average number of passages during a time interval T, the probability Q for non-occurrence of any run during this interval can be estimated approximately by the formula $Q = e^{-\bar{N}_a}$; i.e., the number of passages in the given interval can be considered as obeying approximately a Poisson law.

The formulas for the average number of passages and the average time between successive passages can be generalized for random functions of several variables.

SOLUTION FOR TYPICAL EXAMPLES

Example 33.1 Evaluate the average number of times during $T = 10$ minutes in which the heel angle $\Theta(t)$ of a ship vanishes if $\bar{\theta} = 0$,

$$K_{\theta}(\tau) = ae^{-0.1|\tau|}\left(\cos 0.7\tau + \frac{1}{7}\sin 0.7|\tau|\right),$$

where τ is expressed in seconds and $\Theta(t)$ is a normal random function.

SOLUTION. The average number of passages through zero is

$$p(0) = \frac{1}{2\pi}\frac{\sigma_{\dot{x}}}{\sigma_x} = \frac{1}{2\pi}\sqrt{-\ddot{k}(0)}.$$

Since

$$\ddot{k}(\tau) = -(0.7^2 + 0.1^2)e^{-0.1|\tau|}\left(\cos 0.7\tau - \frac{1}{7}\sin 0.7|\tau|\right),$$

we have

$$p(0) = \frac{1}{2\pi}\sqrt{0.50} = 0.1124 \text{ sec.}^{-1},$$

and the number of passages during 10 minutes, $\bar{N}_0 = 600 \cdot 0.1124 = 67.5$. The required number is $2\bar{N}_0 = 135$.

Example 33.2 The heel angle $\Theta(t)$ and the angle of trim difference $\Psi(t)$ are uncorrelated normal random functions whose correlation functions are given by the formulas

$$K_{\theta}(\tau) = 25e^{-0.07|\tau|}(\cos 0.7\tau + 0.1\sin 0.7|\tau|) \text{ deg.}^2,$$

$$K_{\psi}(\tau) = 12.5e^{-0.02|\tau|}(\cos\sqrt{2}\,\tau + 10^{-2}\sqrt{2}\sin\sqrt{2}\,|\tau|) \text{ deg.}^2,$$

where τ is expressed in seconds, and the expectations $\bar{\theta}$ and $\bar{\psi}$ are equal to zero.

Find the average time that the mast of the ship is outside the cone whose axis is vertical and whose generating angle is 2°, if the deviation of the mast from the vertical ν can be defined by the approximate formula $\nu = \sqrt{\Theta^2 + \Psi^2}$.

SOLUTION. This case differs from the preceding one because the function $\nu(t)$ is not normal. Therefore, one should apply the general formula

$$\bar{\tau}_a = \frac{\int_a^{\infty} f(\nu)\,d\nu}{\int_0^{\infty} vf(a, v)\,dv},$$

where $v(t) = d\nu(t)/dt$.

To find the probability density $f(\nu)$, it is necessary to integrate the probability density of the system of normal random variables $\Theta(t)$, $\Psi(t)$ over the domain $\nu \leqslant \sqrt{\theta^2 + \psi^2} \leqslant \nu + d\nu$, which can be performed easily if we pass from rectangular coordinates θ, ψ to polar coordinates $\nu = \sqrt{\theta^2 + \psi^2}$, $\varphi = \arctan(\psi/\theta)$.

After integration we obtain

$$f(\nu) = \frac{\nu}{\sigma_\theta \sigma_\psi} \exp\left\{-\frac{1}{4}\left(\frac{1}{\sigma_\theta^2} + \frac{1}{\sigma_\psi^2}\right)\nu^2\right\} I_0\left[\frac{1}{4}\left(\frac{1}{\sigma_\theta^2} - \frac{1}{\sigma_\psi^2}\right)\nu^2\right]$$

$$= \frac{\nu}{12.5\sqrt{2}} e^{-0.03\nu^2} I_0(0.01\nu^2),$$

where $I_0(z)$ are the Bessel functions of first kind of an imaginary variable. To obtain $f(\nu, v)$, it is necessary to integrate the probability density of the system of mutually independent random variables $\Theta(t)$, $\dot{\Theta}(t)$, $\Psi(t)$, $\dot{\Psi}(t)$ over the domain of variance of its arguments where the following conditions hold:

$$v \leqslant \frac{d}{dt}\sqrt{\theta^2 + \psi^2} \leqslant v + dv; \qquad \nu \leqslant \sqrt{\theta^2 + \psi^2} \leqslant \nu + d\nu.$$

This integration can be performed easily if one passes from θ, $\dot{\theta}$, ψ, $\dot{\psi}$ to the variables ν, $v = \dot{\nu}$, φ, $\dot{\varphi}$. Using the Jacobian of the transformation, we obtain

$$f(\nu, v) = \frac{\nu^2}{4\pi^2 \sigma_\theta \sigma_\psi \sigma_{\dot{\theta}} \sigma_{\dot{\psi}}} \int_0^{2\pi} \int_{-\infty}^{\infty} \exp\left\{-\frac{1}{2}\left[\left(\frac{\cos^2\varphi}{\sigma_\theta^2} + \frac{\sin^2\varphi}{\sigma_\psi^2}\right)\nu^2 + \frac{(v\cos\varphi - \nu\dot{\varphi}\sin\varphi)^2}{\rho_{\dot{\theta}}^2} + \frac{(v\sin\varphi + \nu\dot{\varphi}\cos\varphi)^2}{\sigma_{\dot{\psi}}^2}\right]\right\} d\dot{\varphi}\, d\varphi.$$

By assumptions, $\sigma_{\dot{\psi}}^2 = \sigma_{\dot{\theta}}^2 = 12.5$ deg.^{2}/sec.^{2} and, consequently, the double integral is simplified and can be computed:

$$f(\nu, v) = \frac{\nu}{62.5\sqrt{2\pi}} \exp\left\{-\frac{v^2}{25} - 0.03\nu^2\right\} I_0(0.01\nu^2).$$

Then

$$\int_0^\infty v f(2, v)\, dv = \frac{25}{62.5\sqrt{2\pi}} e^{-0.12} I_0(0.04).$$

If we substitute the result obtained and the probability density $f(\nu)$ in the formula for $\bar{\tau}_a$, we get

$$\bar{\tau}_a = \frac{\sqrt{\pi}\int_2^\infty e^{-0.03\nu^2}\nu I_0(0.01\nu^2)\, d\nu}{5e^{-0.12} I_0(0.04)}.$$

Since in the theory of Bessel functions it is proved that

$$\int_0^\infty e^{-b\nu^2} I_0(c\nu^2)\nu\, d\nu = \frac{1}{2}\int_0^\infty e^{-bx} I_0(cx)\, dx = \frac{1}{2\sqrt{b^2 - c^2}},$$

the integral in the numerator can be represented as

$$\int_2^\infty e^{-0.03\nu^2} I_0(0.01\nu^2)\nu\, d\nu = \frac{25}{\sqrt{2}} - \int_0^2 e^{-0.03\nu^2} I_0(0.01\nu^2)\nu\, d\nu.$$

In the last integral, the value of the argument of the Bessel function for the upper limit is very small. Therefore, using the expansion of the Bessel function in a series

$$I_0(z) = 1 + \left(\frac{z}{2}\right)^2 + \frac{1}{(2!)^2}\left(\frac{z}{2}\right)^4 + \cdots + \frac{1}{(k!)^2}\left(\frac{z}{2}\right)^{2k} + \cdots,$$

we obtain

$$\int_0^2 e^{-0.03\nu^2}\left[1 + \frac{10^{-4}}{4}\nu^4 + \cdots\right]\nu\, d\nu \approx \frac{1}{0.06}(1 - e^{-0.12});$$

that is,

$$\bar{\tau}_a \approx \frac{\sqrt{\pi}\left(\frac{25}{\sqrt{2}} - \frac{1}{0.06}0.1131\right)}{5e^{-0.12}(1 + 0.0004 + \cdots)} = 6.0 \text{ sec}.$$

Example 33.3 Find the average number of maxima of a normal random function $X(t)$ per unit time if

$$K_{\dot{x}}(\tau) = ae^{-\alpha|\tau|}\left(\cos\beta\tau + \frac{\alpha}{\beta}\sin\beta|\tau|\right), \quad \bar{x} = \text{const}.$$

SOLUTION. The random function $X(t)$ has a maximum if its derivative $\dot{x}(t)$ has a passage through zero from above; that is,

$$\bar{\tau}_a = p_1(a) = \frac{1}{2\pi}\frac{\sigma_{\ddot{x}}}{\sigma_{\dot{x}}} = \frac{\sqrt{\alpha^2 + \beta^2}}{2\pi}.$$

PROBLEMS

33.1 Find the average duration of the passage of a normal random function $X(t)$ through the level $a = 2$ cm. if $\bar{x} = -8$ cm. and $K_x(\tau) = 100e^{-0.1|\tau|}(1 + 0.1|\tau|)$ sq. cm., where τ is expressed in seconds.

33.2 The average number of passages of a normal stationary function through the level $a = \bar{x}$ in one second is 0.01. Find the variance of the rate of change of this function if the variance of the function itself is 64 cm.^{2}.

33.3 The correlation function of the heel angle Θ of a ship is given by the formula

$$K_\theta(\tau) = be^{-\alpha|\tau|}\left(\cos\beta\tau + \frac{\alpha}{\beta}\sin\beta|\tau|\right).$$

If the process of rolling is normal, estimate the average number of times, in 20 minutes, during which the heel angle is outside the bounds $\pm 25°$, if $\bar{\theta} = 0$, $b = 100$ deg.^{2}, $\alpha = 0.1$ sec.$^{-1}$ and $\beta = 0.7$ sec.$^{-1}$.

33.4 The output errors of a dynamical system are normal, with zero expectation and correlation function

$$K(\tau) = ae^{-\alpha|\tau|}(1 + \alpha|\tau|),$$

where $a = 5$ square angular minutes and $\alpha = 1.5$ sec.$^{-1}$. Estimate the average number of times in which the system will be turned off if this occurs automatically in the case of an error whose absolute value exceeds 3′.

33.5 The correlation function of a normal stochastic process is

$$K_x(t_1, t_2) = \sigma^2 t_1 t_2 e^{-\alpha|t_2 - t_1|}(1 + \alpha|t_2 - t_1|).$$

Evaluate the time t at which the average number of passages through the level $a = \bar{x}$ per unit time is less than a prescribed number p_0 ($p_0 > \alpha/2\pi$).

33.6 To remove the damage caused by a random exterior perturbation characterized by a normal random function $X(t)$, it is necessary to use power $W(t)$ proportional to $\dot{X}^2(t)$:

$$W(t) = k\dot{X}^2(t).$$

Estimate the average number of times per unit time in which the power of the motor will be insufficient to remove the damage, if its maximum possible value is w_0, $\bar{x} = 0$,

$$K_{\dot{x}}(\tau) = ae^{-\alpha|\tau|}\left(\cos \beta\tau + \frac{\alpha}{\beta}\sin \beta|\tau|\right),$$

and k, w_0, a, α and β are known constants.

33.7 On an airplane, there is a device (an accelerometer) that measures the accelerations normal to the axis of the fuselage and in the plane of the wing. The automatic pilot is programed for a horizontal rectilinear flight with constant velocity. Because of errors in direction, the angle $\Psi(t)$, made by the velocity vector with the fixed vertical plane, is random. Estimate the average number of times per unit time in which the sensitive element of the accelerometer will go off scale if this event occurs when the instantaneous radius of curvature of the trajectory of the airplane in the horizontal plane becomes equal to the minimal admitted radius of circulation R_0. The velocity of the plane v can be assumed constant and

$$K_{\dot{\psi}}(t_2 - t_1) = ae^{-\alpha|\tau|}\left(\cos \beta\tau + \frac{\alpha}{\beta}\sin \beta|\tau|\right),$$

where $\tau = t_2 - t_1$.

33.8 The altitude $H(t)$ of an airplane directed by an automatic pilot is a random function whose expectation $\bar{h}$ is the given altitude of flight, and whose correlation function is

$$K_h(\tau) = ae^{-\alpha|\tau|}\left(\cos \beta\tau + \frac{\alpha}{\beta}\sin \beta|\tau|\right).$$

Assuming that $H(t)$ is normal, find the minimal altitude $\bar{h}$ that can be established in the system of devices for pilotless flight so that during time T the probability of failure caused by collision with the surface of the earth is less than $\delta = 0.01$ per cent, if $a = 400$ sq. m., $\alpha = 0.01$ sec.$^{-1}$, $\beta = 0.1$ sec.$^{-1}$ and $T = 5$ hours.

33.9 A radio control line insures the transmission of a signal without distortion if the perturbation $X(t)$ at the input of the receiver during transmission does not exceed in absolute value some level a. Find the probability Q for transmission without distortion if

$$\bar{x} = 0, \qquad K_x(\tau) = be^{-\alpha|\tau|}(1 + \alpha|\tau|),$$

and the time of transmission is T.

33.10 Find the distribution law for the ordinates of a normal random $X(t)$ at its points of maxima if

$$\bar{x} = 0, \qquad K_x(\tau) = ae^{-\alpha^2\tau^2}.$$

33.11 Given a normal stochastic process $X(t)$, find the distribution law for the ordinates of its minima if

$$K_x(\tau) = \sigma^2 e^{-\alpha|\tau|}\left(1 + \alpha|\tau| + \frac{1}{3}\alpha^2\tau^2\right).$$

33.12 Estimate the average number of inflexion points of a normal random function $X(t)$ in time T if

$$K_x(\tau) = ae^{-\alpha^2\tau^2}.$$

33.13 Estimate the average number of maxima $\bar{n}$ per unit area of a normal random function of two variables $\zeta(x, y)$ if its two-dimensional correlation function is a function of two variables

$$K_\zeta(\xi, \eta) = \mathbf{M}\{[\zeta^*(x, y) - \bar{\zeta}^*][\zeta(x + \xi, y + \eta) - \bar{\zeta}]\},$$

and its two-dimensional spectral density

$$S_\zeta(\omega_1, \omega_2) = \frac{1}{4\pi^2}\int_{-\infty}^{\infty}\int_{-\infty}^{\infty} e^{-i(\omega_1\xi + \omega_2\eta)}K_\zeta(\xi, \eta)\, d\xi\, d\eta$$

is known.

33.14 Under the assumptions made in the preceding problem, estimate the average number of points $\bar{n}$ per unit area, in which both first partial derivatives $\partial\zeta(x, y)/\partial x$ and $\partial\zeta(x, y)/\partial y$ change their sign from "+" to "−."

34. SPECTRAL DECOMPOSITION OF STATIONARY RANDOM FUNCTIONS

Basic Formulas

Any stationary function $X(t)$ can be written as

$$X(t) - \bar{x} = \int_{-\infty}^{\infty} e^{i\omega t}\, d\Phi(\omega),$$

where, in the case in which

$$\int_{-\infty}^{\infty} |K_x(\tau)|\, d\tau < \infty,$$

the increments $d\Phi(\omega)$ satisfy the relations

$$\mathbf{M}[d\Phi(\omega)] = 0, \qquad \mathbf{M}[d\Phi^*(\omega)\, d\Phi(\omega_1)] = S_x(\omega)\delta(\omega - \omega_1)\, d\omega\, d\omega_1.$$

Here $S_x(\omega)$ is the spectral density of the random function $X(t)$ and $\delta(x)$ denotes the δ-function (see Section 11, page 48).

The correlation function and spectral density are related by mutually inverse Fourier transforms

$$K_x(\tau) = \int_{-\infty}^{\infty} e^{i\omega\tau} S_x(\omega)\, d\omega, \qquad S_x(\omega) = \frac{1}{2\pi}\int_{-\infty}^{\infty} e^{-i\omega\tau} K_x(\tau)\, d\tau,$$

which are the consequence of spectral decomposition of $X(t)$. For $\tau = 0$, the first of the foregoing formulas leads to

$$K_x(0) = \mathbf{D}[X(t)] = \int_{-\infty}^{\infty} S_x(\omega)\, d\omega.$$

The spectral density cannot have negative ordinates; for real functions,

$$S_x(\omega) = S_x(-\omega).$$

The random functions with finite variance have spectral densities vanishing at infinity faster than $1/\omega$.

The spectral density of the derivative $\dot{X}(t)$ is related to $S_x(\omega)$ by the formula

$$S_{\dot{x}}(\omega) = \omega^2 S_x(\omega).$$

The necessary and sufficient condition that a random function be (once) differentiable is

$$\int_{-\infty}^{\infty} \omega^2 S_x(\omega)\, d\omega < \infty,$$

which holds only if $S_x(\omega)$ approaches zero for increasing ω faster than $1/\omega^3$.

If the random functions are stationary and stationarily correlated, then between the mutual correlation function $R_{xy}(\tau)$ and the mutual spectral density $S_{xy}(\omega)$ the following relations hold:

$$R_{xy}(\tau) = \int_{-\infty}^{\infty} e^{i\omega\tau} S_{xy}(\omega)\, d\omega,$$

$$S_{xy}(\omega) = \frac{1}{2\pi}\int_{-\infty}^{\infty} e^{-i\omega\tau} R_{xy}(\tau)\, d\tau.$$

From the definitions of $R_{xy}(\tau)$ and $S_{xy}(\omega)$, it follows that

$$R_{xy}(\tau) = R^*_{yx}(-\tau), \qquad S_{xy}(\omega) = S^*_{yx}(\omega).$$

The spectral density of the product of two normal (real) stationary random functions $X(t)$ and $Y(t)$,

$$Z(t) = X(t)Y(t),$$

is expressed in terms of $S_x(\omega)$, $S_y(\omega)$ and $S_{xy}(\omega)$ by the formula

$$S_z(\omega) = \int_{-\infty}^{\infty} S_x(\omega - \omega_1) S_y(\omega_1)\, d\omega_1 + \int_{-\infty}^{\infty} S_{xy}(\omega - \omega_1) S_{yx}(\omega_1)\, d\omega_1 + \bar{x}^2 S_y(\omega) + \bar{y}^2 S_x(\omega).$$

In the particular case when $Y(t) \equiv X(t)$, $S_y(\omega) = S_{xy}(\omega) = S_x(\omega)$, we have $Z(t) = X^2(t)$ and

$$S_z(\omega) = 2\int_{-\infty}^{\infty} S_x(\omega - \omega_1)S_x(\omega_1)\,d\omega_1 + 2\bar{x}^2 S_x(\omega).$$

The same result can be obtained by using a formula valid for any two normal (stationary) functions:

$$R_{xy}(\tau) = K_x(\tau)K_y(\tau) + R_{xy}(\tau)R_{yx}(\tau) + \bar{x}^2 K_y(\tau) + \bar{y}^2 K_x(\tau),$$

and then applying the Fourier transform to $R_{xy}(\tau)$.

Solution for Typical Examples

To solve Problems 34.1 to 34.10, it is necessary to apply the Fourier transform directly. In determining the correlation function for the case in which the spectral density is the ratio of polynomials in ω, the usual way to obtain the result is by calculations. To find the spectral density when one knows the correlation function, and it involves the modulus of its argument, the infinite domain of integration must be partitioned into two, $(-\infty, 0)$ and $(0, \infty)$. In the rest of the problems it is necessary to find the correlation function or spectral density by using their definitions, and in some problems also by using the properties of normal variables.

Example 34.1 Find the correlation function if

$$S(\omega) = \sum_{j=1}^{n} \frac{a_j}{\omega^2 + \lambda_j^2}.$$

Solution. Using the Fourier transform, we get

$$K(\tau) = \sum_{j=1}^{n} a_j \int_{-\infty}^{\infty} e^{i\omega\tau} \frac{d\omega}{\omega^2 + \lambda_j^2}.$$

For $\tau > 0$, $\int_{-\infty}^{\infty} e^{i\omega\tau}\,d\omega/(\omega^2 + \lambda_j^2)$ is the integral of a function $[e^{i\omega\tau}/(\omega^2 + \lambda_j^2)]$ of a complex variable ω over a contour formed by the real axis and a closed semicircle of infinite radius in the upper half-plane. Thus the integral's value is calculated by multiplying the residue of the function at the unique pole $\omega = i\lambda_j$ (we consider $\operatorname{Re}\lambda_j > 0$) located inside the contour by $2\pi i$, i.e., $\pi/\lambda_j\, e^{-\lambda_j\tau}$, and so

$$K(\tau) = \pi \sum_{j=1}^{n} \frac{a_j}{\lambda_j} e^{-\lambda_j\tau}.$$

Similarly for $\tau < 0$, by closing the real axis through the lower half-plane, we obtain $K(\tau) = \pi \sum_{j=1}^{n} a_j/\lambda_{j_s}\, e^{\lambda_j\tau}$; that is for any sign of τ,

$$K(\tau) = \pi \sum_{j=1}^{n} \frac{a_j}{\lambda_j} e^{-\lambda_j|\tau|}.$$

Example 34.2 Find the spectral density if

$$K(\tau) = \sigma^2 e^{-\alpha|\tau|}\left(1 + \alpha|\tau| + \frac{1}{3}\alpha^2\tau^2\right).$$

SOLUTION. Using the notation

$$J(\alpha, \omega) = \frac{1}{2\pi}\int_{-\infty}^{\infty} e^{-i\omega\tau}\sigma^2 e^{-\alpha|\tau|}\, d\tau,$$

we see that

$$S(\omega) = J - \alpha\frac{\partial J}{\partial\alpha} + \frac{\alpha^2}{3}\frac{\partial^2 J}{\partial\alpha^2}.$$

Since

$$J(\alpha, \omega) = \frac{\sigma^2}{2\pi}\left\{\int_{-\infty}^{0} e^{(\alpha - i\omega)\tau}\, d\tau + \int_{0}^{\infty} e^{(-a-i\omega)\tau}\, d\tau\right\} = \frac{\alpha\sigma^2}{\pi(\omega^2 + \alpha^2)},$$

after differentiation with respect to α and simple transformations, we find that

$$S(\omega) = \frac{8\sigma^2\alpha^5}{3\pi(\omega^2 + \alpha^2)^3}.$$

Example 34.3 Find the spectral density

$$Z(t) = X(t)\dot{X}(t)$$

if $X(t)$ is normal random function and

$$K_x(\tau) = ae^{-\alpha|\tau|}\left(\cos\beta\tau + \frac{\alpha}{\beta}\sin\beta|\tau|\right), \qquad \bar{x} = 0.$$

SOLUTION. Since

$$Z(t) = \frac{1}{2}\frac{d}{dt}X^2(t),$$

$$\begin{aligned} S_z(\omega) &= \frac{1}{2}\omega^2 S_{x^2}(\omega) \\ &= \omega^2\int_{-\infty}^{\infty} S_x(\omega - \omega_1)S_x(\omega_1)S_x(\omega_1)\, d\omega_1 \\ &= \frac{2a\alpha(\alpha^2 + \beta^2)}{\pi}\,\frac{\omega^2(\omega^2 + 20\alpha^2 + 4\beta^2)}{[(\omega^2 + 4\alpha^2 + 4\beta^2)^2 - 16\beta^2\omega^2](\omega^2 + 4\alpha^2)}. \end{aligned}$$

PROBLEMS

34.1 Given the spectral density,

$$S(\omega) = \begin{cases} a, & \text{for} \quad -b \leqslant \omega \leqslant b, \\ 0, & \text{for} \quad b < |\omega|, \end{cases}$$

find the correlation function $K(\tau)$.

34.2 Given the spectral density

$$S(\omega) = \begin{cases} 0, & \text{for} \quad |\omega| < \omega_0, \\ c^2, & \text{for} \quad \omega_0 \leqslant |\omega| \leqslant 2\omega_0, \\ 0, & \text{for} \quad 2\omega_0 < |\omega|. \end{cases}$$

find the correlation function $K(\tau)$.

34.3 Find the spectral density $S(\omega)$ if

$$K(\tau) = ae^{-\alpha|\tau|}(1 + \alpha|\tau|).$$

34.4 Find the spectral density $S(\omega)$ if

$$K(\tau) = \begin{cases} \sigma^2(1 - |\tau|) & \text{for} \quad |\tau| \leqslant 1, \\ 0 & \text{for} \quad |\tau| > 1. \end{cases}$$

34.5 Find the spectral density $S(\omega)$ if

$$K(\tau) = \sigma^2 e^{-\alpha|\tau|} \cos \beta\tau.$$

34.6 Find the spectral density $S(\omega)$ if

$$K(\tau) = \sigma^2 e^{-\alpha|\tau|}\left(\cos \beta\tau + \frac{\alpha}{\beta} \sin \beta|\tau|\right).$$

34.7 Find the spectral density $S(\omega)$ if

$$K(\tau) = \sigma^2 e^{-\alpha|\tau|}\left(1 + \alpha|\tau| - 2\alpha^2\tau^2 + \frac{1}{3}\alpha^3|\tau|^3\right).$$

34.8 Find the spectral density $S(\omega)$ if

$$K(\tau) = ae^{-\alpha|\tau|}\left(\cos \beta\tau - \frac{\alpha}{\beta} \sin \beta|\tau|\right).$$

34.9 According to the form of the spectral density of a random function $X(t)$, determine how many derivatives this function has if

$$K_x(\tau) = \sigma^2 e^{-\alpha|\tau|}\left(1 + \alpha|\tau| + \frac{1}{3}\alpha^2\tau^2\right).$$

34.10 Find the spectral density $S(a)$ if

$$K(\tau) = \sum_{j=1}^{n} a_j e^{-\alpha_j|\tau|}, \qquad \operatorname{Re} \alpha_j > 0.$$

34.11 Find the values of the quotient α/β for which the spectral density

$$S(\omega) = \frac{\alpha\sigma^2}{\pi} \frac{\omega^2 + \alpha^2 + \beta^2}{(\omega^2 + \alpha^2 + \beta^2)^2 - 4\beta^2\omega^2}$$

has a maximum at $\omega = 0$.

34.12 Find the variance of the derivative of a random function $X(t)$ if

$$S_x(\omega) = \frac{\alpha^2}{(\omega^2 + \alpha^2)^2}.$$

34.13 Find the mutual spectral densities $S_{x\dot{x}}(\omega)$ and $S_{\dot{x}x}(\omega)$ if

$$K_x(\tau) = ae^{-\alpha^2\tau^2}.$$

34.14 The control signal $\Delta(t)$ sent to the control units of an automatic system is defined by the formula

$$\Delta(t) = k_1 U(t) + k_2 \dot{U}(t).$$

Find $S_\Delta(\omega)$ if

$$K_u(\tau) = ae^{-\alpha|\tau|}(1 + \alpha|\tau|).$$

34.15 A dynamical system (predictor) is used to obtain the value of the input random function $X(t)$ at time $t + \tau_0$, where τ_0 is the lead time of prediction. Find the mutual spectral density between $X(t)$ and $Y(t) = X(t + \tau_0)$, if $K_x(\tau)$ is known.

36.16 A random function $X(t)$ is fed to the input of a dynamical system. Furthermore, $X(t)$ is the sum of a useful signal $U(t)$ and noise $V(t)$:

$$X(t) = U(t) + V(t).$$

The problem of the dynamical system is the calculation of the function

$$Y(t) = \frac{d^k}{dt^k} U(t + \tau_0).$$

Find the mutual spectral density $S_{xy}(\omega)$ if $S_v(\omega)$, $S_u(\omega)$ and $S_{uv}(\omega)$ are known.

34.17 Find the spectral density $S_z(\omega)$ if

$$Z(t) = \dot{X}(t)\dot{Y}(t)$$

and if $X(t)$ and $Y(t)$ are independent random functions with known correlation functions:

$$K_x(\tau) = a_1 e^{-\alpha_1|\tau|}\left(\cos\beta_1\tau + \frac{\alpha_1}{\beta_1}\sin\beta_1|\tau|\right),$$

$$K_y(\tau) = a_2 e^{-\alpha_2|\tau|}\left(\cos\beta_2\tau + \frac{\alpha_2}{\beta_2}\sin\beta_2|\tau|\right).$$

34.18 Find the spectral density $S_z(\omega)$ if

$$Z(t) = X(t)Y(t),$$

where $X(t)$ and $Y(t)$ are independent random functions, $K_x(\tau) = a_1 e^{-\alpha_1|\tau|}$, $K_y(\tau) = a_2 e^{-\alpha_2|\tau|}$ and $\bar{x}$ and $\bar{y}$ are known.

34.19 The "Cardano error" $\Delta(t)$, which occurs by using a Cardano suspension in some of the stability devices on ships, is related to the heel angle $\Theta(t)$ and trim difference angle $\Psi(t)$ by the formula

$$\Delta(t) = \Theta(t)\Psi(t).$$

Assuming that $\Theta(t)$ and $\Psi(t)$ are independent random functions, find the correlation function, the variance and the spectral density of the error $\Delta(t)$, if $\bar{\theta} = \bar{\psi} = 0$ and

$$K_\theta(\tau) = a_1 e^{-\alpha_1|\tau|}\left(\cos\beta_1\tau + \frac{\alpha_1}{\beta_1}\sin\beta_1|\tau|\right),$$

$$K_\psi(\tau) = a_2 e^{-\alpha_2|\tau|}\left(\cos\beta_2\tau + \frac{\alpha_2}{\beta_2}\sin\beta_2|\tau|\right).$$

34.20 Find the spectral density $S_y(\omega)$ if

$$Y(t) = \dot{X}^2(t),$$

where $X(t)$ is a stationary normal random function and

$$K_x(\tau) = a e^{-\alpha|\tau|}\left(\cos\beta\tau + \frac{\alpha}{\beta}\sin\beta|\tau|\right).$$

34.21 Find the spectral density $S_y(\omega)$ if

$$Y(t) = X^2(t),$$

where $X(t)$ is a normal random function, $\bar{x}$ is known and

$$K_x(\tau) = a e^{-\alpha|\tau|}.$$

34.22 Find the spectral density $S_y(\omega)$ if

$$Y(t) = X(t)\frac{dX(t)}{dt},$$

where $X(t)$ is a normal random function,

$$S_x(\omega) = a e^{-(\omega^2/2\alpha^2)}$$

and $\bar{x}$ is known.

34.23 The correction $\Delta(t)$ for the roll of a ship to the azimuth angle of direction of a navigational radar station is defined by the formula

$$\Delta(t) = -\Phi(t) + \Psi(t)\Theta(t)\cos^2 q - \frac{1}{4}[\Theta^2(t) - \Psi^2(t)]\sin 2q.$$

Find $S_\Delta(\omega)$ if q can be considered constant, and the yaw angle $\Phi(t)$, trim difference angle $\Psi(t)$ and heel angle $\Theta(t)$ are uncorrelated normal random functions with known correlation functions:

$$K_\varphi(\tau) = a_1 e^{-\alpha_1|\tau|}\left(\cos\beta_1\tau + \frac{\alpha_1}{\beta_1}\sin\beta_1|\tau|\right),$$

$$K_\psi(\tau) = a_2 e^{-\alpha_2|\tau|}\left(\cos\beta_2\tau + \frac{\alpha^2}{\beta_2}\sin\beta_2|\tau|\right),$$

$$K_\theta(\tau) = a_3 e^{-\alpha_3|\tau|}\left(\cos\beta_3\tau + \frac{\alpha_3}{\beta_3}\sin\beta_3|\tau|\right), \qquad \bar{\varphi} = \bar{\theta} = \bar{\psi} = 0.$$

34.24 A normal random function $X(t)$ has a correlation function $K_x(\tau) = \sigma_x^2 e^{-\alpha|\tau|}$ and expectation $\bar{x}$. Find the maximum of the spectral density $S_y(\omega)$ if

$$Y(t) = X^2(t).$$

FIGURE 34

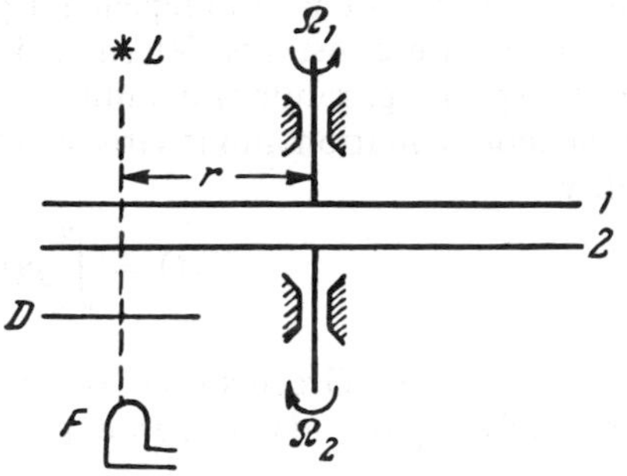

34.25 Two identical disks, whose rotation axes coincide, rotate with different (incommensurable) angular velocities Ω_1 and Ω_2 (Figure 34). In these disks there are holes bounded by two radii making a central angle γ and by the circumferences of radius $r - (1/2)\Delta$ and $r + (1/2)\Delta$. The centers of these holes are selected on the circumference of radius γ according to a uniform distribution law.

On one side of the disks is a point source of light L, and on the other side a photocell F in front of which is placed a diaphragm D; the aperture of the diaphragm has the shape of a sector with angle Γ bounded by the circumferences of radius $r - (1/2)\Delta$ and $r + (1/2)\Delta$. The intensity of the photocurrent J is proportional to the sum of the areas of all the holes within the aperture of the diaphragm. Find the spectral density for the intensity of the current $S_j(\omega)$ if there are n holes in each disk and if it is equally probable that any hole in the first disk, independent of the positions of the other holes, is located opposite a hole in the second disk at any angular distance from the optical axis of the system, light source and the photoelement.[4] (Neglect the case when the size of the aperture is decreased by the diaphragm.)

35. COMPUTATION OF PROBABILITY CHARACTERISTICS OF RANDOM FUNCTIONS AT THE OUTPUT OF DYNAMICAL SYSTEMS

Basic Formulas

For any linear differential equation,

$$\frac{d^n Y(t)}{dt^n} + a_1(t)\frac{d^{n-1}Y(t)}{dt^{n-1}} + \cdots + a_n(t)Y(t) = X(t),$$

the general solution can be represented as

$$Y(t) = \sum_{j=1}^{n} C_j y_j(t) + Y_1(t),$$

[4] Such a device was proposed by V. S. Gytel'son.

where $y_j(t)$ is a system of independent particular integrals of the homogeneous equation, C_j are constants determined by the initial conditions and they are, generally speaking, random quantities, $Y_I(t)$ is a particular integral of the nonhomogeneous equation and it satisfies zero initial conditions and is given by the equality

$$Y_I(t) = \int_0^t p(t, t_1)X(t_1)\,dt_1,$$

where $p(t, t_1)$ is the Green's function of the system (impulse function) expressed in terms of the particular integrals $y_j(t)$ by the formula

$$p(t, t_1) = \begin{vmatrix} y_1(t_1) & \cdots & y_n(t_1) \\ y_1'(t_1) & \cdots & y_n'(t_1) \\ \cdot & \cdots & \cdot \\ y_1^{(n-2)}(t_1) & \cdots & y_n^{(n-2)}(t_1) \\ y_1(t) & \cdots & y_n(t) \end{vmatrix} : \begin{vmatrix} y_1(t_1) & \cdots & y_n(t_1) \\ y_1'(t_1) & \cdots & y_n'(t_1) \\ \cdot & \cdots & \cdot \\ y_1^{(n-2)}(t_1) & \cdots & y_n^{(n-2)}(t_1) \\ y_1^{(n-1)}(t_1) & \cdots & y_n^{(n-1)}(t_1) \end{vmatrix}.$$

In the case in which the coefficients of the equation are constants, the Green's function depends only on the difference of the arguments

$$p(t, t_1) = p(t_1 - t).$$

If the system is stable, $a_j(t) = \text{const.}$, and if $X(t)$ is stationary, then for a sufficiently large t (compared with the time of the transient process), the function $Y(t)$ can also be considered stationary. In this case,

$$\bar{y} = \frac{1}{a_n}\bar{x},$$

$$S_y(\omega) = \frac{S_x(\omega)}{|(i\omega)^n + a_1(i\omega)^{n-1} + \cdots + a_n|^2},$$

and $K_y(\tau)$ can be found by Fourier inversion of $S_y(\omega)$.

If $X(t)$ is related to the stationary random function $Z(t)$ by the formula

$$X(t) = b_0\frac{d^mZ(t)}{dt^m} + b_1\frac{d^{m-1}Z(t)}{dt^{m-1}} + \cdots + b_mZ(t),$$

we have

$$S_y(\omega) = \frac{|b_0(i\omega)^m + b_1(i\omega)^{m-1} + \cdots + b_m|^2}{|(i\omega)^n + a_1(i\omega)^{n-1} + \cdots + a_n|^2}S_z(\omega),$$

the last formula remaining valid even when $Z(t)$ does not have an mth derivative; however, the expression for $S_y(\omega)$ decreases faster than $1/\omega$ when ω increases.

If the elapsed time t from the start of operation of the system is not large, if the function $X(t)$ is nonstationary or if the coefficients of the equation depend on time, then to find the probability characteristics of the solution it is necessary to apply the general formulas for linear operators, which (if, for simplicity, the constants C_j and $X(t)$ are uncorrelated) lead to:

$$\bar{y}(t) = \sum_{j=1}^{n} y_j(t)\bar{c}_j + \int_0^t p(t, t_1)\bar{x}(t_1)\, dt_1,$$

$$K_y(t_1, t_2) = \sum_{j=1}^{n} \sum_{l=1}^{n} y_j^*(t_1)y_l(t_2)k_{jl} + \int_0^{t_1} \int_0^{t_2} p^*(t_1,\xi)p(t_2,\eta)K_x(\xi,\eta)\, d\eta\, d\xi,$$

where $\|k_{jl}\|$ is the correlation matrix of the system of random variables C_j.

For equations with constant coefficients, we replace $p(t_1, t_2)$ by $p(t_2 - t_1)$ in the last formulas.

If $X(t)$ is a stationary function, then

$$Y_I(t) = \int_0^t p(t, t_1)\bar{x}\, dt_1 + \int_{-\infty}^{\infty} y(\omega, t)\, d\Phi(\omega),$$

where $y(\omega, t)$ is a particular integral of the equation with zero intial conditions and where $X(t)$ is replaced by $e^{i\omega t}$.

In this case,

$$K_{y_I}(t_1, t_2) = \int_{-\infty}^{\infty} y^*(\omega, t_1)y(\omega, t_2)S_x(\omega)\, d\omega.$$

A similar formula holds if $X(t)$ is nonstationary, but can be obtained by multiplying a stationary function by a known (nonrandom) function of time, for example:

$$X(t) = b(t)X_1(t),$$

where $X_1(t)$ is stationary. In this case, $y(\omega, t)$ must be looked upon as a particular integral of the equation in which the right-hand side has been replaced by $b(t)e^{i\omega t}$; i.e., as before, the stationary function has been replaced by $e^{i\omega t}$.

Consider a system of differential equations with constant coefficients associated with a stable dynamical system

$$\frac{dY_j(t)}{dt} + \sum_{l=1}^{u} a_{jl}Y_l(t) = X_j(t), \qquad j = 1, 2, \ldots, n,$$

where a_{jl} are constants, $X_j(t)$ are stationary random functions and time t is sufficiently large. Its solutions are stationary random functions, whose spectral densities and mutual spectral densities can be expressed in terms of the spectral densities and mutual spectral densities of the right-hand sides of the equations as follows:

$$S_{y_j}(\omega) = \frac{\sum_{l=1}^{n} \sum_{m=1}^{n} A_{lj}^* A_{mj} S_{x_l x_m}(\omega)}{|\Delta(\omega)|^2},$$

$$S_{y_j y_k}(\omega) = \frac{\sum_{l=1}^{n} \sum_{m=1}^{n} A_{lj}^*(\omega) A_{mk}(\omega) S_{x_l x_m}(\omega)}{|\Delta(\omega)|^2}.$$

Here $\Delta(\omega)$ is the determinant formed from the coefficients appearing on the left-hand sides of the equations:

$$\Delta(\omega) = \begin{vmatrix} a_{11} + i\omega & a_{12} & \cdots & a_{1n} \\ a_{21} & a_{22} + i\omega & \cdots & a_{2n} \\ \cdot & \cdot & \cdot & \cdot \\ a_{n1} & a_{n2} & \cdots & a_{nn} + i\omega \end{vmatrix},$$

where $A_{ij}(\omega)$ is the cofactor of the element located at the intersection of the ith row and the jth column, and $S_{x_j x_j}(\omega) \equiv S_{x_j}(\omega)$.

The distribution law for the solution of a linear equation (system of linear equations), whose right-hand side contains normal random functions and variables, is also normal. If the equation is linear but the distribution law of the random functions on the right-hand side is not normal, the distribution law for the solution also will not be normal. The expectation $\bar{y}$ and the central moments μ_j of this distribution law for any t are determined by the formulas

$$\bar{y} = \int_0^t p(t, t_1)\bar{x}(t_1)\, dt_1,$$

$$\mu_2 = \int_0^t \int_0^t p(t, t_1)p(t, t_2)K_x(t_1, t_2)\, dt_1\, dt_2,$$

$$\mu_3 = \int_0^t \int_0^t \int_0^t p(t, t_1)p(t, t_2)p(t, t_3)K_x(t_1, t_2, t_3)\, dt_1\, dt_2\, dt_3,$$

$$\cdots\cdots\cdots\cdots\cdots,$$

where $X(t)$ is the random function appearing on the right-hand side of the equation and

$$K_x(t_1, t_2, \ldots, t_j) = \mathbf{M}\left\{\prod_{l=1}^{j} [X(t_l) - \bar{x}(t_l)]\right\}.$$

Solution for Typical Examples

Example 35.1 The error $\varepsilon(t)$ in measuring the acceleration of an airplane with the aid of an accelerometer is defined by the equation

$$\ddot{\varepsilon}(t) + 2h\dot{\varepsilon}(t) + n^2\varepsilon(t) = gn^2\gamma(t),$$

where $\gamma(t)$ is a random function characterizing the random perturbation acting on the sensitive element of the accelerometer, and $S_\gamma(\omega) = c^2 \approx \text{const.}$

Find the variance of the velocity of the airplane by integrating the accelerometer readings during time T if no supplementary errors occur during integration and the time for the transient process is much less than T.

SOLUTION. By assumption, the error $\varepsilon(t)$ can be considered a stationary random function of time and, thus,

$$S_\varepsilon(\omega) = \frac{g^2n^4c^2}{|(n^2 - \omega^2) + 2ih\omega|^2} = \frac{g^2n^4c^2}{(\omega^2 - n^2)^2 + 4h^2\omega^2}.$$

The error in velocity $\delta v = \int_0^t \varepsilon(t_1)\,dt_1$ will not be stationary and its variance will be defined by the formula $\mathbf{D}[\delta v(t)] = \int_0^T \int_0^T K_\varepsilon(t_2 - t_1)\,dt_1\,dt_2$. Passing to the new variables $\tau = t_2 - t_1$, $\xi = t_2 + t_1$ and computing the integral with respect to ξ, we obtain

$$\mathbf{D}[\delta v(t)] = 2\int_0^T (T - \tau)K_\varepsilon(\tau)\,d\tau \approx 2T\int_0^\infty K_\varepsilon(\tau)\,d\tau$$
$$= 2\pi T S_\varepsilon(0) = 2\pi g^2c^2T.$$

In a similar way one can solve all the problems in which the required random function is a stationary solution of a linear equation with constant coefficients or the result of application of a linear operator to a stationary solution.

Example 35.2 For time t, find the variance of a particular integral $Y_I(t)$ of the equation $[dY(t)/dt] + aY(t) = tX(t)$, with zero initial conditions, if

$$S_x(\omega) = \frac{\sigma_x^2}{\pi}\frac{\alpha}{\omega^2 + \alpha^2}.$$

SOLUTION. In this particular case, $Y(t)$ is not stationary because on the right-hand side of the equation there is a nonstationary function of time.

We have

$$Y_I(t) = \int_0^t e^{-a(t-t_1)}Z(t_1)\,dt_1,$$

where

$$Z(t) = tX(t).$$

Since

$$K_x(\tau) = \int_{-\infty}^{\infty} e^{i\omega\tau}S_x(\omega)\,d\omega = \sigma_x^2e^{-\alpha|\tau|},$$

then

$$K_z(t_1, t_2) = \sigma_x^2t_1t_2e^{-\alpha|t_2-t_1|},$$

and

$$\mathbf{D}[Y_I(t)] = K_{y_I}(t, t) = \int_0^t\int_0^t K_z(t_1, t_2)e^{-a(2t-t_1-t_2)}\,dt_1\,dt_2,$$

which after integration leads to

$$\mathbf{D}[Y_I(t)] = 2\sigma_x^2\left\{\frac{t^2}{2a(a+\alpha)} - \frac{t(2a+\alpha)}{2a^2(a+\alpha)^2} + \frac{2a+\alpha}{4a^3(a+\alpha)^2}\right.$$
$$\left. + \frac{t(a-\alpha) - 1}{(a^2-\alpha^2)^2}e^{-(a+\alpha)t} + \frac{4a^3 - (2a+\alpha)(a-\alpha)^2}{4a^3(a^2-\alpha^2)}e^{-2at}\right\}.$$

Example 35.3 Find the spectral density and the mutual spectral density of the stationary solutions of the system of equations,

$$\left.\begin{aligned} \frac{d^2 Y(t)}{dt^2} + 2\frac{dY(t)}{dt} + 4Y(t) + Z(t) &= X_1(t) \\ \frac{dZ(t)}{dt} + 9Z(t) &= X_2(t) \end{aligned}\right\},$$

if

$$S_{x_1}(\omega) = \frac{\sigma_1^2}{\pi(\omega^2+1)}, \qquad S_{x_2}(\omega) = \frac{2\sigma_2^2}{\pi(\omega^2+4)},$$

$$S_{x_1x_2}(\omega) = \frac{a}{(\omega^2-2)^2 + i\omega}.$$

SOLUTION. If we replace the differential operator by $i\omega$ on the left-hand sides, the determinant of the resulting system of algebraic equations becomes

$$\Delta(\omega) = \begin{vmatrix} -\omega^2 + 2i\omega + 4 & 1 \\ 0 & i\omega + 9 \end{vmatrix} = [(4-\omega^2) + 2i\omega](i\omega + 9).$$

The cofactors of the elements of the determinant are:

$$A_{11} = i\omega + 9, \qquad A_{12} = 0,$$

$$A_{21} = -1, \qquad A_{22} = (4-\omega^2) + 2i\omega.$$

Consequently, applying the general formula we get:

$$\begin{aligned} S_y(\omega) &= \frac{|A_{11}|^2 S_{x_1}(\omega) + |A_{21}|^2 S_{x_2}(\omega) + A_{11}^* A_{21} S_{x_1x_2}(\omega) + A_{21}^* A_{11} S_{x_2x_1}(\omega)}{|\Delta(\omega)|^2} \\ &= \frac{1}{(\omega^2+81)[(\omega^2-4)^2 + 4\omega^2]} \\ &\quad \times \left\{\frac{\sigma_1^2(\omega^2+81)}{\pi(\omega^2+1)} + \frac{2\sigma_2^2}{\pi(\omega^2+4)} - \frac{2\alpha[\omega^2 - 9(\omega^2-2)^2]}{(\omega^2-2)^4 + \omega^2}\right\}, \end{aligned}$$

$$\begin{aligned} S_{yz}(\omega) &= \frac{A_{11}^* A_{12} S_{x_1}(\omega) + A_{11}^* A_{22} S_{x_1x_2}(\omega) + A_{21}^* A_{12} S_{x_2x_1}(\omega) + A_{21}^* A_{22} S_{x_2}(\omega)}{|\Delta(\omega)|^2} \\ &= \frac{1}{[(\omega^2-4) + 2i\omega](\omega^2+81)}\left\{\frac{a(i\omega-9)}{(\omega^2-2)^2 + i\omega} + \frac{2\sigma_2^2}{\pi(\omega^2+4)}\right\}, \end{aligned}$$

$$S_z(\omega) = \frac{S_{x_2}(\omega)}{|i\omega+9|^2} = \frac{2\sigma_2^2}{\pi(\omega^2+4)(\omega^2+81)}.$$

PROBLEMS

35.1 The input signal of a first-order dynamical system, described by the equation

$$\frac{dY(t)}{dt} + \alpha Y(t) = X(t), \qquad \alpha > 0,$$

is a random function $X(t)$ whose spectral density in the frequency band $|\omega| \leqslant \omega_0$, where $\omega_0 \gg \alpha$ can be considered constant:

$$S_x(\omega) \approx c^2.$$

Find the correlation function of $Y(t)$ for $t \gg 1/\alpha$.

35.2 A dynamical system is described by the equation

$$a_0 \frac{dY(t)}{dt} + a_1 Y(t) = b_0 \frac{dX(t)}{dt} + b_1 X(t),$$

where $\bar{x} = \text{const.}$ is known and $K_x(\tau) = \sigma_x^2 e^{-\alpha|\tau|}$, $a_1/a_0 > 0$.

Find the expectation and variance for the stationary solution of this equation.

35.3 The deviation $U(t)$ of a heel-meter located in the plane of the midship frame is defined by the equation

$$\frac{d^2U(t)}{dt^2} + 2h\frac{dU(t)}{dt} + n^2U(t) = n^2F(t) \qquad (n > h > 0),$$

where $F(t) = 1/g[\ddot{\eta}_c(t) - c\ddot{\Theta}(t)]$. The heel angle $\Theta(t)$ and the velocity of the lateral shift of the center of gravity of the ship $\dot{\eta}_c(t)$, as a result of orbital motion, can be considered uncorrelated random functions

$$K_{\dot{\eta}_c}(\tau) = a_1 e^{-\alpha_1|\tau|}\left(\cos\beta_1\tau + \frac{\alpha_1}{\beta_1}\sin\beta_1|\tau|\right),$$

$$K_\theta(\tau) = a_2 e^{-\alpha_2|\tau|}\left(\cos\beta_2\tau + \frac{\alpha_2}{\beta_2}\sin\beta_2|\tau|\right),$$

and all the constants contained in the formulas are known. Evaluate $S_u(\omega)$.

35.4 An astatic gyroscope with proportional correction is located on a ship in the plane of the midship frame. Find the variance for the deviation α of its axis from the direction given by the physical pendulum if the angle α is determined by the equation

$$\dot{\alpha}(t) + \varepsilon\alpha(t) = \varepsilon U(t) \qquad (\varepsilon > 0).$$

Assume the time elapsed since the start of the gyroscope is sufficiently great so that $\alpha(t)$ can be considered stationary; determine the spectral density $S_u(\omega)$ by use of the result of Problem 35.3, where

$a_1 = 1.24$ sq. m./sec.^{2};	$\alpha_1 = 0.1$ sec.$^{-1}$;	$\beta_1 = 1.20$ sec.$^{-1}$;
$a_2 = 3.8 \cdot 10^{-2}$ rad.^{2};	$\alpha_2 = 0.04$ sec.$^{-1}$;	$\beta_2 = 0.42$ sec.$^{-1}$;
$h = 0.6$ sec.$^{-1}$;	$n = 6.28$ sec.$^{-1}$;	$c = 10$ m.;
		$\varepsilon = 0.01$ sec.$^{-1}$.

35.5 Find the spectral density and correlation function of the stationary solution of the equation

$$\frac{d^2Y(t)}{dt^2} + 2h\frac{dY(t)}{dt} + k^2Y(t) = X(t) \qquad (k \geqslant h > 0)$$

if $X(t)$ has the properties of "white noise," that is, $S_x(\omega) = c^2 = \text{const.}$

35.6 The angular deviation $\Theta(t)$ of the coil of a galvanometer from the equilibrium position in the case of open circuit is defined by the equation

$$I\frac{d^2\Theta(t)}{dt^2} + r\frac{d\Theta(t)}{dt} + D\Theta(t) = M(t), \qquad 4ID \geqslant r^2,$$

where I is the moment of inertia of the coil, r is the friction coefficient, D is the rigidity coefficient of the thread on which the coil is suspended and $M(t)$ is the perturbing moment caused by the impact of molecules from the surrounding medium.

Find the spectral density and the correlation function of the angle $\Theta(t)$ if the spectral density $M(t)$ can be assumed constant and, according to results of statistical mechanics, $\sigma_\theta^2 D = kT$, where k is Boltzmann's constant and T is the absolute temperature of the medium.

35.7 Two random stationary functions $Y(t)$ and $X(t)$ are related by the equation

$$\frac{d^3Y(t)}{dt^3} + 6\frac{d^2Y(t)}{dt^2} + 11\frac{dY(t)}{dt} + 6Y(t) = 5X(t) + 7\frac{d^3X(t)}{dt^3}.$$

Find the spectral density $S_y(\omega)$ for the stationary solution of the equation if $S_x(\omega) = [4/\pi(\omega^2 + 1)]$.

35.8 Does the equation

$$\ddot{Y}(t) - 2\dot{Y}(t) + 3Y(t) = X(t),$$

containing on its right-hand side the stationary function $X(t)$, admit a stationary solution?

35.9 Find the variance of the ordinate of the center of gravity of a ship $\xi_c(t)$ on a wavy sea if

$$\ddot{\zeta}_c(t) + 2h\dot{\zeta}_c(t) + \omega_0^2\zeta_c(t) = \omega_0^2X(t),$$

where the ordinate of the wave front $X(t)$ has the correlation function

$$K_x(\tau) = ae^{-\alpha|\tau|}\left(\cos\beta\tau + \frac{\alpha}{\beta}\sin\beta|\tau|\right);$$

h and ω_0 are constants defined by the parameters of the ship, α is a parameter characterizing the irregularity of waves, β is the dominant frequency of waves and $\omega_0 \geqslant h > 0$.

35.10 The error given by an accelerometer measuring the horizontal acceleration of an airplane is defined by the equation

$$\ddot{\varepsilon}(t) + 2h\dot{\varepsilon}(t) + n^2\varepsilon(t) = gn^2\gamma(t),$$

where $h = 0.6$ sec.$^{-1}$, $n = 6.28$ sec.$^{-1}$, $g = 9.81$ m./sec.^{2} and the heel angle $\gamma(t)$ is a stationary normal random function with a known correlation function:

$$K_\gamma(\tau) = 3\cdot 10^{-4}e^{-0.6|\tau|}(\cos 5\tau + 0.12\sin 5|\tau|).$$

Find the variance of $\varepsilon(t)$ for the stationary operating mode of the accelerometer.

35.11 Prove if the input signal of a linear stable dynamical system described by equations with constant coefficients is a random function $X(t)$ with properties of "white noise" ($S_x(\omega) = c^2$), then for a sufficiently long elapsed time after the start of operations, the correlation function of the output signal $Y(t)$ is defined by the equality

$$K_y(\tau) = 2\pi c^2 \int_0^\infty p(t)p(t-\tau)\,dt,$$

where $p(t)$ is the Green's function of the system.

35.12 Find the variance of the heel angle $\Theta(t)$ of a ship defined by the equation

$$\ddot{\Theta}(t) + 2h\dot{\Theta}(t) + k^2\Theta(t) = k^2F(t) \qquad (k > h > 0),$$

if the wave slope angle $F(t)$ has a zero expectation,

$$K_f(\tau) = ae^{-\alpha|\tau|}\left(\cos\beta\tau + \frac{\alpha}{\beta}\sin\beta|\tau|\right),$$

and the rolling process can be considered stationary.

35.13 A stationary random function $Y(t)$ is related to the stationary function $X(t)$, whose spectral density is known, by the equation

$$\ddot{Y}(t) + 2h\dot{Y}(t) + k^2Y(t) = k^2X(t),$$

where $k \geqslant h > 0$.

Find the mutual spectral density $S_{yx}(\omega)$ and the mutual correlation function $R_{yx}(\tau)$.

35.14 Given

$$\ddot{Y}(t) + 8\dot{Y}(t) + 7Y(t) = X(t),$$

$$K_x(\tau) = 4e^{-\alpha^2\tau^2},$$

find the correlation function $Y(t)$ for times exceeding the time of the transient process.

35.15 The input signal of a dynamical system with Green's function $p(t)$ represents a stationary random function $X(t)$ with zero expectation. Find the variance of the deviation of the output signal $Y(t)$ from some stationary function $Z(t)$ if $K_x(\tau)$ and $R_{xz}(\tau)$ are known, $\bar{z} = 0$ and the transient process of the system can be considered finished.

35.16 Using the spectral decomposition of a stationary random function $X(t)$, find for time $t \gg 1/a$ the variance for the integral of the equation

$$\dot{Y}(t) + aY(t) = tX(t)$$

with zero initial conditions, if

$$S_x(\omega) = \frac{\sigma_x^2}{\pi}\frac{\alpha}{\omega^2 + \alpha^2}.$$

35.17 As a consequence of the random unbalance of the gyromotor placed on a platform with a random vertical acceleration $W(t)$, the direction gyroscope precesses with angular velocity

$$\dot{\alpha}(t) = \frac{PL}{H}\left[1 + \frac{1}{g}W(t)\right].$$

Find the expectation and variance of the azimuthal departure $\alpha(t)$ at time t if $\mathbf{M}[L] = 0$, $\mathbf{D}[L] = \sigma_l^2$, $K_w(\tau)$ and $\bar{w}$ are known, P, H and g are known constants and L and $W(t)$ are uncorrelated.

35.18 Find the correlation function of the particular solution $Y_I(t)$ of the equation

$$\ddot{Y}(t) + 2h\dot{Y}(t) + k^2 Y(t) = e^{-at}X(t)$$

with zero initial conditions, if

$$S_x(\omega) = \frac{\sigma_x^2}{\pi}\frac{\alpha}{\omega^2 + \alpha^2} \qquad (k \geqslant h > 0).$$

35.19 Two random functions $Y(t)$ and $X(t)$ are related by the equation

$$\dot{Y}(t) - tY(t) = X(t).$$

Find $K_y(t_1, t_2)$ if $K_x(\tau) = ae^{-\alpha|\tau|}$ and if for $t = 0$, $Y(t) = 0$.

35.20 Find the expectation and the correlation function of the particular solution of the equation

$$\dot{Y}(t) - a^2 t Y(t) = bX(t)$$

with zero initial conditions, if $\bar{x}(t) = t$,

$$K_x(\tau) = \sigma_x^2 e^{-a^2\tau^2}.$$

35.21 Find the expectation and the correlation function of the solution of the differential equation,

$$\dot{Y}(t) + \frac{1}{t}Y(t) = X(t),$$

if for $t = t_0 \neq 0$, $Y(t) = y_0$, where y_0 is a nonrandom variable and $\bar{x}(t) = 1/t$;

$$K_x(t_1, t_2) = t_1 t_2 e^{-\alpha|t_2 - t_1|}.$$

35.22 Write the general expression for the expectation and correlation function of the solution $Y(t)$ of a differential equation of nth order whose Green's function is $p(t_1, t_2)$, if on the right-hand side of the equation the random function $X(t)$ appears, $\bar{x}(t)$ and $K_x(t_1, t_2)$ are known and the initial values of $Y(t)$ and the first $(n - 1)$ derivatives are random variables uncorrelated with the ordinates of the random function $X(t)$ with known expectations e_j and with correlation matrix $\|k_{jl}\|$ $(l, j = 1, 2, \ldots, n)$.

35.23 Given the system

$$\dot{Y}_1(t) + 3Y_1(t) - Y_2(t) = X(t), \qquad \dot{Y}_2(t) + 2Y_1(t) = 0,$$

find the variance of $Y_2(t)$ for $t = 0.5$ sec. if for $t = 0$, $Y_1(t)$ and $Y_2(t)$ are random variables, uncorrelated to $X(t)$; $\mathbf{D}[Y_1(0)] = 1$, $\mathbf{D}[Y_2(0)] = 2$, $\mathbf{M}\{[Y_1(0) - \bar{y}_1(0)][Y_2(0) - \bar{y}_2(0)]\} = -1$,

$$S_x(\omega) = \frac{2}{\pi(\omega^2 + 1)^2} \text{ sec.}$$

35.24 Find the variance for the solutions of the system of equations

$$\left.\begin{aligned} \dot{Y}_1(t) + 3Y_1(t) - Y_2(t) &= tX(t) \\ \dot{Y}_2(t) + 2Y_1(t) &= 0 \end{aligned}\right\},$$

for time t if the initial conditions are zero and

$$S_x(\omega) = \frac{2}{\pi(\omega^2 + 1)}.$$

35.25 Find the variance for the solutions of the system of equations

$$\left.\begin{aligned} \dot{Y}_1(t) + 3Y_1(t) - Y_2(t) &= tX(t) \\ \dot{Y}_2(t) + 2Y_1(t) &= 0 \end{aligned}\right\},$$

for $t = 0.5$ sec. if $S_x(\omega) = [2/\pi(\omega^2 + 1)]$ and the initial conditions are zero.

35.26 The input signal to an automatic friction clutch serving as a differential rectifier is a random function $X(t)$. Find the variance for the rectified function $Z(t)$ and the variance of the rectified velocity of its variance $Y(t)$ if the operation of the friction clutch is described by the system of equations

$$\left.\begin{aligned} b\dot{Y}(t) + Y(t) &= a\dot{X}(t) \\ b\dot{Z}(t) + Z(t) &= X(t) \end{aligned}\right\},$$

where a and b are constant scale coefficients, and $K_x(\tau) = \sigma_x^2 e^{-\alpha|\tau|}$ and the transient process is finished.

35.27 For $t = 1$, find the distribution law for the solution of the equation

$$\ddot{Y}(t) + 3\dot{Y}(t) + 2Y(t) = X(t)$$

if for $t = 0$, $Y(t) = Y_0$, $\dot{Y}(t) = \dot{Y}_0$, and Y_0, $\dot{Y}_0$ and $X(t)$ are normal and mutually uncorrelated and

$$\mathbf{M}[Y_0] = \mathbf{M}[\dot{Y}_0] = \bar{x} = 0, \qquad \mathbf{D}[Y_0] = 1.5, \qquad \mathbf{D}[\dot{Y}_0] = 0.2,$$
$$K_x(\tau) = 2e^{-|\tau|}.$$

35.28 The deviation $U(t)$ from the vertical position of a plane physical pendulum whose plane of oscillation coincides with the diametral plane of a ship is defined by the equations

$$\ddot{U}(t) + 2h\dot{U}(t) + Y(t)U(t) = X(t),$$

$$X(t) = -\frac{n^2}{g}\{\ddot{\xi}_c(t) + \ddot{\eta}_c(t)\Phi(t) - \rho_x[\dot{\Phi}^2(t) + \dot{\Psi}^2(t) + \Psi(t)\ddot{\Psi}(t)] + \rho_z[\ddot{\Psi}(t) + \ddot{\Phi}(t)\Theta(t) + 2\dot{\Phi}(t)\dot{\Theta}(t)]\},$$

$$Y(t) = n^2\left[1 - \frac{\ddot{\zeta}_c(t) - \rho_x\ddot{\Psi}(t)}{g}\right],$$

where all coefficients are constant, and the yaw angle $\Phi(t)$, the angle of trim difference $\Psi(t)$, the heel angle $\Theta(t)$ and the velocities of the coordinates of the center of gravity of the ship $\dot{\xi}_c(t)$, $\dot{\eta}_c(t)$, $\dot{\zeta}_c(t)$ are normal stationary uncorrelated random functions.

Express the spectral densities $S_x(\omega)$, $S_y(\omega)$ and $S_{xy}(\omega)$, necessary for finding the probability characteristics of $U(t)$ on a simulating system, in terms of spectral densities $S_\varphi(\omega)$, $S_\psi(\omega)$, $S_\theta(\omega)$, $S_{\dot{\xi}_c}(\omega)$, $S_{\dot{\eta}_0}(\omega)$ and $S_{\dot{\zeta}_c}(\omega)$.

35.29 For time $t \gg 1/k$, find the asymmetry S_k and excess Ex of a particular solution of the equation

$$\dot{Y}(t) - kY(t) = X^2(t),$$

with zero initial conditions, if $X(t)$ is a normal stationary function, $\bar{x} = 0$, $K_x(\tau) = ae^{-\alpha|\tau|}$.

35.30 Find the mutual correlation function $R_{yz}(\tau)$ of the stationary solutions of the equations:

$$\frac{d^2Y(t)}{dt^2} + 2h_1\frac{dY(t)}{dt} + k_1^2Y(t) = k_1^2X(t),$$

$$\frac{d^2Z(t)}{dt^2} + 2h_2\frac{dZ(t)}{dt} + k_2^2Z(t) = k_2^2X(t),$$

where the random function $X(t)$ has the properties of "white noise," $(S_x(\omega) \approx c^2)$, $k_1 > h_1 > 0$, $k_2 > h_2 > 0$.

36. OPTIMAL DYNAMICAL SYSTEMS

Basic Formulas

By an optimal dynamical system[5] we mean a system that for an input function $X(t) = U(t) + V(t)$, where $U(t)$ is the useful signal and $V(t)$ is the noise, has an output function $Y(t)$ whose expectation is equal to the expectation of some function $Z(t)$, and

$$\mathbf{D}[\varepsilon(t)] = \mathbf{D}[Y(t) - Z(t)] = \min.$$

The function $Z(t)$ is related to the useful signal $U(t)$ by

$$Z(t) = \mathrm{N}U(t) = \int_0^t n(t, t_1)U(t_1)\,dt_1,$$

where N is a known operator and $n(t, t_1)$ is its Green's function.

[5] There are other possible definitions of the notion of an optimal dynamical system. For example, by optimal system, one can understand a system for which the probability that the difference $Y(t) - Z(t)$ in absolute value does not exceed a prescribed quantity is maximal. The term "dynamical system" is understood in the technical sense of the word, i.e., it means any system whose state (characterized by a function obtained at its output) changes because of the influence of external perturbations (random functions at the "input" of the system).

To find an optimal system is to determine, according to the probability properties of the random functions $U(t)$ and $V(t)$ and the form of the operator N, the form of the operator L or its corresponding Green's function $l(t, t_1)$ so that the function $X(t)$ can be transformed into the function $Y(t)$:

$$Y(t) = \mathrm{L}X(t) = \int_0^t l(t, t_1)X(t_1)\, dt_1.$$

The problem of determination of an optimal dynamical system can be solved if the following hold:

(a) the random functions $U(t)$ and $V(t)$ are stationary and stationarily connected, and N and L are linear operators independent of time;

(b) the spectral density $S_x(\omega) = S_u(\omega) + S_v(\omega) + S_{uv}(\omega) + S_{uv}^*(\omega)$ is a rational function of its argument. It can be expressed as

$$S_x(\omega) = a^2 \frac{|P_m(\omega)|^2}{|Q_n(\omega)|^2},$$

where the polynomials $P_m(\omega)$ and $Q_n(\omega)$ have roots located only in the upper half-plane of the complex variable; i.e., they can be represented as

$$P_m(\omega) = \prod_{j=1}^{\beta} (\omega - \mu_j)^{m_j}, \qquad Q_n(\omega) = \prod_{l=1}^{\gamma} (\omega - \nu_l)^{n_l},$$

where the complex numbers μ_j and ν_l have positive imaginary parts, m_j and n_l are the multiplicities of the corresponding roots, $\sum_{j=1}^{\beta} m_j = m$, $\sum_{l=1}^{\gamma} n_l = n$;

(c) in the determination of the ordinates of the function $Y(t)$, one can use the values of the ordinates of the function $X(t)$ for an infinitely long time previous to the current time t. In this case, the transmission function $L(i\omega)$ of the optimal dynamical system related to the Green's function by

$$l(\tau) = \frac{1}{2\pi} \int_{-\infty}^{\infty} e^{i\omega\tau} L(i\omega)\, d\omega,$$

$$L(i\omega) = \int_{-\infty}^{\infty} e^{-i\omega\tau} l(\tau)\, d\tau,$$

is defined in the following way (we assume that $\bar{u} = \bar{v} = 0$):

If the system operates without delay (that is, $Z(t)$ is the result of application of some operator to the present or future values of the ordinates of the function $U(t)$), then

$$L(i\omega) = \frac{1}{a^2} \frac{Q_n(\omega)}{P_m(\omega)} \chi(\omega),$$

where

$$\chi(\omega) = \sum_{r=1}^{a} \sum_{k=1}^{l_r} \frac{c_{kr}}{(\omega - \lambda_r)^k},$$

$$c_{kr} = \frac{1}{(l_r - k)!} \frac{d^{l_r - k}}{d\omega^{l_r - k}} \left[(\omega - \lambda_r)^{l_r} \frac{Q_n^*(\omega)}{P_m^*(\omega)} S_{xz}(\omega) \right] \bigg|_{\omega = \lambda_r},$$

and λ_r $(r = 1, 2, \ldots, \alpha)$ is the pole of multiplicity lr (of the expression $[Q_n^*(\omega)/P_m^*(\omega)]S_{xz}(\omega))$ located in the upper half-plane.

If the optimal dynamical system must operate with delay (that is, the function $Z(t)$ is the result of application of some operator to the ordinates of the function $U(t)$ at an instant preceding the present time t by τ_0 seconds), then

$$L(i\omega) = \frac{S_{xz}(\omega)}{S_x(\omega)} - \frac{1}{a^2}\frac{Q_n(\omega)}{P_m(\omega)}\Psi(\omega),$$

where

$$\Psi(\omega) = \sum_{r=1}^{a'}\sum_{k=1}^{l'_r}\frac{c'_{kr}}{(\omega - \kappa_r)^k},$$

$$c'_{kr} = \frac{1}{(l'_r - k)!}\frac{d^{l'_r - k}}{d\omega^{l'_r - k}_r}\left[(\omega - \kappa_r)^{l'_r}\frac{Q_n^*(\omega)}{P_m^*(\omega)}S_{xz}(\omega)\right]\Bigg|_{\omega=\kappa_r},$$

and κ_r $(r = 1, 2, \ldots, \alpha')$ is the pole of multiplicity l'_r (of the expression $[Q_n^*(\omega)/P_m^*(\omega)]S_{xz}(\omega))$, located in the lower half-plane.

The variance $\mathbf{D}[\varepsilon(t)]$ for the optimal dynamical system is

$$\mathbf{D}[\varepsilon(t)] = \mathbf{D}[Z(t)] - \mathbf{D}[Y(t)].$$

If the dynamical system makes use of the ordinates of the random function during a finite interval of time $(t - T, t)$ preceding the present time t ("system with finite memory") and the useful signal is the sum of the polynomial $R_k(t)$ of a preassigned degree k (the coefficients of the polynomial being arbitrary constants) and a stationary random function $U(t)$, that is, the input function $X(t)$ is

$$X(t) = R_k(t) + U(t) + V(t),$$

then under the same assumptions about the form of the spectral density $S_x(\omega)$, the Green's function $l(\tau)$ of the optimal dynamical system is defined by the formulas

$$l(t) = \sum_{j=0}^{k} D_j t^j + \sum_{r=1}^{2m} c_r e^{\alpha_r t} + \int_{-\infty}^{\infty}\frac{|Q_n(\omega)|^2}{|P_m(\omega)|^2}N(i\omega)S_u(\omega)e^{i\omega t}\,d\omega$$
$$+ \sum_{l=1}^{n-m} A_l\delta^{(l-1)}(t) + \sum_{l=1}^{n-m} B_l\delta^{(l-1)}(t - T) \qquad (0 \leqslant t \leqslant T).$$

Here α_r are the roots of the equation $|P_m(i\alpha)|^2 = 0$, $N(i\omega)$ is the transmission function of the operator N and the constants on the right side of the equality are determined by substituting the expression for $l(\tau)$ in the equation

$$\int_{-0}^{T+} l(\tau)K_x(t - \tau)\,d\tau - R_{xz}(t) = \sum_{j=0}^{k}\lambda_j t^j$$
$$(0 \leqslant t \leqslant T),$$

satisfied by the Green's function $l(\tau)$ of the optimal dynamical system and, then, equating the coefficients of equal powers in t as well as those of equal exponential functions. To the $2n + k + 1$ equations thus obtained, should be added the $k + 1$ equations formed by equating the moments of the function

$l(\tau)$ and the Green's function $n(\tau)$ associated with the operator N; i.e., the equations

$$\int_{0-}^{T+} l(\tau)\tau^j\,d\tau = \mu_j \qquad (j = 0, 1, 2, \ldots, k),$$

where

$$\mu_j = \int_{-\infty}^{\infty} n(\tau)\tau^j\,d\tau.$$

The system of equations thus obtained completely defines all the constants contained in the expression for $l(\tau)$. The transmission function $L(i\omega)$ can be found from $l(\tau)$ by a Fourier transform

$$L(i\omega) = \int_{-\infty}^{\infty} e^{-i\omega\tau} l(\tau)\,d\tau,$$

and the variance of error $\varepsilon(t)$ for the optimal system in the present case is

$$\mathbf{D}[\varepsilon(t)] = \mathbf{D}[Z(t)] - R_{yz}(0) + \sum_{j=0}^{k} \lambda_j \mu_j.$$

In a similar way one can solve the problem of finding the Green's function of an optimal dynamical system if the nonrandom part of the useful signal contains a linear combination (with constant but unknown parameters) of trigonometric or exponential functions of time. The only difference is that in the expression for $l(\tau)$ a similar linear combination will appear, whose coefficients can be determined by substitution in the initial integral equation.

In some problems one prefers not to form optimal dynamical systems because of difficulties connected with their practical realization and, instead, one forms systems that are not optimal in the strict meaning of the word but that give the minimal variance $\mathbf{D}[\varepsilon(t)]$ among systems whose realization in the particular case presents no special difficulties. For example, to find the value of the function $U(t)$ at time $t + \tau$ one can take as $Y(t)$

$$Y(t) = a_1 U(t) + a_2 \dot{U}(t)$$

and determine a_1 and a_2 so that for $\bar{y}(t) = \bar{z}(t)$,

$$\mathbf{D}[Y(t) - Z(t)] = \min.$$

For such a statement of the problem, the determination of the form of operator L (the values of the constants appearing in the expression for this operator) reduces to the determination of the extremum of a function of several variables.

Solution for Typical Examples

Example 36.1 A dynamical system is designed to give the best approximation of the random function $Z(t) = NU(t + \tau_0)$. Find the mutual spectral density $S_{xz}(\omega)$ if $X(t) = U(t) + V(t)$, and the transmission function $N(i\omega)$ of operator N, the prediction time τ_0, the spectral densities $S_u(\omega)$, $S_v(\omega)$ and the mutual spectral density $S_{uv}(\omega)$ are known.

SOLUTION. Setting $U + V$ (instead of $X(t)$) in the expression

$$R_{xz}(\tau) = \mathbf{M}\{[X^*(t) - \bar{x}^*][Z(t + \tau) - \bar{z}]\},$$

replacing $U(t)$ and $V(t)$ by their spectral decompositions and taking into account that $Z(t) = \int_{-\infty}^{\infty} n(\tau)U(t - \tau)\,d\tau$, after simple transformations we obtain

$$S_{xz}(\omega) = [S_u(\omega) + S_{vu}(\omega)]N(i\omega)e^{i\omega\tau_0}.$$

Similarly one can solve Problems 36.1 and 36.2.

Example 36.2 The random function $X(t) = U(t) + V(t)$ is fed into the input of a dynamical system, where the spectral density of the useful signal $S_u(\omega) = \alpha^2/(\omega^2 + \beta^2)$, $S_{uv}(\omega) = 0$ and the spectral density of noise can be considered constant: $S_v(\omega) = c^2$. Find the transmission function $L(i\omega)$ of the optimal dynamical system if the job of the system is to produce the function $Z(t) = U(t + \tau)$, where (a) $\tau \geqslant 0$, (b) $\tau < 0$.

SOLUTION. In this case,

$$S_x(\omega) = \frac{c^2\omega^2 + \alpha^2 + c^2\beta^2}{\omega^2 + \beta^2} = c^2\frac{|P_1(\omega)|^2}{|Q_1(\omega)|^2}, \qquad P_1(\omega) = \omega - i\gamma,$$

$$Q_1(\omega) = \omega - i\beta, \qquad \gamma = \frac{1}{c}\sqrt{\alpha^2 + c^2\beta^2}.$$

(a) For $\tau \geqslant 0$, the expression $[Q_1^*(\omega)/P_1^*(\omega)]S_{xz}(\omega)$ has one pole in the upper half-plane: $\omega = i\beta$; consequently,

$$L(i\omega) = \frac{1}{c^2}\frac{\omega - i\beta}{\omega - i\gamma}\frac{1}{\omega - i\beta}\left[(\omega - i\beta)\frac{\omega + i\beta}{\omega + i\gamma}e^{i\omega\tau}\frac{\alpha^2}{\omega^2 + \beta^2}\right]_{\omega = i\beta}$$

$$= \frac{\alpha^2}{c^2}\frac{e^{-\beta\tau}}{(\beta + \gamma)(\gamma + i\omega)}.$$

(b) For $\tau < 0$, $[Q_1^*(\omega)/P_1^*(\omega)]S_{xz}(\omega)$ has one pole in the lower half-plane: $\omega = -i\gamma$; consequently,

$$L(i\omega) = \frac{\alpha^2}{\omega^2 + \beta^2}e^{i\omega\tau}\frac{\omega^2 + \beta^2}{c^2(\omega^2 + \gamma^2)} - \frac{1}{c^2}\frac{\omega - i\beta}{\omega - i\gamma}\frac{1}{\omega + i\gamma}\left[(\omega + i\gamma)\frac{\omega + i\beta}{\omega + i\gamma}\frac{\alpha^2}{\omega^2 + \beta^2}e^{i\omega\tau}\right]_{\omega = -i\gamma}$$

$$= \frac{\alpha^2}{c^2}\frac{1}{\omega^2 + \gamma^2}\left[e^{i\omega\tau} + \frac{\omega - i\beta}{i(\beta + \gamma)}e^{\gamma\tau}\right].$$

Example 36.3 The distance $D(t)$ to an airplane, measured with the aid of a radar device with error $V(t)$, is the input to a dynamical system that estimates the present value of the velocity by taking into account only its values during time $(t - T, t)$. Determine the optimal Green's function $l(\tau)$ if $K_v(\tau) = \sigma_v^2 e^{-\alpha|\tau|}$; the correct value of the distance can be quite accurately approximated by a polynomial of third degree in t, $\sigma_v = 30$ m., $\alpha = 0.5$ sec.$^{-1}$, $\beta = 2.0$ sec.$^{-1}$ and $T = 20$ sec.

SOLUTION. Since to the correlation function $K_v(\tau)$ there corresponds the spectral density $S_v(\omega) = [\alpha\sigma_v^2/\pi(\omega^2 + \alpha^2)]$, and the useful part of the random signal $U(t) = 0$, then, in the notations assumed in this example, we have $k = 3$, $n - m = 1$, $S_x(\omega) = S_v(\omega)$, the numerator of $S_v(\omega)$ contains no ω and, consequently, it has no roots.

Green's function of the optimal system will be

$$l(\tau) = A_1\delta(\tau) + B_1\delta(\tau - T) + D_0 + D_1\tau + D_2\tau^2 + D_3\tau^3.$$

To determine the constants after substituting $l(\tau)$ in the equation

$$\int_{0-}^{T+} l(\tau)K_x(t - \tau)\,d\tau = \sum_{j=0}^{3} \lambda_j t^j,$$

we equate the coefficients of equal exponential functions:

$$-\alpha A_1 + D_0 - \frac{1}{\alpha} D_1 + \frac{2}{\alpha^2} D_2 - \frac{6}{\alpha^3} D_3 = 0,$$

$$-\alpha B_1 + D_0 + \frac{1}{\alpha}(1 + \alpha T)D_1 + \frac{1}{\alpha^2}(2 + 2\alpha T + \alpha^2T^2)D_2 + \frac{1}{\alpha^3}(6 + 6\alpha T + 3\alpha^2T^2 + \alpha^3T^3)D_3 = 0.$$

Adding to these equations the equalities obtained by equating the moments of $l(\tau)$ and $n(\tau) = \delta^{(1)}(\tau)$:

$$A_1 + B_1 + TD_0 + \frac{1}{2}T^2D_1 + \frac{1}{3}T^3D_2 + \frac{1}{4}T^4D_3 = 0,$$

$$B_1 + \frac{1}{2}TD_0 + \frac{1}{3}T^2D_1 + \frac{1}{4}T^3D_2 + \frac{1}{5}T^4D_3 = -1,$$

$$B_1 + \frac{1}{3}TD_0 + \frac{1}{4}T^2D_1 + \frac{1}{5}T^3D_2 + \frac{1}{6}T^4D_3 = 0,$$

$$B_1 + \frac{1}{4}TD_0 + \frac{1}{5}T^2D_1 + \frac{1}{6}T^3D_2 + \frac{1}{7}T^4D_3 = 0,$$

we obtain a complete system of linear equations, which determine the required constants. Solving this system, we find:

$$D_0 = 5.948\cdot 10^{-1}, \quad D_2 = 9.618\cdot 10^{-2}, \quad A_1 = 6.138.$$
$$D_1 = -7.803\cdot 10^{-1}, \quad D_3 = -0.2896\cdot 10^{-2}, \quad B_1 = -2.582.$$

PROBLEMS

36.1 At the output of a dynamical system,

$$X(t) = U(t) + V(t)$$

emerges, where $U(t)$ is a useful signal and $V(t)$ is the noise. Find $S_x(\omega)$ if $S_u(\omega)$, $S_v(\omega)$ and $S_{uv}(\omega)$ are known.

36.2 At the output of a dynamical system designed to receive a function $Z(t) = \dot{U}(t)$, a function $X(t) = U(t) + V(t)$ emerges, where

$V(t)$ denotes the noise added in the reception of the ordinates of function $U(t)$. Find the mutual spectral density $S_{xz}(\omega)$ if $S_u(\omega)$, $S_{uv}(\omega)$ and $S_v(\omega)$ are known.

36.3 Find the transmission function $L(i\omega)$ of an optimal dynamical system designed to receive the derivative of the random function $X(t)$ during τ seconds before the last observation of the ordinate of $X(t)$ if

$$S_x(\omega) = \frac{a^2}{(\omega^2 + \alpha^2)^2}.$$

Find the variance of the error in the estimate of the velocity.

36.4 Find the transmission function $L(i\omega)$ of an optimal differentiable system if the system serves to determine the derivative of a random function $U(t)$ at time $t - \tau$ ($\tau > 0$) and, if at the output the signal is a random function $X(t)$ that is the sum of a useful signal $U(t)$ and noise $V(t)$ not related to $U(t)$. Assume that

$$S_u(\omega) = \frac{a^2}{(\omega^2 + \alpha^2)^2}, \qquad S_v(\omega) = \frac{b^2}{(\omega^2 + \beta^2)^2}.$$

36.5 Find the transmission function of an optimal filter designed to receive the present value of a useful signal if its input signal consists of the sum of the useful signal $U(t)$ and the noise signal $V(t)$; $U(t)$ and $V(t)$ are mutually uncorrelated and

$$S_u(\omega) = \frac{a^2}{\omega^2 + \alpha^2}, \qquad S_v(\omega) = \frac{b^2}{\omega^2 + \beta^2}.$$

36.6 Express the variance of the error of an optimal dynamical system in terms of the spectral densities $S_u(\omega)$, $S_v(\omega)$ and $S_{uv}(\omega)$, ($U(t)$ denotes a useful signal and $V(t)$, the noise) if the transmission function of the optimal system is $L(i\omega)$ and N is the operator that, applied to the function $U(t)$, minimizes the error in the system.

36.7 At the output of a dynamical system, designed to receive the derivative $\dot{U}(t)$, $X(t) + U(t) + V(t)$ emerges, where the noise $V(t)$ and the signal $U(t)$ are uncorrelated,

$$S_u(\omega) = \frac{\alpha^2}{(\omega^2 - 2\beta^2)^2 + 4\gamma^4}, \qquad S_v(\omega) = c^2 = \text{const}.$$

Find the optimal transmission function of the system and the variance of the error in the estimate of the derivative $\dot{U}(t)$.

36.8 Find the optimal transmission function of a dynamical system designed to receive the values of the ordinate of $U(t + \tau)$ if the input signal is represented by a random function $U(t)$,

$$S_u(\omega) = \frac{a^2}{\omega^2 + \alpha^2}, \qquad \alpha > 0, \quad \tau \geqslant 0.$$

36.9 The spectral density of the input signal is $S_x(\omega) = 1/(\omega^2 + 1)^2$ and $\tau \geqslant 0$ is the prediction time. Find the optimal transmission function of the dynamical system.

36.10 The spectral density of the input signal is

$$S_x(\omega) = \frac{a^2(\omega^2 + \alpha^2)}{\omega^4 + 2\beta^4}.$$

Find the optimal transmission function of a dynamical system designed to produce $X(t + \tau)$ and the variance of the error in the estimate of $X(t + \tau)$ for $\tau \geqslant 0$.

36.11 The input to a dynamical system consists of the sum of two uncorrelated functions, useful signal $U(t)$ and noise $V(t)$. Determine the optimal transmission function for the evaluation of the signal at time $t + \tau$ if $\tau \geqslant 0$,

$$S_u(\omega) = \frac{a^2}{\omega^2 + \alpha^2}, \qquad S_v(\omega) = \frac{b^2}{\omega^2 + \beta^2}.$$

36.12 The input to a delay filter consists of the sum of two uncorrelated functions, signal $U(t)$ and noise $V(t)$, whose correlation functions are known:

$$K_u(\tau) = \sigma_u^2 e^{-\alpha|\tau|}, \qquad K_v(\tau) = \sigma_v^2 e^{-\beta|\tau|}.$$

Find the optimal transmission function of the dynamical system and the error in filtering if the delay is τ_0 ($\tau_0 \geqslant 0$).

36.13 The spectral density of the input signal is $S_x(\omega) = \alpha^2/(\omega^4 + 4\alpha^4)$ and the prediction time is τ ($\tau \geqslant 0$). Find the optimal transmission function of the dynamical system designed for the determination of $X(t + \tau)$.

36.14 On a rolling ship it is necessary to determine a time t so that τ_0 seconds later the linear function of the heel angle $\Theta(t)$ and its derivative $n_1\Theta(t) + n_2\dot{\Theta}(t)$ (where n_1 and n_2 are known constants) will assume a prescribed value c. Find the optimal transmission function of the predictor and the variance σ_ε^2 of the error if $\bar{\theta} = 0$,

$$K_\theta(\tau) = \sigma_\theta^2 e^{-\alpha|\tau|}\left(\cos \beta\tau + \frac{\alpha}{\beta} \sin \beta|\tau|\right).$$

36.15 The coordinate of a ship moving on a rectilinear course with a constant velocity is estimated with an error $V(t)$ characterized by the correlation function

$$K_v(\tau) = \sigma_v^2 e^{-\alpha|\tau|},$$

where $\sigma_v = 25$ m. and $\alpha = 0.25$ sec.$^{-1}$.

Find the maximal accuracy attained in estimating the velocity of variation of the coordinate for the observation times $T = 20$, 40 and 240 seconds.

36.16 Under the assumptions of the preceding problem, find the maximal accuracy attained in the estimate of the velocity of variation of the ship coordinate if

$$K_v(\tau) = \sigma_v^2 e^{-\alpha|\tau|}(1 + \alpha|\tau|),$$

and all the other conditions are the same.

36.17 To estimate the present values of the angular rolling velocity $\dot{\Theta}(t)$ of a ship, one uses a dynamical system; the input to this system is the present value of the heel angle $\Theta(t)$, distorted by an error of measurement $V(t)$. Find the variance of the error $\varepsilon(t)$ in the estimate of the angular velocity if this system can be considered optimal with $\bar{\theta} = 0$, $\bar{v} = 0$, $K_v(\tau) = \sigma_v^2 e^{-\alpha_v|\tau|}$, $R_{\theta v}(\tau) \equiv 0$, $K_\theta(\tau) = \sigma_\theta^2 e^{-\alpha|\tau|}[\cos \beta\tau + (\alpha/\beta) \sin \beta|\tau|]$, $\sigma_\theta = 0.1$ rad., $\alpha = 0.1$ sec.$^{-1}$, $\beta = 0.75$ sec.$^{-1}$, $\sigma_v = 2.10^{-2}$ rad. and $\alpha_v = 0.5$ sec.$^{-1}$.

36.18 A dynamical system has been designed to determine the values of a random function $X(t)$ at time $t + \tau_0$ according to the values of the ordinates of this function during the interval $(t - T, t)$. Find the optimal transmission function of the system and the variance of the error in the determination of $X(t + \tau_0)$ if the measurements of the ordinates of function $X(t)$ are performed practically without errors:

$$X(t) = c_1 + c_2 t + U(t),$$

where c_1 and c_2 are unknown constants and $U(t)$ is a random function whose correlation function is

$$K_u(\tau) = \sigma_u^2 e^{-\alpha|\tau|}(1 + \alpha|\tau|),$$

$$\sigma_u = 1, \qquad \alpha = 0.1 \text{ sec.}^{-1}, \qquad \tau_0 = 10 \text{ sec.}, \qquad T = 40 \text{ sec}.$$

36.19 A dynamical system obtains the derivative of a random function $X(t)$ at time $t + \tau_0$. Find the optimal transmission function of the system if

$$X(t) = c_1 + c_2 t + U(t), \qquad K_u(\tau) = \sigma_u^2 e^{-\alpha|\tau|}(1 + \alpha|\tau|),$$

where c_1 and c_2 are unknown constants and the system has a "finite memory" (that is, uses only the values of $X(t)$ during the interval $(t - T, T)$), $\sigma_u = 1$, $\alpha = 0.1$ sec.$^{-1}$, $\tau_0 = 10$ sec. and $T = 40$ sec.

36.20 Find the Green's function $l(\tau)$ of an optimal dynamical system with "finite memory" T designed for the differentiation of the function $X(t) = R_1(t) + U(t)$, and find the error in the determination of $\dot{X}(t)$, where R_1 is a polynomial of first degree and

$$K_u(\tau) = \sigma_u^2 e^{-\alpha|\tau|}(1 + \alpha|\tau|).$$

36.21 For automatic control of airplanes one can use an inertial control system consisting of devices of two types: in the first case, during the operation of the system the following signal is determined:

$$u_1(t) = c_1 + c_2 t + c_3 \sin \Omega t + c_4 \cos \Omega t,$$

where c_1, c_2, c_3, c_4 are some (unknown) constants, and $\Omega = 1.25 \cdot 10^{-2}$ sec.$^{-1}$; in the second case, the signal has the form

$$u_2(t) = c_3 \sin \Omega t + c_4 \cos \Omega t.$$

Find the optimal transmission functions of the dynamical systems used for the determination of the signal in both cases if the systems have a "finite memory" T, $T = 20$ sec. and the useful input signal is distorted by an error $V(t)$,

$$K_v(\tau) = \sigma_v^2 e^{-\alpha|\tau|}\left(\cos \beta\tau + \frac{\alpha}{\beta} \sin \beta|\tau|\right), \qquad \bar{v} = 0,$$

$$\alpha = 0.5 \text{ sec.}^{-1}, \qquad \beta = 3 \text{ sec.}^{-1}, \qquad \sigma_v^2 = 4 \cdot 10^{-4}.$$

36.22 The predicting value of the random function $X(t + \tau_0)$ is $Y(t) = aX(t)$. Find the value of the constant a that minimizes the variance of the error $\varepsilon(t) = aX(t) - X(t + \tau_0)$, and the minimal value of the variance if $\bar{x} = 0$,

$$S_x(\omega) = \frac{\alpha\sigma_x^2}{\pi(\omega^2 + \alpha^2)}.$$

36.23 The predicting value of the random function $X(t + \tau)$ is the linear combination $Z(t) = aX(t) + b\dot{X}(t)$. Find the values of constants a and b that minimize the variance of the error

$$\varepsilon(t) = aX(t) + b\dot{X}(t) - X(t + \tau),$$

and the minimal variance of this error if $\bar{x} = 0$,

$$K_x(\tau) = \sigma_x^2 e^{-\alpha|\tau|}\left(\cos \beta\tau + \frac{\alpha}{\beta} \sin \beta|\tau|\right).$$

36.24 The predicting value of the random function $U(t + \tau_0)$ is $Y(t) = a[U(t) + V(t)]$, where $V(t)$ is the error in the estimate of the present value of the useful signal $U(t)$. Find the value of the constant a that minimizes the variance of

$$\varepsilon(t) = Y(t) - U(t + \tau_0) \quad \text{and} \quad \mathbf{D}[\varepsilon(t)]_{\min},$$

if

$$S_{uv}(\omega) = 0, \qquad S_u(\omega) = \frac{\alpha\sigma_u^2}{\pi(\omega^2 + \alpha^2)}, \qquad S_v(\omega) = \frac{\beta\sigma_v^2}{\pi(\omega^2 + \beta^2)},$$

$$\bar{u} = \bar{v} = 0.$$

36.25 A signal must be sent to predict the zero value of the derivative $\dot{\Theta}(t)$ by τ_0 seconds. Actually the signal is sent at the instant in which the following linear combination becomes zero:

$$Y(t) = a\Theta(t) + b\dot{\Theta}(t) + c.$$

Find the optimal values of constants a, b and c and the magnitude of the variance of $\dot{\Theta}(t + \tau_0)$ if $\bar{\theta} = 0$,

$$K_\theta(\tau) = \sigma_\theta^2 e^{-\alpha|\tau|}\left(\cos \beta\tau + \frac{\alpha}{\beta} \sin \beta|\tau|\right),$$

$\sigma_\theta = 5°$, $\beta = 0.7$ sec.$^{-1}$, $\alpha = 0.042$ sec.$^{-1}$ and $\tau_0 = 0.2$ sec.

36.26 Under the assumptions made in the preceding problem, find the optimal values of the constants a, b and c for which

$$\mathbf{D}[\Theta(t - \tau_0) - Y(t)] = \min.$$

37. THE METHOD OF ENVELOPES

Basic Formulas

Any normal stationary function $X(t)$ can be represented for $\bar{x} = 0$ as

$$X(t) = A(t) \cos \Phi(t),$$

where the random functions $A(t)$ and $\Phi(t)$ are mutually uncorrelated.

The functions $X(t)$ and $Y(t) = A(t) \sin \Phi(t)$ have a mutual correlation function that can be expressed in terms of $S_x(\omega)$ by the relation

$$R_{xy}(\tau) = 2 \int_0^\infty S_x(\omega) \sin \omega\tau \, d\omega \equiv \sigma_x^2 r(\tau),$$

where $R_{xy}(\tau)$ vanishes for $\tau = 0$. Consequently, for equal times the functions $X(t)$ and $Y(t)$ are uncorrelated and, being normal, they also are independent.

The distribution laws for the ordinates of the functions $A(t)$ and $\Phi(t)$ are uniquely defined by the correlation function $K_x(\tau) = \sigma_x^2 k(\tau)$ according to the following formulas:

the one-dimensional distribution densities,

$$f(a) = \frac{a}{\sigma_x^2} \exp\left\{-\frac{a^2}{2\sigma_x^2}\right\},$$

$$f(\varphi) = \frac{1}{2\pi}, \qquad 0 \leqslant \varphi \leqslant 2\pi;$$

the two-dimensional distribution densities,

$$f(a_1, a_2) = \frac{a_1 a_2}{\sigma_x^4 q^2} \exp\left\{-\frac{a_1^2 + a_2^2}{2\sigma_x^2 q^2}\right\} I_0\left(\frac{a_1 a_2 \sqrt{1 - q^2}}{\sigma_x^2 q^2}\right),$$

$$f(\varphi_1, \varphi_2) = \frac{q^2}{4\pi^2}\left[\frac{1}{1 - \kappa^2} + \frac{\kappa}{(1 - \kappa^2)^{3/2}}\left(\frac{\pi}{2} + \arcsin \kappa\right)\right],$$

where a_1, φ_1 and a_2, φ_2 are the values for the amplitude and the phase of the envelope at times t and $t + \tau$, $q^2 = 1 - k^2(\tau) - r^2(\tau)$, $\kappa = \kappa(\tau) = \sqrt{1 - q^2}$ $\cos(\varphi_2 - \varphi_1 - \gamma)$, $\gamma = \gamma(\tau) = \arctan[r(\tau)/k(\tau)]$ and $I_0(z)$ is the Bessel function of the first kind, of zero order and of an imaginary argument.

The preceding formulas lead to the conditional distribution laws,

$$f(a_2 \mid a_1) = \frac{a_2}{\sigma_x^2 q^2} \exp\left\{-\frac{a_2^2}{2\sigma_x^2 q^2}\right\} I_0\left(\frac{a_1 a_2 \sqrt{1 - q^2}}{\sigma_x^2 q^2}\right) \exp\left\{-\frac{a_1^2(1 - q^2)}{2\sigma_x^2 q^2}\right\},$$

$$f(\varphi_2 \mid \varphi_1) = \frac{q^2}{2\pi}\left[\frac{1}{1 - \kappa^2} + \frac{\kappa}{(1 - \kappa^2)^{3/2}}\left(\frac{\pi}{2} + \arcsin \kappa\right)\right]$$

and the formula for the correlation function,

$$K_a(\tau) = \sigma_x^2\left[2E(1 - q^2) - q^2 K(1 - q^2) - \frac{\pi}{2}\right],$$

where $K(k^2)$ and $E(k^2)$ denote the total elliptic integrals of first and second kinds:

$$K(k^2) = \int_0^{\pi/2} \frac{d\varphi}{\sqrt{1 - k^2 \sin^2 \varphi}},$$

$$E(k^2) = \int_0^{\pi/2} \sqrt{1 - k^2 \sin^2 \varphi}\, d\varphi.$$

The four-dimensional and two-dimensional distribution laws for the amplitude of the envelope, its phase and the corresponding velocities have the form:

$$f(a, \dot{a}, \varphi, \dot{\varphi}) = \frac{a^2}{4\pi^2(\omega_2^2 - \omega_1^2)\sigma_x^2} \times \exp\left\{-\frac{1}{2\sigma_x^2(\omega_2^2 - \omega_1^2)}[\dot{a}^2 + a^2(\omega_2^2 - 2\omega_1\dot{\varphi} + \dot{\varphi}^2)]\right\},$$

$$f(a, \dot{\varphi}) = \frac{a^2}{\sigma_x^2\sqrt{2\pi}\sqrt{\omega_2^2 - \omega_1^2}} \times \exp\left\{-\frac{a^2}{2\sigma_x^2(\omega_2^2 - \omega_1^2)}(\omega_2^2 - 2\omega_1\dot{\varphi} + \dot{\varphi}^2)\right\},$$

$$f(\dot{a}, \varphi) = \frac{1}{\sigma_x(2\pi)^{3/2}\sqrt{\omega_2^2 - \omega_1^2}} \exp\left\{-\frac{\dot{a}^2}{2\sigma_x^2(\omega_2^2 - \omega_1^2)}\right\},$$

$$f(a, \dot{a}) = f(a)f(\dot{a}),$$

$$f(\varphi, \dot{\varphi}) = f(\varphi)f(\dot{\varphi}),$$

$$f(\dot{a}) = \frac{1}{\sigma_x\sqrt{2\pi}\sqrt{\omega_2^2 - \omega_1^2}} \exp\left\{-\frac{\dot{a}^2}{2\sigma_x^2(\omega_2^2 - \omega_1^2)}\right\},$$

$$f(\dot{\varphi}) = \frac{\omega_2^2 - \omega_1^2}{2[(\dot{\varphi} - \omega_1)^2 + (\omega_2^2 - \omega_1^2)]^{3/2}},$$

where

$$\omega_1^2 = \frac{2}{\sigma_x^2}\int_0^\infty S_x(\omega)\omega\, d\omega,$$

$$\omega_2^2 = \frac{2}{\sigma_x^2}\int_0^\infty S_x(\omega)\omega^2\, d\omega.$$

The probability that $\dot{\varphi}$ is greater than zero is defined by

$$\mathbf{P}(\dot{\varphi} \geqslant 0) = \frac{1}{2}\left(1 + \frac{\omega_1}{\omega_2}\right).$$

Similarly,

$$\mathbf{P}(\dot{\varphi} \leqslant 0) = \frac{1}{2}\left(1 - \frac{\omega_1}{\omega_2}\right).$$

For a narrow-band spectrum of the random variable $X(t)$ the quantity $\Delta^2 = \omega_2^2 - \omega_1^2$ is small compared to ω_1^2, and some of the foregoing formulas can be simplified by expanding the corresponding expressions in powers of the small quotient Δ/ω_1. In particular, for a narrow-band spectrum the variances

$\mathbf{D}[\dot{A}(t)]$ and $\mathbf{D}[\dot{\Phi}(t)]$ become small and, since $\mathbf{M}[A(t)] = 0$, $\mathbf{M}[\Phi(t)] = \omega_1$, by differentiating the random function $X(t) = A(t)\cos\Phi(t)$ one may consider in some cases that $\dot{A}(t)$ vanishes, and replace $\dot{\Phi}(t)$ by ω_1.

In the case of a narrow-band spectrum, the probability density of the time τ during which the random function is above (below) the zero level ("the distribution law of the half-period") has the following approximate expression

$$f(\tau) \approx \frac{\pi\Delta^2\tau}{2[(\pi - \omega_1\tau)^2 + \Delta^2\tau^2]^{3/2}},$$

whose accuracy increases with the decrease of the quotient Δ/ω_1.

Solution for Typical Examples

Example 37.1 Find the average number of passages per unit time for the random function

$$\Theta(t) = \Phi(t) - \omega_1 t,$$

where $\Phi(t)$ is the phase of the normal random function $X(t)$ if

$$K_x(\tau) = \sigma_x^2 e^{-\alpha|\tau|}(1 + \alpha|\tau|),$$

$$\omega_1 = \frac{2}{\sigma_x^2}\int_0^\infty S_x(\omega)\omega\,d\omega, \qquad \alpha = 0.1\text{ sec.}^{-1}.$$

Solution. We determine the spectral density

$$S_x(\omega) = \frac{1}{2\pi}\int_{-\infty}^{\infty} K_x(\tau)e^{-i\omega\tau}\,d\tau = \frac{2\alpha^3\sigma_x^2}{\pi(\omega^2 + \alpha^2)^2}.$$

Consequently,

$$\omega_1 = \frac{4\alpha^3}{\pi}\int_0^\infty \frac{\omega\,d\omega}{(\omega^2 + \alpha^2)^2} = \frac{2\alpha}{\pi}.$$

Applying the general formula for the number of passages per unit time, we obtain

$$p = \int_0^\infty f(\theta, \dot{\theta})_{\theta=0}\mid \dot{\theta}\,d\dot{\theta}.$$

Since $\Theta(t) = \Phi(t) - \omega_1 t$, Θ has a uniform distribution law in the interval $(0, 2\pi)$, and the distribution law $f(\theta, \dot{\theta})$ can be obtained easily if we replace $\dot{\varphi}$ by $\dot{\theta} + \omega_1$ in the distribution law $f(\varphi, \dot{\varphi})$; that is,

$$f(\theta, \dot{\theta}) = \frac{\omega_2^2 - \omega_1^2}{4\pi[\dot{\theta}^2 + (\omega_2^2 - \omega_1^2)]^{3/2}},$$

where

$$\omega_2^2 = \frac{2}{\sigma_x^2}\int_0^\infty S_x(\omega)\omega^2\,d\omega = \alpha^2.$$

Setting $f(\theta, \dot{\theta})$ in the formula for p, we get

$$p = \frac{1}{4\pi}\sqrt{\omega_2^2 - \omega_1^2} = \frac{\alpha\sqrt{\pi^2 - 4}}{4\pi^2} = 0.0061\text{ sec.}^{-1}.$$

PROBLEMS

37.1 The correlation function is defined by the formula

$$K_x(\tau) = \sigma_x^2 e^{-\alpha|\tau|}.$$

Considering $X(t)$ normal ($\bar{x} = 0$), find the correlation function for the amplitude of the envelope of this function.

37.2 What is the probability that the phase of the envelope of a normal random function $X(t)$ will decrease if

$$K_x(\tau) = \sigma_x^2 e^{-\alpha|\tau|}(1 + \alpha|\tau|), \qquad \bar{x} = 0,$$
$$\alpha = 0.01;\ 0.10;\ 0.50\ \text{sec.}^{-1}?$$

37.3 For a stationary normal random function $X(t)$, find the probability that the phase will increase (decrease) if

$$K_x(\tau) = \sigma_x^2 e^{-\alpha|\tau|}\left(\cos \beta\tau + \frac{\alpha}{\beta} \sin \beta|\tau|\right).$$

37.4 Find the probability P that the velocity of variation of the phase of the envelope will be greater than

$$\omega_1 = \frac{2}{\sigma_x^2} \int_0^\infty S_x(\omega)\omega\, d\omega$$

if

$$K_x(\tau) = \sigma_x^2 e^{-\alpha|\tau|}\left(\cos \beta\tau + \frac{\alpha}{\beta} \sin \beta|\tau|\right), \qquad \bar{x} = 0.$$

37.5 For a normal random function $X(t)$, find the distribution law for the velocity of variation of the phase if

$$K_x(\tau) = \sigma_x^2 e^{-\alpha|\tau|}(1 + \alpha|\tau|), \qquad \bar{x} = 0.$$

37.6 Find the distribution law for the phase of a normal random function $X(t) - \bar{x}$, for which

$$K_x(\tau) = \sigma_x^2 e^{-\alpha|\tau|}.$$

37.7 Find the distribution law for the velocity of phase variation of a normal random function $X(t)$ with spectral density

$$S_x(\omega) = \frac{c^2}{(\omega^2 - \alpha^2 - \beta^2)^2 + 4\alpha^2\omega^2}, \qquad \bar{x} = 0.$$

37.8 Find the distribution law for the envelope and the velocity of variation of the envelope of a normal random function $X(t)$ if

$$S_x(\omega) = \frac{2\alpha^3\sigma_x^2}{\pi(\omega^2 + \alpha^2)^2}, \qquad \bar{x} = 0.$$

37.9 Under the assumptions made in the preceding problem, find the conditional distribution law of the envelope at time $t + \tau$ if at time t

$$A(t) = \sigma_x, \qquad \tau = 2\ \text{sec.}, \qquad \alpha = 0.1\ \text{sec.}^{-1}.$$

37.10 Find an approximate expression for the distribution law of the time during which a random function is below the zero level if

$$K_x(\tau) = \sigma_x^2 e^{-0.01|\tau|}\left(\cos 0.7\tau + \frac{1}{70}\sin 0.7|\tau|\right), \qquad \bar{x} = 0.$$

37.11 Assuming that the formulas for the envelope of a random function with a narrow-band spectrum are applicable, find the distribution law for the intervals between successive moments during which the deck of a ship passes through equilibrium if the heel angle $\Theta(t)$ is a normal random function, whose correlation function

$$K_\theta(\tau) = \sigma_\theta^2 e^{-0.1|\tau|}\left(\cos 0.7\tau + \frac{1}{7}\sin 0.7|\tau|\right), \qquad \bar{\theta} = 0,$$

and there is no pitching.

37.12 Find the average number of passages beyond the level $2\sigma_x$ per unit time for a random function $A(t)$ if $A(t)$ is the envelope of the normal random function $X(t)$ and

$$K_x(\tau) = \sigma_x^2 e^{-\alpha|\tau|}(1 + \alpha|\tau|), \qquad \bar{x} = 0.$$

37.13 Find the average number of passages beyond the level $2\sigma_x$ for the amplitude of the envelope of a normal stochastic process $X(t)$ if

$$K_x(\tau) = \sigma_x^2 e^{-\alpha|\tau|}\left(1 + \alpha|\tau| + \frac{1}{3}\alpha^2\tau^2\right), \qquad \bar{x} = 0.$$

37.14 Find the conditional distribution law for the phase of a normal function $X(t)$ at time $t + \tau$ if at time t, the phase is zero and

$$K_x(\tau) = \sigma_x^2 e^{-\alpha|\tau|}\left(\cos \beta\tau + \frac{\alpha}{\beta}\sin \beta|\tau|\right), \qquad \bar{x} = 0.$$

Neglecting the variance of the amplitude of the envelope, determine the variance of $X(t)$ at time $(t + \pi/\omega_1)$, where

$$\omega_1 = \frac{2}{\sigma_x^2}\int_0^\infty S_x(\omega)\omega\, d\omega, \qquad \alpha = 0.01 \text{ sec.}^{-1}, \qquad \beta = 0.70 \text{ sec.}^{-1}.$$

37.15 Find the mutual correlation function for two normal stationary random functions $X(t)$ and $Y(t)$ if

$$X(t) = A(t)\cos\Phi(t), \qquad Y(t) = A(t)\sin\Phi(t),$$

$$K_x(\tau) = \sigma_x^2 e^{-0.01|\tau|}\left(\cos 0.7\tau + \frac{1}{70}\sin 0.7|\tau|\right), \qquad \bar{x} = 0.$$

VIII MARKOV PROCESSES

38. MARKOV CHAINS

Basic Formulas

Let S be a finite sample space consisting of outcomes $Q_1, Q_2, \ldots, Q_m$. A sequence of trials of the underlying experiment is called a finite Markov chain if $p_{ij}(k)$, the conditional probability at the kth trial of Q_j under the assumption that Q_i occurred at the $(k-1)$st trial, is independent of the outcomes at the $(k-2)$nd, $(k-3)$rd, ... trials. The events $Q_1, Q_2, \ldots, Q_m$ are called states of the Markov chain and the kth trial can be considered as the change of state at time t_k.

In each column of matrix $\mathscr{P}_k = \|p_{ij}(k)\|$, there is at least one element different from zero, and the transition probabilities $p_{ij}(k)$ $(i, j = 1, 2, \ldots, m)$ for any k satisfy the relation

$$\sum_{j=1}^{m} p_{ij}(k) = 1 \qquad (i = 1, 2, \ldots, m).$$

A Markov chain is called irreducible if any state can be reached from any other state and periodic if the return to any state can be made through a number of steps, which are a multiple of some $\kappa > 1$.

A Markov chain is called homogeneous if the transition probabilities $p_{ij}(k)$ are independent of k; that is, $p_{ij}(k) = p_{ij}$ $(i, j = 1, 2, \ldots, m)$.

The column $p(n) = \{p_1(n); p_2(n); \ldots; p_m(n)\}$, which is formed of the unconditional probabilities that at the nth trial the system will pass respectively to states $Q_1, Q_2, \ldots, Q_m$, is defined by the formula

$$p(n) = (\mathscr{P}_1\mathscr{P}_2\cdots\mathscr{P}_n)'p(0),$$

and for a homogeneous chain by

$$p(n) = (\mathscr{P}')^n p(0),$$

where the accent means transposed matrix; that is, if $\mathscr{P} = \|p_{ij}\|$, then $\mathscr{P}' = \|p_{ji}\|$.

For any n, but relatively small m, to calculate $\mathscr{P}^n$ we can use the Lagrange-Sylvester formula, which in the case of simple eigenvalues $\lambda_1, \lambda_2, \ldots, \lambda_m$ (roots of the equation $|\lambda\mathscr{E} - \mathscr{P}| = 0$, where $\mathscr{E}$ is the unit matrix) has the form

$$\mathscr{P}^n = \sum_{k=1}^{m} \frac{(\mathscr{P} - \lambda_1\mathscr{E})\cdots(\mathscr{P} - \lambda_{k-1}\mathscr{E})(\mathscr{P} - \lambda_{k+1}\mathscr{E})\cdots(\mathscr{P} - \lambda_m\mathscr{E})}{(\lambda_k - \lambda_1)\cdots(\lambda_k - \lambda_{k-1})(\lambda_k - \lambda_{k+1})\cdots(\lambda_k - \lambda_m)} \lambda_k^n.$$

In the general case, for finding $\mathscr{P}^n$ it is convenient to reduce $\mathscr{P}$ to normal form $\mathscr{P} = HJH^{-1}$, where J is a diagonal or a quasidiagonal matrix, depending only on the eigenvalues of matrix $\mathscr{P}$. For simple eigenvalues, $J = \|\delta_{ik}\lambda_k\|$, where $\delta_{ik} = 0$ for $i \neq k$ and $\delta_{kk} = 1$. The elements of matrices H and H^{-1} are the solutions of algebraic equations of the form $\mathscr{P}H = HJ$, $H^{-1}\mathscr{P} = JH^{-1}$, $HH^{-1} = \mathscr{E}$. Then $\mathscr{P}^n = HJ^nH^{-1}$, where for simple eigenvalues $J^n = \|\delta_{ik}\lambda_k^n\|$.

The elements $p_{ij}^{(n)}$ of matrix $\mathscr{P}^n$ are also determined by the Perron formula

$$p_{ij}^{(n)} = \sum_{s=1}^{r} \frac{1}{(\nu_s - 1)!} \left\{ \frac{d^{\nu_s - 1}}{d\lambda^{\nu_s - 1}} \left[\frac{\lambda^n A_{ji}(\lambda)(\lambda - \lambda_s)^{\nu_s}}{|\lambda\mathscr{E} - \mathscr{P}|} \right] \right\}_{\lambda = \lambda_s},$$

where r is the number of distinct eigenvalues, ν_s is their multiplicity ($\sum_{s=1}^{r} \nu_s = m$) and $A_{ji}(\lambda)$ is the cofactor of the element $\lambda\delta_{ji} - p_{ji}$ in the determinant $|\lambda\mathscr{E} - \mathscr{P}|$. The matrix $\mathscr{P}^{\infty} = \|p_{ij}^{(\infty)}\|$ of the limiting transition probabilities and the column $p(\infty) = (\mathscr{P}^{\infty})p(0)$ of the limiting unconditional probabilities can be obtained from the corresponding expression by passage to the limit for $n \to \infty$. The limits exist only if $|\lambda_s| < 1$ for $s = 2, 3, \ldots, r$ (for the transition probability matrices $|\lambda_s| \leqslant 1$ always obtains and one eigenvalue λ_1 equals unity). For this,

$$\mathscr{P}^{\infty} = \nu_1 \left\{ \frac{\dfrac{d^{\nu_1 - 1}}{d\lambda^{\nu_1 - 1}} [(\lambda\mathscr{E} - \mathscr{P})^{-1} |\lambda\mathscr{E} - \mathscr{P}|]}{\dfrac{d^{\nu_1}}{d\lambda^{\nu_1}} |\lambda\mathscr{E} - \mathscr{P}|} \right\}_{\lambda = 1},$$

where ν_1 is the multiplicity of the eigenvalue $\lambda_1 = 1$.

For $\nu_1 = 1$, all m rows of matrix $\mathscr{P}^{\infty}$ are equal and the elements of the column $p(\infty)$ coincide with the corresponding elements of any row; that is,

$$p_j(\infty) = p_{ij}^{(\infty)} = p_j^{(\infty)} \qquad (i, j = 1, 2, \ldots, m).$$

In this case, the probabilities $p_j^{(\infty)}$ can also be determined from the solution of the algebraic system

$$\sum_{i=1}^{m} p_{ij} p_i^{(\infty)} = p_j^{(\infty)} \qquad (j = 1, 2, \ldots, m), \text{ where } \sum_{j=1}^{m} p_j^{(\infty)} = 1.$$

If the finite Markov chain is irreducible and nonperiodic, then to find the probabilities $p_j^{(\infty)}$ one can use the last equations. If the number of states $m = \infty$, the Markov chain is irreducible and nonperiodic and the system of linear equations $\sum_{i=1}^{\infty} u_i p_{ij} = u_j$ $(j = 1, 2, \ldots)$ has a nontrivial solution for which $\sum_{i=1}^{\infty} |u_i| < \infty$, and probabilities $p_j^{(\infty)}(p_j^{(\infty)} > 0, j = 1, 2, \ldots)$ are the solutions of the system $\sum_{i=1}^{\infty} p_{ij} p_i^{(\infty)} = p_j^{(\infty)}$ $(j = 1, 2, \ldots)$, where $\sum_{j=1}^{\infty} p_j^{(\infty)} = 1$.

If one can separate a group of states of the system so that a transition from any state of this group to any of the remaining states is impossible, the group can be considered an independent Markov chain. A group may consist of one state Q_k so that $p_{kk} = 1$; Q_k is called an absorbing state.

In the general case, from the states $Q_1, Q_2, \ldots, Q_m$ one can select mutually disjoint groups $C_1, C_2, \ldots, C_h$ called essential states; the rest form a group T

of inessential states. For a proper numbering of states, the matrix $\mathscr{P}$ is reduced to the form

$$\mathscr{P} = \left\| \begin{matrix} R_1 & 0 & \cdots & 0 & \cdots & 0 \\ 0 & R_2 & \cdots & 0 & \cdots & 0 \\ . & . & . & . & . & . \\ 0 & 0 & \cdots & R_h & \cdots & 0 \\ & U & & & W & \end{matrix} \right\|,$$

where $R_1, R_2, \ldots, R_h$ are the matrices of transition probabilities of the groups $C_1, C_2, \ldots, C_h$, W is a square matrix associated with the inessential states of group T, and U is a nonzero (if there are inessential states) not necessarily square matrix.

If all the eigenvalues of matrices $R_1, R_2, \ldots, R_h$, except those equal to unity, are less than unity in absolute value, then

$$\mathscr{P}^\infty = \left\| \begin{matrix} R_1^\infty & 0 & \cdots & 0 & \cdots & 0 \\ 0 & R_2^\infty & \cdots & 0 & \cdots & 0 \\ . & . & . & . & . & . \\ 0 & 0 & \cdots & R_h^\infty & \cdots & 0 \\ & U_\infty & & & 0 & \end{matrix} \right\|,$$

where U_∞ is some rectangular matrix.

Let $h = 1$ in the matrix $\mathscr{P}$; i.e., there is one group C of absorbing states. If the Markov chain formed from the states of this group is nonperiodic, the probabilities p_{*j} of transition from an inessential state Q_j to the group C of essential states is determined from the equation

$$p_{*j} = \sum_T p_{j\nu} p_{*\nu} + \sum_C p_{jr},$$

where in the first term the summation is extended over inessential states and in the second over the essential states.

Let κ_j $(j = 1, 2, \ldots, h)$ be the number of eigenvalues (considering their multiplicity) of the matrix R_j that are not exactly equal to unity, but equal in modulus to unity. The minimal common multiplicity of these eigenvalues is the period κ of the Markov chain. If the chain is irreducible, all states of the periodic chain can be divided into groups $G_0, G_1, \ldots, G_{\kappa-1}$ so that a transition from a state contained in G_r always leads in one step to a state in G_{r+1} $(G_\kappa = G_0)$. In the Markov chain with matrix $\mathscr{P}^\kappa$, each group G_r can be considered an independent chain. The following limits for $r = 0, 1, \ldots, \kappa - 1$ exist:

$$\lim_{n\to\infty} p_{ik}^{(n\kappa+r)} = \begin{cases} p_{k\kappa'} & \text{if } Q_j \text{ is from } G_\nu, \text{ and } Q_k \text{ is from } G_{\nu+r}, \\ 0 & \text{otherwise;} \end{cases}$$

the probabilities $p_{k,\kappa}$ are determined as in the case $\kappa = 0$.

In the general case, there also exists a matrix $(\mathscr{P}^\kappa)^\infty$ and matrices

$\mathscr{P}_r = \mathscr{P}^r(\mathscr{P}^\kappa)^\infty$ $(r = 0, 1, \ldots, \kappa - 1)$. The matrix $\hat{\mathscr{P}} = \|\hat{p}_{ij}\|$ of mean limiting transition probabilities is defined by the formula

$$\hat{\mathscr{P}} = \frac{1}{\kappa}(\mathscr{E} + \mathscr{P} + \cdots + \mathscr{P}^{\kappa-1})(\mathscr{P}^\kappa)^\infty.$$

The column $\hat{p}$ of mean limiting unconditional probabilities is given by $\hat{p} = \hat{\mathscr{P}}'p(0)$. If $h = 1$ in the matrix $\mathscr{P}$, then the mean limiting unconditional probabilities $\hat{p}_j$ $(j = 1, 2, \ldots, m)$ are uniquely defined by the equalities: $\mathscr{P}'\hat{p} = \hat{p}$, $\sum_{j=1}^m \hat{p}_j = 1$.

Solution for Typical Examples

Example 38.1 Some numbers are selected at random from a table of random numbers containing integers 1 to m inclusive. The system is in state Q_j if the largest of the selected numbers is j $(j = 1, 2, \ldots, m)$. Find the probabilities $p_{ik}^{(n)}$ $(i, k = 1, 2, \ldots, m)$ that after selecting n random numbers from this table, the largest number will be k if before it was i.

Solution. Any integer 1 to m appears equally probable in the table of random numbers and, thus, any transition from state Q_1 (the largest selected number is 1) to any state Q_j is equally probable. Then $p_{1j} = 1/m$ $(j = 1, 2, \ldots, m)$. The transition from Q_2 to Q_1 is impossible and, consequently, $p_{21} = 0$. The system can remain in state Q_2 in two cases: if the selected number is 1 or 2 and, consequently, $p_{22} = 2/m$, $p_{2j} = 1/m$ $(j = 3, 4, \ldots, m)$. In the general case, we find

$$p_{ii} = \frac{i}{m}; \qquad p_{ij} = 0 \quad \text{for} \quad i > j; \qquad p_{ij} = \frac{1}{m} \quad \text{for} \quad i < j$$
$$(i, j = 1, 2, \ldots, m).$$

The matrix of transition probabilities can be written as

$$\mathscr{P} = \begin{Vmatrix} \frac{1}{m} & \frac{1}{m} & \frac{1}{m} & \cdots & \frac{1}{m} & \frac{1}{m} \\ 0 & \frac{2}{m} & \frac{1}{m} & \cdots & \frac{1}{m} & \frac{1}{m} \\ 0 & 0 & \frac{3}{m} & \cdots & \frac{1}{m} & \frac{1}{m} \\ \cdot & \cdot & \cdot & \cdot & \cdot & \cdot \\ 0 & 0 & 0 & \cdots & \frac{m-1}{m} & \frac{1}{m} \\ 0 & 0 & 0 & \cdots & 0 & 1 \end{Vmatrix}.$$

The characteristic equation

$$|\lambda\mathscr{E} - \mathscr{P}| \equiv \prod_{k=1}^{m}\left(\lambda - \frac{k}{m}\right) = 0$$

has roots $\lambda_k = k/m$ ($k = 1, 2, \ldots, m$). To find the probabilities $p_{ik}^{(n)}$, representing the elements of the matrix $\mathscr{P}^n$, let us apply Perron's formula. The cofactors $A_{ki}(\lambda)$ of the elements of the determinant $|\lambda\mathscr{E} - \mathscr{P}|$ are the following:

$$\text{for } i > k \quad A_{ki}(\lambda) = 0; \qquad A_{kk}(\lambda) = \frac{|\lambda\mathscr{E} - \mathscr{P}|}{\lambda - \dfrac{k}{m}};$$

$$\text{for } i < k \quad A_{ki}(\lambda) = \frac{1}{m}\prod_{\nu=1}^{k-2}\left(\lambda - \frac{\nu}{m}\right)\prod_{r=k+1}^{m}\left(\lambda - \frac{r}{m}\right) = \frac{|\lambda\mathscr{E} - \mathscr{P}|}{m\left(\lambda - \dfrac{k}{m}\right)\left(\lambda - \dfrac{k-1}{m}\right)}.$$

Substituting these expressions in Perron's formula, we obtain

$$p_{ik}^{(n)} = \begin{cases} 0 & \text{for } i > k, \\ \left(\dfrac{k}{m}\right)^n & \text{for } i = k, \\ \left(\dfrac{k}{m}\right)^n - \left(\dfrac{k-1}{m}\right)^n & \text{for } i < k. \end{cases}$$

In a similar way one can solve Problems 38.3 to 38.10.

Example 38.2 A vending machine that sells tokens in a subway station can be operated with nickels and dimes. If a nickel is inserted, the machine releases one token if the container, which can hold m nickels, is not full; otherwise the machine releases no token. If a dime is inserted, the machine releases one token and a nickel change if there is at least one nickel in the container; if not, the machine turns off. One knows that a nickel and a dime are inserted with probabilities p and $q = p - 1$. Find the probabilities $p_{ik}^{(n)}$ ($i, k = 0, 1, \ldots, m$) that after n demands for tokens, the machine will contain k nickels if initially it held i nickels.

SOLUTION. Let the state Q_j mean that the container has j nickels ($j = 0, 1, \ldots, m$). For $1 \leqslant j \leqslant m - 1$, a transition from Q_j to Q_{j+1} is possible with probability p and to Q_{j-1} with probability q. When the states Q_0 or Q_m, representing absorbing states, are reached, the machine turns off. Therefore,

$$p_{00} = 1,\ p_{mm} = 1,\ p_{j,j+1} = p,\ p_{j,j-1} = q \qquad (j = 1, 2, \ldots, m - 1).$$

The matrix of transition probabilities has the form

$$\mathscr{P} = \begin{Vmatrix} 1 & 0 & 0 & 0 & \cdots & 0 & 0 & 0 \\ q & 0 & p & 0 & \cdots & 0 & 0 & 0 \\ 0 & q & 0 & p & \cdots & 0 & 0 & 0 \\ \cdot & \cdot & \cdot & \cdot & \cdot & \cdot & \cdot & \cdot \\ 0 & 0 & 0 & 0 & \cdots & q & 0 & p \\ 0 & 0 & 0 & 0 & \cdots & 0 & 0 & 1 \end{Vmatrix} = \begin{Vmatrix} 1 & 0 & 0 \\ U & W & V \\ 0 & 0 & 1 \end{Vmatrix},$$

where W is a square matrix of order $m - 1$, and U and V are two columns of order $m - 1$;

$$W = \begin{Vmatrix} 0 & p & 0 & \cdots & 0 & 0 \\ q & 0 & p & \cdots & 0 & 0 \\ 0 & q & 0 & \cdots & 0 & 0 \\ \cdot & \cdot & \cdot & \cdot & \cdot & \cdot \\ 0 & 0 & 0 & \cdots & 0 & p \\ 0 & 0 & 0 & \cdots & q & 0 \end{Vmatrix}, \qquad U = \begin{Vmatrix} q \\ 0 \\ 0 \\ \cdot\ \cdot\ \cdot \\ 0 \\ 0 \end{Vmatrix}, \qquad V = \begin{Vmatrix} 0 \\ 0 \\ 0 \\ \cdot\ \cdot\ \cdot \\ 0 \\ p \end{Vmatrix},$$

where the matrix W is associated with the inessential states $Q_1, Q_2, \ldots, Q_{m-1}$.

The required probabilities are the elements of the matrix

$$\mathscr{P}^n = \begin{Vmatrix} 1 & 0 & 0 \\ U_n & W^n & V_n \\ 0 & 0 & 1 \end{Vmatrix},$$

and, consequently,

$$p_{00}^{(n)} = 1, \; p_{mn}^{(n)} = 1, \; p_{0r}^{(n)} = 0 \quad (r = 1, 2, \ldots, m),$$

$$p_{mj}^{(n)} = 0 \quad (j = 0, 1, \ldots, m - 1).$$

To find the elements of matrix W^n, form the characteristic equation $\Delta_{m-1} = |\lambda\mathscr{E} - W| = 0$. For determinants of this type there obtains the following recursion relation:

$$\Delta_k = \lambda\Delta_{k-1} - pq\Delta_{k-2} \qquad (k = 2, 3, \ldots, m - 1),$$

with $\Delta_0 = 1, \Delta_1 = \lambda$. Then,

$$\Delta^{m-1} = \lambda^{m-1} - C_{m-2}^{1}pq\lambda^{m-3} + C_{m-3}^{2}(pq)^2\lambda^{m-5} - \cdots = 0.$$

The last term of the equation is $(-1)^{(m-1)/2}(pq)^{(m-1)/2}$ for odd m and $(-1)^{(m-2)/2}(m/2)(pq)^{(m-2)/2}$ for even m.

Making the substitution $\lambda = \sqrt{pq}\,[\mu + (1/\mu)]$, we can write the equation $\Delta_{m-1} = 0$ in the form

$$\Delta_{m-1} = \left(\frac{pq}{\mu^2}\right)^{(m-1)/2} \frac{1 - \mu^{2m}}{1 - \mu^2}.$$

From this it follows that $\mu_k = \exp\{i(k\pi/m)\}$ $(k = 1, 2, \ldots, m - 1)$. Therefore the eigenvalues will be $\lambda_k = 2\sqrt{pq}\cos(k\pi/m)$ $(k = 1, 2, \ldots, m - 1)$.

The matrix W can be reduced to the form $W = HJH^{-1}$, where $J = \|2\sqrt{pq}\cos(k\pi/m)\,\delta_{jk}\|$, and $H = \|h_{jk}\|$ is to be determined.

The matrix equation $WH = HJ$ is equivalent to the following equations:

$$ph_{2,k} = h_{1,k}\lambda_k, \qquad qh_{m-2,k} = h_{m-1,k}\lambda_k, \qquad qh_{j-1,k} + ph_{j+1,k} = h_{j,k}\lambda_k$$

$$(j = 2, 3, \cdots, m - 2; \quad k = 1, 2, \ldots, m - 1).$$

Up to a factor, the solutions of this system are the elements

$$h_{jk} = \left(\frac{q}{p}\right)^{j/2} \sin\frac{kj\pi}{m} \qquad (k, j = 1, 2, \ldots, m-1).$$

Thus, $H = \|(q/p)^{j/2} \sin (kj\pi/m)\|$.

The inverse matrix H^{-1} can be written in the form

$$H^{-1} = \|h_{jk}^{(-1)}\| = \left\| C_k \left(\frac{p}{q}\right)^{k/2} \sin\frac{jk\pi}{m} \right\|.$$

From $HH^{-1} = \mathscr{E}$, we find $C_k = 2/m$ $(k = 1, 2, \ldots, m-1)$. Using the equality $W^n = HJ^nH^{-1}$, we obtain

$$\begin{aligned} p_{ik}^{(n)} &= \sum_{j=1}^{m-1} h_{ij}\lambda_j^n h_{jk}^{(-1)} \\ &= \frac{2^{n+1}}{m} p^{(n+k-i)/2} q^{(n-k+i)/2} \sum_{j=1}^{m-1} \cos\frac{j\pi}{m} \sin\frac{ij\pi}{m} \sin\frac{jk\pi}{m} \end{aligned}$$

$$(i, k = 1, 2, \ldots, m-1).$$

To determine the elements $p_{j0}^{(n)}$ $(j = 1, 2, \ldots, m-1)$ of the column U_n, we shall use Perron's formula. The characteristic polynomial of the matrix $\mathscr{P}$ will be $|\lambda\mathscr{E} - \mathscr{P}| = (\lambda - 1)^2 \prod_{k=1}^{m-1} (\lambda - \lambda_k)$. For the cofactors of $A_{0j}(\lambda)$ of the elements of the determinant $|\lambda\mathscr{E} - \mathscr{P}|$, we get the following expressions:

$$A_{0j}(\lambda) = q^j(\lambda - 1) \prod_{k=1}^{m-j-1} \left(\lambda - 2\sqrt{pq}\cos\frac{k\pi}{m-j}\right) \quad (j = 1, 2, \ldots, m-2),$$

$$A_{0,m-1}(\lambda) = q^{m-1}(\lambda - 1).$$

Then,

$$p_{j0}^{(n)} = \frac{A'_{0j}(1)}{\prod_{k=1}^{m-1} (1 - \lambda_k)} + \sum_{k=1}^{m-1} \frac{\lambda_k^n A_{0j}(\lambda_k)}{(\lambda_k - 1)^2 \prod_{\nu=1}^{m-1 *} (\lambda_k - \lambda_\nu)},$$

where the asterisk means that the factor with $k = \nu$ must be eliminated from the product.

The probabilities $p_{jm}^{(n)}$ $(j = 1, 2, \ldots, m-1)$ can be calculated similarly. To evaluate them, we can also use the equalities

$$p_{jm}^{(n)} = 1 - \sum_{k=0}^{m-1} p_{jk}^{(n)} \qquad (j = 1, 2, \ldots, m-1).$$

Problems 38.11 to 38.14 may be solved similarly.

Example 38.3 A substance is irradiated by a stream of radioactive elements during equal time intervals Δt. The probability that during irradiation the substance will absorb r radioactive particles is determined by the formula $\beta_r = a^r/r!\, e^{-a}$. Each radioactive particle contained in the substance may decay during two successive irradiations with probability q. Find the limiting probabilities for the number of particles in the substance.

SOLUTION. Let state Q_i mean that after an irradiation the substance will contain i $(i = 0, 1, \ldots)$ radioactive particles. During the interval Δt the transition from Q_i to Q_k will occur if $i - \nu$ particles $(\nu = 0, 1, \ldots, i)$ decay and

$k - \nu$ $(k \geqslant \nu)$ are absorbed by the substance. The transition probabilities are

$$p_{ik} = \sum_{\nu=0}^{i(k)} C_i^\nu p^\nu q^{i-\nu} \frac{a^{k-\nu}}{(k-\nu)!} e^{-a} \qquad (i, k = 0, 1, \ldots),$$

where $p = 1 - q$, and summation is extended up to i if $i \leqslant k$, and up to k if $k < i$.

The substance can contain any number of particles; i.e., all the states of the system are attainable. Therefore the Markov chain is irreducible. Since probabilities p_{ii} are different from zero, the chain is nonperiodic.

Let us consider the system of equations

$$\sum_{i=0}^{\infty} u_i p_{ij} = u_j \qquad (j = 0, 1, \ldots).$$

We set

$$G(z) = \sum_{j=0}^{\infty} u_j z^j,$$

and multiply both sides of the system by z^j, sum over j from 0 to ∞, and then apply the formula $n - 1$ times. Hence,

$$\begin{aligned} G(z) &= \exp\{a(z-1)\}G[1 + (z-1)p] \\ &= \exp\{a(z-1)(1 + p + p^2 + \cdots + p^{n-1})\}G[1 + (z-1)p^n]. \end{aligned}$$

From this, we find that

$$G(z) = \exp\left\{\frac{a}{q}(z-1)\right\}G(1) = \exp\left\{-\frac{a}{q}\right\}G(1) \sum_{j=0}^{\infty} \frac{\left(\frac{a}{q}z\right)^j}{j!}.$$

Comparing the two expressions for $G(z)$, we obtain

$$u_j = \exp\left\{-\frac{a}{q}\right\}G(1)\frac{\left(\frac{a}{q}\right)^j}{j!} \qquad (j = 0, 1, \ldots).$$

Since $\sum_{j=0}^{\infty} |u_j| = G(1)$ and the arbitrary constant $G(1)$ can be taken different from zero and infinity, the algebraic system has a nontrivial solution and the series $\sum_{j=0}^{\infty} |u_j|$ is convergent. Consequently, $p_i^{(\infty)}$ can be determined from the system $\sum_{i=0}^{\infty} p_{ij} p_i^{(\infty)} = p_j^{(\infty)}$ $(j = 0, 1, \ldots)$. The system for $p_i^{(\infty)}$ is similar to the preceding system solved for u_j and, therefore,

$$p_j^{(\infty)} = \exp\left\{-\frac{a}{q}\right\}G(1)\frac{\left(\frac{a}{q}\right)^j}{j!} \qquad (j = 0, 1, \ldots).$$

Since $\sum_{j=0}^{\infty} p_j^{(\infty)} = 1$, $G(1) = 1$ and, thus, the required probabilities are

$$p_j^{(\infty)} = \frac{\left(\frac{a}{q}\right)^j}{j!} \exp\left\{-\frac{a}{q}\right\} \qquad (j = 0, 1, \ldots).$$

One can solve Problems 38.16 to 38.22 in a similar way.

Example 38.4 The number X of defective items in each independent sample of size N selected from an infinitely large lot obeys a binomial distribution law; that is, $\mathbf{P}(X = k) = p_k = C_N^k p^k q^{N-k}$ $(k = 0, 1, \ldots, N)$, $q = 1 - p$. If a sample contains r defective items, then according to the acceptance criteria, one considers the lot as changing its preceding state Q_ν to $Q_{\nu+r-1}$. The lot is rejected if $\nu + r - 1 \geqslant m$ and accepted if $\nu + r - 1 = 0$. Find the probability that the lot will be accepted if its initial state is Q_j $(j = 1, 2, \ldots, m - 1)$.

SOLUTION. There are $m + 1$ states Q_i $(i = 0, 1, \ldots, m - 1)$ possible. If the state Q_0 is reached, the lot is accepted; if Q_m is reached, it is rejected. Since these two are absorbing states, $p_{00} = 1$, $p_{mm} = 1$. If $i \neq 0$ and $i \neq m$, $p_{i,i+j-1} = p_j$ $(j = 0, 1, \ldots, m - i)$, $p_{im} = 1 - \sum_{j=0}^{m-i} p_j$ $(i = 1, 2, \ldots, m - 1)$.

The matrix of transition probabilities is

$$\mathscr{P} = \left\| \begin{matrix} 1 & 0 & 0 & 0 & \cdots & 0 & 0 & 0 \\ p_0 & p_1 & p_2 & p_3 & \cdots & p_{m-2} & p_{m-1} & p_{1,m} \\ 0 & p_0 & p_1 & p_2 & \cdots & p_{m-3} & p_{m-2} & p_{2,m} \\ 0 & 0 & p_0 & p_1 & \cdots & p_{m-4} & p_{m-3} & p_{3,m} \\ \cdot & \cdot & \cdot & \cdot & \cdot & \cdot & \cdot & \cdot \\ 0 & 0 & 0 & 0 & \cdots & p_1 & p_2 & p_{m-2,m} \\ 0 & 0 & 0 & 0 & \cdots & p_0 & p_1 & p_{m-1,m} \\ 0 & 0 & 0 & 0 & \cdots & 0 & 0 & 1 \end{matrix} \right\|.$$

The required probabilities p_{*j} $(j = 1, 2, \ldots, m - 1)$ are the probabilities of transition from inessential states $Q_1, Q_2, \ldots, Q_{m-1}$ to the essential state Q_0 and can be determined from the algebraic system

$$p_{*j} = \sum_{\nu=1}^{m-1} p_{j\nu} p_{*\nu} + p_{j0} \qquad (j = 1, 2, \ldots, m - 1),$$

which can be written in the form

$$(p_1 - 1)p_{*1} + \sum_{k=2}^{m-1} p_k p_{*k} = -p_0,$$

$$p_0 p_{*,r-1} + (p_1 - 1)p_{*r} + \sum_{k=r+1}^{m-1} p_{k-r+1} p_{*k} = 0 \qquad (r = 2, 3, \ldots, m - 1).$$

The determinant Δ_{m-1} of this system can be found by the recursion formula

$$\Delta_{m-r} = (p_1 - 1)\Delta_{m-r-1} - \sum_{j=2}^{m-r} (-1)^j p_j p_0^{j-1} \Delta_{m-r-j}$$

$$(r = 1, 2, \ldots, m - 1),$$

where $\Delta_0 = 1$. The required probabilities are determined by the equations

$$p_{*j} = (-1)^j p_0^j \frac{\Delta_{m-j-1}}{\Delta_{m-1}} \qquad (j = 1, 2, \ldots, m - 1).$$

Problems 38.23 to 38.25 can be solved in a similar way.

Example 38.5 A truck transports goods among $2m$ points located on a circular route. These goods are carried only from one point to the next with probability p, or to the preceding point with probability $q = 1 - p$. Find the probabilities $p_{jk}^{(n)}$ $(j, k = 1, 2, \ldots, 2m)$ that after n transports the truck will pass from the jth point to the kth point. Evaluate these probabilities for $n \to \infty$ and compute the mean limiting probabilities of transition.

SOLUTION. Let state Q_j $(j = 1, 2, \ldots, 2m)$ mean that the truck is at the kth point. The transition probabilities are

$$p_{j,j+1} = p \qquad (j = 1, 2, \ldots, 2m - 1),$$

$$p_{j,j-1} = q \qquad (j = 2, 3, \ldots, 2m - 1),$$

$$p_{2m,1} = p, \qquad p_{1,2m} = q.$$

The matrix of transition probabilities is

$$\mathscr{P} = \left\| \begin{array}{cccccccc} 0 & p & 0 & 0 & \cdots & 0 & 0 & q \\ q & 0 & p & 0 & \cdots & 0 & 0 & 0 \\ 0 & q & 0 & p & \cdots & 0 & 0 & 0 \\ . & . & . & . & . & . & . & . \\ 0 & 0 & 0 & 0 & \cdots & 0 & p & 0 \\ 0 & 0 & 0 & 0 & \cdots & q & 0 & p \\ p & 0 & 0 & 0 & \cdots & 0 & q & 0 \end{array} \right\|.$$

Let us introduce the matrix $H = \|h_{jk}\| = \|\varepsilon^{(j-1)(k-1)}\|$ of order $2m$, in which $\varepsilon = e^{\pi i/m}$. By direct multiplication, we find that $\mathscr{P}H = H\|(p\varepsilon^{k-1} + q\varepsilon^{-(k-1)})\delta_{jk}\|$ and, consequently, the eigenvalues of $\mathscr{P}$ will be $\lambda_k = p\varepsilon^{k-1} + q\varepsilon^{-(k-1)}$ $(k = 1, 2, \ldots, 2m)$.

The eigenvalues with maximal absolute value are $\lambda_1 = 1$ and $\lambda_{m+1} = -1$; they have multiplicity one and, thus, the chain is periodic with period $\kappa = 2$. The inverse matrix

$$H^{-1} = \|h_{jk}^{(-1)}\| = \frac{1}{2m}\|\varepsilon^{-(j-1)(k-1)}\|.$$

From the equality $\mathscr{P}^n = HJ^nH^{-1}$, where $J^n = \|\lambda_k\delta_{jk}\|$, we find

$$p_{jk}^{(n)} = \frac{1}{2m}\sum_{\nu=1}^{2m}[p\varepsilon^{(\nu-1)} + q\varepsilon^{-(\nu-1)}]^n\varepsilon^{(\nu-1)(j-k)},$$

which can be written as

$$p_{jk}^{(n)} = \frac{1}{2m}[1 + (-1)^{n+j-k}]\sum_{\nu=1}^{m}[p\varepsilon^{\nu-1} + q\varepsilon^{-(\nu-1)}]^n\varepsilon^{(\nu-1)(j-k)}$$

$$(j, k = 1, 2, \ldots, 2m).$$

All terms in the sum except the first are smaller than unity in modulus so that for $n \to \infty$,

$$p_{jk}^{(n)} \to \frac{1}{2m}[1 + (-1)^{n+j-k}].$$

This implies that

$$\lim_{n\to\infty} p_{jk}^{(2n)} = \begin{cases} \dfrac{1}{m}, & \text{for } j+k - \text{even}, \\ 0, & \text{for } j+k - \text{odd}, \end{cases}$$

$$\lim_{n\to\infty} p_{jk}^{(2n+1)} = \begin{cases} \dfrac{1}{m}, & \text{for } j+k - \text{odd}, \\ 0, & \text{for } j+k - \text{even}. \end{cases}$$

The last equalities can be written without using the expression for $p_{jk}^{(n)}$ as an irreducible chain, and the transition in one step from the group C_0 of states with odd numbers always leads to the group C_1 of states with even numbers and conversely.

The mean limiting transition probabilities are

$$\hat{p}_{jk} = \frac{1}{2} \lim_{n\to\infty} (p_{jk}^{(2n)} + p_{jk}^{(2n+1)}) = \frac{1}{2m} \qquad (j, k = 1, 2, \ldots, 2m).$$

Using this solution, one can solve Problems 38.26 and 38.27.

Example 38.6 In discussing the fundamental statements of kinetic theory of matter, Ehrenfest proposed the following model: m molecules distributed in two containers are randomly removed one by one from one container to the other. Find the mean limiting unconditional probabilities for the number of molecules in the first container.

SOLUTION. Let the state Q_i mean that there are i molecules in the first container $(i = 0, 1, \ldots, m)$. Then $p_{i,i-1} = i/m$, $p_{i,i+1} = 1 - i/m$ $(i = 0, 1, \ldots, m)$. The matrix of transition probabilities can be written as follows:

$$\mathscr{P} = \left\| \begin{matrix} 0 & 1 & 0 & 0 & \cdots & 0 & 0 \\ \dfrac{1}{m} & 0 & 1 - \dfrac{1}{m} & 0 & \cdots & 0 & 0 \\ 0 & \dfrac{2}{m} & 0 & 1 - \dfrac{2}{m} & \cdots & 0 & 0 \\ \cdot & \cdot & \cdot & \cdot & \cdot & \cdot & \cdot \\ 0 & 0 & 0 & 0 & \cdots & 0 & \dfrac{1}{m} \\ 0 & 0 & 0 & 0 & \cdots & 1 & 0 \end{matrix} \right\|.$$

From any state Q_i, a return to Q_i is possible only in a number of steps that is a multiple of two. Therefore, in the present case, the Markov chain is periodic with period $\kappa = 2$. The chain is irreducible because each state can be reached from any other state.

The column $\hat{p}$ of mean limiting unconditional probabilities can be determined from the condition $\hat{\mathscr{P}}'\hat{p} = \hat{p}$; that is,

$$\frac{1}{m}\hat{p}_1 = \hat{p}_0, \quad \left(1 - \frac{k-1}{m}\right)\hat{p}_{k-1} + \frac{k+1}{m}\hat{p}_{k+1} = \hat{p}_k, \quad \hat{p}_m = \frac{1}{m}\hat{p}_{m-1}$$

$$(k = 1, 2, \ldots, m-1).$$

From this it follows that $\hat{p}_k = \hat{p}_0 C_m^k$. Using the equality $\sum_{k=0}^m \hat{p}_k = 1$, we find that $1/\hat{p}_0 = \sum_{k=0}^m C_m^k = 2^m$; consequently, the required probabilities are

$$\hat{p}_k = \frac{1}{2^m} C_m^k \qquad (k = 0, 1, \ldots, m).$$

Similarly one can solve Problems 38.28 and 38.29.

PROBLEMS

38.1 Show that for a homogeneous Markov chain the transition probabilities $p_{ij}^{(n)}$ are correlated by the equality

$$p_{ij}^{(\alpha+\beta)} = \sum_{\nu=1}^{m} p_{i\nu}^{(\alpha)} p_{\nu j}^{(\beta)} \qquad (i, j = 1, 2, \ldots, m).$$

38.2 Given the column of initial probabilities $p(0) = \{\alpha, \beta, \gamma\}$ and the matrices of transition probabilities for times t_1, t_2, t_3,

$$\mathscr{P}_1 = \begin{Vmatrix} \alpha_1 & \alpha_2 & \alpha_3 \\ \alpha_3 & \alpha_1 & \alpha_2 \\ \alpha_2 & \alpha_3 & \alpha_1 \end{Vmatrix}, \quad \mathscr{P}_2 = \begin{Vmatrix} \alpha_2 & \alpha_3 & \alpha_1 \\ \alpha_1 & \alpha_2 & \alpha_3 \\ \alpha_3 & \alpha_1 & \alpha_2 \end{Vmatrix}, \quad \mathscr{P}_3 = \begin{Vmatrix} \alpha_3 & \alpha_1 & \alpha_2 \\ \alpha_2 & \alpha_3 & \alpha_1 \\ \alpha_1 & \alpha_2 & \alpha_3 \end{Vmatrix},$$

determine the column of unconditional probabilities $p(3)$.

38.3 According to the rules of a competition, a contestant quits a match if he loses two points in one game or if there are two ties. A contestant without ties can win at each game with probability α, can tie with probability β and can lose with probability $1 - \alpha - \beta$. In case of one tie, the probability of winning at each game is γ. Find the probability of losing various numbers of points in n games for the contestant whose outcomes in the previous games are known.

38.4 If the current in an electric circuit increases, the blocking system of a certain device fails with probability α and the entire device ceases to operate with probability β. If the blocking system fails, then at the next increase of current the device ceases to operate with probability γ. Find the probabilities that no failure will occur in the circuit, that only the blocking system will fail and that the device will cease to operate after n increases in current, if the initial state of the device is known.

38.5 There are several teams in a certain competition. During each round only three members of a team can compete with another team. According to the rules of the competition no ties can occur and the one who loses once is eliminated from this competition. Let α, β and γ be the probabilities that, in the next round in turn, among one, two, and three members remaining respectively, from a team, none loses; let β_1 and γ_1 be the probabilities that in the next round, in turn, among two and three remaining team members, respectively, one loses and let γ_2 be the probability that two among three members of this team lose. Determine the probabilities $p_{ik}^{(n)}$ $(i, k = 0, 1, 2, 3)$ that after n rounds, k members of this team compete if before these rounds i members of the same team competed.

38.6 An automatic system can operate if from N identical units, $m - 1$ fail; each unit can fail only during an operation cycle. The probabilities p_{ik} of transition of the system during one cycle from state Q_i to state Q_k are known, where the index of a state represents the number of units that failed so that for $k < i$, $p_{ik} = 0$ $(i, k = 0, 1, \ldots, m)$, $p_{mm} = 1$. Prove that the transition probabilities $p_{ik}^{(n)}$ for n cycles, during which the defective units are not replaced with probabilities $p_k = p_{kk}$ $(k = 0, 1, \ldots, m)$, are determined by the formulas:

$$p_{kk}^{(n)} = p_k^n \qquad (k = 0, 1, \ldots, m),$$

for $i > k$, $p_{ik}^{(n)} = 0$ $(i, k = 0, 1, \ldots, m)$ and for $k > i$

$$p_{ik}^{(n)} = \sum_{v=i}^{k} \frac{p_v^n D_{ki}(p_v)}{(p_v - p_i)(p_v - p_{i+1})\cdots(p_v - p_{v-1})(p_v - p_{v+1})\cdots(p_v - p_{k-1})(p_v - p_k)},$$

where

$$D_{ki}(\lambda) = \begin{Vmatrix} p_{i,i+1} & p_{i,i+2} & p_{i,i+3} & \cdots & p_{i,k-1} & p_{ik} \\ p_{i+1} - \lambda & p_{i+1,i+2} & p_{i+1,i+3} & \cdots & p_{i+1,k-1} & p_{i+1,k} \\ 0 & p_{i+2} - \lambda & p_{i+2,i+3} & \cdots & p_{i+2,k-1} & p_{i+2,k} \\ \cdot & \cdot & \cdot & \cdot & \cdot & \cdot \\ 0 & 0 & 0 & \cdots & p_{k-2,k-1} & p_{k-2,k} \\ 0 & 0 & 0 & \cdots & p_{k-1} - \lambda & p_{k-1,k} \end{Vmatrix}.$$

38.7 Prove that if under the assumptions made in the preceding problem $p_{kk} = p$ $(k = 0, 1, \ldots, m - 1)$, then

$$p_{mm}^{(n)} = 1, \qquad p_{kk}^{(n)} = p^n \qquad (k = 0, 1, \ldots, m - 1);$$

for $i > k$, $p_{ik}^{(n)} = 0$ $(i, k = 0, 1, \ldots, m)$, and for $k > i$

$$p_{ik}^{(n)} = \frac{(m - k + i - 1)!}{(m - 1)!} \left\{\frac{d^{k-i}}{d\lambda^{k-i}} [\lambda^n D_{ki}(\lambda)]\right\}_{\lambda=p},$$

$$p_{im}^{(n)} = \frac{(i - 1)!}{(m - 1)!} \left\{\frac{d^{m-i}}{d\lambda^{m-i}} \left[\frac{\lambda^n}{\lambda - 1} D_{mi}(\lambda)\right]\right\}_{\lambda=p} + \frac{D_{mi}(1)}{(1 - p)^{m-i}},$$

where $D_{ki}(\lambda)$ is determined by the formula of the preceding problem for $p_k = p$ $(k = 0, 1, \ldots, m - 1)$.

38.8 From an urn containing N white and black balls, m balls are drawn simultaneously. The black balls are used to replace the white balls that are drawn. Initially the urn contains m white balls and after several drawings it contains i white balls. Determine the probabilities $p_{ik}^{(n)}$ $(i, k = 0, 1, \ldots, m)$ that after n additional drawings there will be k white balls in the urn. Evaluate these probabilities for $N = 6$, $m = 3$.

38.9 For a given series of shots, each marksman from one group scores any number of points ranging from $N + 1$ to $N + m$ with equal probabilities. Determine the probability that among the next n marksmen of this group at least one will score $N + k$ points, if the maximal number of points scored by the previous marksmen is $N + i$ $(k \geqslant i = 1, 2, \ldots, m)$.

38.10 Along a straight line AB in a horizontal plane, there are placed identical, vertical cyclinders of radius r whose centers are a distance l apart. Perpendicular to this line spheres of radius R are thrown, and the path of a moving sphere crosses AB with equal probability at any point of the interval L on which there stand m cylinders. The distance between the centers of the cylinders is $l > 2(r + R)$; each time a sphere hits a cylinder the number of cylinders decreases by one. Determine the probabilities $p_{ik}^{(n)}$ $(i, k = 0, 1, \ldots, m)$ that after n throws k cylinders will remain if before this there were i cylinders.

38.11 In a domain D, partitioned into m equal parts, points are placed successively so that their positions are equally probable throughout the domain. Determine the probabilities $p_{ik}^{(n)}$ $(i, k = 1, 2, \ldots, m)$ that after placing a new series of n points, the number of parts of D containing at least one point will increase from i to k.

38.12 At times $t_1, t_2, t_3, \ldots$ a ship can change its direction by selecting one out of m possible courses, $Q_1, Q_2, \ldots, Q_m$. The probability p_{ij} that at time t_r the ship changes from Q_i to Q_j is $p_{ij} = \alpha_{m-i+j+1}$, and $\alpha_{m+k} = \alpha_k \neq 0$ $(k = 1, 2, \ldots, m)$, $\sum_{k=1}^{m} \alpha_k = 1$. Determine the probability $p_{jk}^{(n)}$ that for $t_n < t < t_{n+1}$, the direction of the ship will be Q_k if the initial direction was Q_j $(j, k = 1, 2, \ldots, m)$. Find this probability for $n = \infty$.

38.13 Consider the following model of the diffusion process with central force. A particle can lie only on the segment AB at points with coordinates $x_k = x_A + k\Delta$ $(k = 0, 1, \ldots, m)$, where $x_m = x_B$. It shifts stepwise from x_j to the next point toward A with probability j/m, and to the next point toward B with probability $1 - j/m$. Determine the probabilities $p_{ik}^{(n)}$ $(i, k = 0, 1, \ldots, m)$ that after n steps the particle will be at point x_k if initially it was at x_i.

38.14 The assumptions here are the same as in Example 38.2, but the machine does not turn off. When there are no nickels in the container and a dime is inserted, or there are m nickels and a nickel is inserted, the machine returns the last coin inserted without releasing a token. Find the probabilities $p_{ik}^{(n)}$ $(i, k = 0, 1, \ldots, m)$ that after n demands for tokens, there will be k nickels in the container if initially there were i nickels.

38.15 Two marksmen A and B fire shots in turn so that after each hit A fires, and after each failure B fires. The right for the first shot is determined on the same basis, by reference to the outcome of a preliminary shot fired by a randomly chosen marksman. Determine the probability of failure at the nth trial independent of the previous hits if the probabilities of failure at each trial for these two marksmen are α and β, respectively.

38.16 Given the matrix $\mathscr{P} = \|p_{ij}\|$ of transition probabilities, that is irreducible, nonperiodic and twice-stochastic, i.e., the sum of elements of each column and of each row is unity, find the limiting probabilities $p_j^{(\infty)}$ $(j = 1, 2, \ldots, m)$.

38.17 There are m white and m black balls, that are mixed thoroughly and then equally distributed in two urns. From each urn one ball is randomly drawn and placed in the other. Find the probabilities p_{ik} $(i, k = 0, 1, \ldots, m)$ that after an infinite number of such interchanges, the first urn will contain k white balls if initially it contained i white balls.

38.18 A segment AB is divided into m equal intervals. A particle can lie only on the midpoint of some interval and shifts stepwise by an amount equal to the length of one interval toward point B with probability p and toward point A with probability $q = 1 - p$. At the endpoints of AB, reflecting screens are placed so that upon reaching A or B, the particle is reflected toward its initial position. Find the limiting unconditional probabilities $p_k^{(\infty)}$ ($k = 1, 2, \ldots, m$) that the particle is in each of the m intervals.

38.19 Given the following transition probabilities for a Markov chain with an infinite number of states,

$$p_{i1} = \frac{i}{i+1}, \qquad p_{i,i+1} = \frac{1}{i+1} \qquad (i = 1, 2, \ldots),$$

determine the limiting probabilities $p_j^{(\infty)}$ ($j = 1, 2, \ldots$).

38.20 The transition probabilities for a Markov chain with an infinite number of states is defined by $p_{i1} = q$, $p_{i,i+1} = p = 1 - q$ ($i = 1, 2, \ldots$). Find the limiting probabilities $p_j^{(\infty)}$ ($j = 1, 2, \ldots$).

38.21 A Markov chain with an infinite number of states has the following transition probabilities:

$$p_{11} = \frac{1}{2}, \quad p_{12} = \frac{1}{2}, \quad p_{i1} = \frac{1}{i}, \quad p_{i,i+1} = \frac{i-1}{i} \qquad (i = 2, 3, \ldots).$$

Find the limiting probabilities $p_{ik}^{(\infty)}$ ($i, k = 1, 2, \ldots$).

38.22 A particle makes a random walk on the positive portion of the x-axis. The particle can shift by one step Δ to the right with probability α, to the left with probability β or it can remain fixed; it can reach only points with coordinates x_j ($j = 1, 2, \ldots$). From the point with coordinate $x_1 = \Delta$, the particle can move to the right with probability α or remain fixed with probability $1 - \alpha$. Find the limiting transition probabilities $p_k^{(\infty)}$ ($k = 1, 2, \ldots$).

38.23 The matrix of transition probabilities is given in the form

$$\mathscr{P} = \begin{Vmatrix} R & 0 \\ U & W \end{Vmatrix},$$

where R is the matrix associated with the irreducible nonperiodic group C of essential states $Q_1, Q_2, \ldots, Q_s$, and the square matrix W is associated with the inessential states $Q_{s+1}, Q_{s+2}, \ldots, Q_m$. Determine the limiting probabilities p_{*j} ($j = s + 1, s + 2, \ldots, m$) that the system will pass into a state belonging to group C.

38.24 The matrix of transition probabilities is given in the form

$$\mathscr{P} = \begin{Vmatrix} R & 0 & 0 \\ 0 & R_1 & 0 \\ U & U_1 & W \end{Vmatrix},$$

where R is the matrix corresponding to the nonperiodic group C of essential states $Q_1, Q_2, \ldots, Q_s$, and the square matrix W corresponds to the inessential states $Q_{r+1}, Q_{r+2}, \ldots, Q_m$. Find the probabilities p_{*j} $(j = r+1, r+2, \ldots, m)$ that the system will pass into a state belonging to the group C if all the elements of W are equal to α and the sum of elements of any row of matrix U is β.

38.25 Two players A and B continue a game until the complete financial ruin of one. Their probabilities of winning at each play are, respectively, p and q $(p + q = 1)$. At each play the win of one player (loss for the other) is one dollar, and the total capital of the players is m dollars. Determine the probabilities of financial ruin for each if A has j dollars $(j = 1, 2, \ldots, m-1)$ before the game begins.

38.26 Given the transition probabilities $p_{j,j+1} = 1$ $(j = 1, 2, \ldots, m-1)$, $p_{m1} = 1$, determine the transition probabilities $p_{jk}^{(n)}$ and the mean limiting transition probabilities $\hat{p}_{jk}$ $(j, k = 1, 2, \ldots, m)$.

38.27 The matrix of transition probabilities is

$$\mathscr{P} = \begin{Vmatrix} \alpha & \beta & \gamma & \delta \\ 0 & 0 & 1 & 0 \\ 0 & 0 & 0 & 1 \\ 0 & 1 & 0 & 0 \end{Vmatrix},$$

where $\alpha \neq 1$. Determine the transition probabilities $p_{jk}^{(n)}$ and the mean limiting transition probabilities $\hat{p}_{jk}$ $(j, k = 1, 2, 3, 4)$.

38.28 Given the elements of the matrix of transition probabilities,

$$p_{j,j+1} = p \qquad (j = 1, 2, \ldots, 2m-1),$$

$$p_{j,j-1} = q = 1 - p \qquad (j = 2, 3, \ldots, 2m),$$

without evaluating the eigenvalues of the matrix $\mathscr{P}$, find the limiting transition probabilities and the mean limiting unconditional probabilities.

38.29 A particle is displaced on a segment AB by random impacts, and can be at the points with coordinates $x_j = x_A + j\Delta$ $(j = 0, 1, \ldots, m)$. Reflecting screens are placed at the endpoints A and B. Each impact can shift the particle to the right with probability p and to the left with probability $q = 1 - p$. If the particle is next to a screen, any impact shifts it to the screen in question. Find the mean limiting unconditional probabilities that the particle is at each division point of the segment AB.

39. THE MARKOV PROCESSES WITH A DISCRETE NUMBER OF STATES

Basic Formulas

The behavior of a system with possible states $Q_0, Q_1, Q_2, \ldots, Q_m$ can be described by a random function $X(t)$ assuming the value k, if at time t the system is in state Q_k. If the passage from one state to another is possible at any time t, and the probabilities $P_{ik}(t, \tau)$ of transition from state Q_i at time t to state Q_k at time τ ($\tau \geqslant t$) are independent of the behavior of the system before the time t, then $X(t)$ is a Markov stochastic process with a discrete number of states. (The number of states can be finite or infinite.)

The transition probabilities $P_{ik}(t, \tau)$ satisfy the relation

$$P_{ik}(t, \tau) = \sum_{j=0}^{m} P_{ij}(t, s)P_{jk}(s, \tau), \qquad t \leqslant s \leqslant \tau.$$

The process is homogeneous if

$$P_{ik}(t, \tau) = P_{ik}(\tau - t).$$

In this case, for the Markov process

$$P_{ik}(\tau - t) = \sum_{j=0}^{m} P_{ij}(s - t)P_{jk}(\tau - s), \qquad t \leqslant s \leqslant \tau.$$

A Markov process is called regular if:

(a) for each state Q_k there exists a limit

$$c_k(t) = \lim_{\Delta t \to 0} \frac{1}{\Delta t}[1 - P_{kk}(t, t + \Delta t)];$$

(b) for each pair of states Q_i and Q_k there exists a temporal transition probability density $p_{ik}(t)$, continuous in t, defined by

$$p_{ik}(t) = \frac{1}{c_i(t)} \lim_{\Delta t \to 0} \frac{P_{ik}(t, t + \Delta t)}{\Delta t},$$

where the limit exists uniformly with respect to t, and for fixed k uniformly with respect to i.

For regular Markov processes the probabilities $P_{ik}(t, \tau)$ are determined by two systems of differential equations:

$$\frac{\partial P_{ik}(t, \tau)}{\partial \tau} = -c_k(t)P_{ik}(t, \tau) + \sum_{j \neq k} P_{ij}(t, \tau)c_j(\tau)p_{jk}(\tau) \qquad \text{(the first system)},$$

$$\frac{\partial P_{ik}(t, \tau)}{\partial t} = c_i(t)P_{ik}(t, \tau) - c_i(t) \sum_{j \neq 1} P_{jk}(t, \tau)p_{ij}(t) \qquad \text{(the second system)}$$

$$(i, j, k = 0, 1, 2, \ldots, m),$$

with initial conditions

$$P_{ik}(t, t) = \delta_{ik},$$

where

$$\delta_{ik} = \begin{cases} 1, & \text{for } i = k, \\ 0, & \text{for } i \neq k. \end{cases}$$

For a homogeneous Markov process, $c_i(t)$ and $p_{ij}(t)$ are independent of time, $P_{ik}(t, \tau) = P_{ik}(\tau - t)$ and the systems of differential equations become

$$\frac{dP_{ik}(t)}{dt} = -c_k P_{ik}(t) + \sum_{j \neq k} c_j p_{jk} P_{ij}(t) \qquad \text{(the first system)},$$

$$\frac{dP_{ik}(t)}{dt} = -c_i P_{ik}(t) + c_i \sum_{j \neq k} p_{ij} P_{jk}(t) \qquad \text{(the second system)}$$

$$(i, j, k = 0, 1, 2, \ldots, m)$$

with initial conditions

$$P_{ik}(0) = \delta_{ik}.$$

The probabilities $P_k(t)$ that the system is in state Q_k at time t is given by the system of equations

$$\frac{dP_k(t)}{dt} = -c_k(t)P_k(t) + \sum_{j \neq k} c_j(t)p_{jk}(t)P_j(t) \qquad (j, k = 0, 1, 2, \ldots, m),$$

with corresponding initial conditions for $P_j(t)$. If the initial state Q_i is given, the initial conditions are

$$P_k(t) = \delta_{ik} \quad \text{for} \quad t = 0.$$

For homogeneous Markov processes, the last system becomes

$$\frac{dP_k(t)}{dt} = -c_kP_k(t) + \sum_{j \neq k} c_jp_{jk}P_j(t) \qquad (j, k = 0, 1, 2, \ldots, m),$$

and the initial conditions are

$$P_k(t) = \delta_{ik} \quad \text{for} \quad t = 0.$$

If for a homogeneous Markov process there exists a time interval $t > 0$ such that $P_{ik}(t) > 0$ for all i and k, then the process is called transitive and for it the limit

$$\lim_{t \to \infty} P_{ik}(t) = \lim_{t \to \infty} P_k(t) = p_k$$

exists independent of the index of the initial state. The limiting probabilities p_k in this case are determined from the system of algebraic equations

$$c_kp_k = \sum_{j \neq k} c_jp_{jk}p_j \qquad (j, k = 0, 1, 2, \ldots, m).$$

The equations for probabilities $P_{ik}(t, \tau)$ and $P_i(t)$ can be obtained either by applying the foregoing general formulas or finding the variations of probabilities for different states of the system during a small time interval Δt and passing to the limit as $\Delta t \to 0$.

An example of a Markov process is the simple flow of events with the following properties:

stationarity, that is, for any $\Delta t > 0$ and integer $k \geqslant 0$, the probability that k events will occur during the interval $(t, t + \Delta t)$ is the same for all $t \geqslant 0$;

absence of after effect, that is, the probability of occurrence of k events during the interval $(t, t + \Delta t)$ is independent of the number of occurrences before the time t;

ordinarity, that is,

$$\lim_{\Delta t \to 0} \frac{R_2(\Delta t)}{\Delta t} = 0,$$

where $R_2(\Delta t)$ is the probability that at least two events occur during interval Δt.

Solution for Typical Examples

Example 39.1 A system can be in one of the states Q_0, Q, $Q_2, \ldots$, and it passes, during time Δt, into a state whose index is higher by one with probability $\lambda\Delta t + o(\Delta t)$. Find the probabilities $P_{ik}(t)$ of transition from state Q_i to state Q_k $(k \geqslant i)$ during time t.

SOLUTION. The process is Markovian by assumption. Moreover, it is regular, since

$$c_i = \lim_{\Delta t \to 0} \frac{1}{\Delta t} [\lambda\Delta t + o(\Delta t)] = \lambda = \text{const.},$$

$$p_{i,i+1}(t) = \frac{1}{c_i} \lim_{\Delta t \to 0} \frac{\lambda\Delta t + o(\Delta t)}{\Delta t} = 1,$$

and otherwise $p_{ik} = 0$.

Consequently, the equations for homogeneous Markov processes are applicable;

$$\frac{dP_{ii}(t)}{dt} = -\lambda P_{ii}(t),$$

$$\frac{dP_{ik}(t)}{dt} = -\lambda P_{ik}(t) + \lambda P_{i,k-1}(t) \qquad (k \geqslant i + 1),$$

with initial conditions $P_{ik}(0) = \delta_{ik}$. Multiplying both sides of the obtained equations by u^k and summing over k from i to ∞, we get

$$\frac{\partial G(t, u)}{\partial t} = \lambda(u - 1)G(t, u),$$

where $G(t, u) = \sum_{k=i}^{\infty} P_{ik}(t)u^k$.

The solution of the last equation has the form

$$\ln G(t, u) = \lambda(u - 1)t + \ln G(0, u).$$

Since by definition,

$$G(0, u) = \sum_{k=i}^{\infty} P_{ik}(0)u^k = u^i,$$

we have

$$G(t, u) = u^i e^{\lambda(u-1)t} = u^i e^{-\lambda t} \sum_{m=0}^{\infty} \frac{(\lambda t)^m}{m!} u^m = e^{-\lambda t} \sum_{k=i}^{\infty} \frac{(\lambda t)^{k-i}}{(k - i)!} u^k.$$

Comparing the last expression with the definition of $G(t, u)$, we obtain

$$P_{ik}(t) = \frac{(\lambda t)^{k-i}}{(k - i)!} e^{-\lambda t}.$$

The initial system of differential equations for $P_{ik}(t)$ can also be obtained in another way: the probability $P_{ik}(t + \Delta t)$ is the sum of the probability $P_{ik}(t)[1 - \lambda\Delta t - o(\Delta t)]$ that the passage from state Q_i to state Q_k $(k > i)$ occurred during time T and the probability $P_{i,k-1}(t)[\lambda\Delta t + o(\Delta t)]$ that this passage occurs in the interval $(t, t + \Delta t)$; that is,

$$P_{ik}(t + \Delta t) = P_{ik}(t)[1 - \lambda\Delta t - o(\Delta t)] + P_{i,k-1}(t)[\lambda\Delta t + o(\Delta t)].$$

Transposing $P_{ik}(t)$ to the left side of the equality, dividing both sides by Δt and passing to the limit as $\Delta t \to 0$, we obtain the required equation. In this manner the equation for $k = i$ can be deduced.

Problem 39.6 may also be solved in a similar way.

Example 39.2 A queuing system consists of a large (practically infinite) number of identical devices, each device servicing only one call at a time and spending on it a random time that obeys an exponential distribution law with probability density $\mu e^{-\mu t}$. The incoming calls for service form a simple queue with parameter λ. Evaluate (a) the probability $P_n(t)$ that at time t exactly n devices will be busy ($n \leqslant m$) if initially all devices were free; (b) the limiting probabilities $p_n = \lim_{t\to\infty} P_n(t)$; (c) the expected number of devices busy at time t.

SOLUTION. Since the queue of calls is simple and the servicing time obeys an exponential distribution, during the time interval $(t, t + \Delta t)$ the system will change its state more than once with a probability whose order of magnitude is higher than Δt.

Therefore, considering only the first-order terms during time interval Δt, we obtain:

$$P_{n,n+1}(t, t + \Delta t) = \lambda\Delta t(1 - n\mu\Delta t) + o(\Delta t) = \lambda\Delta t + o(\Delta t),$$

$$P_{n,n-1}(t, t + \Delta t) = (1 - \lambda\Delta t)n\mu\Delta t + o(\Delta t) = n\mu\Delta t + o(\Delta t),$$

$$\begin{aligned} P_{n,n}(t, t + \Delta t) &= (1 - \lambda\Delta t)(1 - n\mu\Delta t) + o(\Delta t) \\ &= 1 - (\lambda + n\mu)\Delta t + o(\Delta t). \end{aligned}$$

The system is regular because

$$c_n = \lim_{\Delta t\to 0} \frac{1 - P_{n,n}(t, t + \Delta t)}{\Delta t} = \lambda + n\mu = \text{const.},$$

$$p_{n,n+1} = \frac{1}{c_n}\lim_{\Delta t\to 0}\frac{P_{n,n+1}(t, t + \Delta t)}{\Delta t} = \frac{\lambda}{\lambda + n\mu},$$

$$p_{n,n-1} = \frac{1}{c_n}\lim_{\Delta t\to 0}\frac{P_{n,n-1}(t, t + \Delta t)}{\Delta t} = \frac{n\mu}{\lambda + n\mu}.$$

(a) We substitute the calculated values for c_n, $p_{n,n+1}$ and $p_{n,n-1}$ in the system of differential equations for $P_n(t)$:

$$\frac{dP_n(t)}{dt} = -(\lambda + n\mu)P_n(t) + \lambda P_{n-1}(t) + (n + 1)\mu P_{n+1}(t),$$

for $n \geqslant 1$ and

$$\frac{dP_0(t)}{dt} = -\lambda P_0(t) + \mu P_1(t).$$

If one assumes that at time $t = 0$ all devices are free, the initial conditions are

$$P_n(0) = \delta_{n0}.$$

The resulting system can be solved with the aid of the generating function

$$G(t, u) = \sum_{n=0}^{\infty} P_n(t)u^n.$$

Multiplying both sides of differential equations by u^n and summing, after simple transformations we find

$$\frac{\partial G(t, u)}{\partial t} = (1 - u)\left\{-\lambda G(t, u) + \mu \frac{\partial G(t, u)}{\partial u}\right\}.$$

The initial condition is $G(0, u) = 1$.

The resulting linear nonhomogeneous partial differential equation is replaced by an equivalent homogeneous one,[1]

$$\frac{\partial V}{\partial t} - \mu(1 - u)\frac{\partial V}{\partial u} - \lambda(1 - u)G\frac{\partial V}{\partial G} = 0,$$

with initial condition $V = G - 1$ for $t = 0$.

To solve the last equation it is necessary first to solve the system of ordinary differential equations

$$\frac{dt}{1} = -\frac{du}{\mu(1 - u)} = -\frac{dG}{\lambda(1 - u)G},$$

whose independent integrals are

$$t - \frac{1}{\mu}\ln(1 - u) = c_1,$$

$$\frac{\lambda}{\mu}u - \ln G = c_2.$$

Using the initial conditions $t = 0$, $u = u_0$, $G = G_0$, we obtain the Cauchy integrals of the system

$$u_0 = 1 - (1 - u)e^{-\mu t},$$

$$G_0 = G\exp\left\{\frac{\lambda}{\mu}(1 - u)(1 - e^{-\mu t})\right\}.$$

The right-hand sides are the principal solutions of the homogeneous partial differential equation. Using these solutions, we form the solution of Cauchy's problem for the homogeneous partial differential equation

$$V = G\exp\left\{\frac{\lambda}{\mu}(1 - u)(1 - e^{-\mu t})\right\} - 1.$$

The solution of the Cauchy problem for the initial equation is the function G for which $V = 0$; hence,

$$G = \exp\left\{-\frac{\lambda}{\mu}(1 - u)(1 - e^{-\mu t})\right\}.$$

The probabilities $P_n(t)$ are related to the generating function $G(t, u)$ by the equality

$$P_n(t) = \frac{1}{n!}\left.\frac{\partial G(t, u)}{\partial u}\right|_{\mu = 0},$$

[1] Weinberger, H. F.: First Course in Partial Differential Equations. Waltham, Mass., Blaisdell Publishing Company, 1965; and Petrovskii, I. G.: Partial Differential Equations. Philadelphia, W. B. Saunders Company, 1967.

which leads to

$$P_n(t) = \frac{1}{n!}\left\{\frac{\lambda}{\mu}(1 - e^{-\mu t})\right\}^n \exp\left\{-\frac{\lambda}{\mu}(1 - e^{-\mu t})\right\};$$

that is, a Poisson law with parameter

$$a = \frac{\lambda}{\mu}(1 - e^{-\mu t}).$$

(b) The limiting probabilities p_n are obtained from the initial ones by passage to the limit:

$$p_n = \lim_{t\to\infty} P_n(t) = \frac{1}{n!}\left(\frac{\lambda}{\mu}\right)^n \exp\left\{-\frac{\kappa}{\mu}\right\};$$

that is, p_n obey a Poisson distribution law with parameter $a = \lambda/\mu$. (The same result can be obtained if we solve the system of algebraic equations obtained from the differential system for $P_n(t)$ after replacing $P_n(t)$ by p_n and $[dP_n(t)/dt]$ by zero.)

(c) The expected number of busy devices is

$$M(t) = \sum_{n=0}^{\infty} nP_n(t).$$

For $M(t)$, write the differential equation:

$$\begin{aligned}\frac{dM(t)}{dt} &= \sum_{n=0}^{\infty} n\frac{dP_n(t)}{dt}\\ &= \sum_{n=0}^{\infty} n\{-(\lambda + n\mu)P_n(t) + \lambda P_{n-1}(t) + (n + 1)\mu P_{n+1}(t)\} = \lambda - \mu M(t).\end{aligned}$$

Since initially all devices are free,

$$M(t) = \frac{\lambda}{\mu}(1 - e^{-\mu t}).$$

Problems 39.17 to 39.19 may be solved in a similar way.

Example 39.3 A queuing system consists of m devices, each of which, at any given time, can service only one call. It services for a random time, obeying an exponential distribution law with parameter μ. The incoming calls form a simple queue with parameter λ. A call is serviced immediately after it is received if there is at least one free device at that time; otherwise, the call is rejected and does not return to the system. Determine the limiting probability for a rejected call.

SOLUTION. Let Q_i denote a state of the system in which i devices are busy, then, $P_{ik}(t) > 0$ for a finite time interval. Consequently, we can apply Markov's theorem, stating that there exist limiting probabilities such that

$$p_n = \lim_{t\to\infty} P_n(t),$$

and determined by the formula

$$c_n p_n = c_{n-1} p_{n-1,n} p_{n-1} + c_{n+1} p_{n+1,n} p_{n+1}.$$

As in the preceding example, we have

$$c_n = \lambda + n\mu, \qquad p_{n,n+1} = \frac{\lambda}{\lambda + n\mu},$$

$$p_{n,n-1} = \frac{n\mu}{\lambda + n\mu} \qquad (0 \leqslant n \leqslant m - 1),$$

$$c_m = m\mu, \qquad p_{m,m+1} = 0,$$

and the other probabilities $p_{jk} = 0$. Substituting these values for p_{jk} in the equations for p_n, we get

$$(\lambda + n\mu)p_n = \lambda p_{n-1} + (n + 1)\mu p_{n+1} \qquad (0 \leqslant n \leqslant m - 1;\ p_{-1} = 0),$$

$$m\mu p_m = \lambda p_{m-1}.$$

If we set $z_n = \lambda p_{n-1} - n\mu p_n$, the system becomes

$$z_1 = 0, \quad z_n - z_{n+1} = 0 \qquad (0 < n < m), \quad z_m = 0,$$

hence, it follows that $z_n = 0$ for all n and this means that

$$p_n = \frac{\lambda}{n\mu} p_{n-1} \quad \text{or} \quad p_n = \frac{1}{n!}\left(\frac{\lambda}{\mu}\right)^n p_0.$$

The system is certainly in one of states Q_n ($n = 0, 1, 2, \ldots, m$), therefore,

$$\sum_{n=0}^{m} p_n = 1;$$

from this, the probability p_0 that all devices are free is

$$p_0 = \frac{1}{\sum_{n=0}^{m} \frac{1}{n!}\left(\frac{\lambda}{\mu}\right)^n}.$$

The probability that the service is refused is

$$p_m = \frac{\frac{1}{m!}\left(\frac{\lambda}{\mu}\right)^m}{\sum_{n=0}^{m} \frac{1}{n!}\left(\frac{\lambda}{\mu}\right)^n} \qquad \text{(Erlang's formula).}$$

Following this solution, Problems 39.8, 39.10, 39.11 and 39.14 may be solved.

PROBLEMS

39.1 The particles emitted by a radioactive substance in the disintegration process form a simple flow with parameter λ. Each particle can independently reach a counter with probability p. Determine the probability that during time t n particles will be recorded by the counter.

39.2 Two communication channels feed two independent simple flows of telegrams to a given point. Find the probability that n telegrams will arrive during time t if the parameters of the component flows are λ_1 and λ_2.

39.3 The electronic emission of the cathode of an electronic tube represents a simple flow of electrons with parameter λ. The flight times for different electrons are independent random variables with the same distribution function $F(x)$. Determine the probability that at time t after the start of emission there will be exactly n electrons between the electrodes of the tube, and determine the limiting probability of the same event.

39.4 For a simple flow of events, determine the correlation coefficient between the number of occurrences in the intervals $(0, t)$ and $(0, t + \tau)$.

39.5 For a random time T_n of occurrence of the nth event in a simple flow with parameter λ, determine the distribution function $F_n(t)$, the probability density $f_n(t)$ and the moments m_k.

39.6 Find the transition probabilities of a system from state Q_i to state Q_k in time t in a homogeneous Markov process if, in a single change of state, the system can pass only from state Q_n to state Q_{n+1}, and the probability that the system will change its state during time interval $(t, t + \Delta t)$ is $[\lambda\Delta t + o(\Delta t)]$.

39.7 The customers of a repair shop form a simple queue with parameter λ. Each customer is serviced by one repairman during a random time, obeying an exponential distribution law with parameter μ. If there are no free repairmen, the customer leaves without service. How many repairmen should there be so that the probability that a customer will be refused immediate service is at most 0.015, if $\mu = \lambda$.

39.8 One repairman services m automatic machines, which need no care during normal operation. The failures of each machine form an independent simple flow with parameter λ. To remove the defects, a repairman spends a random time distributed according to an exponential law with parameter μ. Find the limiting probabilities that k machines do not run (are being repaired or are waiting for repairs), and the expected number of machines waiting for repairs.

39.9 Solve Problem 39.8 under the assumption that the number of repairmen is r $(r < m)$.

39.10 A computer uses either units of type A or units of type B. The failures of these units form a simple flow with parameters $\lambda_A = 0.1$ units/hour and $\lambda_B = 0.01$ units/hour. The total cost of all units of type A is a, and that of all units of type B is b $(b > a)$. A defective unit causes a random delay, obeying an exponential distribution law with an average time of two hours. The cost per hour of delay is c. Find the expectation for the saving achieved by using more reliable elements during 1000 hours of use.

39.11 The incoming calls for service in a system consisting of n homogeneous devices form a simple queue with parameter λ. The service starts immediately if there is at least one free device, and each call requires a single free device whose servicing time is a random variable obeying an exponential distribution with parameter μ $(\mu n > \lambda)$. If a call finds no free device, it waits in line.

Determine the limiting values for (a) the probabilities p_k that there are exactly k calls in the system (being serviced and waiting in line); (b) the probability p^* that all devices are busy; (c) the distribution function $F(t)$ and the expected time t spent by a device waiting in line; (d) the expected number m_1 of calls waiting in line, the expected number m_2 of calls in the servicing system and the expected number of working devices m_3 that need no service.

39.12 The machines arriving at a repair shop that gives guaranteed service form a simple queue with parameter $\lambda = 10$ units/hour. The servicing time for one unit is a random variable obeying an exponential distribution law with parameter $\mu = 5$ units/hour. Determine the average time elapsed from the moment a machine arrives until it is repaired if there are four repairmen, each servicing only one machine at a time.

39.13 How many positions should an experimental station have so that an average of one per cent of items wait more than 2/3 of a shift to start, if the duration of the experiments is a random variable obeying an exponential distribution law with a mean shift of 0.2, and the incoming devices used in these experiments form a simple queue with an average number of 10 units per shift?

39.14 A servicing system consists of n devices, each servicing only one call at a time. The servicing time is an exponentially distributed random variable with parameter μ. The incoming calls for service form a simple queue with parameter λ ($\mu n > \lambda$). A call is serviced immediately if at least one device is free. If all devices are busy and the number of calls in the waiting line is less than m, the calls line up in the waiting line; if there are m calls in the waiting line, a new call is refused service.

Find the limiting values for (a) the probabilities p_k that there will be exactly k calls in the servicing system; (b) the probability that a call will be denied service; (c) the probabilities that all servicing devices will be busy; (d) the distribution function $F(t)$ for the time spent in the waiting line; (e) the expected number of calls m_1 in the waiting line, the expected number of calls m_2 in the servicing system, and the expected number of devices m_3 freed from service.

39.15 A barbershop has three barbers. Each barber spends an average of 10 minutes with each customer. The customers form a simple queue with an average of 12 customers per hour. The customers stand in line if when they arrive there are fewer than three persons in the waiting line; otherwise, they leave.

Determine the probability p_0 for no customers, the probability p that a customer will leave without having his hair cut, the probability p^* that all barbers will be busy working, the average number of customers m_1 in the waiting line, and the average number of customers m_2 in the barbershop in general.

39.16 An electric circuit supplies electric energy to m identical machines, which need service independently. The probability that during the interval $(t, t + \Delta t)$ a machine stops using electric energy is $\mu\Delta t + o(\Delta t)$, and the probability that it will need energy during the same interval is $[\lambda\Delta t + o(\Delta t)]$. Determine the limiting probability that there will be n machines connected in the circuit.

39.17 A shower of cosmic particles is caused by one particle reaching the atmosphere at some given moment. Determine the probability that at time t after the first particle reaches the atmosphere there will be n particles, if each particle during the time interval $(t, t + \Delta t)$ can produce with probability $[\lambda t\Delta + o(\Delta t)]$ a new particle with practically the same reproduction probability.

39.18 A shower of cosmic particles is produced by one particle reaching the atmosphere at some given moment. Estimate the probability that at time t after the first particle reaches the atmosphere there will be n particles, if each particle during the time interval $(t, t + \Delta t)$ can produce a new particle with probability $[\lambda\Delta t + o(\Delta t)]$ or disappear with probability $[\mu\Delta t + o(\Delta t)]$.

39.19 In a homogeneous process of pure birth (birth without death), a number of n particles at time t can change into $n + 1$ particles during the interval $(t, t + \Delta t)$ with probability $\lambda_n(t)\Delta t + 0(\Delta t)$, where

$$\lambda_n(t) = \frac{1 + an}{1 + at},$$

or they can fail to increase in number. Determine the probability that at time t there will be exactly n particles.

40. CONTINUOUS MARKOV PROCESSES

Basic Formulas

A continuous stochastic process $U(t)$ is called a Markov process if the distribution function $F(u_n \mid u_1, \ldots, u_{n-1})$ of the ordinate of $U(t)$ at time t_n, computed under the assumption that the values of the ordinates $u_1, u_2, \ldots, u_{n-1}$ at times $t_1, t_2, \ldots, t_{n-1}$ are known $(t_1 < t_2 < \cdots < t_{n-1} < t_n)$, depends only on the value of the last ordinate; i.e.,

$$F(u_n \mid u_1, \ldots, u_{n-1}) = F(u_n \mid u_{n-1}).$$

The conditional probability density $f(u_n \mid u_{n-1})$ is a function $f(t, x; \tau, y)$ of four variables, where for the sake of brevity one uses the notations:

$$U(t) = X, \qquad U(\tau) = Y, \qquad t \leqslant \tau.$$

The function $f(t, x; \tau, y)$ satisfies the Kolmogorov equations[2]

$$\frac{\partial f}{\partial \tau} + a(t, x)\frac{\partial f}{\partial x} + \frac{1}{2}b(t, x)\frac{\partial^2 f}{\partial x^2} = 0 \qquad \text{(the first equation)},$$

$$\frac{\partial f}{\partial \tau} + \frac{\partial}{\partial y}[a(\tau, y)f] - \frac{1}{2}\frac{\partial^2}{\partial y^2}[b(\tau, y)f] = 0 \qquad \text{(the second equation)},$$

[2] The second Kolmogorov equation is sometimes called the Fokker-Planck equation or Fokker-Planck-Kolmogorov equation since before it was rigorously proved by Kolmogorov it had appeared in the works of these physicists.

where

$$a(t, x) = \lim_{\tau \to t} \frac{1}{\tau - t} \mathbf{M}\{[Y - X] \mid X = x\},$$

$$b(t, x) = \lim_{\tau \to t} \frac{1}{\tau - t} \mathbf{M}\{[Y - X]^2 \mid X = x\}.$$

The function $f(t, x; \tau, y)$ has the general properties of the probability density:

$$f(t, x; \tau, y) \geqslant 0, \qquad \int_{-\infty}^{\infty} f(t, x; \tau, y)\, dy = 1$$

and satisfies the initial condition

$$f(t, x; \tau, y) = \delta(y - x) \quad \text{for} \quad \tau = t.$$

If the range for the ordinates of the random function is bounded, that is,

$$\alpha \leqslant U(t) \leqslant \beta,$$

then in addition to the previously mentioned conditions, the function

$$G(\tau, y) \equiv a(\tau, y)f - \frac{1}{2} \frac{\partial}{\partial y} [b(\tau, y)f]$$

should also be constrained by the following boundary conditions:

$$G(\tau, \alpha) = G(\tau, \beta) = 0 \quad \text{for any} \quad \tau.$$

($G(\tau, y)$ may be regarded as a "probability flow.")

A set of n random functions $U_1(t), \ldots, U_n(t)$ forms a Markov process if the probability density (distribution function) f for the ordinates $Y_1, Y_2, \ldots, Y_n$ of these functions at time τ, calculated under the assumption that at time t the ordinates of the random functions assumed the values $X_1, X_2, \ldots, X_n$, is independent of the values of the ordinates of $U_1(t), U_2(t), \ldots, U_n(t)$ for times previous to t. In this case, the function f satisfies the system of multidimensional Kolmogorov equations

$$\frac{\partial f}{\partial t} + \sum_{j=1}^{n} a_j(t, x_1, x_2, \ldots, x_n) \frac{\partial f}{\partial x_j} + \frac{1}{2} \sum_{j,l=1}^{n} b_{jl}(t, x_1, x_2, \ldots, x_n) \frac{\partial^2 f}{\partial x_j \partial x_l} = 0 \qquad \text{(first equation)},$$

$$\frac{\partial f}{\partial \tau} + \sum_{j=1}^{n} \frac{\partial}{\partial y_j} [a_j(\tau, y_1, \ldots, y_n)f] - \frac{1}{2} \sum_{j,l=1}^{n} \frac{\partial^2}{\partial y_j \partial y_l} [b_{jl}(\tau, y_1, \ldots, y_n)f] = 0 \qquad \text{(second equation)},$$

where the coefficients a_j and b_{jl} are determined by the equations

$$a_j(t, x_1, \ldots, x_n) = \lim_{\tau \to t} \frac{1}{\tau - t} \mathbf{M}[(Y_j - X_j) \mid X_1 = x_1, \ldots, X_n = x_n],$$

$$b_{jl}(t, x_1, \ldots, x_n) = \lim_{\tau \to t} \frac{1}{\tau - t} \mathbf{M}[(Y_j - X_j)(Y_l - X_l) \mid X_1 = x_1, \ldots, X_n = x_n],$$

and the initial conditions

$$\tau = t, f(t, x_1, \ldots, x_n; \tau, y_1, \ldots, y_n) = \delta(y_1 - x_1)\delta(y_2 - x_2)\cdots\delta(y_n - x_n).$$

Given the differential equation for the components of a Markov process $U_1(t), U_2(t), \ldots, U_n(t)$, to determine the coefficients a_j and b_{jl} (a and b in the linear case) one must compute the ratio of the increments of the ordinates of $U_j(t)$ during a small time interval to $(\tau - t)$, find the conditional expectations of these increments and of their products and pass to the limit as $\tau \to t$.

To any multidimensional Kolmogorov equation, there corresponds a system of differential equations for the components of the process

$$\frac{\partial U_l}{\partial t} = \psi_l(t, U_1, \ldots, U_n) + \sum_{m=1}^{n} g_{lm}(t, U_1, \ldots, U_n)\xi_m(t),$$
$$l = 1, 2, \ldots, n,$$

where $\xi_m(t)$ are mutually independent random functions with independent ordinates ("white noise") whose correlation functions are $K_m(\tau) = \delta(\tau)$, and the function ψ_l and g_{lm} are uniquely determined by the system

$$\left.\begin{aligned} \sum_{j=1}^{n} g_{lj}g_{jm} = b_{lm};\ g_{lj} = g_{jl} \\ \psi_l = a_l \end{aligned}\right\}, \qquad (l, m = 1, 2, \ldots, n).$$

To solve the Kolmogorov equations one can use the general methods of the theory of parabolic differential equations (see, for example Koshlyakov, Gliner and Smirnov, 1964). When a_l and b_{lm} are linear functions of the ordinates $U_l(t)$, the solution can be obtained by passing from the probability density $f(t, x_1, \ldots, x_n; \tau, y_1, \ldots, y_n)$ to the characteristic function

$$E(z_1, \ldots, z_n) = \int_{-\infty}^{\infty} \cdots \int_{-\infty}^{\infty} \exp\{i(z_1y_1 + \cdots + z_ny_n)\}$$
$$\times f(t, x_1, \ldots, x_n; \tau, y_1, \ldots, y_n)\, dy_1 \cdots dy_n,$$

obeying a partial differential equation of first order, which can be solved by general methods.[1]

If the coefficients a_l, b_{lm} are independent of t, then the problem of finding the stationary solutions of the Kolmogorov equations makes sense. To find the stationary solution of the second Kolmogorov equation, set $\partial f/\partial\tau = 0$ and look for the solution of the resulting equation as a function of $y_1, y_2, \ldots, y_n$ only. In the particular case of a one-dimensional Markov process, the solution is obtained by quadratures.

Any stationary normal process with a rational spectral density can be considered as a component of a multidimensional Markovian process.

The probability $W(T)$ that the ordinate of a one-dimensional Markov process during a time $T = \tau - t$ after a time t will, with known probability density $f_0(x)$ for the ordinates of the random function, remain within the limits of the interval (α, β) is

$$W(T) = \int_{\alpha}^{\beta} w(\tau, y)\, dy, \qquad \tau = t + T,$$

where the probability density $w(\tau, y)$ is the solution of the second Kolmogorov equation with conditions:

$$w(\tau, y) = f_0(y) \qquad \text{to} \quad \tau = t,$$

$$w(\tau, \alpha) = w(\tau, \beta) = 0 \quad \text{to} \quad \tau \geqslant t.$$

When the initial value of the ordinate is known, $f_0(y) = \delta(y - x)$. The probability density $f(T)$ of the sojourn time of a random function in the interval (α, β) is defined by the equality

$$f(T) = -\int_\alpha^\beta \frac{\partial w(\tau, y)}{\partial \tau}\, dy, \qquad T = \tau - t.$$

The average sojourn time $\overline{T}$ of the random function in the interval (α, β) is related to $w(\tau, y)$ by $\overline{T} = \int_0^\infty W(T)\, dT$. For $\alpha \neq \infty$, $\beta = \infty$, the last formulas give the probability $W(T)$ of sojourn time above a given level, the probability density $f(T)$ of the passage time and the average passage time T.

The average number of passages beyond the level α per unit time for a one-dimensional Markov process is infinity. However, the average number $n(\tau_0)$ of passages per unit time for passages with duration greater than $\tau_0 > 0$ is finite, and for a stationary process it is defined by the formula

$$n(\tau_0) = f(\alpha) \int_\alpha^\infty v(\tau_0, y)\, dy,$$

where $f(\alpha)$ is the probability density for the ordinate (corresponding to argument α) of the process and $v(\tau, y)$ is the solution of the second Kolmogorov equation for a stochastic process with conditions:

$$\tau < t,\ v(\tau, y) = 0;\ \tau \geqslant t,\ v(\tau, \alpha) = \delta(\tau - t),$$

which is equivalent to the solution of the equation for the Laplace-Carson transform $\bar{v}(p, y)$. For a stationary process

$$\frac{\partial}{\partial y}\left[\frac{1}{2}\frac{\partial}{\partial y}(b\bar{v}) - a\bar{v}\right] = p\bar{v}, \qquad \bar{v} = p \quad \text{for} \quad y = \alpha, \qquad \bar{v} = 0 \quad \text{for} \quad y = \infty.$$

The transform of $n(\tau_0)$ is

$$\bar{n}(p) = -\frac{1}{p} f(\alpha) \left.\frac{\partial(b\bar{v})}{\partial y}\right|_{y=\alpha} + f(\alpha)a(\alpha).$$

The probability $W(T)$ that the ordinate $U_1(t)$ of a component of a multidimensional Markov process will remain within the interval (α, β) if initially the distribution law for the components $U_1(t), U_2(t), \ldots, U_n(t)$ is known, is defined by the equation

$$W(T) = \int_{-\infty}^\infty \cdots \int_{-\infty}^\infty \int_\alpha^\beta w(\tau, y_1, \ldots, y_n)\, dy_1 \cdots dy_{n-1}\, dy_n, \qquad T = \tau - t,$$

where $w(\tau, y_1, \ldots, y_n)$ is the probability density that the components of the process reach a volume element $dy_1 \cdots dy_n$ at time τ under the assumption that during the interval (t, τ) the ordinate $U_1(t)$ has never left the limits of the interval (α, β). The function $w(\tau, y_1, \ldots, y_n)$ is the solution of the second Kolmogorov equation with the conditions

$$w(\tau, y_1, \ldots, y_n) = f_0(y_1, \ldots, y_n) \quad \text{for} \quad \tau = t;$$

$$w(\tau, \alpha, y_2, \ldots, y_n) = w(\tau, \beta, y_2, \ldots, y_n) = 0 \quad \text{for} \quad \tau \geqslant t.$$

The probability density $f(T)$ of the sojourn time of $U_1(t)$ in the interval (α, β) is defined by the formula

$$f(T) = -\int_{-\infty}^{\infty} \cdots \int_{-\infty}^{\infty} \int_{\alpha}^{\beta} \frac{\partial w(\tau, y_1, \ldots, y_n)}{\partial \tau} \, dy_1 \cdots dy_{n-1} \, dy_n,$$

$$T = \tau - t.$$

In the last formula, α can be $-\infty$, or β can be $+\infty$, which correspond to probabilities of sojourn time neither above nor below a given level.

Solution for Typical Examples

Example 40.1 Write the Kolmogorov equations for a multidimensional Markov process whose components $U_1(t), U_2(t), \ldots, U_n(t)$ satisfy the system of differential equations

$$\frac{dU_j(t)}{dt} = \psi_j(t, U_1, \ldots, U_n) + c_j \xi_j(t), \qquad j = 1, 2, \ldots, n,$$

where ψ_j are known continuous functions, c_j are known constants and $\xi_j(t)$ are independent random functions with the property of "white noise"; that is,

$$\bar{\xi}_j(t) = 0, \qquad K_{\xi_j}(\tau) = \delta(\tau).$$

Solution. To write the Kolmogorov equations, it suffices to determine the coefficients a_j and b_{jl} of these equations.

Denoting by X_j the ordinate of the random function $U_j(t)$ at time t and by Y_j its ordinate at time τ, and integrating the initial equations, we obtain

$$Y_j - X_j = \int_t^{\tau} \psi_j[t_1, U_1(t_1), \ldots, U_n(t_1)] \, dt_1 + c_j \int_t^{\tau} \xi_j(t_1) \, dt_1.$$

Considering the difference $\tau - t$ small, we can carry ψ_j outside the first integral with a precision up to second order terms and set $t_1 = T$, $U_1 = X_1$, $U_2 = X_2$, $\ldots$, $U_n = X_n$, which leads to

$$Y_j - X_j = \psi_j(t, X_1, \ldots, X_n)(\tau - t) + c_j \int_t^{\tau} \xi(t_1) \, dt_1;$$

that is,

$$\frac{Y_j - X_j}{\tau - t} = \psi_j(t, X_1, \ldots, X_n) + \frac{c_j}{\tau - t} \int_t^{\tau} \xi(t_1) \, dt_1.$$

Assuming that the random variables $X_1, \ldots, X_n$ are equal to $x_1, \ldots, x_n$, finding the expectation of the last equality and passing to the limit as $\tau \to t$, we obtain

$$a_j(t, x_1, \ldots, x_n) = \psi_j(t, x_1, \ldots, x_n).$$

Multiplying the expression for $(Y_j - X_j)$ by that for $(Y_l - X_l)$ and finding the expectation of the product obtained, we get

$$\begin{aligned}
&\mathbf{M}[(Y_j - X_j)(Y_l - X_l) \mid x_1, \ldots, x_n] \\
&\quad = \psi_j(t, x_1, \ldots, x_n)\psi_l(t, x_1, \ldots, x_n)(\tau - t)^2 + c_j c_l \int_t^\tau \int_t^\tau \delta(t_2 - t_1)\, dt_1\, dt_2 \\
&\quad = \psi_j(t, x_1, \ldots, x_n)\psi_l(t, x_1, \ldots, x_n)(\tau - t)^2 + c_j c_l(\tau - t),
\end{aligned}$$

which after division by $(\tau - t)$ and passage to the limit gives

$$b_{jl}(t, x_1, \ldots, x_n) = c_j c_l.$$

Example 40.2 Given the first Kolmogorov equation for the conditional probability density $f(t, x_1, x_2; \tau, y_1, y_2)$ of a normal Markov process,

$$\frac{\partial f}{\partial t} + x_2 \frac{\partial f}{\partial x_1} - (k^2 x_1 + 2h x_2)\frac{\partial f}{\partial x_2} + \frac{1}{2}\sigma^2 \frac{\partial^2 f}{\partial x_2^2} = 0,$$

determine the system of differential equations satisfied by the components $U_1(t)$ and $U_2(t)$.

SOLUTION. According to the notations, for the coefficients of the Kolmogorov equations we have

$$a_1 = x_2, \qquad a_2 = -k^2 x_1 - 2h x_2, \qquad b_{11} = b_{12} = 0, \qquad b_{22} = \sigma^2.$$

The required system of equations has the form:

$$\frac{\partial U_l}{\partial t} = \psi_l(t, U_1, U_2) + \sum_{m=1}^{2} g_{lm}(t, U_1, U_2)\xi_m(t), \qquad l = 1, 2,$$

where $\xi_m(t)$ is the "white noise" with zero expectation and unit variance. By the general formula given in the introductory section, p. 258, we have

$$g_{11}^2 + g_{12}^2 = 0, \qquad g_{11}g_{12} + g_{12}g_{22} = 0,$$
$$g_{12}^2 + g_{22}^2 = \sigma^2, \qquad \psi_l = a_l.$$

Consequently,

$$g_{11} = g_{12} = 0, \qquad g_{22} = \sigma,\ \psi_1 = x_2,\ \psi_2 = -k^2 x_1 - 2h x_2,$$

and the required system has the form

$$\frac{\partial U_1(t)}{\partial t} = U_2(t),$$

$$\frac{\partial U_2(t)}{\partial t} = -[k^2 U_1(t) + 2h U_2(t)] + \sigma\xi(t).$$

Eliminating $U_2(t)$ from the last equation, we obtain for $U_1(t)$ a second-order equation

$$\frac{\partial^2 U_1}{\partial t^2} + 2h\frac{\partial U_1}{\partial t} + k^2 U_1 = \sigma\xi(t).$$

Example 40.3 A normal stationary process $U(t)$ has the spectral density

$$S_u(\omega) = \frac{|P_m(i\omega)|^2}{|Q_n(i\omega)|^2},$$

where

$$P_m(x) = \beta_0 x^m + \beta_1 x^{m-1} + \cdots + \beta_m,$$

$$Q_n(x) = x^n + \alpha_1 x^{n-1} + \cdots + \alpha_n, \qquad n > m,$$

and α_j and β_j are known constants. Considering $U(t)$ as a component of a multidimensional Markov process, determine the coefficients of the Kolmogorov equations of this process.

SOLUTION. A stationary normal random function with rational spectral density is the solution of a linear differential equation, containing "white noise" on the right side. In the present case the equation has the form

$$\frac{d^n U}{dt^n} + \alpha_1 \frac{d^{n-1}U}{dt^{n-1}} + \cdots + \alpha_n U = \beta_0 \frac{d^m \xi}{dt^m} + \beta_1 \frac{d^{m-1}\xi}{dt^{m-1}} + \cdots + \beta_m \xi.$$

We turn from the nth order equation containing the derivatives of $\xi(t)$ on its right side to a system of equations of first order without derivatives of $\xi(t)$ on the right side. Let $U(t) = U_1(t)$, and introduce new variables defined by the equalities:

$$U_2 = \dot{U}_1,\ U_3 = \dot{U}_2, \ldots, U_{n-m} = \dot{U}_{n-m-1},$$

$$U_{n-m+1} = \dot{U}_{n-m} - c_{n-m}\xi,\ U_{n-m+2} = \dot{U}_{n-m+1} - c_{n-m+1}\xi,$$

$$\cdots\cdots\cdots\cdots\cdots\cdots$$

$$U_n = \dot{U}_{n-1} - c_{n-1}\xi,$$

where c_l are arbitrary constants for the time being. The foregoing equations form a system of $n - 1$ equations of first order. To get the last (nth) equation, in the initial nth-order differential equation it is necessary to express all derivatives of U in terms of U_j and their first derivatives. Performing these transformations, we obtain

$$\begin{aligned}\frac{dU_n}{dt} &+ \alpha_1 U_n + \cdots + \alpha_n U_1 + c_{n-m}\xi^{(m)} \\ &+ (c_{n-m+1} + \alpha_1 c_{n-m})\xi^{(m-1)} + (c_{n-m+2} + \alpha_1 c_{n-m+1} \\ &+ \alpha_2 c_{n-m})\xi^{(m-2)} + \cdots + (c_{n-1} + \alpha_1 c_{n-2} + \alpha_2 c_{n-3} + \cdots \\ &+ \alpha_{m-1}c_{n-m})\xi' + (\alpha_1 c_{n-1} + \alpha_2 c_{n-2} + \cdots + \alpha_m c_{n-m})\xi \\ &= \beta_0 \xi^{(m)} + \beta_1 \xi^{(m-1)} + \cdots + \beta_{m-1}\xi' + \beta_m \xi.\end{aligned}$$

Determining the coefficients c_j so that the derivatives of $\xi(t)$ disappear in the equation, we find the recursion relations

$$c_l = \beta_{l+m-n} - \sum_{j=n-m}^{l-1} \alpha_{l-j}c_j, \qquad l = n - m, \ldots, n - 1,$$

which for the last equation of the system gives

$$\frac{dU_n}{dt} + \alpha_1 U_n + \cdots + \alpha_n U_1 = c_n \xi, \qquad c_n = \beta_m - \sum_{j=n-m}^{n-1} \alpha_{n-j}c_j.$$

Since the components of an n-dimensional process satisfy a system of first-order equations on the right sides of which there is the "white noise," the process is an n-dimensional Markov process. The coefficients of the Kolmogorov equations are determined as in Example 40.1.

Example 40.4 The conditional probability density $f(t, x_1, x_2; \tau, y_1, y_2)$ of a two-dimensional stochastic process $U_1(t)$, $U_2(t)$ satisfies the equation

$$\frac{\partial f}{\partial \tau} - \frac{\partial}{\partial y_1}(\beta y_2 f) - \frac{\partial}{\partial y_2}(\alpha y_1 f) - \frac{1}{2}\frac{\partial^2}{\partial y_1^2}(\beta\sqrt{1 + y_2^2} f) - \frac{1}{2}\frac{\partial^2}{\partial y_2^2}(\alpha\sqrt{1 + y_1^2} f) = 0,$$

where α and β are constants.

Determine the system of differential equations satisfied by $U_1(t)$ and $U_2(t)$.

SOLUTION. The given equation represents the second Kolmogorov equation and, consequently, the process is a two-dimensional Markov process.

The coefficients of the equation are

$$a_1 = -\beta y_2, \qquad a_2 = -\alpha y_1, \qquad b_{11} = \beta\sqrt{1 + y_2^2}$$

$$b_{22} = \alpha\sqrt{1 + y_1^2}, \qquad b_{12} = 0.$$

The required system of equations has the form

$$\frac{dU_1}{dt} = \psi_1(t, U_1, U_2) + \sum_{m=1}^{2} g_{1m}(t, U_1, U_2)\xi_m(t),$$

$$\frac{dU_2}{dt} = \psi_2(t, U_1, U_2) + \sum_{m=1}^{2} g_{2m}(t, U_1, U_2)\xi_m(t),$$

where $\xi_1(t)$ and $\xi_2(t)$ are uncorrelated random functions of the "white noise" type with unit variance. According to the general theory, to determine g_{lm} one should solve the algebraic system of equations

$$g_{11}^2 + g_{12}^2 = \beta\sqrt{1 + y_2^2}, \qquad g_{21}^2 + g_{22}^2 = \alpha\sqrt{1 + y_1^2},$$

$$g_{11}g_{12} + g_{12}g_{22} = 0.$$

Hence, it follows that

$$g_{12} = 0, \qquad g_{11} = \sqrt{\beta\sqrt{1 + y_2^2}}, \qquad g_{22} = \sqrt{\alpha\sqrt{1 + y_1^2}},$$

$$\psi_1 = -\beta y_2, \qquad \psi_2 = -\alpha y_1.$$

Consequently, the required system has the form

$$\frac{dU_1}{dt} + \beta U_2 = \sqrt{\beta\sqrt{1 + U_2^2}}\,\xi_1(t),$$

$$\frac{dU_2}{dt} + \alpha U_1 = \sqrt{\alpha\sqrt{1 + U_1^2}}\,\xi_2(t).$$

Example 40.5 Determine the asymmetry Sk and the excess Ex of the ordinate of a random function $Z(t)$ defined by the equality

$$\frac{dZ(t)}{dt} + k^2 Z(t) = \zeta^2(t),$$

if $\zeta(t)$ is a normal random function, $\bar{\zeta} = 0$, $K_\zeta(\tau) = \sigma^2 e^{-\alpha|\tau|}$ and the transient phase of the process is assumed to have ended (compare with Problem 35.29).

SOLUTION. Since the spectral density

$$S_\zeta(\omega) = \frac{\alpha\sigma^2}{\pi}\frac{1}{\omega^2 + \alpha^2}$$

is a rational function of frequency, $\zeta(t)$ satisfies the equation

$$\frac{d\zeta}{dt} + \alpha\zeta = \sigma\sqrt{\frac{\alpha}{\pi}}\,\xi(t),$$

where $\xi(t)$ is "white noise" with zero expectation and unit variance. Therefore, considering a two-dimensional stochastic process with components $U_1(t) = Z(t)$, $U_2(t) = \zeta(t)$, for the conditional probability density $f(t, x_1, x_2; \tau, y_1, y_2)$ of this process we obtain the second Kolmogorov equation in the following form:

$$\frac{\partial f}{\partial \tau} + \frac{\partial}{\partial y_1}[(y_2^2 - k^2 y_1)f] - \alpha\frac{\partial}{\partial y_2}(y_2 f) - \alpha\sigma^2\frac{\partial^2 f}{\partial y_2^2} = 0.$$

For the stationary mode $f(t, x_1, x_2; \tau, y_1, y_2) = f(y_1, y_2)$ and the Kolmogorov equation becomes

$$\frac{\partial}{\partial y_1}[(y_2^2 - k^2 y_1)f] - \alpha\frac{\partial}{\partial y_2}(y_2 f) - \alpha\sigma^2\frac{\partial^2 f}{\partial y_2^2} = 0.$$

According to the assumption of this problem, it is necessary to determine the moments m_l of the ordinate of $Y_1(t)$ up to the fourth, inclusive. The required moments relate to the two-dimensional probability density $f(y_1, y_2)$ by

$$m_l = \int_{-\infty}^{\infty}\int_{-\infty}^{\infty} y_1^l f(y_1, y_2)\,dy_1\,dy_2 = \int_{-\infty}^{\infty} \chi_l(y_2)\,dy_2,$$

where

$$\chi_l(y_2) = \int_{-\infty}^{\infty} y_1^l f(y_1, y_2)\,dy_1.$$

Multiplying both sides of the Kolmogorov equation by y_1^l, integrating the result with respect to y_1 with infinite limits, and taking into account that

$$\int_{-\infty}^{\infty} y_1^l \frac{\partial}{\partial y_1}[(y_2^2 - k^2 y_1)f(y_1, y_2)]\,dy_1$$
$$= y_1^l(y_2^2 - k^2 y_1)f(y_1, y_2)\Big|_{-\infty}^{\infty} - l\int_{-\infty}^{\infty} y_1^{l-1}(y_2^2 - k^2 y_1)f(y_1, y_2)\,dy_1$$
$$= k^2 l\chi_l(y_2) - l y_2^2\chi_{l-1}(y_2);$$

we obtain a recursion relation between $\chi_l(y_1)$ and $\chi_{l-1}(y_2)$:

$$\alpha\sigma^2\frac{d^2\chi_l}{dy_2^2} + \alpha\frac{d}{dy_2}(y_2\chi_l) - k^2 l\chi_l = -l y_2^2\chi_{l-1}.$$

Multiplying both sides of the last equality successively by 1, y_2^2, y_2^4 and y_2^6, integrating by parts and eliminating all zero terms that appear outside the integral, we get a series of equations:

$$m_l = \frac{1}{k^2}\int_{-\infty}^{\infty} y_2^2\chi_{l-1}(y_2)\,dy_2,$$

$$m_{l+1} = \frac{1}{k^2(k^2l + 2\alpha)}\left\{l\int_{-\infty}^{\infty} y_2^4\chi_{l-1}(y_2)\,dy_2 + 2\alpha\sigma^2 m_l\right\},$$

$$m_{l+2} = \frac{l+1}{k^2(k^2l + 4\alpha)[(l+1)k^2 + 2\alpha]} \times \left\{l\int_{-\infty}^{\infty} y_2^6\chi_{l-1}(y_2)\,dy_2 + \frac{2\alpha\sigma^2(6k^2 + 7k^2l + 4\alpha)}{l+1}m_{l+1}\right\},$$

$$m_{l+3} = \frac{(l+1)(l+2)}{k^2(k^2l + 6\alpha)(k^2l + k^2 + 4\alpha)[k^2(l+2) + 2\alpha]} \times \left\{l\int_{-\infty}^{\infty} y_2^8\chi_{l-1}(y_2)\,dy_2 + \frac{2\sigma^2\alpha[(13k^2 + 7k^2l + 4\alpha)(k^2l + 6\alpha) + 15k^2(k^2l + k^2 + 2\alpha)(l+2)]m_{l+2}}{(l+1)(l+2)} - \frac{4\alpha^2}{l+1}m_{l+1}\right\}.$$

Setting $l = 1$ in these equalities, we can express the four moments in terms of $\chi_0(y_2)$. Because of the normality of function $Y_2(\tau) = \zeta(\tau)$,

$$\chi_0(y_2) = \int_{-\infty}^{\infty} f(y_1, y_2)\,dy_1 = f(y_2) = \frac{1}{\sigma\sqrt{2\pi}}\exp\left\{-\frac{y_2^2}{2\sigma^2}\right\}.$$

Consequently, all the integrals appearing in the preceding equalities can be computed and the result coincides with that of Problem 35.29, which is solved in a more complicated manner.

Example 40.6 Determine the conditional probability density $f(t, x_1, \ldots, x_n; \tau, y_1, \ldots, y_n)$ of a multidimensional Markov process if in the second Kolmogorov equation

$$\frac{\partial f}{\partial \tau} + \sum_{k=1}^{n}\frac{\partial}{\partial y_k}(a_k f) - \frac{1}{2}\sum_{k,j=1}^{n}\frac{\partial^2}{\partial y_j \partial y_k}(b_{jk}f) = 0,$$

the coefficients b_{jk} are constants, the coefficients a_l are linear functions of y_j:

$$a_l = \alpha_l + \sum_{j=1}^{n}\alpha_{lj}y_j, \qquad l, j = 1, 2, \ldots, n,$$

and the range of y_j is $(-\infty, \infty)$.

SOLUTION. By assumption, the solution must satisfy the initial condition

$$f = \prod_{l=1}^{n}\delta(y_l - x_l) \quad \text{for} \quad \tau = t$$

and the condition that f vanishes as $|y_l| \to \infty$, and $\int_{-\infty}^{\infty} \cdots \int_{-\infty}^{\infty} f\, dy_1 \cdots dy_n = 1$ for any τ.

We now turn from the probability density f of the system of random variables $Y_1, Y_2, \ldots, Y_n$ to the characteristic function

$$E(z_1, z_2, \ldots, z_n) = \int_{-\infty}^{\infty} \cdots \int_{-\infty}^{\infty} \exp\left\{i \sum_{l=1}^{n} z_l y_l\right\} f\, dy_1 \cdots dy_n.$$

For this purpose, we multiply both sides of the second Kolmogorov equation by $\exp\{i \sum_{l=1}^{n} z_l y_l\}$ and integrate with respect to $y_1, y_2, \ldots, y_n$ between infinite limits.

Since

$$\int_{-\infty}^{\infty} \cdots \int_{-\infty}^{\infty} \exp\left\{i \sum_{j=1}^{n} z_j y_j\right\} \frac{\partial}{\partial y_l}(a_l f)\, dy_1 \cdots dy_n$$

$$= -i \int_{-\infty}^{\infty} \cdots \int_{-\infty}^{\infty} \exp\left\{i \sum_{j=1}^{n} z_j y_j\right\} z_l a_l f\, dy_1 \cdots dy_n$$

$$= -i z_l \alpha_l E - z_l \sum_{j=1}^{n} \alpha_{lj} \frac{\partial E}{\partial z_j},$$

$$\int_{-\infty}^{\infty} \cdots \int_{-\infty}^{\infty} \frac{\partial^2 f}{\partial y_j \partial y_l} \exp\left\{i \sum_{m=1}^{n} z_m y_m\right\} dy_1 \cdots dy_n = -z_j z_l E,$$

the equation for E has the form

$$\frac{\partial E}{\partial \tau} - \sum_{l,m=1}^{n} \alpha_{ml} z_m \frac{\partial E}{\partial z_l} = \left(i \sum_{m=1}^{n} z_m \alpha_m - \frac{1}{2} \sum_{m,l=1}^{n} z_m z_l b_{ml}\right) E.$$

Letting $E = \exp\{-V\}$, we get for V the equation

$$\frac{\partial V}{\partial \tau} - \sum_{j,l=1}^{n} \alpha_{jl} z_j \frac{\partial V}{\partial z_l} = -i \sum_{j=1}^{n} z_j \alpha_j + \frac{1}{2} \sum_{j,l=1}^{n} z_j z_l b_{jl},$$

which according to the initial conditions for f must be solved under the conditions

$$\tau = t, \qquad V = i \sum_{j=1}^{n} z_j x_j.$$

From the general theory, it is known that the distribution law for the process considered is normal. Therefore, we seek the solution for V in the form of a second-degree polynomial of z_j, that is,

$$V = \frac{1}{2} \sum_{j,l=1}^{n} k_{jl} z_j z_l - i \sum_{j=1}^{n} \bar{y}_j z_j,$$

where k_{jl} and $\bar{y}_j$ are real functions of τ. To determine these functions, we substitute the last expression in the differential equation for V and equate the coefficients of equal powers of z_i in the left and right sides. We find

$$\begin{aligned} &\frac{d\bar{y}_j}{d\tau} - \sum_{l=1}^{n} \alpha_{jl} \bar{y}_l = \alpha_j, \\ &\frac{dk_{jl}}{d\tau} - \sum_{m=1}^{n} \alpha_{jm} k_{ml} = b_{jl} \end{aligned} \qquad (j, l = 1, 2, \cdots, n).$$

The system of equations for $\bar{y}_j$ is independent of k_{jl} and should be solved with the initial conditions: $\tau = t$, $\bar{y}_j = x_j$. The system of equations for k_{jl} is independent of y_j and should be solved with the initial conditions: $\tau = t$, $k_{jl} = 0$. From the general theory of linear differential equations, it follows that $\bar{y}_j$ and k_{jl} are linear combinations of exponential functions of form $e^{\lambda(\tau - t)}$, where λ are the roots of the corresponding characteristic equation (in case of multiple roots, the coefficients in the exponent can be polynomials of τ). The general formulas can be obtained by matrix operations.

Example 40.7 Find the conditional probability density $f(t, x; \tau, y)$ for a process defined by the equation

$$\frac{\partial f}{\partial \tau} - \frac{\partial}{\partial y}\left[\left(\beta y - \frac{\alpha^2}{2y}\right)f\right] - \frac{\alpha^2}{2}\frac{\partial^2 f}{\partial y^2} = 0, \qquad 0 < y < \infty$$

if α and β are constants.

SOLUTION. We apply the Fourier method; i.e., first we seek two functions $\psi(\tau)$ and $\chi(y)$ whose product satisfies the given equation independent of the form of the initial conditions. Substituting them in the equation, we get

$$\frac{1}{\psi}\frac{d\psi}{d\tau} = \frac{1}{\chi}\left\{\frac{d}{dy}\left[\left(\beta y - \frac{\alpha^2}{2y}\right)\chi\right] + \frac{\alpha^2}{2}\frac{d^2\chi}{dy^2}\right\}.$$

Since the left side of the equality is independent of y and the right side is independent of τ, both sides must be equal to a constant, which we denote by λ, obtaining

$$\frac{d\psi}{d\tau} + \lambda\psi = 0,$$

$$\frac{\alpha^2}{2}\frac{d^2\chi}{dy^2} + \frac{d}{dy}\left[\left(\beta y - \frac{\alpha^2}{2y}\right)\chi\right] + \lambda\chi = 0.$$

The first equation has the obvious solution

$$\psi(\tau) = e^{-\lambda(\tau - t)}.$$

The second equation has a solution vanishing at infinity only for discrete values of $\lambda = 2n\beta$, $n = 0, 1, \ldots$. In this case, the equation for $\chi(y)$ has the solution

$$\chi(y) = \frac{1}{n!}\frac{y}{\sigma^2}\exp\left\{-\frac{y^2}{2\sigma^2}\right\}L_n\left(\frac{y^2}{2\sigma^2}\right),$$

where $L_n(x) = e^x\, d^n/dx^n\,(e^{-x}x^n)$ are the orthogonal Laguerre polynomials, and $\sigma^2 = \alpha^2/2\beta^2$. Since the functions $\psi(\tau)$ and $\chi(y)$ depend on the integer n, the solution f of the initial differential equation can be found as a linear combination of the products of these functions; that is,

$$f(t, x; \tau, y) = \sum_{n=0}^{\infty} c_n \exp\{-2n\beta(\tau - t)\}\frac{1}{n!}\frac{y}{\sigma^2}\exp\left\{-\frac{y^2}{2\sigma^2}\right\}L_n\left(\frac{y^2}{2\sigma^2}\right),$$

where the coefficients c_n should be such that for $\tau = t$, the function $f(t, x; \tau, y)$ becomes $\delta(y - x)$, that is,

$$\sum_{n=0}^{\infty} c_n \frac{1}{n!}\frac{y}{\sigma^2}\exp\left\{-\frac{y^2}{2\sigma^2}\right\}L_n\left(\frac{y^2}{2\sigma^2}\right) = \delta(y - x).$$

To determine the constants c_n, it is sufficient to multiply the last equation by

$$y \exp\left\{-\frac{y^2}{2\sigma^2}\right\}L_n\left(\frac{y^2}{2\sigma^2}\right)$$

and to integrate with respect to y between the limits $(0, \infty)$. Using the orthogonality of the Laguerre polynomials and the properties of the δ-function, we find that

$$c_n = \frac{1}{n!}\frac{x}{\sigma^2}\exp\left\{-\frac{x^2}{2\sigma^2}\right\}L_n\left(\frac{x^2}{2\sigma^2}\right);$$

that is,

$$f(t, x; \tau, y) = \frac{xy}{\sigma^4}\exp\left\{-\frac{x^2 + y^2}{2\sigma^2}\right\}$$
$$\times \sum_{n=0}^{\infty}\frac{1}{(n!)^2}L_n\left(\frac{x^2}{2\sigma^2}\right)L_n\left(\frac{y^2}{2\sigma^2}\right)\exp\{-2n\beta(\tau - t)\}.$$

Example 40.8 Find the probability $W(\tau)$ that the ordinate of the process $U(t)$ defined by equation $dU/dt + \alpha U = \xi(t)$, where $S_\xi(\omega) = c^2 = \text{const.}$, $\bar{\xi} = 0$ at time τ never exceeds the level $y = 0$, if for $t = 0$, $U(t) = -\beta$; $\beta > 0$.

SOLUTION. The probability density $w(\tau, y)$ that at time τ the ordinate of the stochastic process, which never exceeds the zero level, will lie on the interval $(y, y + dy)$ is defined by the second Kolmogorov equation

$$\frac{\partial w}{\partial \tau} - \alpha\frac{\partial}{\partial y}(yw) - \frac{c^2}{2}\frac{\partial^2 w}{\partial y^2} = 0,$$

which in the present case should be solved for $y \leqslant 0$ with the conditions: $w(\tau, y) = \delta(y + \beta)$ for $\tau = 0$; $w(\tau, 0) = 0$ for any τ. The required probability is

$$W(\tau) = \int_{-\infty}^{\infty} w(\tau, y)\,dy.$$

To simplify the coefficients of the equation, let us introduce nondimensional variables

$$\tau_1 = \alpha\tau, \qquad y_1 = \frac{\sqrt{\alpha}}{c}y,$$

after which the equation becomes

$$\frac{1}{2}\frac{\partial^2 w}{\partial y_1^2} + \frac{\partial}{\partial y_1}(y_1 w) - \frac{\partial w}{\partial \tau_1} = 0;$$

$$w(\tau_1, y_1) = \frac{\sqrt{\alpha}}{c}\delta(y_1 + \beta_1) \quad \text{for} \quad \tau_1 = 0,$$

$w(\tau_1, 0) = 0$ for $\tau_1 > 0$, where $\beta_1 = \sqrt{\alpha}/c\,\beta$.

Solving this equation by the Fourier method and setting $w(\tau_1, y_1) = \psi(\tau_1)\chi(y_1)$, we obtain for $\psi(\tau_1)$ and $\chi(y_1)$ the equations

$$\frac{1}{\psi}\frac{d\psi}{d\tau_1} = -\lambda^2, \qquad \frac{d^2\chi}{dy_1^2} + 2y_1\frac{d\chi}{dy_1} + 2(\lambda^2 + 1)\chi = 0.$$

The first equation has the obvious solution $\psi(\tau_1) = \exp\{-\lambda^2\tau_1\}$ and the second one has finite solutions at infinity only if $\lambda^2 = n$ ($n = 0, 1, 2, \ldots$), when

$$\chi(y_1) = \exp\{-y_1^2\}H_n(y_1),$$

where

$$H_n(y_1) = (-1)^n \exp\{y_1^2\} \frac{d^n}{dy_1^n} (\exp\{-y_1^2\})$$

is the Hermite polynomial. Consequently, the solution must be sought in the form

$$w = \exp\{-y_1^2\} \sum_{n=0}^{\infty} a_n \exp\{-n\tau_1\}H_n(y_1).$$

Since for $y_1 = 0$, w must vanish for any τ_1, the series can contain only polynomials $H_n(y_1)$ with odd indices ($H_{2k+1}(0) = 0$, $H_{2k}(0) \neq 0$ for any integer $k > 0$). Therefore, the solution should be of the form

$$w = \exp\{-y_1^2\} \sum_{k=0}^{\infty} a_{2k+1} \exp\{-(2k + 1)\tau_1\}H_{2k+1}(y_1).$$

To find the coefficients a_{2k+1}, it is necessary to fulfill the initial condition; that is,

$$\exp\{-y_1^2\} \sum_{k=0}^{\infty} a_{2k+1}H_{2k+1}(y_1) = \frac{\sqrt{\alpha}}{c}\,\delta(y_1 + \beta_1), \qquad y_1 \leqslant 0.$$

This condition is equivalent, for the range $(-\infty, +\infty)$ of y_1, to the condition

$$\exp\{-y_1^2\} \sum_{k=0}^{\infty} a_{2k+1}H_{2k+1}(y_1) = \frac{\sqrt{\alpha}}{c}\,\delta(y_1 + \beta_1) - \frac{\sqrt{\alpha}}{c}\,\delta(y_1 - \beta_1).$$

Multiplying both sides of the last equality by $H_{2k+1}(y_1)$, integrating with respect to y_1 from $-\infty$ to $+\infty$ and considering that

$$\int_{-\infty}^{\infty} e^{-x^2}H_n(x)H_m(x)\,dx = 2^n n!\sqrt{\pi}\,\delta_{mn}$$

($\delta_{nn} = 1$, $\delta_{mn} = 0$ for $n \neq m$), we obtain

$$a_{2k+1} = -\frac{1}{2^{2k+1}(2k + 1)!\sqrt{\pi}}\,\frac{\sqrt{\alpha}}{c}\cdot 2H_{2k+1}(\beta_1).$$

Thus,

$$w = -\exp\{-y_1^2\} \sum_{k=0}^{\infty} \exp\{-(2k + 1)\tau_1\} \times \frac{\sqrt{\alpha}}{2^{2k}(2k + 1)!\sqrt{\pi}\,c} H_{2k+1}(\beta_1)H_{2k+1}(y_1).$$

Returning to variables y and τ, we find

$$w(\tau, y) = -\frac{\sqrt{\alpha}}{c\sqrt{\pi}} \sum_{k=0}^{\infty} \exp\{-(2k + 1)\tau\alpha\} \exp\left\{-\frac{\alpha}{c^2}y^2\right\} \times \frac{1}{2^{2k}(2k + 1)!} H_{2k+1}\left(\frac{\sqrt{\alpha}\,\beta}{c}\right)H_{2k+1}\left(\frac{\sqrt{\alpha}\,y}{c}\right).$$

Substituting the resulting series in the formula for $W(\tau)$ and considering that

$$\int_{-\infty}^{0} \exp\left\{-\frac{\alpha y^2}{c^2}\right\} H_{2k+1}\left(\frac{\sqrt{\alpha}\, y}{c}\right) \frac{\sqrt{\alpha}}{c}\, dy = \int_{-\infty}^{0} \exp\{-y_1^2\} H_{2k+1}(y_1)\, dy_1 = (-1)^{k+1} \frac{(2k)!}{k!},$$

we obtain that

$$W(\tau) = \frac{1}{\sqrt{\pi}} \sum_{k=0}^{\infty} \exp\{-(2k+1)\tau\alpha\} \frac{(-1)^k}{2^{2k}(2k+1)k!} H_{2k+1}\left(\frac{\sqrt{\alpha}\, \beta}{c}\right).$$

PROBLEMS

40.1 Find the coefficients of the Kolmogorov equations for an n-dimensional Markov process if its components $U_1(t), U_2(t), \ldots, U_n(t)$ are determined by the system of equations

$$\frac{dU_j}{dt} = \psi_j(t, U_1, \ldots, U_n) + \varphi_j(t, U_1, \ldots, U_n)\xi_j(t),$$
$$j = 1, 2, \ldots, n,$$

where ψ_j and φ_j are known continuous functions of their variables and $\xi_j(t)$ are independent random functions with the properties of "white noise":

$$\bar{\xi}_j = 0, \qquad K_{\xi j}(\tau) = \delta(\tau).$$

40.2 Given the system of differential equations

$$\frac{dU_j}{dt} = \psi_j(t, U_1, \ldots, U_n, Z), \qquad j = 1, 2, \ldots, n,$$

where ψ_j are known functions of their arguments and $Z(t)$ is a normal stationary stochastic process with spectral density,

$$S_z(\omega) = \frac{c^2}{(\omega^2 + \alpha^2)^3},$$

add to the multidimensional process $U_1(t), \ldots, U_n(t)$ the necessary number of components so that the process obtained is Markovian. Write the Kolmogorov equations for it.

40.3 Suppose $U(t)$, a stationary normal process, is given with spectral density

$$S_u(\omega) = \frac{c^2\omega^2}{[(\omega^2 + \alpha^2 + \beta^2)^2 - 4\beta^2\omega^2]},$$

where c, α and β are constants.

Show that $U(t)$ can be considered as a component of a multidimensional Markov process. Determine the number of dimensions of this process and the coefficients of the Kolmogorov equations.

40.4 Determine the coefficients of the Kolmogorov equations of a multidimensional Markov process defined by the system of equations

$$\frac{dU_j}{dt} = \varphi_j(t, U_1, \ldots, U_n) + Z_j(t),$$

where

$$\bar{z}_j(t) = 0, \qquad \mathbf{M}[Z_j(t)Z_l(t + \tau)] = \psi_{jl}(t)\delta(\tau),$$
$$j, l = 1, 2, \ldots, n,$$

and φ_j and ψ_{jl} are known continuous functions of their arguments.

40.5 The random functions $U_j(t)$ satisfy the system of differential equations

$$\frac{dU_j}{dt} = \varphi_j(t, U_1, \ldots, U_r, Z), \qquad j = 1, 2, \ldots, r,$$

where φ_j are known continuous functions of their arguments and $Z(t)$ is a stationary normal random function with rational density

$$\bar{z} = 0, \qquad S_z(\omega) = \frac{|P_m(i\omega)|^2}{|Q_n(i\omega)|^2}, \qquad n > m,$$

where the polynomials

$$P_m(x) = \beta_0 x^m + \beta_1 x^{m-1} + \cdots + \beta_m,$$
$$Q_n(x) = x^n + \alpha_1 x^{n-1} + \cdots + \alpha_n$$

have roots only in the upper half-plane.

Show that $U_1(t), \ldots, U_r(t)$ can be considered as components of a multidimensional Markov process, determine the number of dimensions and the coefficients of the Kolmogorov equations of this process.

40.6 Show that if the Kolmogorov equations

$$\frac{\partial f}{\partial \tau} + \sum_{j=1}^{n} \frac{\partial}{\partial y_j}\left[\left(\alpha_j - \sum_{m=1}^{n} \alpha_{jm} y_m\right) f\right] - \frac{1}{2} \sum_{j,m=1}^{n} \frac{\partial^2}{\partial y_j \partial y_m}(b_{jm} f) = 0,$$

where $\alpha_j, \alpha_{jm}, b_{jm}$ $(j, m = 1, 2, \ldots, n)$ are constants, hold for a multidimensional Markov process, then the stochastic process satisfies the system of differential equations:

$$\frac{dU_j}{dt} + \sum_{m=1}^{n} \alpha_{jm} U_m = \eta_j(t), \qquad j = 1, 2, \ldots, n,$$

where

$$\bar{\eta}_j = \alpha_j, \qquad K_{\eta_j}(\tau) = b_{jj}\delta(\tau), \qquad R_{\eta_j \eta_m}(\tau) = b_{jm}\delta(\tau).$$

40.7 Derive the system of differential equations for the components of a two-dimensional Markov process $U_1(t)$, $U_2(t)$ if the conditional probability density $f(t, x_1, x_2; \tau, y_1, y_2)$ satisfies the equation

$$\frac{\partial f}{\partial \tau} + \frac{1}{\mu}\left[y_2 \frac{\partial f}{\partial y_1} + \varphi(y_1) \frac{\partial f}{\partial y_2}\right] - \frac{1}{\mu^2}\left[\frac{\partial^2}{\partial y_1^2}(y_2 f) + \frac{\kappa^2}{2} \frac{\partial^2 f}{\partial y_2^2}\right] = 0.$$

40.8 Determine the distribution law for the ordinate of a random function $U(t)$ for the stationary mode if

$$\frac{d^2U}{dt^2} + \frac{\alpha^2}{2}\frac{dU}{dt} + \varphi(U) = \xi(t),$$

where α is a constant, $\varphi(U)$ is a given function that ensures the existence of a stationary mode and

$$\bar{\xi}(t) = 0, \qquad K_{\xi}(\tau) = \sigma^2\delta(\tau).$$

Solve the problem for the particular case when $\varphi(U) = \beta^2 U^3$.

40.9 Determine the stationary distribution law for the ordinate of a random function $U(t)$ if

$$\frac{dU}{dt} = \varphi(U) + \psi(U)\xi(t),$$

where $\varphi(U)$ and $\psi(U)$ are known functions and $\xi(t)$ represents "white noise" with zero expectation and unit variance.

40.10 A diode detector consists of a nonlinear element with volt-ampere characteristic $F(V)$ connected in series with a parallel RC circuit. A random input signal $\zeta(t)$ is fed to the detector. Determine the stationary distribution law of the voltage $U(t)$ in the RC circuit if the equation of the detector has the form

$$\frac{dU}{dt} + \frac{1}{RC}U = \frac{1}{C}F(\zeta - U),$$

where R and C are constants and $\zeta(t)$ is a normal stationary function for which

$$\bar{\zeta}(t) = 0, \qquad K_{\zeta}(\tau) = \sigma^2 \exp\left\{-\frac{|\tau|}{RC}\right\}.$$

Solve the problem for the particular case in which

$$F(v) = \begin{cases} kv, & v \geqslant 0, \\ 0, & v < 0. \end{cases}$$

40.11 Determine the distribution law for the ordinate of a random function $U(t)$ for time $\tau > 0$ if

$$\alpha\frac{dU}{dt} - \frac{\sigma^2}{2U} = \sigma\xi(t), \qquad \bar{\xi} = 0, \qquad K_{\xi}(\tau) = \alpha\delta(\tau),$$

$$f_0(x) = \frac{x}{a^2}\exp\left\{-\frac{x^2}{2a^2}\right\}, \qquad U(t) = X \quad \text{for} \quad t = 0 \quad (x \geqslant 0).$$

40.12 An input signal representing a normal stochastic process $\zeta(t)$ with a small correlation time is received by an exponential detector whose voltage $U(t)$ is defined by the equation

$$\frac{dU}{dt} + \frac{1}{RC}U = \frac{i_0}{C}e^{a\zeta - aU},$$

where R, C, a, i_0 are the constants of the detector, $\bar{\zeta} = 0$ and

$$K_\zeta(\tau) = \sigma^2 e^{-\alpha|\tau|}.$$

Using the approximate representation

$$e^{a\xi(t)} = \mathbf{M}[e^{a\xi(t)}] + \xi(t)$$

and considering that

$$\xi(t) = e^{a\xi(t)} - \mathbf{M}[e^{a\xi(t)}]$$

is a δ-correlated process,

$$K_\xi(\tau) \approx \sigma_\zeta^2 \delta(\tau),$$

where

$$\sigma_\zeta^2 = \int_{-\infty}^{\infty} K_\zeta(\tau)\, d\tau,$$

determine the stationary distribution law for the ordinate of $U(t)$.

40.13 A stochastic process $U(t)$ satisfies the equation

$$\frac{dU}{dt} = -\varphi(U) + \sigma\xi(t),$$

where $\varphi(U)$ is a given function, $\zeta(t)$ represents "white noise" with zero expectation and unit variance, and for a given form of the function $\varphi(U)$ a stationary mode is possible. Determine the probability density $f(y)$ of the stationary mode.

40.14 A random function $U(t)$ satisfies the equation

$$\frac{dU}{dt} = \alpha(t) + \beta(t)U + \gamma(t)\xi(t)$$

with initial conditions $\tau = t$, $U(t) = x$.

Find the distribution law for the ordinates of this random function for time $\tau \geqslant t$ if $\alpha(t)$, $\beta(t)$ and $\gamma(t)$ are known functions of time and $\xi(t)$ is "white noise" with zero expectation and unit variance.

40.15 The deviation of the elevator of an airplane is communicated to the automatic pilot to eliminate the effect of wind pulsations characterized by a random function $\varepsilon(t)$. The signal is approximately described by the differential equation

$$T_0 \frac{d\Delta}{dt} + \Delta = i_0\varepsilon(t),$$

where T_0 and i_0 are constants.

Determine the conditional probability density $f(t, x; \tau, y)$ of the ordinate of the random function $\Delta(t)$ if the expectation $\bar{\varepsilon}(t) = 0$ and one may approximately consider that $K_\varepsilon(\tau) = \sigma_\varepsilon^2\delta(\tau)$, and $\Delta = x$ for $\tau = t$.

40.16 The incoming random perturbation at the input of a system of second order is described by $\zeta(t)$:

$$\frac{d^2U}{dt^2} + 2h\frac{dU}{dt} + k^2U = \zeta(t), \qquad k > h > 0.$$

Determine the conditional distribution law of the ordinate of the stochastic process $U(t)$ at time $\tau \geqslant t$ if at time t, $U(t) = x$, $\dot{U}(t) = 0$, $\bar{\zeta}(t) = 0$, $K_\zeta(\tau) = c^2\delta(\tau)$; c, h, k are known constants.

40.17 The equation defining the operation of an element of a system of automatic control has the form

$$\frac{dU}{dt} = -\alpha \operatorname{sgn} U + c\xi(t),$$

where α and c are constants and

$$\bar{\xi}(t) = 0, \qquad K_\xi(\tau) = \delta(\tau).$$

Write the Kolmogorov equation for the determination of the conditional probability density $f(t, x; \tau, y)$.

40.18 A moving charged particle is under the influence of three forces directed parallel to the velocity vector $U(t)$: the forces created by the electric field of intensity $\xi(t)$, the accelerating force created by the field whose intensity can be taken inversely proportional to the velocity of the particle and the friction forces proportional to the velocity. The motion equation has the form

$$\frac{dU}{dt} = -\beta U + \frac{\gamma}{U} + \alpha\xi(t).$$

Find the probability density $f(t, x; \tau, y)$ for the magnitude of the velocity $U(t)$ if α, β and γ are constants and $\bar{\xi}(t) = 0$, $K_\xi(\tau) = \delta(\tau)$; the mass of the particle is m.

40.19 A radio receiver can detect a random input noise $U(t)$ only if the absolute value of the signal is greater than the sensitivity level of the receiver u_0. Determine the probability $W(T)$ that during time T no false signal will be received if $U(t)$ is a normal stochastic process with zero expectation and with correlation function

$$K_u(\tau) = \sigma^2 e^{-\alpha|\tau|},$$

where u_0, α and σ are constants and $U(t) = 0$ for $t = 0$.

40.20 A radio receiver can detect a random input noise $U(t)$ if the signal (not its absolute value) is greater than the sensitivity level u_0 of the receiver. Determine the probability $W(T)$ that during time T no false signal will be received if $U(t)$ is a normal stochastic process with zero expectation and with correlation function

$$K_u(\tau) = \sigma^2 e^{-\alpha|\tau|},$$

where u_0, α and σ are constants and $U(t) = 0$ for $t = 0$.

IX METHODS OF DATA PROCESSING

41. DETERMINATION OF THE MOMENTS OF RANDOM VARIABLES FROM EXPERIMENTAL DATA

Basic Formulas

The approximate values of the moments of random variables obtained by processing the experimental data are called estimates (fitting values) of these variables and are denoted by the same symbols as the estimated numerical characteristics of random variables, but with a tilde above (for example, $\tilde{\mathbf{M}}[X] = \tilde{x}$, $\tilde{\mathbf{D}}[X]$, $\tilde{k}_{xy}$, $\tilde{\sigma}_x$, and so forth). The set of values $(x_1, x_2, \ldots, x_n)$ for a random variable X, obtained in n experiments, is called a sample of size n. It is assumed that the experiments are performed independently under the same conditions. If the sample size n tends to infinity, the estimate should converge in probability to the parameter being estimated. The estimate is called unbiased if, for any sample size, its expectation coincides with the required parameter. The unbiased estimate for the expectation is the arithmetic mean,

$$\tilde{x} = \frac{1}{n}\sum_{j=1}^{n} x_j = C + \frac{1}{n}\sum_{j=1}^{n}(x_j - C),$$

where C is an arbitrary number introduced for convenience in computations ("false zero").

If the expected value is unknown, the unbiased estimate of the variance will be

$$\tilde{\mathbf{D}}[X] = \frac{1}{n-1}\sum_{j=1}^{n}(x_j - \tilde{x})^2 = \frac{1}{n-1}\sum_{j=1}^{n}(x_j - C)^2 - \frac{n}{n-1}(\tilde{x} - C)^2.$$

If the random variable considered is normally distributed, then the unbiased estimate of the standard deviation is given by the formula

$$\tilde{\sigma} = k_n\sqrt{\frac{1}{n-1}\sum_{j=1}^{n}(x_j - \tilde{x})^2} \approx \sqrt{\frac{1}{n-1.45}\sum_{j=1}^{n}(x_j - \tilde{x})^2},$$

where

$$k_n = \sqrt{\frac{n-1}{2}}\,\frac{\Gamma\left(\frac{n-1}{2}\right)}{\Gamma\left(\frac{n}{2}\right)}.$$

TABLE 23

n	k_n	n	k_n	n	k_n
3	1.1284	10	1.0280	30	1.0087
4	1.0853	12	1.0230	35	1.0072
5	1.0640	15	1.0181	40	1.0064
6	1.0506	20	1.0134	45	1.0056
7	1.0423	25	1.0104	50	1.0051

The values of the coefficient k_n are included in Table 23.

If the expectation is known, the unbiased estimate of the variance is

$$\tilde{\mathbf{D}}[X] = \frac{1}{n} \sum_{j=1}^{n} (x_j - \bar{x})^2.[1]$$

If $x_1, y_1, \ldots, x_n, y_n$ are the values of the random variables X and Y, obtained as a result of n independent experiments that are performed under identical conditions, the unbiased estimate of the covariance of these random variables is

$$\tilde{k}_{xy} = \frac{1}{n-1} \sum_{j=1}^{n} (x_j - \tilde{x})(y_j - \tilde{y}),$$

for unknown expectations X and Y;

$$\tilde{k}_{xy} = \frac{1}{n} \sum_{j=1}^{n} (x_j - \bar{x})(y_j - \bar{y}),$$

for known expectations.

The estimate of the correlation coefficient can be found from the formula

$$\tilde{r}_{xy} = \frac{\tilde{k}_{xy}}{\tilde{\sigma}_x \tilde{\sigma}_y}.$$

For a large sample size the elements of the statistical series are combined in groups (classes) by representing the experimental data in the form of an ordered array (Table 24).

TABLE 24

Class No.	1	2	...	k
Limits of the class $x_{j-1} - x_j$	$x_0 - x_1$	$x_1 - x_2$	...	$x_{k-1} - x_k$
The mean value x_j^* of the class	x_1^*	x_2^*	...	x_k^*
Size m_j of the class	m_1	m_2	...	m_k
Frequency $p_j^* = m_j/n$ of the class	p_1^*	p_2^*	...	p_k^*

[1] If the variable considered is normal, then the unbiased estimate for the standard deviation is determined from the formula

$$\tilde{\sigma} = k_{n+1} \sqrt{\frac{1}{n} \sum_{j=1}^{n} (x_j - \bar{x})^2}.$$

In this case, the estimates for the expectation, variance and moments of higher order are approximately determined by the formulas:

$$\tilde{x} \approx \sum_{j=1}^{k} x_j^* p_j^*,$$

$$\tilde{\mathbf{D}}[X] \approx \sum_{j=1}^{k} (x_j^* - \tilde{x})^2 p_j^*,$$

$$\tilde{\mu}_s[X] \approx \sum_{j=1}^{k} (x_j^* - \tilde{x})^s p_j^*,$$

$$\tilde{m}_s[X] \approx \sum_{j=1}^{k} (x_j^*)^s p_j^*,$$

or more precisely (taking into account the Sheppard corrections) by:

$$\tilde{m}_1[X] \approx \sum_{j=1}^{k} x_j^* p_j^*,$$

$$\tilde{m}_2[X] \approx \sum_{j=1}^{k} (x_j^*)^2 p_j^* - \frac{h^2}{12},$$

$$\tilde{m}_3[X] \approx \sum_{j=1}^{k} (x_j^*)^3 p_j^* - \frac{h^2}{4} \sum_{j=1}^{k} x_j^* p_j^*,$$

$$\tilde{m}_4[X] \approx \sum_{j=1}^{k} (x_j^*)^4 p_j^* - \frac{h^2}{2} \sum_{j=1}^{k} (x_j^*)^2 p_j^* + \frac{7h^4}{240},$$

$$\tilde{\mu}_2[X] \approx \sum_{j=1}^{k} (x_j^* - \tilde{x})^2 p_j^* - \frac{h^2}{12},$$

$$\tilde{\mu}_3[X] \approx \sum_{j=1}^{k} (x_j^* - \tilde{x})^3 p_j^*,$$

$$\tilde{\mu}_4[X] \approx \sum_{j=1}^{k} (x_j^* - \tilde{x})^4 p_j^* - \frac{h^2}{2} \sum_{j=1}^{k} (x_j^* - \tilde{x})^2 p_j^* + \frac{7h^4}{240},$$

where h is the class interval length.

Solution for Typical Examples

Example 41.1 To determine the precision of a measuring instrument whose systematic error is practically zero, one performs five independent measurements whose results are given in Table 25.

Determine the unbiased estimate for the variance of errors if the value of the quantity being measured is (a) known to be 2800 m., (b) unknown.

TABLE 25

Experiment No.	1	2	3	4	5
x_j, m.	2781	2836	2807	2763	2858

SOLUTION. The value of the quantity being measured is $\bar{x}$. Therefore in (a) the unbiased estimate of the variance is determined by the formula

$$\tilde{D}[X] = \frac{1}{n}\sum_{j=1}^{n}(x_j - \bar{x})^2 = \frac{6439}{5} = 1287.8 \text{ sq. m.}$$

If the value of the measured quantity is unknown, its estimate is

$$\tilde{x} = \frac{1}{n}\sum_{j=1}^{n} x_j = 2809 \text{ m.}$$

Thus, in (b), the unbiased estimate is

$$\tilde{D}[X] = \frac{1}{n-1}\sum_{j=1}^{n}(x_j - \tilde{x})^2 = \frac{6034}{4} = 1508.5 \text{ sq. m.}$$

In a similar way one can solve Problems 41.1 to 41.14 and 41.13 to 41.16.

Example 41.2 To determine the estimates of the standard deviation of the errors given by a measuring instrument whose systematic errors are practically zero, one performs five independent experiments whose results are included in Table 26.

TABLE 26

Experiment No.	1	2	3	4	5
x_j, m.	92	101	103	98	96

To process the data obtained in measurements, the following formulas for the unbiased estimates are used:

$$\tilde{\sigma}_1 = k_n\sqrt{\frac{1}{n-1}\sum_{j=1}^{n}(x_j - \tilde{x})^2},$$

$$\tilde{\sigma}_2 = \sqrt{\frac{\pi}{2n(n-1)}}\sum_{j=1}^{n}|x_j - \tilde{x}| \equiv k\sum_{j=1}^{n}|x_j - \tilde{x}|.$$

Find $\tilde{\sigma}_1$ and $\tilde{\sigma}_2$ and determine the variance of these estimates, if the errors obey a normal distribution law.

SOLUTION. Filling in Table 27 and summing by columns, we obtain:

$$A_1 = \sum_j x_j = 490, \qquad \tilde{x} = \frac{A_1}{n} = 98 \text{ m.},$$

$$A_2 = \sum_j |x_j - \tilde{x}| = 16, \qquad \tilde{\sigma}_1 = k_n\sqrt{\frac{A_3}{n-1}} = 1.064\sqrt{\frac{74}{4}} = 4.57 \text{ m.},$$

$$A_3 = \sum_j (x_j - \tilde{x})^2 = 74, \qquad \tilde{\sigma}_2 = \sqrt{\frac{\pi}{2n(n-1)}}A_2 = 4.48 \text{ m.}$$

TABLE 27

x_j	$\lvert x_j - \tilde{x}\rvert$	$(x_j - \tilde{x})^2$
92	6	36
101	3	9
103	5	25
98	0	0
96	2	4

The obtained estimates $\tilde{\sigma}_1$ and $\tilde{\sigma}_2$ are random variables whose expectation is $\mathbf{M}[\tilde{\sigma}_1] = \mathbf{M}[\tilde{\sigma}_2] = \sigma$. To find the variance $\tilde{\sigma}_1$, we have

$$\mathbf{D}[\tilde{\sigma}_1] = \mathbf{M}[(\tilde{\sigma}_1)^2] - \{\mathbf{M}[\tilde{\sigma}_1]\}^2 = \mathbf{M}\left[k_n^2 \frac{\sum_{j=1}^n (x_j - \tilde{x})^2}{n-1}\right] - \sigma^2$$

$$= k_n^2\mathbf{M}\{\tilde{\mathbf{D}}[X]\} - \sigma^2 = k_n^2\sigma^2 - \sigma^2 = (k_n^2 - 1)\sigma^2.$$

For the variance of the random variable $\tilde{\sigma}_2$, we have

$$\mathbf{D}[\tilde{\sigma}_2] = \mathbf{M}[(\tilde{\sigma}_2)^2] - \{\mathbf{M}[\tilde{\sigma}_2]\}^2,$$

where $\mathbf{M}[(\tilde{\sigma}_2)^2] = k^2\mathbf{M}[(\sum_{i=1}^m |x_i - \tilde{x}|)^2]$.

Let $z_i = x_i - (1/n)\sum_{j=1}^n x_j$. Since z_i is a linear function of normal random variables, it also obeys a normal distribution law with parameters

$$\mathbf{M}[z_i] = \mathbf{M}\left[x_i - \frac{1}{n}\sum_{j=1}^n x_j\right] = 0,$$

$$\mathbf{D}[z_i] = \mathbf{D}\left[x_i - \frac{1}{n}\sum_{j=1}^n x_j\right] = \mathbf{D}\left[\frac{n-1}{n}x_i - \frac{1}{n}\sum_{\substack{j=1\\ j\neq i}}^n x_j\right]$$

$$= \left(\frac{n-1}{n}\right)^2\sigma^2 + \frac{n-1}{n^2}\sigma^2 = \frac{n-1}{n}\sigma^2.$$

Therefore,

$$\mathbf{M}[(\tilde{\sigma}_2)^2] = k^2\mathbf{M}\left[\left(\sum_{j=1}^n |z_j|\right)^2\right]$$

$$= k^2\mathbf{M}\left[\sum_{i=1}^n z_i^2 + 2\sum_{i=1}^{n-1}\sum_{j=i+1}^n |z_i|\,|z_j|\right]$$

$$= k^2\{n\mathbf{D}[z_i] + n(n-1)\mathbf{M}[|z_i|\,|z_j|]\},$$

where $(j \neq i)$

$$\mathbf{M}[|z_i|\,|z_j|] = \frac{1}{2\pi\sigma_z^2\sqrt{1-r^2}}\int_{-\infty}^{\infty}\int_{-\infty}^{\infty} |z_1|\,|z_2| \exp\left\{-\frac{z_1^2 + z_2^2 - 2rz_1z_2}{2\sigma_z^2(1-r^2)}\right\} dz_1\, dz_2$$

Passing to polar coordinates, we find

$$\mathbf{M}[|z_i|\,|z_j|] = \frac{1}{2\pi\sigma_z^2\sqrt{1-r^2}}\int_0^{2\pi}\int_0^{\infty} R^3|\sin\varphi|\,|\cos\varphi| \times \exp\left\{-\frac{R^2(1-r\sin 2\varphi)}{2\sigma_z^2(1-r^2)}\right\} dR\,d\varphi$$

$$= \frac{2\sigma_z^2}{\pi}[\sqrt{1-r^2} + r\arcsin r].$$

Here,

$$r = \frac{\mathbf{M}[z_i z_j]}{\sigma_z^2} = \frac{1}{\sigma_z^2}\mathbf{M}\left[\left(\frac{n-1}{n}x_i - \frac{1}{n}\sum_{\substack{l=1\\ i\neq l}}^{n} x_l\right) \times \left(\frac{n-1}{n}x_j - \frac{1}{n}\sum_{\substack{k=1\\ k\neq j}}^{n} x_k\right)\right] = -\frac{1}{n-1}.$$

Finally we get

$$\mathbf{D}[\tilde{\sigma}_2] = \mathbf{M}[(\tilde{\sigma}_2)^2] - \sigma^2 = \frac{\sigma^2}{n}\left(\frac{\pi}{2} + \sqrt{n(n-2)} - n + \arcsin\frac{1}{n-1}\right).$$

The ratio between the variances for the random variables $\tilde{\sigma}_1$ and $\tilde{\sigma}_2$ for different n are shown in Table 28.

TABLE 28

n	5	10	20	50
$\frac{\mathbf{D}[\tilde{\sigma}_2]}{\mathbf{D}[\tilde{\sigma}_1]}$	1.053	1.096	1.150	1.170

The solution for this example implies that the estimate of σ given by the formula

$$\tilde{\sigma}_1 = k_n\sqrt{\frac{\sum_{j=1}^{n}(x_j - \tilde{x})^2}{n-1}}$$

has a smaller variance than the result obtained from the formula

$$\tilde{\sigma}_2 = k\sum_{j=1}^{n}|x_j - \tilde{x}|;$$

that is, the estimate $\tilde{\sigma}_1$ is more efficient.

Similarly one can solve Problems 41.7, 41.12 and 41.20.

Example 41.3 From the current production of an automatic boring machine, a sample of 200 cylinders is selected. The measured deviations of the diameters of these cylinders from the rated value are given in Table 29.

Determine the estimates for the expectation, variance, asymmetry and the excess of these deviations.

SOLUTION. To simplify the intermediary calculations, we introduce the random variable

$$z_j = \frac{x_j^* - C}{h},$$

where as "false zero" we take $C = 2.5$ microns and the class width is $h = 5$ microns.

TABLE 29

Class No.	1	2	3	4	5	6	7	8	9	10
Limits (in microns) from to	−20 −15	−15 −10	−10 −5	−5 0	0 +5	+5 +10	+10 +15	+15 +20	+20 +25	+25 +30
The mean value x_j^* (in microns)	−17.5	−12.5	−7.5	−2.5	+2.5	+7.5	+12.5	+17.5	+22.5	+27.5
Size	7	11	15	24	49	41	26	17	7	3
Frequency	0.035	0.055	0.075	0.120	0.245	0.205	0.130	0.085	0.035	0.015

Let us determine the estimates of the first four moments of the random variable by considering the Sheppard corrections. The calculations are summarized in Table 30.

TABLE 30

Class No.	p_j^*	x_j^*	$z_j = \frac{x_j^* - C}{h}$	z_j^2	z_j^3	z_j^4	$p_j^* z_j$	$p_j^* z_j^2$	$p_j^* z_j^3$	$p_j^* z_j^4$
1	2	3	4	5	6	7	8	9	10	11
1	0.035	−17.5	−4	16	−64	256	−0.140	0.560	−2.240	8.960
2	0.055	−12.5	−3	9	−27	81	−0.165	0.495	−1.485	4.455
3	0.075	−7.5	−2	4	−8	16	−0.150	0.300	−0.600	1.200
4	0.120	−2.5	−1	1	−1	1	−0.120	0.120	−0.120	0.120
5	0.245	+2.5	0	0	0	0	0	0	0	0
6	0.205	+7.5	1	1	1	1	0.205	0.205	0.205	0.205
7	0.130	+12.5	2	4	8	16	0.260	0.520	1.040	2.080
8	0.085	+17.5	3	9	27	81	0.255	0.765	2.295	6.885
9	0.035	+22.5	4	16	64	256	0.140	0.560	2.240	8.960
10	0.015	+27.5	5	25	125	625	0.075	0.375	1.875	9.375
Σ							A	B	D	E

$A = 0.36$; $D = 3.21$;
$B = 3.90$; $E = 42.24$.

Taking into account the Sheppard corrections, we obtain:

$$\tilde{x} \approx hA + C = 4.30\,\mu,$$

$$\tilde{\mathbf{D}}[X] \approx h^2\left(B - A^2 - \frac{1}{12}\right) = 92.25\,\mu^2,$$

$$\tilde{\mu}_3[X] \approx h^3(D - 3AB + 2A^3) = -113.75\,\mu^3,$$

$$\tilde{\mu}_4[X] \approx h^4\left[E - A\left(4D - \frac{A}{2}\right) + B\left(6A^2 - \frac{1}{2}\right) - 3A^4 + \frac{7}{240}\right] = 24{,}215.62\,\mu^4,$$

$$\widetilde{\mathrm{Sk}}[X] \approx \frac{\tilde{\mu}_3[X]}{\tilde{\sigma}_x^3} = \frac{-113.75}{885.97} = -0.128,$$

$$\widetilde{\mathrm{Ex}}[X] \approx \frac{\tilde{\mu}_4[X]}{\sigma_x^4} - 3 = \frac{24{,}215.62}{8510.06} - 3 = -0.16.$$

For the same variables, but without considering the Sheppard corrections, we have (see Examples 43.2 and 43.4):

$$\tilde{x} \approx 4.30\,\mu, \qquad \tilde{\mu}_4[X] \approx 25{,}375.00\,\mu^4,$$

$$\widetilde{\mathrm{Sk}}[X] \approx -0.125, \qquad \widetilde{\mathrm{Ex}}[X] \approx -0.145.$$

$$\tilde{\mathbf{D}}[X] \approx 94.26\,\mu^2,$$

Problems 41.5, 41.8, 41.18 and 41.19 can be solved in a similar manner.

PROBLEMS

41.1 In 12 independent measurements of a base of length 232.38 m., which were performed with the same instrument, the following results were obtained: 232.50, 232.48, 232.15, 232.53, 232.45, 232.30, 232.48, 232.05, 232.45, 232.60, 232.47 and 232.30 m. Assuming that the errors obey a normal distribution and do not contain systematic errors, determine the unbiased estimate for the standard deviations.

41.2 The following are the results of eight independent measurements performed with an instrument with no systematic error: 369, 378, 315, 420, 385, 401, 372 and 383 m. Determine the unbiased estimate for the variance of the errors in measurements if (a) the length of the base that is being measured is known: $\bar{x} = 375$ m., (b) the length of the measured base is unknown.

41.3 In processing the data obtained in 15 tests performed with a model airplane, the following values for its maximal velocity were obtained: 422.2, 418.7, 425.6, 420.3, 425.8, 423.1, 431.5, 428.2, 438.3, 434.0, 411.3, 417.2, 413.5, 441.3 and 423.0 m./sec. Determine the unbiased estimates for the expectation and standard deviation of the maximal velocity assumed to obey a normal distribution law.

41.4 In processing the data of six tests performed with a motorboat, the following values for its maximal velocity were obtained: 27, 38, 30, 37, 35 and 31 m./sec. Determine the unbiased estimates for the expectation and standard deviation of the maximal velocity, assuming that the maximal velocity of the boat obeys a normal distribution law.

41.5 The sensitivity of a television set to video signals is characterized by data in Table 31.

TABLE 31

x_j^*, μv	m_j	x_j^*, μv	m_j	x_j^*, μv	m_j
200	10	350	20	550	3
225	1	375	10	600	19
250	26	400	29	625	3
275	8	425	5	650	1
300	23	450	26	700	6
325	9	500	24	800	4

Find the estimates for the expectation and standard deviation of the sensitivity of the set.

41.6 A number n of independent experiments are performed to determine the frequency of an event A. Determine the value of $\mathbf{P}(A)$ that maximizes the variance of the frequency.

41.7 A number n of independent measurements of the same unknown constant quantity are performed. The errors obey a normal distribution law with zero expectation.

To determine the estimates of the variance by using the experimental data, the following formulas are applied:

$$\tilde{\sigma}_1^2 = \sum_{j=1}^{n} \frac{(x_j - \tilde{x})^2}{n}, \qquad \tilde{\sigma}_2^2 = \sum_{j=1}^{n} \frac{(x_j - \tilde{x})^2}{n - 1}.$$

Find the variance of the random variables $\tilde{\sigma}_1^2$ and $\tilde{\sigma}_2^2$.

41.8 The experimental values of a random variable X are divided into groups. The average value x_j^* for the jth group and the number of elements m_j in the jth group are in Table 32.

TABLE 32

x_j^*	m_j	x_j^*	m_j	x_j^*	m_j
44	7	47	48	50	1
45	18	48	33	52	1
46	120	49	5	58	2

Find the estimates for the asymmetry coefficient and the excess.

41.9 A sample, $x_1, x_2, \ldots, x_n$, selected from a population is processed by differences in order to determine the estimates for the variance. The formula used for processing the results of the experiment is

$$\tilde{\sigma}_x^2 = k \sum_{j=1}^{n-1} (x_{j+1} - x_j)^2.$$

How large should k be so that $\tilde{\sigma}_x^2$ is an unbiased estimate of σ_x^2 if the random variable X is normal?

41.10 Let $x_1, x_2, \ldots, x_n$ be the outcomes of independent measurements of an unknown constant. The errors in measurements obey the same normal distribution law. The standard deviation is determined by the formula

$$\tilde{\sigma} = k \sum_{j=1}^{n} |x_j - \tilde{x}|,$$

where

$$\tilde{x} = \frac{1}{n} \sum_{j=1}^{n} x_j.$$

Determine the value of k for which $\tilde{\sigma}$ is an unbiased estimate of σ.

41.11 Independent measurements of a known constant x are $x_1, x_2, \ldots, x_n$. The errors obey the same normal distribution law. For processing the results of these observations in order to obtain the estimates for the standard deviation of errors the following formula is used:

$$\tilde{\sigma} = k \sum_{j=1}^{n} |x_j - \bar{x}|.$$

How large should k be so that the estimates are unbiased for (a) the standard deviation of the errors, (b) the variance of the errors?

41.12 Independent measurements $x_1, x_2, \ldots, x_n$ with different accuracies of the same unknown constant are made. The estimate of the quantity being measured is determined from the formula

$$\tilde{x} = \frac{\sum_{j=1}^{n} A_j x_j}{\sum_{j=1}^{n} A_j}.$$

How large should A_j be so that the variance of $\tilde{x}$ is minimal if the standard deviation of the errors of the jth measurement is σ_j?

41.13 A system of two random variables with a normal distribution in the plane is subjected to n independent experiments, in which the values (x_k, y_k) $(k = 1, 2, \ldots, n)$ of these variables are determined. The principal dispersion axes are parallel to the coordinate axes. Determine the unbiased estimates for the expectation and the standard deviations of these variables.

41.14 Solve Problem 41.13 for the results of the independent trials given in Table 33.

41.15 Under the conditions of Problem 41.13 find the estimates for the parameters of the unit dispersion ellipse if before the experiments the direction of the principal axes is unknown.

41.16 Solve Problem 41.15 for the results of 16 independent trials given in Table 34.

TABLE 33

Experiment No. k	x_k, m.	y_k, m.	Experiment No. k	x_k, m.	y_k, m.
1	55	77	9	41	31
2	43	46	10	36	60
3	63	34	11	56	48
4	57	61	12	72	78
5	44	84	13	48	62
6	26	54	14	16	49
7	59	53	15	49	31
8	72	21	16	36	64

TABLE 34

Experiment No.	Deviations, m.		Experiment No.	Deviations, m.		Experiment No.	Deviations, m.		Experiment No.	Deviations, m.	
	x_i	y_i		x_i	y_i		x_i	y_i		x_i	y_i
1	+2	+59	5	+2	+72	9	+1	+7	13	+4	+103
2	+3	+88	6	0	+34	10	−2	+57	14	0	+65
3	+2	+32	7	+2	−12	11	−1	+42	15	+1	+16
4	−2	−24	8	+3	+50	12	+2	+23	16	+1	+28

41.17 A sample $x_1, x_2, \ldots, x_n$ selected from a normal population is processed to determine the estimates for the standard deviation by the formula

$$\tilde{\sigma} = k\sqrt{\frac{\sum_{j=1}^{n}(x_j - \tilde{x})^2}{n}},$$

where

$$\tilde{x} = \frac{1}{n}\sum_{j=1}^{n} x_j.$$

How large should k be so that $\tilde{\sigma}$ is an unbiased estimate of the standard deviation σ.

41.18 From a table of random numbers, 150 two-digit numbers (00 is taken for 100) are selected. These numbers are divided into intervals of 10 (Table 35).

TABLE 35

1–10	11–20	21–30	31–40	41–50	51–60	61–70	71–80	81–90	91–100
16	15	19	13	14	19	14	11	13	16

Construct the histogram and the graph of the frequency count. Find the estimates for the expectation and variance.

41.19 With the aid of a table of random one-digit numbers 250 sums of five numbers each are formed. The numbers are distributed into classes as indicated in Table 36 (if the number coincides with the limit of a class, 1/2 is added to the two adjacent classes). Construct the histogram and find the estimates for the expectation and variance.

TABLE 36

0–3	3–6	6–9	9–12	12–15	15–18	18–21	21–24
0	0.5	1.5	10.0	17.5	28.5	39	41
24–27	27–30	30–33	33–36	36–39	39–42	42–45	
45	30.5	27.0	7.5	1	1	0	

41.20 To determine the value of an unknown constant, n independent measurements are performed. The systematic errors in measurements are zero and the random errors are normally distributed. The following two formulas are used to find the estimated variances:

$$\tilde{\sigma}_1^2 = \frac{\sum_{j=1}^{n} (x_j - \tilde{x})^2}{n - 1}, \qquad \tilde{\sigma}_2^2 = \frac{1}{2(n-1)} \sum_{j=1}^{n-1} (x_{j+1} - x_j)^2.$$

Are $\tilde{\sigma}_1^2$ and $\tilde{\sigma}_2^2$ unbiased estimates of the variance? Which of these two formulas gives a more accurate value for the variance?

42. CONFIDENCE LEVELS AND CONFIDENCE INTERVALS

Basic Formulas

A confidence interval is an interval that with a given confidence level α covers a parameter Θ to be estimated.

The width of a symmetrical confidence interval 2ε is determined by the condition

$$\mathbf{P}\{|\Theta - \tilde{\Theta}| \leqslant \varepsilon\} = \alpha,$$

where $\tilde{\Theta}$ is the estimate of parameter Θ, and the probability $\mathbf{P}\{|\Theta - \tilde{\Theta}| \leqslant \varepsilon\}$ is determined by the distribution law for $\tilde{\Theta}$.

If $x_1, x_2, \ldots, x_n$ is a sample from a normal population, then the confidence level is determined by the formulas:

(a) for the expectation in the case when σ is known,

$$\alpha = \mathbf{P}\{|\tilde{x} - \bar{x}| \leqslant \varepsilon\} = \Phi\left(\frac{\varepsilon\sqrt{n}}{\sigma}\right);$$

for unknown σ,

$$\alpha = \int_{-t_\alpha}^{t_\alpha} S_n(t)\, dt,$$

where

$$S_n(t) = \frac{\Gamma\left(\frac{n}{2}\right)\left(1 + \frac{t^2}{n-1}\right)^{-n/2}}{\sqrt{\pi(n-1)}\,\Gamma\left(\frac{n-1}{2}\right)}$$

is Student's distribution law and

$$t_\alpha = \frac{\varepsilon\sqrt{n}}{\tilde{\sigma}} = \frac{\varepsilon\sqrt{n}}{\sqrt{\frac{1}{n-1}\sum_{j=1}^{n}(x_j - \tilde{x})^2}}.$$

The values of t_α are given in Table 16T,[2] whose entries are the number of degrees of freedom $k = n - 1$ and the confidence level α.

(b) For the standard deviation,

$$\alpha = \mathbf{P}\{|\tilde{\sigma} - \sigma| \leqslant \varepsilon\} = \mathbf{P}\left\{\frac{\sqrt{k}}{1+q} < \chi < \frac{\sqrt{k}}{1-q}\right\} = \int_{\sqrt{k}/(1+q)}^{\sqrt{k}/(1-q)} P_k(\chi)\,d\chi,$$

where

$$P_k(\chi) = \frac{\chi^{k-1}e^{-\chi^2/2}}{2^{(k-2)/2}\Gamma\left(\frac{k}{2}\right)}, \qquad k = n - 1, \qquad q = \frac{\varepsilon}{\tilde{\sigma}}.$$

The values of the integral $\int_{\sqrt{k}/(1+q)}^{\sqrt{k}/(1-q)} P_k(\chi)\,d\chi$ are given in Table 20T.

The confidence interval for σ $(\gamma_1\tilde{\sigma}, \gamma_2\tilde{\sigma})$, where the probabilities of its lying entirely to the right and entirely to the left of the true value are both $(1 - \alpha)/2$, is determined by the formula

$$\mathbf{P}\left\{\frac{\sigma}{\tilde{\sigma}} \leqslant \gamma_1\right\} = \mathbf{P}\left\{\frac{\sigma}{\tilde{\sigma}} \geqslant \gamma_2\right\} = \mathbf{P}\left\{\chi^2 \leqslant \frac{n-1}{\gamma_2^2}\right\}$$
$$= \mathbf{P}\left\{\chi^2 \geqslant \frac{n-1}{\gamma_1^2}\right\} = \frac{1-\alpha}{2}.$$

In order to find γ_1 and γ_2 for a given confidence level α and $k = n - 1$ degrees of freedom, one may use Table 19T or 18T.

For an exponential distribution law, the confidence interval for expectation $(\nu_1\tilde{x}, \nu_2\tilde{x})$ is given by the expression

$$\mathbf{P}\left\{\frac{\bar{x}}{\tilde{x}} \leqslant \nu_1\right\} = \mathbf{P}\left\{\frac{\bar{x}}{\tilde{x}} \geqslant \nu_2\right\} = \mathbf{P}\left\{\chi^2 \leqslant \frac{2n}{\nu_2} = \chi_{1-\delta}^2\right\}$$
$$= \mathbf{P}\left\{\chi^2 \geqslant \frac{2n}{\nu_1} = \chi_\delta^2\right\} = \frac{1-\alpha}{2} = \delta.$$

From this $\nu_1 = 2n/\chi_\delta^2$, $\nu_2 = 2n/\chi_{1-\delta}^2$.

The values for χ_δ^2 and $\chi_{1-\delta}^2$ are determined from Table 18T for the probabilities δ and $1 - \delta$, respectively, and for $k = 2n$ degrees of freedom.

[2] References for the table numbers followed by T are found on pages 471–473.

For a sufficiently large sample size ($n > 15$) the limits of the confidence interval for $\bar{x}$ are calculated approximately by the formulas,

$$\frac{4\sum_{j=1}^{n} x_j}{(\sqrt{4n-1}+\varepsilon_0)^2}, \qquad \frac{4\sum_{j=1}^{n} x_j}{(\sqrt{4n-1}-\varepsilon_0)^2},$$

where ε_0 is the solution of the equation $\alpha = \Phi(\varepsilon_0)$.

If, from the same population, there are selected N samples each of size n, if the event whose probability of realization obeys a Poisson distribution occurs m_j times ($j = 1, 2, \ldots, N$) in the jth sample and the expected value of the parameter is given by the formula $\tilde{a} = (\sum_{j=1}^{n} m_j)/N$, then for $\tilde{a} > 0$ the limits of the confidence interval are determined from the relation

$$\mathbf{P}\{2N\tilde{a} > \chi_\delta^2\} = \mathbf{P}\{2N\tilde{a} < \chi_{1-\delta}^2\} = \frac{1-\alpha}{2} = \delta,$$

that is, the upper and lower limits are equal to

$$\frac{\chi_{1-\delta}^2}{2N}, \qquad \frac{\chi_\delta^2}{2N},$$

respectively, where $\chi_{1-\delta}^2$ and χ_δ^2 for a given δ are chosen from Table 18T, $\chi_{1-\delta}^2$ being taken for $k = 2\sum_{j=1}^{N} m_j$ degrees of freedom and χ_δ^2 for $k = 2(\sum_{j=1}^{N} m_j + 1)$ degrees of freedom.

For $\tilde{a} = 0$, the lower limit becomes zero and the upper limit is $\chi_{2\delta}^2/2N$, where $\chi_{2\delta}^2$ is found from Table 18T for $k = 2$ and level $\mathbf{P}\{2N\tilde{a} > \chi_{2\delta}^2\} = 2\delta$.

For a sufficiently large k (practically greater than 30), the limits of the confidence interval are determined approximately by the formulas:

$$\text{(lower limit)} \quad \frac{\left(\sqrt{4\sum_{j=1}^{N} m_j - 1} - \varepsilon_0\right)^2}{4N},$$

$$\text{(upper limit)} \quad \frac{\left(\sqrt{4\sum_{j=1}^{N} m_j - 1} + \varepsilon_0\right)^2}{4N},$$

where ε_0 is the solution of the equation $\alpha = \Phi(\varepsilon_0)$.

If in n independent trials a certain event occurs exactly m times ($0 < m < n$), the limits p_1, p_2 of the confidence interval, if the probability of occurrence of this event is p, are determined from the equations

$$\sum_{j=m}^{n} C_n^j p_1^j (1-p_1)^{n-j} = \frac{1-\alpha}{2},$$

$$\sum_{j=0}^{m} C_n^j p_2^j (1-p_2)^{n-j} = \frac{1-\alpha}{2}.$$

These equations can be solved approximately with the aid of the incomplete β-function. In Table 30T the values of p_1 and p_2 are given for different m and n and two values of the level α, 0.95 and 0.99.

For n sufficiently large, one can write approximately

$$p_1 = \tilde{p} - \varepsilon, \qquad p_2 = \tilde{p} + \varepsilon,$$

where $\tilde{p} = m/n$, and ε is the solution of the equation

$$\alpha = \Phi\left(\frac{\varepsilon\sqrt{n}}{\sqrt{\tilde{p}(1-\tilde{p})}}\right).$$

A better approximation is given by the formulas

$$\left.\begin{matrix} p_2 \\ p_1 \end{matrix}\right\} = \frac{1}{n+\varepsilon_0^2}\left[n\tilde{p} + \frac{\varepsilon_0^2}{2} \pm \frac{1}{2} \pm \varepsilon_0\sqrt{\frac{\left(n\tilde{p} \pm \frac{1}{2}\right)\left(n - n\tilde{p} \mp \frac{1}{2}\right)}{n} + \frac{\varepsilon_0^2}{4}}\right]$$

and

$$\left.\begin{matrix} 2\arcsin\sqrt{p_2} \\ 2\arcsin\sqrt{p_1} \end{matrix}\right\} = 2\arcsin\sqrt{\frac{n\tilde{p} \pm \frac{1}{2}}{n}} \pm \frac{\varepsilon_0}{2\sqrt{n}},$$

one of which underestimates the interval, while the other overestimates it by a quantity of the same order of magnitude; ε_0 is the solution of $\alpha = \Phi(\varepsilon_0)$.

If $m = 0$, then $p_1 = 0$ and $p_2 = 1 - \sqrt[n]{1-\alpha}$.

If $m = n$, then $p_2 = 1$ and $p_1 = \sqrt[n]{1-\alpha}$.

The confidence interval for the correlation coefficient whose estimate is obtained from a normal sample of size n can be expressed approximately in terms of auxiliary random variable $Z = (1/2)\ln[(1+\tilde{r})/(1-\tilde{r})]$, whose limits (z_H, z_B) of the confidence interval are determined by the formulas

$$z_H = \bar{z} - \varepsilon_0\sigma_z, \qquad z_B = \bar{z} + \varepsilon_0\sigma_z,$$

where $\sigma_z = 1/\sqrt{n-3}$, ε_0 is the solution of the equation $\alpha = \Phi(\varepsilon_0)$, $\bar{z} = \Delta_1 + \Delta_2$, $\Delta_1 = (1/2)\ln[(1+\tilde{r})/(1-\tilde{r})]$ (the value of this quantity is determined from Table 31T) and $\Delta_2 = \tilde{r}/2(n-1)$.

By the values z_H and z_B, found from Table 31T, or the formula $r = \tanh z$, one can find the limits of the confidence interval for r. In the case of large n $(n > 50)$ and small $\tilde{r}$ $(\tilde{r} < 0.5)$, the limits r_H, r_B of the confidence interval for r are given approximately by

$$\left.\begin{matrix} r_H \\ r_B \end{matrix}\right\} = \tilde{r} \mp \varepsilon_0\sigma_{\tilde{r}},$$

where ε_0 is the solution of the quation $\alpha = \Phi(\varepsilon_0)$,

$$\sigma_{\tilde{r}} = \frac{1-\tilde{r}^2}{\sqrt{n-1}}.$$

Solution for Typical Examples

Example 42.1 The average distance measured from a reference point in four independent trials is 2250 m. The mean error of the measuring instrument is $E = 40$ m. Given the confidence level, 95 per cent, find the confidence interval for the quantity measured.

SOLUTION. The probability of covering the true value $\bar{x}$ of the measured quantity by the interval $(\tilde{x} - \varepsilon, \tilde{x} + \varepsilon)$ with random endpoints for a known E is determined by the formula

$$\mathbf{P}\{|\tilde{x} - \bar{x}| \leqslant \varepsilon\} = \frac{\rho}{\sqrt{\pi}\, E_1} \int_{-\varepsilon}^{\varepsilon} \exp\left\{-\rho^2 \frac{z^2}{E_1^2}\right\} dz = \hat{\Phi}\left(\frac{\varepsilon}{E_1}\right),$$

where $E_1 = E/\sqrt{n}$ is the standard deviation of the random variable $\tilde{x} = (1/n) \sum_{j=1}^{n} x_j$.

Solving the equation $\hat{\Phi}(\varepsilon\sqrt{n}/E) = 0.95$ by means of Table 11T, we find that

$$\frac{\varepsilon\sqrt{n}}{E} = 2.91,$$

$$\varepsilon = \frac{2.91}{\sqrt{n}} E = \frac{2.91 \cdot 40}{2} = 58.2 \text{ m}.$$

From this, the limits of the confidence interval will be:

the upper limit, 2250 m. + 58.2 m. = 2308.2 m.;

the lower limit, 2250 m. − 58.2 m. = 2191.8 m.

Similarly one can solve Problems 42.1, 42.6 and 42.13.

Example 42.2 The standard deviation of an altimeter is $\sigma = 15$ m. How many altimeters should there be on an airplane so that with confidence level 0.99, the mean error in altitude $\tilde{x}$ is not greater than −30 m., if the errors given by the altimeters are normally distributed, and there are no systematic errors?

SOLUTION. The assumptions of the problem can be written as

$$\mathbf{P}\{-30 < \tilde{x} - \bar{x} < \infty\} = 0.99.$$

The random variable

$$Z = \tilde{x} - \bar{x} = \frac{1}{n} \sum_{j=1}^{n} x_j - \bar{x}$$

is a linear function of normally distributed random variables and, hence, it also obeys a normal distribution with parameters

$$\mathbf{M}[Z] = \mathbf{M}\left[\frac{1}{n} \sum_{j=1}^{n} x_j - \bar{x}\right] = 0,$$

$$\mathbf{D}[Z] = \frac{\sigma^2}{n}.$$

Then

$$\mathbf{P}\{-30 < Z < \infty\} = \frac{1}{\sigma_z\sqrt{2\pi}} \int_{-30}^{\infty} \exp\left\{-\frac{z^2}{2\sigma_z^2}\right\} dz$$

$$= \frac{1}{2}\left[1 + \Phi\left(\frac{30}{\sigma_z}\right)\right] = 0.99.$$

Solving the equation

$$\Phi\left(\frac{30\sqrt{n}}{\sigma}\right) = 0.98,$$

we find from Table 8T that $(30\sqrt{n})/\sigma = 2.33$,

$$n = \left(\frac{2.33\sigma}{30}\right)^2 = 1.36.$$

Thus, the number of altimeters on the airplane should be at least two. Problems 42.7 and 42.11 can be solved similarly.

Example 42.3 In control tests performed with 16 light bulbs, estimates for their expected lifetime and the standard deviation were found: $\tilde{x} = 3000$ hours and $\tilde{\sigma} = 20$ hours. If the lifetime of each bulb is a normal random variable, determine (a) the confidence interval for the expectation and the standard deviation if the confidence level is 0.9, (b) the probability with which one can assert that the absolute value of the error $\bar{x}$ will be at most 10 hours and the error will be less than two hours.

SOLUTION. (a) To determine the limits of the confidence interval for the expectation, we make use of the equation

$$\alpha = \int_{-t_\alpha}^{t_\alpha} S_n(t)\,dt.$$

In Table 16T for $k = n - 1$ and $\alpha = 0.9$, we find that $t_\alpha = \varepsilon\sqrt{n}/\tilde{\sigma} = 1.753$, hence, it follows that $\varepsilon = 1.753\tilde{\sigma}/\sqrt{n} = 8.765$ hours.

Therefore, the upper and lower limits of the confidence interval for $\bar{x}$ are $3000 + 8.765 = 3008.765$ hours and $3000 - 8.765 = 2991.235$ hours, respectively.

To determine the limits of the confidence interval for σ, we make use of Table 19T. The entries to this table are $k = n - 1$ and the confidence level is α. For $k = 15$ and $\alpha = 0.9$, we have

$$\gamma_1 = 0.775, \qquad \gamma_2 = 1.437.$$

Consequently, for a confidence level 0.9, the values of σ compatible with the experimental data lie within the limits $0.775\tilde{\sigma} = 15.50$ hours to $1.437\tilde{\sigma} = 28.74$ hours.

(b) The probability for the inequality -10 hours $< \tilde{x} - \bar{x} < 10$ hours is determined by Student's distribution,

$$\alpha = \mathbf{P}\{|\bar{x} - \tilde{x}| < \varepsilon\} = \int_{-t_\alpha}^{t_\alpha} S_n(t)\,dt.$$

From Table 16T for $t_\alpha = \varepsilon\sqrt{n}/\tilde{\sigma} = 10\sqrt{16}/20 = 2$ and $k = n - 1 = 15$, we find that $\alpha = 0.93$.

The chi-square distribution permits us to determine the probability for the existence of inequality -2 hours $< \tilde{\sigma} - \sigma < 2$ hours;

$$\alpha = \mathbf{P}\{|\tilde{\sigma} - \sigma| < \varepsilon\} = \int_{\sqrt{k}/(1+q)}^{\sqrt{k}/(1-q)} P_k(\chi)\,d\chi.$$

For $q = \varepsilon/\tilde{\sigma} = 2/20 = 0.1$ and $k = n - 1 = 15$ degrees of freedom, we find from Table 20T that $\alpha = 0.41$.

Following this solution one can solve Problems 42.2 to 42.5 and 42.8 to 42.10.

Example 42.4 A random variable T obeys an exponential distribution law with a probability density $f(t) = 1/\bar{t}\exp\{-t/\bar{t}\}$.

The estimate for the parameter $\bar{t}$ is determined by the formula

$$\tilde{t} = \frac{1}{n}\sum_{j=1}^{n} t_j.$$

Express, in terms of $\tilde{t}$, the limits of the confidence interval for $\bar{t}$ so that $\mathbf{P}\{\nu_1\tilde{t} > \bar{t}\} = \mathbf{P}\{\nu_2\tilde{t} < \bar{t}\} = (1 - \alpha)/2$, if the confidence level $\alpha = 0.9$ and n equals 3, 5, 10, 20, 30, 40.

Solution. By the assumptions made in this example,

$$\mathbf{P}\{\nu_1\tilde{t} > \bar{t}\} = \mathbf{P}\{\nu_2\tilde{t} < \bar{t}\} = \frac{1 - \alpha}{2} = \delta.$$

Rewriting the inequalities in this expression leads to

$$\mathbf{P}\left\{\frac{2n\tilde{t}}{\bar{t}} > \frac{2n}{\nu_1} = \chi_\delta^2\right\} = \mathbf{P}\left\{\frac{2n\tilde{t}}{\bar{t}} < \frac{2n}{\nu_2} = \chi_{1-\delta}^2\right\} = \delta.$$

The random variable $U = 2n\tilde{t}/\bar{t}$ has a chi-square distribution with $2n$ degrees of freedom and, for $n > 15$, the random variable $Z = \sqrt{2U}$ has an approximately normal distribution with $\bar{z} = \sqrt{2n - 1}$ and $\sigma_z = 1$. Therefore, in the first case (for $n < 15$) we have

$$\nu_1 = \frac{2n}{\chi_\delta^2}, \qquad \nu_2 = \frac{2n}{\chi_{1-\delta}^2}.$$

After determining χ_δ^2 and $\chi_{1-\delta}^2$ from Table 18T (for $2n$ degrees of freedom and probabilities δ and $1 - \delta$), we calculate ν_1 and ν_2 (see Table 37).

Table 37

n	3	5	10
χ_δ^2	12.60	18.30	31.40
$\chi_{1-\delta}^2$	1.63	3.94	10.90
ν_1	0.48	0.55	0.64
ν_2	3.68	2.54	1.83

In the second case ($n > 15$) according to the formulas at the beginning of this solution, we have (see Table 38)

$$\nu_1 = \frac{4n}{(\sqrt{4n - 1} + \varepsilon_0)^2}, \qquad \nu_2 = \frac{4n}{(\sqrt{4n - 1} - \varepsilon_0)^2}.$$

TABLE 38

n	20	30	40
ε_0	1.65	1.65	1.65
ν_1	0.72	0.76	0.79
ν_2	1.53	1.40	1.33

The quantity ε_0 is determined from Table 8T for the level $\alpha = 0.9$.

In Figure 35, there is given the graph representing ν_1 and ν_2 as functions of n for the confidence level $\alpha = 0.9$.

Example 42.5 Three types of devices (A, B and C) are subjected to 50 independent trials during a certain time interval; the numbers of failures are recorded as in Table 39. Find the limits of the confidence intervals for the expected number of failures of each type during a selected time interval if the confidence level $\alpha = 0.9$, and the number of failures for each type obeys a Poisson distribution law during this interval.

TABLE 39

Number of failures	*Number of observations in which such number of failures occurred*		
	for type A	*for type B*	*for type C*
0	38	4	50
1	12	16	0
2	0	20	0
3	0	6	0
4	0	4	0

FIGURE 35

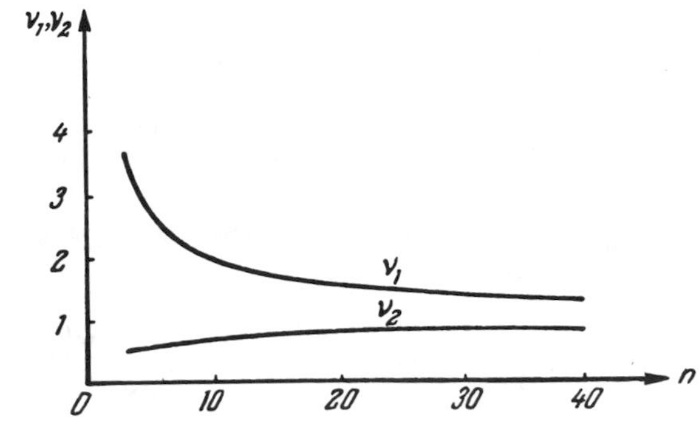

SOLUTION. To determine the limits of the confidence interval for the devices of type A, we make use of a chi-square distribution. From Table 18T, for $k = 24$ degrees of freedom and probability $(1 + \alpha)/2 = 0.95$, we find $\chi^2_{1-\delta} = 13.8$; for $k = 26$ and probability $\delta = (1 - \alpha)/2 = 0.05$, we find $\chi^2_\delta = 38.9$.

The upper limit a_2 and the lower limit a_1 of the confidence interval for $\bar{a}$ devices of type A are equal to

$$a_2 = \frac{\chi^2_\delta}{2N} = \frac{38.9}{100} = 0.389, \qquad a_1 = \frac{\chi^2_{1-\delta}}{2N} = \frac{13.8}{100} = 0.138.$$

To determine the limits of the confidence interval for the expected number of devices of type B that failed, one also should use the chi-square distribution for $k = 180$ and $k = 182$ degrees of freedom. Table 18T contains the data only for $k = 30$. Therefore, considering that for a number of degrees of freedom greater than 30, a chi-square distribution practically coincides with a normal one, we have

$$a_1 \approx \frac{\left(\sqrt{4 \sum_{j=1}^N m_j - 1} - \varepsilon_0\right)^2}{4N} = \frac{(\sqrt{4 \cdot 90 - 1} - 1.64)^2}{200} = 1.50,$$

$$a_2 \approx \frac{\left(\sqrt{4 \sum_{j=1}^N m_j - 1} + \varepsilon_0\right)^2}{4N} = \frac{(\sqrt{4 \cdot 90 - 1} + 1.64)^2}{200} = 2.12.$$

For devices of type C, $\sum_{j=1}^N m_j = 0$ and, therefore, the lower limit of the confidence interval is certainly zero. From Table 18T, for $k = 2$ and probability $1 - \alpha = 0.1$, we determine $\chi^2_{2\delta} = 4.6$ and calculate the value for the upper limit: $a_2 = \chi^2_{2\delta}/2N = 4.6/100 = 0.046$.

Example 42.6 Ten items out of thirty tested are defective. Determine the limits of the confidence interval for the probability of a defect if the confidence level is 0.95 and the number of defective items obeys a binomial distribution. Compare the results of the exact and approximate solutions.

SOLUTION. The exact solution can be obtained directly from Table 30T. For $x = 10$, $n - x = 20$ and a confidence level equal to 95 per cent, we have $p_1 = 0.173$, $p_2 = 0.528$.

For large $np(1 - p)$, the equations from which we determine the limits of the confidence interval for p can be written approximately by using the normal distribution:

$$\sum_{x=m}^{n} C_n^x p_1^x (1 - p_1)^{n-x} \approx \frac{1}{\sigma_1\sqrt{2\pi}} \int_{m-(1/2)}^{n+(1/2)} \exp\left\{-\frac{(z - \bar{z}_1)^2}{2\sigma_1^2}\right\} dz = \frac{1 - \alpha}{2},$$

$$\sum_{x=0}^{m} C_n^x p_2^x (1 - p_2)^{n-x} \approx \frac{1}{\sigma_2\sqrt{2\pi}} \int_{-1/2}^{m+(1/2)} \exp\left\{-\frac{(z - \bar{z}_2)^2}{2\sigma_2^2}\right\} dz = \frac{1 - \alpha}{2}.$$

From this,

$$\left.\begin{matrix} p_2 \\ p_1 \end{matrix}\right\} \approx \frac{1}{n + \varepsilon_0^2}\left[n\tilde{p} + \frac{\varepsilon_0^2}{2} \pm \frac{1}{2} \pm \varepsilon_0 \sqrt{\frac{\left(n\tilde{p} \pm \frac{1}{2}\right)\left(n - n\tilde{p} \mp \frac{1}{2}\right)}{n} + \frac{\varepsilon_0^2}{4}}\,\right],$$

where $\tilde{p} = m/n = 1/3$ and the quantity ε_0 can be determined from Table 8T for level $\alpha = 0.95$,

$$p_1 \approx \frac{1}{33.84}\left[10 - 0.5 + 1.92 - 1.96\sqrt{\frac{9.5 \cdot 20.5}{30} + 0.96}\right] = 0.180,$$

$$p_2 \approx \frac{1}{33.84}\left[10 + 0.5 + 1.92 + 1.96\sqrt{\frac{10.5 \cdot 19.5}{30} + 0.96}\right] = 0.529.$$

An approximation of the same kind gives the formula

$$\left.\begin{matrix}\arcsin\sqrt{p_2}\\ \arcsin\sqrt{p_1}\end{matrix}\right\} \approx \arcsin\sqrt{\frac{n\tilde{p} \pm (1/2)}{n}} \pm \frac{\varepsilon_0}{2\sqrt{n}},$$

which, when applied, leads to

$$p_2 \approx 0.526, \qquad p_1 \approx 0.166.$$

By a rougher approximation p_1 and p_2 can be found if one considers that the frequency $\tilde{p}$ is approximately normally distributed about p with variance $\tilde{p}(1 - \tilde{p})/\sqrt{n}$.

In this case,

$$\left.\begin{matrix}p_2\\ p_1\end{matrix}\right\} \approx \tilde{p} \pm \varepsilon,$$

where ε is the solution of the equation $\alpha = \Phi[\varepsilon\sqrt{n}/\sqrt{\tilde{p}}(1 - \tilde{p})]$.

Using Table 8T for $\alpha = 0.95$, we get

$$\frac{\varepsilon\sqrt{n}}{\sqrt{\tilde{p}(1 - \tilde{p})}} = 1.96, \qquad \varepsilon = 1.96\sqrt{\frac{2}{9 \cdot 30}} = 0.169,$$

hence, it follows that $p_1 \approx 0.333 - 0.169 = 0.164$, $p_2 \approx 0.333 + 0.169 = 0.502$.

Example 42.7 To study the mechanical properties of steel, 30 independent experiments are performed; based on their outcomes, estimates are determined for the correlation coefficients $\tilde{r}_{12} = 0.88$ and $\tilde{r}_{13} = 0.40$ characterizing the relation of the endurance level to the resistance and fluidity levels, respectively. Determine the limits of the confidence interval for r_{12} and r_{13} if the confidence level is 0.95.

SOLUTION. For a large sample size n and small values of the correlation coefficient r, its estimate $\tilde{r}$ has a distribution approximately normal with expectation $\mathbf{M}[\tilde{r}] = r$ and standard deviation $\sigma_{\tilde{r}} = (1 - r)/(\sqrt{n - 1})$. Taking $r \approx \tilde{r}$, we have:

$$r_{12} \approx 0.88, \; r_{13} \approx 0.40,$$

$$\sigma_{\tilde{r}_{12}} \approx \frac{1 - 0.88}{\sqrt{0.29}} = 0.022,$$

$$\sigma_{\tilde{r}_{13}} \approx \frac{1 - 0.40}{\sqrt{0.29}} = 0.111.$$

From Table 8T, for the confidence level $\alpha = 0.95$, we find $\varepsilon_0 = 1.96$ (ε_0 being the solution of the equation $\alpha = \Phi(\varepsilon_0)$) and the confidence interval:

(0.84, 0.92) for r_{12},

(0.18, 0.62) for r_{13}.

The confidence interval obtained can be corrected if we transform $\tilde{r}$ so that $\sigma_{\tilde{r}}$ is independent of r. This leads to a new random variable

$$Z = \frac{1}{2} \ln \frac{1 + \tilde{r}}{1 - \tilde{r}},$$

whose distribution is approximately normal even for small n.

In this case

$$\mathbf{M}[Z] = \frac{1}{2} \ln \frac{1 + r}{1 - r} + \frac{r}{2(n - 1)} \approx \frac{1}{2} \ln \frac{1 + \tilde{r}}{1 - \tilde{r}} + \frac{\tilde{r}}{2(n - 1)}$$

and

$$\sigma_z = \frac{1}{\sqrt{n - 3}}.$$

Using Table 31T, we determine the confidence interval for the random variable Z:

0.88 (1.014, 1.768) for $\tilde{r}_{12}$,

0.40 (0.053, 0.808) for $\tilde{r}_{13}$.

Using Table 31T, we find the confidence interval:

(0.77, 0.94) for $\tilde{r}_{12}$,

(0.05, 0.67) for $\tilde{r}_{13}$.

PROBLEMS

42.1 A constant quantity is measured 25 times with an instrument whose systematic error is zero and random errors are normally distributed with standard deviation $E = 10$ m. Determine the limits of the confidence interval for the values of the quantity being measured if the confidence level is 0.99 and $\tilde{x} = 100$ m.

42.2 The results of measurements not containing systematic errors are written in the form of a statistical series (Table 40). The errors obey a normal distribution. Determine the estimate of the quantity being measured and the limits of the confidence interval for a confidence level 0.95.

TABLE 40

j	x_j^*, m.	m_j	j	x_j^*, m.	m_j
1	114	2	4	117	4
2	115	5	5	118	3
3	116	8			

42.3 From the results of 40 measurements of a base of constant length, estimates of the length and the standard deviation are found: $\tilde{x} = 10{,}400$ m. and $\tilde{\sigma}_x = 85$ m. The errors obey a normal distribution law. Find the probabilities that the confidence intervals with random limits $(0.999\tilde{x}, 1.001\tilde{x})$ and $(0.95\tilde{\sigma}, 1.05\tilde{\sigma})$ will cover the parameters $\bar{x}$ and σ_x, respectively.

42.4 The results of 11 measurements of a constant quantity are included in Table 41. The errors are normally distributed and the systematic errors are absent.

TABLE 41

Measurement No.	x_j, m.	*Measurement No.*	x_j, m.	*Measurement No.*	x_j, m.
1	9.9	5	6.0	9	11.6
2	12.5	6	10.9	10	9.8
3	10.3	7	10.3	11	14.0
4	9.2	8	11.8		

Determine (a) the estimates of the length being measured and the standard deviation, (b) the probability that the absolute value of the error in finding the exact value is less than 2 per cent of $\tilde{x}$, (c) the probability that the absolute value of the error in the standard deviation is less than 1 per cent of $\tilde{\sigma}$.

42.5 As a result of 100 experiments, it has been established that the average time necessary for the production of one item is $\tilde{\omega} = 0.5$ sec., and that $\tilde{\sigma}_\omega = 1.7$ sec. Assuming that the time to produce an item is a normal random variable, find the limits of the true values of $\tilde{\omega}$ and σ_ω for confidence levels 85 per cent and 90 per cent, respectively.

42.6 The estimate for the velocity of an airplane obtained from the results of five trials is $\tilde{v} = 870.3$ m./sec. Find the 95 per cent-confidence interval if it is known that the dispersion of the velocity obeys a normal distribution with standard deviation $E_v = 2.1$ m./sec.

42.7 The depth of a sea is measured with an instrument whose systematic error is zero and the random errors are normally distributed with standard deviation $E = 20$ m. How many independent measurements should be performed to find the depth with an error of at most 15 m. if the confidence level is 90 per cent?

42.8 Find for confidence level 0.9 the confidence limits for the distance to a reference point $\bar{x}$ and the standard deviation E if in 10 independent measurements, the results given in Table 42 were obtained, and the errors obey a normal distribution.

TABLE 42

Experiment No.	1	2	3	4	5	6	7	8	9	10
x_j, m.	25,025	24,970	24,780	25,315	24,907	24,646	24,717	25,354	24,911	25,374

42.9 Assume that five independent measurements with equal accuracy are performed to determine the charge of an electron. The experiments give the following results (in absolute electrostatic units):

$$4.781 \cdot 10^{-10}, \qquad 4.792 \cdot 10^{-10},$$
$$4.795 \cdot 10^{-10}, \qquad 4.779 \cdot 10^{-10}.$$
$$4.769 \cdot 10^{-10},$$

Find the estimate for the magnitude of the charge and the confidence limits of a confidence level of 99 per cent.

42.10 From the results of 15 independent equally accurate measurements, there were derived the following values for the estimate of the expected magnitude and the standard deviation of the maximal velocity of an airplane: $\tilde{v} = 424.7$ m./sec. and $\tilde{\sigma}_v = 8.7$ m./sec.

Determine (a) the confidence limits for the expectation and the standard deviation if the confidence level is 0.9, (b) the probabilities with which one may assert that the absolute value of the error in $\tilde{v}$ and σ_v does not exceed 2 m./sec.

42.11 The arithmetic mean of the results of independent measurements performed with n range finders is taken as an estimate of the distance to a navigational marker. The measurements contain no systematic errors and the random errors are normally distributed with standard deviation $E = 10$ m. How many range finders should there be so that the absolute value of the error in determining the distance will be at most 15 m. with probability 0.9?

42.12 It is known that a measuring instrument has no systematic errors and the random errors of each measurement obey the same normal distribution law. How many measurements should be performed to find the estimates for the standard deviation so that with confidence level of 70 per cent the absolute value of the error is at most 20 per cent of $\tilde{\sigma}$?

42.13 The systematic errors of an instrument are practically zero and the random errors are normally distributed with standard deviation $E = 20$ m. It is necessary that the difference between the estimate of the measured quantity and its true value is at most 10 m. What is the probability with which this condition will be satisfied if the number of observations is 3, 5, 10, 25 (construct the graph)?

42.14 The estimate for a measured quantity is given by the formula

$$\tilde{x} = \frac{1}{n} \sum_{j=1}^{n} x_j.$$

The results of individual measurements obey the same normal distribution law. Find the limits of the confidence interval for level 0.9 with the following conditions: (a) $\sigma = 20$ m., $n = 3, 5, 10, 25$; (b) $\tilde{\sigma} = 20$ m., $n = 3, 5, 10, 25$.

42.15 Ten identical devices are tested. The instants when each of them failed are recorded. The results of the observations are in Table 43.

Determine the estimate for the expected time t for nonstop

TABLE 43

Device No.	1	2	3	4	5	6	7	8	9	10
t_i, hours	200	350	600	450	400	400	500	350	450	550

operation of a device and the confidence interval for t if the confidence level is 0.9 and the random variable T obeys an exponential distribution.

42.16 A randomly selected sample of eight devices is subjected to reliability tests. The numbers of hours during which the devices operate without failures are 100, 170, 400, 250, 520, 680, 1500 and 1200. Determine the 80 per cent confidence interval for the average duration of operation if the reliable operation time obeys an exponential distribution law.

42.17 The probability density for the time between successive failures of a radio-electronic device is given by the formula

$$f(t) = \frac{1}{\bar{t}} \exp\left\{-\frac{t}{\bar{t}}\right\},$$

where t is the operating time between two successive failures, $\bar{t}$ is the expected value of the random variable T, which is the expected time during which the device operates in good condition (called in reliability theory "the expected lifetime").

For the determination of the estimates of the parameter $\bar{t}$ 25 failures were observed and the total duration of the reliable time from the start of the tests to the last failure turned out to be $\sum_{j=1}^{25} t_j = 1600$ hours.

Find the limits of the confidence interval for the parameter $\bar{t}$ according to the results of these experiments if the confidence level is $\alpha = 0.8$.

42.18 To determine the toxic dose, a certain poison is administered to 30 rats, eight of which die. Determine the limits of the confidence interval for the probability that the given dose will be fatal if the confidence level is 0.95 and if the number of fatal outcomes in this experiment obeys a binomial distribution law.

42.19 In 100 independent trials, a certain event A is observed 40 times. Determine the limits of the confidence interval for the probability of occurrence of this event if the confidence levels are 0.95 and 0.99 and the number of occurrences of A has a binomial distribution.

42.20 Ten devices are tested and no failures are observed. Determine the limits of the confidence interval in the cases in which the confidence levels are 0.8, 0.9 and 0.99 if the failures have a binomial distribution.

42.21 A marksman A scores five hits in 10 shots and B scores 50 hits in 100 shots; both marksmen fire shots at the same target. Determine the limits of the confidence interval for the probabilities that each marksman scores a hit in one shot if the confidence level is 0.99 and the hits obey a binomial distribution.

42.22 Six identical devices are tested. During 15 hours of tests 12 failures are recorded. Find the limits of the confidence interval for the expected number of failures during 15 hours operation of such a device if the confidence level is 0.9 and the tested devices obey a Poisson distribution.

42.23 The number of particles recorded by a counter in the Rutherford-Chadwick-Ellis experiment during every one of 2608 intervals of 7.5 sec. each is given in Table 44. Assuming that the number of particles obeys a Poisson distribution, determine the limits of the confidence interval for the parameter of this distribution, corresponding to an interval of 7.5 sec. and to the confidence level 0.9999.

TABLE 44

No. of particles reaching the counter	*No. of observations in which this number occurred*	*No. of particles reaching the counter*	*No. of observations in which this number occurred*
0	57	6	273
1	203	7	139
2	383	8	45
3	525	9	27
4	532	10	16
5	408		

42.24 In analyzing the amount of dodder in seeds of clover, it has been established that a sample of 100 g. contains no dodder seeds. Find the 99 per cent-confidence interval for the average amount of dodder in a sample that weighs 100 g. if the amount of dodder obeys a Poisson distribution.

42.25 From the results of 190 experiments performed with items made of Type A steel, estimates were found for the correlation coefficients $\tilde{r}_{12} = 0.55$, $\tilde{r}_{13} = 0.30$, $\tilde{r}_{14} = 0.37$, characterizing the dependence of the coercive force on the grain and carbon and sulphur compositions, respectively. Determine the limits of the confidence intervals for the correlation coefficients if the confidence levels are 0.99 and 0.95 and if the random variables have a normal distribution.

42.26 In a certain experiment, 25 pairs of values were obtained for a system of random variables (X, Y) with a normal distribution. With these experimental data, the parameters of this system were estimated: $\tilde{x} = 10.5, \tilde{y} = 74, \tilde{\sigma}_x = 2.0, \tilde{\sigma}_y = 10.0, \tilde{r}_{xy} = 0.62$. Find the limits of the confidence intervals for the parameters of the system (X, Y) if the confidence level is 0.9.

43. TESTS OF GOODNESS-OF-FIT

Basic Notions

The tests of goodness-of-fit permit estimation of the probability that a certain sample does not contradict the assumption made regarding the form of the distribution law of a considered random variable. For this purpose, one selects some quantity κ representing the discrepancy measure of the statistical and theoretical distribution laws, and determines for it a value κ_α such that $\mathbf{P}(\kappa > \kappa_\alpha) = \alpha$, where α is a sufficiently small quantity (significance level) whose value is determined by the nature of the problem. If the experimental value of the discrepancy measure κ_q is greater than κ_α, the deviation from the theoretical distribution law is considered significant and the assumption regarding the form of the distribution is disproved (the probability of disproving a correct assumption with regard to the form of the distribution in this case is equal to α). If $\kappa_q \leqslant \kappa_\alpha$, then the experimental data agree with the hypothesis made about the form of the distribution law.

The test of the hypothesis about the character of the distribution by means of goodness-of-fit procedures can be performed in another order: according to the value κ_q, one determines the probability $\alpha_q = \mathbf{P}(\kappa < \kappa_q)$. If $\alpha_q < \alpha$, the deviations are significant; if $\alpha_q \geqslant \alpha$, the deviations are insignificant. The values α_q, very close to 1 (very good fit), correspond to an event with very small probability of occurrence and indicate that the sample is defective (for example, elements with large deviations from the average are eliminated from the initial sample without further reason).

In different tests of goodness-of-fit, different quantities are taken as measures of discrepancy between the statistical and theoretical distributions.

In the chi-square tests (the Pearson tests), the discrepancy measure is the quantity χ^2, whose experimental value χ_q^2 is given by the formula

$$\chi_q^2 = \sum_{i=1}^{l} \frac{(m_i - np_i)^2}{np_i},$$

where l is the number of classes into which all experimental values of X are divided, n is the sample size, m_i is the number in the ith class and p_i is the probability, computed from the theoretical distribution law, that the random variable X is in the ith class interval.

For $n \to \infty$, the distribution of χ_q^2, regardless of the distribution of the random variable X, tends to a chi-square distribution with $k = l - r - 1$ degrees of freedom, where r is the number of parameters, computed according to the given sample, of the theoretical distribution law.

The values of the probabilities $\mathbf{P}(\chi^2 \geqslant \chi_q^2)$ as functions of χ_q^2 and k are given in Table 17T.

To apply the chi-square test in the general case it is necessary that the sample size n and class numbers m_i be sufficiently large (practically, it is considered sufficient that $n \sim 50$–60, $m_i \sim 5$–8).

The Kolmogorov test of goodness-of-fit is applicable only if the parameters of the theoretical distribution law are not determined by the data of the sample. The biggest value D of the absolute value of the difference between the statistical and theoretical distribution functions is selected as the discrepancy measure of the statistical and theoretical distribution laws. The experimental value D_q of D is determined by the formula

$$D_q = \max |\tilde{F}(x) - F(x)|,$$

where $\tilde{F}$ and F are the statistical and the theoretical distribution functions, respectively.

As $n \to \infty$, the distribution law for $\lambda = \sqrt{N}D$, regardless of the form of the distribution of the random variable X, tends to the Kolmogorov distribution. The values of the probabilities $\alpha_q = \mathbf{P}(D \geqslant D_q) = P(\lambda) = 1 - K(\lambda)$ are included in Table 25T.

The Kolmogorov test is also a statistical test of the hypothesis that two samples of size n_1 and n_2 arise from a single population. In this case, $\alpha_q = \mathbf{P}(\lambda)$, where $\mathbf{P}(\lambda)$ is given in Table 25T, but

$$\lambda = \sqrt{\frac{n_1 n_2}{n_1 + n_2}} \cdot \max |\tilde{F}_1(x) - \tilde{F}_2(x)|,$$

where $\tilde{F}_1(x)$ and $\tilde{F}_2(x)$ are the statistical distribution functions for the first and second samples.

The form of the theoretical distribution is chosen either on the basis of data about the random variables selected or by qualitative analysis of the form of the distribution histogram. If the form of the distribution cannot be established from general considerations, then it is approximated by a distribution whose first few moments are the same as the estimates obtained from the sample. For approximating expressions, one can use Pearson's curves (Gnedenko and Khinchin, 1962), which consider the four first moments or the infinite Edgeworth series (Gnedenko and Khinchin, 1962). Here, for a small deviation of the statistical distribution from the normal, one can retain only the first terms, forming a Charlier-A series,

$$F(z) = 0.5 + 0.5\Phi(z) - \frac{\widetilde{\mathrm{Sk}}}{6}\varphi_2(z) + \frac{\widetilde{\mathrm{Ex}}}{24}\varphi_3(z),$$

where $\varphi_2(z)$, $\varphi_3(z)$ are the second and third derivatives of the normal probability density $\varphi(z)$; $z = (x - M[x])/\tilde{\sigma}$, $\widetilde{\mathrm{Sk}} = \tilde{\mu}_3/\tilde{\sigma}_3$ is the estimate for asymmetry, $\widetilde{\mathrm{Ex}} = (\tilde{\mu}_4/\tilde{\sigma}_4) - 3$ is the estimate for excess and $\tilde{\sigma}^2$, $\tilde{\mu}_3$ and $\tilde{\mu}_4$ are the estimates for the second, third and fourth central moments, respectively.

The values of $\Phi(z)$, $\varphi_2(z)$, $\varphi_3(z)$ are given in Tables 8T and 10T.

The chi-square test also permits us to test hypotheses about the independence of two random variables X and Y. In this case, χ_q^2 is determined by the formula

$$\chi_q^2 = \sum_{i=1}^{l} \sum_{j=1}^{m} \frac{(h_{ij} - m_{ij})^2}{m_{ij}},$$

where h_{ij} is the number of cases in which the values $X = x_i$, $Y = y_j$ are observed simultaneously,

$$m_{ij} = \frac{h_{i0} h_{0j}}{n},$$

h_{i0} being the number of cases in which the value $X = x_i$ is observed, h_{0j} is the total number of cases in which the value $Y = y_j$ is observed and l and m are the numbers of values assumed by the variables X and Y.

The number k of degrees of freedom necessary for the calculation of the probability $\mathbf{P}(\chi^2 \geqslant \chi_q^2)$ is given by the formula

$$k = (l - 1)(m - 1).$$

Solution for Typical Examples

Example 43.1 A radioactive substance is observed during 2608 equal time intervals (each, 7.5 sec.). For each interval the number of particles reaching a counter is recorded. The numbers m_i of time intervals during which exactly i particles reached the counter are given in Table 45:

Table 45

i	m_i	i	m_i
0	57	6	273
1	203	7	139
2	383	8	45
3	525	9	27
4	532	10 or more	16
5	408	Total	$n = \sum m_i = 2608$

Test, using the chi-square test, the hypothesis that the data agree with the Poisson distribution law

$$P(i, a) = \frac{e^{-a}a^i}{i!}.$$

The significance level should be taken as 5 per cent.

Solution. Using the data, we compute the estimate $\tilde{a}$ of the parameter a of the Poisson distribution by the formula

$$\tilde{a} = \frac{\sum_{i=0}^{\infty} im_i}{n},$$

where $n = \sum_{i=0}^{\infty} m_i = 2608$, $\tilde{a} = 3.870$.

For the function $P(i, \tilde{a}) = p_i$, we compute, using Table 6T, the theoretical probabilities p_i that i particles with Poisson distribution reach the counter. As a result of interpolation between $a = 3$ and $a = 4$, we obtain the values p_i and np_i, which are given in Table 46.

We compute the values of χ_q^2 by performing the calculations in Table 46:

$$\chi_q^2 = \sum_{i=0}^{10} \frac{(m_i - np_i)^2}{np_i} = 13.05.$$

Since the number of degrees of freedom is $k = l - r - 1$, where the total number of intervals is $l = 11$ and the number of parameters determined from the data is $r = 1$ (the parameter a), we have

$$k = 11 - 1 - 1 = 9.$$

From Table 17T, we find for $k = 9$ and $\chi_q^2 = 13.05$ the probability $\mathbf{P}(\chi^2 \geqslant \chi_q^2)$ that the quantity χ^2 will exceed χ_q^e. We obtain

$$\alpha_q = \mathbf{P}(\chi^2 \geqslant \chi_q^2) = 0.166.$$

TABLE 46

i	p_i	np_i	$m_i - np_i$	$(m_i - np_i)^2$	$\frac{(m_i - np_i)^2}{np_i}$
0	0.021	54.8	2.2	4.84	0.088
1	0.081	211.2	−8.2	67.24	0.318
2	0.156	406.8	−23.8	566.44	1.392
3	0.201	524.2	0.8	0.64	0.001
4	0.195	508.6	23.4	547.56	1.007
5	0.151	393.8	14.2	201.64	0.512
6	0.097	253.0	20.0	400.00	1.581
7	0.054	140.8	−1.8	3.24	0.023
8	0.026	67.8	−22.8	519.84	7.667
9	0.011	28.7	−1.7	2.89	0.101
10	0.007	18.3	−2.3	5.29	0.289
	1.000				$\chi_q^2 = 13.049$

Since $\alpha_q > \alpha = 0.05$, the deviations from the Poisson distribution are insignificant.

Similarly one can solve Problems 43.1 to 43.4.

Example 43.2 A sample of 200 items is selected from the current output of a precision automatic lathe. The dimension of each item is measured with a precision of 1 micron. The deviations x_i (from the nominal dimension) divided into classes and the numbers m_i in the classes and their frequencies p^* are given in Table 47.

TABLE 47

Class No. i	Limits of the interval x_i to x_{i+1}	m_i	p_i^*	Class No. i	Limits of the interval x_i to x_{i+1}	m_i	p_i^*
1	−20 to −15	7	0.035	6	5 to 10	41	0.205
2	−15 to −10	11	0.055	7	10 to 15	26	0.130
3	−10 to −5	15	0.075	8	15 to 20	17	0.085
4	−5 to 0	24	0.120	9	20 to 25	7	0.035
5	0 to 5	49	0.245	10	25 to 30	3	0.015

Estimate with the aid of the chi-square test the hypothesis that the sample distribution obeys a normal distribution law for a significance level $\alpha = 5$ per cent.

SOLUTION. We determine the values x_i^* of the midpoints of the intervals and find the estimates for the expectation and variance by the formulas:

$$\tilde{x} = \sum_{i=1}^{10} x_i^* p_i^* = 4.30\ \mu, \qquad \tilde{\sigma}^2 = \tilde{m}_2 - \tilde{x}^2 = 94.26\ \mu^2,$$

$$\tilde{m}_2 = \sum_{i=1}^{10} x_i^{*2} p_i^* = 112.75\ \mu^2, \qquad \tilde{\sigma} = 9.71\ \mu.$$

The computations are summarized in Table 48.

TABLE 48

i	x_i^*	z_i	$\frac{1}{2}\Phi(z_i)$	p_i	np_i	$\frac{(m_i - np_i)^2}{np_i}$
1	−17.5	$-\infty$	−0.5000	0.0239	4.78	1.04
2	−12.5	−1.99	−0.4761	0.0469	9.38	0.28
3	−7.5	−1.47	−0.4292	0.0977	19.54	1.05
4	−2.5	−0.96	−0.3315	0.1615	32.30	2.13
5	2.5	−0.44	−0.1700	0.1979	39.58	2.24
6	7.5	0.07	0.0279	0.1945	38.90	0.11
7	12.5	0.59	0.2224	0.1419	28.38	0.20
8	17.5	1.10	0.3643	0.0831	16.62	0.01
9	22.5	1.62	0.4474	0.0526	10.52	0.03
10	27.5	2.13	0.4834			
11	—	∞	0.5000	—	200	$\chi_q^2 = 7.09$

The theoretical probabilities p_i that the deviations lie on the intervals (x_i, x_{i+1}) are computed by the formula

$$p_i = \frac{1}{2}\Phi(z_{i+1}) - \frac{1}{2}\Phi(z_i),$$

where z_i is the left limit of the ith interval measured with respect to $\tilde{x}$ in $\tilde{\sigma}$ units:

$$z_i = \frac{x_i - \tilde{x}}{\tilde{\sigma}}.$$

Here the smallest $z_i = z_0 = 2.06$ is replaced by $-\infty$, and the largest $z_{11} = 3.09$ by $+\infty$.

The value of the Laplace function $\Phi(z)$ is found from Table 8T. The interval $i = 10$, because of its small number, is attached to the interval $i = 9$. The results of the computations are given in Table 48.

We find that

$$\chi_q^2 = \sum_{i=1}^{9} \frac{(m_i - np_i)^2}{np_i} = 7.09.$$

The number of degrees of freedom is

$$k = l - r - 1 = 9 - 2 - 1 = 6,$$

since, because of the small numbers in the last two classes, the 9th and 10th classes are united.

From Table 17T, for the values χ_q^2 and k we find $\alpha_q = \mathbf{P}(\chi^2 \geqslant \chi_q^2) = 0.313$. The hypothesis on the normality of the deviations from the nominal dimension does not contradict the observations.

Problems 43.6, 43.7, 43.9, 43.11, 43.13 to 43.21, 43.24 and 43.25 can be solved in a similar manner.

Example 43.3 The results x_i of several measurements (rounded-off to 0.5 mm.) of 1000 items are given in Table 49.

TABLE 49

i	x_i	m_i	i	x_i	m_i
1	98.0	21	6	100.5	201
2	98.5	47	7	101.0	142
3	99.0	87	8	101.5	97
4	99.5	158	9	102.0	41
5	100.0	181	10	102.5	25

(m_i is the number of measurements giving the result x_i.)

By using the Kolmogorov test of goodness-of-fit, verify that the observations agree with the assumption that the variable X obeys a normal distribution law with expectation $\bar{x} = 100.25$ mm. and standard deviation $\sigma = 1$ mm., if the influence of round-off errors may be neglected.

SOLUTION. The theoretical distribution function $F(x)$ is defined by the formula

$$F(x) = \frac{1}{2} + \frac{1}{2}\Phi(x - \bar{x}).$$

The statistical distribution function $\tilde{F}^*(x)$ can be calculated by the formula

$$\tilde{F}^*(x_k) = \frac{1}{1000}\left[\sum_{i=1}^{k} m_i + 0.5m_k\right].$$

The computations are performed in Table 50.

TABLE 50

i	$x_i - \bar{x}$	$\frac{1}{2}\Phi(x_i - \bar{x})$	$F(x_i)$	$\tilde{F}^*(x_i)$	$\lvert\tilde{F}^*(x_i) - F(x_i)\rvert$
1	−2.25	−0.4877	0.0123	0.0105	0.0018
2	−1.75	−0.4599	0.0401	0.0445	0.0044
3	−1.25	−0.3944	0.1056	0.1115	0.0059
4	−0.75	−0.2734	0.2266	0.2340	0.0074
5	−0.25	−0.0987	0.4013	0.4035	0.0022
6	0.25	0.0987	0.5987	0.5945	0.0042
7	0.75	0.2734	0.7734	0.7660	0.0074
8	1.25	0.3944	0.8954	0.8855	0.0089
9	1.75	0.4599	0.9599	0.9545	0.0054
10	2.25	0.4877	0.9877	0.9875	0.0002

For each value x_i, forming the differences $\tilde{F}^*(x_i) - F(x_i)$ and selecting from them the largest in absolute value, according to Table 50 we find $D_q = 0.0089$.

Computing

$$\lambda = \sqrt{n}\, D_q = \sqrt{1000}\cdot 0.0089 = 0.281,$$

we find the value of $P(\lambda)$ from Table 25T:

$$\alpha_q = P(\lambda) = 1000.$$

The value of $P(\lambda)$ is large. Consequently, the deviations are insignificant and it can be assumed that the hypothesis that the data obey a normal distribution with parameters $\bar{x} = 100.25$, $\sigma = 1$ is valid; however, a large value of α leads to doubts about the high quality of the sample.

Following this solution, one can solve Problems 43.5, 43.8, 43.10, 43.12, 43.22 and 43.23.

Example 43.4 According to the data of Example 43.2, select the distribution law by using a Charlier-*A* series, and test by means of the chi-square test whether the goodness-of-fit of the data with the resulting distribution law will be improved by use of the normal distribution.

From Example 43.2, we take the estimates of the expectation $\tilde{x}$ and standard deviation $\tilde{\sigma}$:

$$\tilde{x} = 4.30\ \mu, \qquad \tilde{\sigma} = 9.71\ \mu.$$

Moreover, using the data of Tables 47 and 48, we estimate the third central moment $\tilde{\mu}_3$ and the fourth central moment $\tilde{\mu}_4$ of the random variable X.

$$\tilde{\mu}_3 = \sum_{i=1}^{10} (x_i^* - \tilde{x})^3 p_i^* = -113.86\ \mu^3,$$

$$\tilde{\mu}_4 = \sum_{i=1}^{10} (x_i^* - \tilde{x})^4 p_i^* = 25{,}375\ \mu^4.$$

The computations are performed in Table 51.

TABLE 51

i	x_i^*	$x_i - \tilde{x}$	$(x_i - \tilde{x})^3$	$(x_i - \tilde{x})^4$	$(x_i - \tilde{x})^3 p_i^*$	$(x_i - \tilde{x})^4 p_i^*$
1	−17.5	−21.8	−10,360	225,853	−362.6	7,904.9
2	−12.5	−16.8	−4,742	79,659	−260.8	4,381.2
3	−7.5	−11.8	−1,643	19,388	−123.2	1,454.1
4	−2.5	−6.8	−314	2,138	−37.7	256.6
5	2.5	−1.8	−6	10	−1.5	2.4
6	7.5	3.2	33	105	6.8	21.5
7	12.5	8.2	551	4,125	71.6	587.7
8	17.5	13.2	2,300	30,360	195.5	2,580.6
9	22.5	18.2	6,029	109,720	211.0	3,840.2
10	27.5	23.2	12,487	289,702	187.3	4,345.5
				Total	−113.86	25,375

Furthermore, we compute the estimates for the asymmetry $\widetilde{\mathrm{Sk}}$ and excess $\widetilde{\mathrm{Ex}}$ by the formulas

$$\widetilde{\mathrm{Sk}} = \frac{\tilde{\mu}_3}{\tilde{\sigma}^3} = -0.1247,$$

$$\widetilde{\mathrm{Ex}} = \frac{\tilde{\mu}_4}{\tilde{\sigma}^4} - 3 = -0.1455.$$

Using the first three terms of the distribution function for the Charlier-A series,

$$F(z) = 0.5 + 0.5\Phi(z) - \frac{\widetilde{\text{Sk}}}{6}\varphi_2(z) + \frac{\widetilde{\text{Ex}}}{24}\varphi_3(z),$$

where

$$z = \frac{x - \tilde{x}}{\tilde{\sigma}},$$

we find

$$F(z) = 0.5 + 0.5\Phi(z) + 0.0208\varphi_2(z) - 0.0061\varphi_3(z).$$

We now compute the values $F(z_i)$, and use tables 8T, 10T for the determination of the values of $\Phi(z)$, $\varphi_2(z)$, $\varphi_3(z)$; here z_i are the coordinates with respect to $\tilde{x}$ in $\tilde{\sigma}$ units of the limits of the intervals. The values of z_i and the subsequent computations of $F(z_i)$ are given in Table 52.

TABLE 52

i	z_i to z_{i+1}	$0.5\Phi(z_i)$ to $0.5\Phi(z_{i+1})$	$\varphi_2(z_i)$ to $\varphi_2(z_{i+1})$
1	$-\infty$ to -1.99	-0.5 to -0.4761	0 to 0.1630
2	-1.99 to -1.47	-0.4761 to -0.4292	0.1630 to 0.1572
3	-1.47 to -0.96	-0.4292 to -0.3315	0.1572 to -0.0197
4	-0.96 to -0.44	-0.3315 to -0.1700	-0.0197 to -0.2920
5	-0.44 to 0.07	-0.1700 to 0.0279	-0.2920 to -0.3960
6	0.07 to 0.59	0.0279 to 0.2224	-0.3960 to -0.2185
7	0.59 to 1.10	0.2224 to 0.3643	-0.2185 to 0.0458
8	1.10 to 1.62	0.3643 to 0.4474	0.0458 to 0.1745
9	1.62 to 2.13	0.4474 to 0.4834	0.1745 to 0.1460
10	2.13 to ∞	0.4834 to 0.5000	0.1460 to 0

i	$\varphi_3(z_i)$ to $\varphi_3(z_{i+1})$	$F(z_i)$ to $F(z_{i+1})$	p_i	np_i	$m_i - np_i$	$\frac{(m_i - np_i)^2}{np_i}$
1	0 to 0.1062	0 to 0.0267	0.0267	5.34	-1.66	0.516
2	0.1062 to -0.1670	0.0267 to 0.0751	0.0484	9.68	-1.32	0.180
3	-0.1670 to -0.5021	0.0751 to 0.1712	0.0961	19.22	4.22	0.926
4	-0.5021 to -0.4472	0.1712 to 0.3266	0.1554	28.08	4.08	0.593
5	-0.4472 to 0.0834	0.3266 to 0.5192	0.1926	38.52	10.48	2.852
6	0.0834 to 0.5245	0.5192 to 0.7147	0.1955	39.10	-1.90	0.092
7	0.5245 to 0.4290	0.7147 to 0.8627	0.1480	29.60	3.60	0.438
8	0.4290 to 0.0654	0.8627 to 0.9506	0.0879	17.58	0.58	0.019
9	0.0654 to -0.1351	0.9506 to 0.9872	0.0366	7.32	-0.12	0.001
10	-0.1351 to 0	0.9872 to 1.0000	0.0128	2.56		$\chi_q^2 = 5.615$

The theoretical probabilities p_i based on the distribution law defined by the Charlier-A series are computed by the formula

$$p_i = F(z_{i+1}) - F(z_i).$$

Using these values and noting that $n = \sum_{i=1}^{10} m_i = 200$, we compute (see Table 52)

$$\chi_q^2 = \sum_{i=1}^{10} \frac{(m_i - np_i)^2}{np_i} = 5.615.$$

The number of degrees of freedom is $k = l - r - 1 = 4$ since the number of classes is $l = 9$ (the last two intervals, because of their small number, are united into a single interval; the number of parameters determined on the basis of the data is $r = 4(\tilde{x}, \tilde{\sigma}, \widetilde{\mathrm{Sk}}, \widetilde{\mathrm{Ex}})$. From Table 17T, for $k = 4$ and $\chi_q^2 = 5.615$, we find that $\alpha_q = \mathbf{P}(\chi^2 \geqslant \chi_q^2) = 0.208$.

The hypothesis on the goodness-of-fit of the experimental data with the distribution law $F(z)$ specified by a Charlier-A series is not disproved. However, there are no reasons to assert that the goodness-of-fit is better than what is provided by the normal distribution law mentioned in the assumption of the problem.

In a similar way one can solve Problems 43.26 and 43.27.

Example 43.5 There are two groups of 60 identical items produced by two machines. The data obtained from several measurements of some specific dimension x of the items are given in Table 53.

TABLE 53

Item No.	Dimension x		Item No.	Dimension x		Item No.	Dimension x	
	Group I	*Group II*		*Group I*	*Group II*		*Group I*	*Group II*
1	72.58	72.50	21	72.50	72.35	41	72.30	72.31
2	72.35	72.35	22	72.69	72.16	42	72.28	72.46
3	72.33	72.69	23	72.54	72.51	43	72.51	72.36
4	72.54	72.60	24	72.48	72.50	44	72.37	72.39
5	72.24	72.54	25	72.36	72.50	45	72.14	72.30
6	72.42	72.42	26	72.50	72.48	46	72.42	72.30
7	72.58	72.68	27	72.43	72.53	47	72.36	72.38
8	72.47	72.54	28	72.46	72.25	48	72.28	72.55
9	72.54	72.55	29	72.56	72.48	49	72.20	72.36
10	72.24	72.33	30	72.48	72.36	50	72.48	72.24
11	72.38	72.56	31	72.43	72.53	51	72.66	72.23
12	72.70	72.36	32	72.56	72.23	52	72.64	72.16
13	72.47	72.36	33	72.34	72.55	53	72.73	72.17
14	72.49	72.15	34	72.38	72.51	54	72.43	72.37
15	72.28	72.48	35	72.56	72.25	55	72.28	72.38
16	72.47	72.46	36	72.32	72.11	56	72.64	72.46
17	71.95	72.36	37	72.41	72.44	57	72.72	72.12
18	72.18	72.38	38	72.14	72.51	58	72.35	72.28
19	72.66	72.40	39	72.29	72.55	59	72.60	72.23
20	72.35	72.38	40	72.31	72.24	60	72.46	72.38

Test, by means of the Kolmogorov test, the hypothesis that both samples belong to a single population; i.e., that both machines give the same distribution of the dimension x at a significance level $\alpha = 8$ per cent.

SOLUTION. We divide the items into groups according to the increasing dimension x and compute the statistical distribution functions $\tilde{F}_1(x)$ and $\tilde{F}_2(x)$ for each group (see Table 54).

TABLE 54

x_i	Number of values $x \leqslant x_i$ Group I	Group II	$\tilde{F}_1(x_i)$	$\tilde{F}_2(x_i)$	$\tilde{F}_1(x_i) - \tilde{F}_2(x_i)$
71.95	1	0	0.0167	0	0.0167
72.11	1	1	0.0167	0.0167	0.0000
72.12	1	2	0.0167	0.0333	0.0167
72.14	3	2	0.0500	0.0333	0.0167
...	...	...	...	...	...
72.53	43	50	0.7167	0.8333	0.1167
72.54	46	52	0.7667	0.8667	0.1000
72.55	46	56	0.7667	0.9333	0.1667
72.56	49	57	0.8167	0.9500	0.1333
72.58	51	57	0.8500	0.9500	0.1000
72.60	52	58	0.8667	0.9667	0.1000
...	...	...	...	...	...
72.69	57	60	0.9500	0.0000	0.0500
72.70	58	60	0.9667	1.0000	0.0333
72.72	59	60	0.9833	1.0000	0.0167
72.73	60	60	1.0000	1.0000	0.0000

We find the largest absolute value D_{n_1,n_2} of the difference $\tilde{F}_1(x) - \tilde{F}_2(x)$:

$$D_{n_1,n_2} = 1.667 \quad \text{(see Table 54)}.$$

Determining

$$\lambda = \sqrt{\frac{n_1 n_2}{n_1 + n_2}}\, D_{n_1,n_2},$$

where in our case $n_1 = n_2 = 6$, we obtain $\lambda = 0.9130$. Using Table 25T, for λ, we have $\mathbf{P}(\lambda) = 0.375 = \alpha_q$.

The value of α_q is large; consequently, the deviations are insignificant and the hypothesis that both samples belong to the same main population is not contradicted.

Example 43.6 Six-hundred items are measured and for each item the dimensions X and Y are checked. The results are given in Table 55, where h_{ij} denotes the number of items with dimensions $X = x_i$, $Y = y_j$.

TABLE 55

x_i \ x_j	Under-estimated: $j = 1$	Within tolerance limits: $j = 2$	Over-estimated: $j = 3$	h_{i0}
Underestimated: $i = 1$	6	48	8	62
Within tolerance limits: $i = 2$	52	402	36	490
Overestimated: $i = 3$	6	38	4	48
h_{0i}	54	488	48	600

For X: $i = 1$ if the dimension is underestimated, $i = 2$ if the dimension is within the tolerance limits, $i = 3$ if the dimension is overestimated; for Y: $j = 1, 2, 3$ if the dimension is underestimated, within the tolerance limits, or overestimated, respectively.

Test by using the chi-square test whether the deviations of dimensions X and Y from admissible dimensions are independent at a significance level $\alpha = 5$ per cent.

SOLUTION. We find the estimates m_{ij} of the expected number of observations in which $X = x_i$, $Y = y_j$ by starting from the hypothesis on the independence of X and Y:

$$m_{ij} = \frac{h_{i0}h_{0j}}{n}.$$

The values m_{ij} are given in Table 56.

TABLE 56

i \ j	1	2	3
1	6.61	50.43	4.96
2	52.26	308.57	39.20
3	5.12	39.06	3.86

We compute χ_q^2 by the formula

$$\chi_q^2 = \sum_{i=1}^{3} \sum_{j=1}^{3} \frac{(h_{ij} - m_{ij})^2}{m_{ij}}.$$

The computations are performed in Table 57, in which the values of $(h_{ij} - m_{ij})^2/m_{ij}$ are given.

TABLE 57

i \ j	1	2	3	$\sum_j$
1	0.0563	0.1171	1.8632	2.0366
2	0.0013	0.0295	0.2612	0.2920
3	0.1513	0.0316	0.0077	0.1906
$\sum_j$	0.2089	0.1782	2.1321	$\sum_{i,j} = \chi_q^2 = 2.5192$

We get $\chi_q^2 = 2.519$. Then, we determine the number of degrees of freedom

$$k = (l - 1)(m - 1),$$

where l is the number of groups according to the dimension X, m is the number of groups according to Y, $l = 3$, $m = 3$, $k = 4$. Using Table 17T for $k = 4$ and $\chi_q^2 = 2.519$, we find $\alpha_q = \mathbf{P}(\chi^2 \geqslant \chi_q^2) = 0.672$.

The value of α_q is large; consequently, the hypothesis on the independence of the deviations of the dimensions of an item according to the test of X and Y against the admissible dimensions is not disproved.

Problem 43.28 can be solved similarly.

PROBLEMS

43.1 In Table 58 are listed the numbers m_i of plots of equal area (0.25 km.^{2}) of the southern part of London. During the Second World War each of the plots was hit by i buzz bombs. Test with the aid of the chi-square test that the experimental data agree with the Poisson distribution law

$$P(i, a) = \frac{a^i e^{-a}}{i!},$$

if the significance level is 6 per cent.

TABLE 58

i	0	1	2	3	4	5 or more	Total
m_i	229	211	93	35	7	1	$n = \sum m_i = 576$

43.2 For a thin layer of gold solution, there is recorded the number of particles of gold reaching the field of view of a microscope during equal time intervals. The results of these observations are given in Table 59.

TABLE 59

No. of particles, i	0	1	2	3	4	5	6	7	Total
m_i	112	168	130	68	32	5	1	1	$\sum m_i = 517$

Test with the chi-square test the goodness-of-fit of the Poisson distribution: use the 5 per cent significance level.

43.3 Ten shots are fired from a rifle at each of 100 targets and the hits and the misses are recorded. The results appear in Table 60.

Test, by using the chi-square test, that the probabilities of hitting the targets are the same for all shots; in other words, test that the outcomes obey a binomial distribution law: use the 10 per cent significance level.

TABLE 60

No. of hits	m_i	*No. of hits*	m_i	*No. of hits*	m_i
0	0	4	22	8	4
1	2	5	26	9	2
2	4	6	18	10	0
3	10	7	12		

43.4 Seven coins are tossed simultaneously 1536 times and each time the number X of heads is recorded. Table 61 lists the number m_i of cases in which the number of heads is X_i.

TABLE 61

X_i	0	1	2	3	4	5	6	7
m_i	12	78	270	456	386	252	69	13

Using the chi-square test, test the hypothesis that the experimental data obey a binomial distribution law. Assume that the probability of occurrence of a head is 0.5 for each coin. The significance level should be 5 per cent.

43.5 Each of 100 machines produces a lot of 40 first-grade and second-grade items during one shift. Samples of 10 items from each lot are selected and for each sample the number of second grade items is recorded. The results of the tests are given in Table 62.

TABLE 62

i	0	1	2	3	4	5	6 or more
m_i	1	10	27	36	25	1	0

The m_i denote the numbers of samples with i second-grade items. The number of second-grade items produced during a long operation period of the plant is 30 per cent ($p = 0.30$) of all production.

Test, by using the Kolmogorov test, that the experimental results obey the hypergeometric and binomial distribution laws if one uses the 5 per cent significance level.

For the quantity i distributed according to a hypergeometric law, there obtains the formula

$$p_{i,n} = \frac{C_L^i C_{N-L}^{n-i}}{C_N^n},$$

where N is the number of items in the lot, L is the number of second-grade items in the lot and n is the sample size.

For a binomial distribution,

$$p_{i,n} = C_n^i p^i (1 - p)^{n-i}.$$

43.6 Table 63 contains the deviations from a given dimension of the diameters of several cylinders produced by a machine.

TABLE 63

The limits of the interval in microns	0 to 5	5 to 10	10 to 15	15 to 20	20 to 25
No. m_i in the class	15	75	100	50	20
Frequency p_i^*	0.06	0.30	0.40	0.20	0.04

Test, with the chi-square test, the hypothesis that the observations obey a normal distribution law if the 5 per cent significance level is used.

43.7 Suppose that 250 numbers are generated by summing the digits of five-digit numbers selected from a table of random numbers. The resulting sums are divided into 15 intervals as shown in Table 64.

TABLE 64

Interval	m_i	*Interval*	m_i	*Interval*	m_i
0 to 3	0	15 to 18	28.5	30 to 33	27.0
3 to 6	0.5	18 to 21	39.0	33 to 36	7.5
6 to 9	1.5	21 to 24	41.0	36 to 39	1.0
9 to 12	10.0	24 to 27	45.0	39 to 42	1.0
12 to 15	17.5	27 to 30	30.5	42 to 45	0

Sums representing multiples of three are equally divided between two adjacent intervals. Using the chi-square test, test whether the given statistical distribution obeys a normal distribution whose parameters are the expectation and variance determined from the data if the significance level is 5 per cent.

43.8 Solve the preceding problem by using the Kolmogorov test. Assume (because of the narrowness of the interval in Table 64) that it is possible to take all elements in each interval to be the value at the mid-point of the interval. To establish the hypothetical normal distribution law, consider that any value 0 to 9 for the individual digits of a random five-digit number has probability $p = 0.1$.

43.9 The digits 0, 1, 2, . . ., 9 among the first 800 decimals of the number π occur 74, 92, 83, 79, 80, 73, 77, 75, 76 and 91 times, respectively. Using the chi-square test, test the hypothesis that these data obey a uniform distribution law if the 10 per cent significance level is used.

43.10 Solve the preceding problem by using the Kolmogorov test, and by assuming that the probability that any digit appears at any decimal place is 0.10.

43.11 From a table of random numbers, 150 two-digit numbers (00 is also considered a two-digit number) are selected. The results appear in Table 65.

TABLE 65

Interval	*No.* m_i *in the class*	*Frequency* p_i^*	*Interval*	*No.* m_i *in the class*	*Frequency* p_i^*
0 to 9	16	0.107	50 to 59	19	0.127
10 to 19	15	0.100	60 to 69	14	0.093
20 to 29	19	0.127	70 to 79	11	0.073
30 to 39	13	0.087	80 to 89	13	0.087
40 to 49	14	0.093	90 to 99	16	0.107

Using the chi-square test, verify the hypothesis that the observations obey a uniform distribution law for a 5 per cent significance level.

43.12 Solve the preceding problem by applying the Kolmogorov test. Assume (because of the narrowness of an interval in Table 65) that all the elements in one interval may be taken equal to the midpoint of the interval.

43.13 The readings on the scale of a measuring instrument are estimated approximately in fractions of one division. Theoretically, any value of the last digit is equally probable, but in some cases certain digits are favored over others. In Table 66, 200 readings of the last digit between adjacent divisions of the scale are listed. Using the chi-square test, establish whether there is a systematic error in readings, i.e., whether the observations obey a uniform distribution law if the probability of appearance of any digit is $p_i = 0.10$ and the significance level is 5 per cent.

TABLE 66

Digit, i	0	1	2	3	4	5	6	7	8	9
m_i	35	16	15	17	17	19	11	16	30	24

43.14 The observed daily mean temperature of the air during 320 days is given in Table 67.

Establish, with the aid of the chi-square test, which of the two distributions, normal or Simpson (triangular), agrees with the data better, if the significance level is 3 per cent.

TABLE 67

x_i, °C	m_i	x_i, °C	m_i
−40 to −30	5	10 to 20	81
−30 to −20	11	20 to 30	36
−20 to −10	25	30 to 40	20
−10 to 0	42	40 to 50	8
0 to 10	88	50 to 60	4

43.15 In Table 68 there are listed the observed time periods necessary to find and remove the failure of a certain electronic device; these periods are expressed in hours with a precision of one minute.

TABLE 68

Interval No., i	*Limits of the interval* y_i to y_{i+1}	*No. in the class* m_i	*Interval No.,* i	*Limits of the interval* y_i to y_{i+1}	*No. in the class* m_i
1	1/60 to 3/60	2	8	1.8 to 3.2	10
2	3/60 to 6/60	5	9	3.2 to 5.6	7
3	6/60 to 10/60	7	10	5.6 to 10	4
4	10/60 to 18/60	11	11	10 to 18	2
5	18/60 to 35/60	15	12	18 to 30	1
6	35/60 to 1.0	21	13	more than 30	0
7	1.0 to 1.8	15			

Using the chi-square test, test that the data obey a logarithmically normal distribution in which $x = \log y$ obeys a normal distribution if the significance level is 5 per cent.

43.16 The data of the Vorontsov-Vel'yaminov catalog, the distribution of distances to planetary nebulae, are exhibited in Table 69, where X_i is the distance (in kiloparsecs) and m_i the number of cases (number in the class).

TABLE 69

X_i	m_i	X_i	m_i	X_i	m_i	X_i	m_i
0 to 0.5	9	3.0 to 3.5	12	6.0 to 6.5	3	9.0 to 9.5	2
0.5 to 1.0	11	3.5 to 4.0	7	6.5 to 7.0	2	9.5 to 10.0	0
1.0 to 1.5	8	4.0 to 4.5	10	7.0 to 7.5	1	10.0 to 10.5	0
1.5 to 2.0	12	4.5 to 5.0	8	7.5 to 8.0	0		
2.0 to 2.5	13	5.0 to 5.5	5	8.0 to 8.5	0	$n = \sum_{i=1}^{21} m_i = 119$	
2.5 to 3.0	16	5.5 to 6.0	0	8.5 to 9.0	0		

Using the chi-square test, test the hypothesis that the data agree with the distribution law whose distribution function $F(|x|)$ has the form

$$F(|x|) = \frac{1}{2}\left[\Phi\left(\frac{x - \bar{x}}{\sigma}\right) + \Phi\left(\frac{x + \bar{x}}{\sigma}\right)\right],$$

where $\bar{x}$ and σ are the expectation and the standard deviation of the random variable X obeying a normal distribution law and are related to the expectation $\mathbf{M}[|X|]$ and the second moment m_2 of the absolute $|X|$ by the formulas:

$$\sigma = \sqrt{\frac{m_2}{1 + \nu^2}}, \qquad \bar{x} = \nu\sqrt{\frac{m_2}{1 + \nu^2}}.$$

Here, ν is the root of the equation

$$\frac{2[\varphi(\nu) + 0.5\nu\Phi(\nu)]}{\sqrt{1 + \nu^2}} = \frac{\mathbf{M}[|X|]}{\sqrt{m_2}},$$

where $\varphi(\nu)$ and $\Phi(\nu)$ are determined from tables 9T and 8T. The significance level is 5 per cent.

43.17 In Table 70 the results of several measurements of a quantity X are given.

TABLE 70

Limits of the interval x_i	m_i	*Limits of the interval* x_i	m_i	*Limits of the interval* x_i	m_i
75 to 77	2	85 to 87	32	95 to 97	8
77 to 79	4	87 to 89	24	97 to 99	3
79 to 81	12	89 to 91	23	99 to 101	1
81 to 83	24	91 to 93	22		
83 to 85	25	93 to 95	20	$n = \sum_{i=1}^{13} m_i = 200$	

Using the chi-square test, test that the data agree with the normal distribution law, and with the convolution of the normal and uniform distributions whose parameters are to be determined from the results of measurements.

Remember that for the random variable $X = Y + Z$, where Y and Z are independent and Y obeys a normal distribution law with zero expectation and variance σ^2 and Z obeys a uniform distribution law in the interval (α, β), the probability density $\psi(x)$ is given by the expression

$$\psi(x) = \frac{1}{2(\beta - \alpha)}\left[\Phi\left(\frac{x - \alpha}{\sigma}\right) - \Phi\left(\frac{x - \beta}{\sigma}\right)\right].$$

To determine the estimates of the parameters σ, α, β, appearing in the formula for $\psi(x)$, it is necessary to derive, from the data, the estimates for the expectation $\tilde{x}$ and the second and fourth central moments $\tilde{\mu}_2$ and $\tilde{\mu}_4$, after which the estimates of σ, α, β are given by the equations:

$$\tilde{\sigma}^2 = \tilde{\mu}_2 - \sqrt{\frac{5}{2}\tilde{\mu}_2^2 - \frac{5}{6}\tilde{\mu}_4},$$

$$\frac{(\tilde{\beta} - \tilde{\alpha})^2}{12} = \sqrt{\frac{5}{2}\tilde{\mu}_2^2 - \frac{5}{6}\tilde{\mu}_4},$$

$$\frac{\tilde{\beta} + \tilde{\alpha}}{2} = \tilde{x}.$$

43.18 For 602 samples, the distance r (in microns) of the center of gravity of an item to the axis of its exterior cylindrical surface is measured with the aid of a control instrument. The results of the measurements appear in Table 71.

TABLE 71

Intervals of values r_i	m_i	*Intervals of values* r_i	m_i
0 to 16	40	80 to 96	45
16 to 32	129	96 to 112	19
32 to 48	140	112 to 128	8
48 to 64	126	128 to 144	3
64 to 80	91	144 to 160	1

Using the chi-square test, verify that the data obey a Rayleigh distribution

$$f(r) = \frac{1}{a^2} re^{-r^2/2a^2};$$

the estimate of the parameter a should be determined in terms of the estimate $\tilde{r}$ for the expectation by the formula

$$\mathbf{M}[r] = a\sqrt{\frac{\pi}{2}}.$$

Use the 5 per cent significance level.

43.19 Table 72 gives the results of 228 measurements of the sensitivity X of a television set (in microvolts).

TABLE 72

x_k	m_k	x_k	m_k	x_k	m_k
200	1	450	33	650	19
250	2	500	34	700	13
300	11	550	31	750	8
350	20	600	25	800	3
400	28				

Using the chi-square test, determine the better fit between the normal and the Maxwell distribution whose probability density is defined by the formula

$$f(x) = \sqrt{\frac{2}{\pi}} \frac{(x - x_0)^2}{a^3} \exp\left\{-\frac{(x - x_0)^2}{2a^2}\right\}, \qquad x \geqslant x_0.$$

Assume the expectation $\mathbf{M}[X]$ of X and a are related by the formula $\mathbf{M}[X] = x_0 + 1.596a$. For simplicity, select as x_0 the smallest observed value of X.

43.20 A lot of 200 light bulbs is tested for lifetime T (in hours) and gives results as in Table 73.

TABLE 73

Class No. i	*Limits of the class* t_i to t_{i+1}	*No. in the class* m_i	*Class No.* i	*Limits of the class* t_i to t_{i+1}	*No. in the class* m_i
1	0 to 300	53	7	1800 to 2100	9
2	300 to 600	41	8	2100 to 2400	7
3	600 to 900	30	9	2400 to 2700	5
4	900 to 1200	22	10	2700 to 3000	3
5	1200 to 1500	16	11	3000 to 3300	2
6	1500 to 1800	12	12	more than 3300	0

Using the chi-square test, test that the data obey an exponential distribution law whose probability density is expressed by the formula

$$f(t) = \lambda e^{-\lambda t}.$$

The significance level should be taken equal to 5 per cent.

Consider the fact that the parameter λ of the exponential distribution law is related to the expectation of the random variable T by the formula

$$\lambda = \frac{1}{\mathbf{M}[T]}.$$

43.21 A lot of 1000 electronic tubes is tested for lifetime. Table 74 gives the lifetime intervals $(t_i; t_{i+1})$ before breakdowns occur and the corresponding sizes m_i of the classes; t_i are expressed in hours.

Using the chi-square test, verify the hypothesis that the experimental data agree with the Weibull distribution law. The distribution function $F(t)$ for this law is given by the formula

$$F(t) = 1 - \exp\left\{-\left(\frac{b_m t}{\bar{t}}\right)^m\right\},$$

where

$$b_m = \Gamma\left(\frac{1}{m} + 1\right);$$

$\Gamma(x)$ is the Γ-function.

TABLE 74

No. of the interval i	1	2	3	4	5	
Limits of the interval t_i to t_{i+1}	0 to 100	100 to 200	200 to 300	300 to 400	400 to 500	
No. in the class m_i	78	149	174	165	139	
No. of the interval i	6	7	8	9	10	11
Limits of the interval t_i to t_{i+1}	500 to 600	600 to 700	700 to 800	800 to 900	900 to 1000	more than 1000
No. in the class m_i	107	77	50	32	27	2

The parameters $\bar{t}$ (the expected value of T) and m should be computed from the data. Take into account that m is related to the standard deviation σ by the formula

$$\sigma = v_m \bar{t},$$

where

$$v_m = \sqrt{\frac{\Gamma\left(\frac{2}{m} + 1\right)}{\Gamma^2\left(\frac{1}{m} + 1\right)} - 1};$$

$v_m = \sigma/\bar{t}$ is the coefficient of variation.

In Table 32T, there are given the values of b_m and v_m as functions of m. Knowing v_m, we can find n and b_m from this table. The following is a section of this table (Table 75).

TABLE 75

m	b_m	v_m
1.7	0.892	0.605
1.8	0.889	0.575

43.22 The position of a point M in the plane is defined by rectangular coordinates X and Y. An experiment consists of measuring the angle φ made by the radius-vector of a point M with the y-axis (Figure 36). The results of 1000 measurements of φ rounded-off to the nearest multiple of 15 degrees and the numbers m_i of appearances of a given value φ_i are shown in Table 76.

FIGURE 36

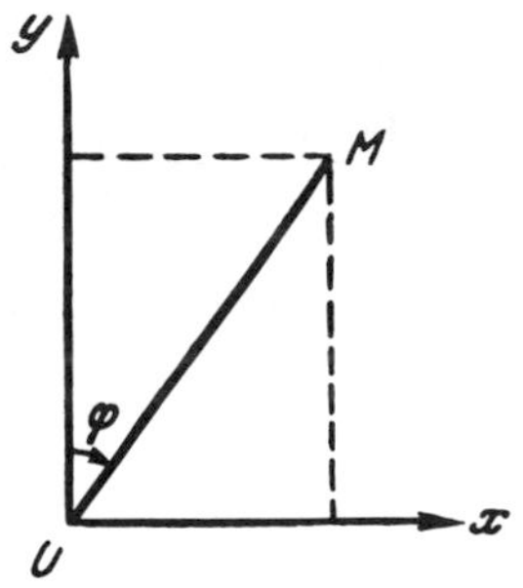

TABLE 76

φ_i, degrees	m_i	φ_i, degrees	m_i	φ_i, degrees	m_i
−82.5	155	−22.5	49	37.5	67
−67.5	118	−7.5	48	52.5	66
−52.5	73	7.5	48	67.5	111
−37.5	59	22.5	53	82.5	153

If X and Y are independent normal variables with zero expectations and variances equal to σ^2 and $(1/4)\sigma^2$, respectively, then $z = \tan \varphi$ must obey the Cauchy distribution (the arctan law),

$$f(z) = \frac{2}{\pi(z^2 + 4)}.$$

Assuming that there are no errors in the measurements of φ and that the round-off errors may be discounted, test, by using the Kolmogorov test, the validity of the preceding assumptions made about X and Y if the significance level is 5 per cent.

43.23 To check the precision of a special pendulum clock at random times, one records the angles made by the axis of the pendulum and the vertical. The amplitude of oscillation is constant and equal to $\alpha = 15°$.

The results of 1000 such measurements, rounded-off to the nearest multiple of 3°, appear in Table 77.

TABLE 77

α_i measured in degrees	*m_i—no. of occurrences of α_i*	*α_i measured in degrees*	*m_i—no. of occurrences of α_i*
−13.5	188	1.5	74
−10.5	88	4.5	76
−7.5	64	7.5	81
−4.5	86	10.5	100
−1.5	62	13.5	181

Assuming that the round-off errors may be discounted, test, using the Kolmogorov test, the hypothesis that the data agree with the arcsine distribution law if the significance level is 5 per cent.

43.24 To check the stability of a certain machine, the following test is conducted every hour: a sample of 20 items selected at random is measured and, using the results of the measurements, one computes in the ith sample the unbiased estimate of the variance $\tilde{\sigma}_i^2$. The values of $\tilde{\sigma}_i^2$ for 47 such samples are given in Table 78.

TABLE 78

i	$\tilde{\sigma}_i^2$	i	$\tilde{\sigma}_i^2$	i	$\tilde{\sigma}_i^2$	i	$\tilde{\sigma}_i^2$
1	0.1225	13	0.1444	25	0.1681	37	0.1089
2	0.1444	14	0.1600	26	0.1369	38	0.1089
3	0.1296	15	0.1521	27	0.1681	39	0.0784
4	0.1024	16	0.1444	28	0.0676	40	0.1369
5	0.1369	17	0.1024	29	0.1024	41	0.0729
6	0.0961	18	0.0961	30	0.1369	42	0.1089
7	0.1296	19	0.1156	31	0.0576	43	0.0784
8	0.1156	20	0.1024	32	0.1024	44	0.5121
9	0.1764	21	0.1521	33	0.0841	45	0.1600
10	0.0900	22	0.1024	34	0.1521	46	0.1681
11	0.1225	23	0.1600	35	0.0676	47	0.1089
12	0.1156	24	0.1296	36	0.1225		

Using the chi-square test, test at a 5 per cent significance level the hypothesis of proportionality of the variances; that is, test the assumption that there is no disorder, which means that the dispersion varies with the measured dimension of an item. Take into account the fact that if this hypothesis is valid, the quantity

$$q_i = \frac{(n_i - 1)\tilde{\sigma}_i^2}{\tilde{\sigma}^2}$$

obeys approximately a chi-square distribution law with $(n_i - 1)$ degrees of freedom, where $\tilde{\sigma}^2$ is the unbiased estimate for the variance σ^2 of the entire main population and can be computed by the formula

$$\tilde{\sigma}^2 = \frac{\sum_{i=1}^{m} \tilde{\sigma}_i^2(n_i - 1)}{N - m},$$

where $n_i = n = 20$ is the number of items in each sample, $m = 47$ is the number of samples and $N = \sum_{i=1}^{m} n_i = 940$ is the total number of items in all samples.

43.25 There are $m = 40$ samples of $n = 20$ items each, and for the ith group there is given as an estimate for the expectation $\tilde{x}_i$, a randomly selected value x_{i1} from the ith sample x_{i1} (for example, the first in each sample), and for the variance, the unbiased estimate of the variance $\tilde{\sigma}_i^2$ for the dimension x of an item. The values of $\tilde{x}_i$, x_{i1}, $\tilde{\sigma}_i^2$ for the 40 samples appear in Table 79.

TABLE 79

i	x_{i1}	$\tilde{x}_i$	$\tilde{\sigma}_i^2$	i	x_{i1}	$\tilde{x}_i$	$\tilde{\sigma}_i^2$
1	148	132	24	21	114	112	39
2	182	152	38	22	112	108	32
3	195	145	40	23	49	97	52
4	81	134	32	24	116	106	36
5	149	124	37	25	138	124	36
6	143	144	31	26	120	149	37
7	133	142	31	27	120	129	41
8	132	143	34	28	104	120	26
9	111	109	42	29	121	105	26
10	156	121	30	30	99	110	32
11	103	93	35	31	123	105	37
12	61	118	45	32	109	123	24
13	149	116	38	33	100	116	32
14	209	123	40	34	115	123	29
15	124	106	39	35	108	109	27
16	52	181	46	36	125	138	35
17	147	102	32	37	170	126	33
18	145	124	31	38	132	132	33
19	128	125	34	39	114	131	28
20	98	119	32	40	155	115	37

Using the Kolmogorov test, verify for the 10 per cent significance level the hypothesis that the normal distribution obtains for the dimension x.

Note that in this case (for $n \neq 4$)

$$\eta_i = \frac{\tau_i\sqrt{n-2}}{\sqrt{n-1-\tau_i^2}},$$

where

$$\tau_i = \frac{x_{ij} - \tilde{x}_i}{\tilde{\sigma}_i}$$

obey a Student's distribution law with $k = n - 2 = 18$ degrees of freedom, where x_{ij} is a randomly selected value from the ith sample (in our case x_{i1}).

43.26 The results of 300 measurements of some quantity x are included in Table 80.

TABLE 80

Limits of the interval x_i	m_i	*Limits of the interval* x_i	m_i	*Limits of the interval* x_i	m_i
50 to 60	1	100 to 110	56	140 to 150	19
60 to 70	2	110 to 120	61	150 to 160	16
70 to 80	9	120 to 130	49	160 to 170	4
80 to 90	23	130 to 140	25	170 to 180	2
90 to 100	33				

Using the chi-square test, test that the data agree with the normal distribution whose parameter estimates should be computed from the experimental data. Smooth the data with the aid of a distribution specified by a Charlier-A series, and, using the chi-square test, verify that the data agree with the obtained distribution.

43.27 The measurements of light velocity c in the Michelson-Pease-Pearson experiment gave the results shown in Table 81. For brevity, the first three digits of c_i (in km./sec.) are omitted (299 000).

TABLE 81

Limits of the interval c_i	m_i	*Limits of the interval* c_i	m_i	*Limits of the interval* c_i	m_i	*Limits of the interval* c_i	m_i
735 to 740	3	755 to 760	17	775 to 780	40	795 to 800	5
740 to 745	7	760 to 765	23	780 to 785	17	800 to 805	2
745 to 750	4	765 to 770	29	785 to 790	16	805 to 810	3
750 to 755	8	770 to 775	45	790 to 795	10	810 to 815	4

The following estimates for the expected value $\tilde{c}$ and the standard deviation $\tilde{\sigma}$ were obtained from the data:

$$\tilde{c} = 299733.85 \text{ km./sec.}, \qquad \tilde{\sigma} = 14.7 \text{ km./sec.}$$

The chi-square test of the hypothesis that the data agree with a normal distribution law with parameters $\tilde{c}$ and $\tilde{\sigma}$ gives the value $\chi_q^2 = \chi_{qH}^2 = 18.52$; the number of degrees of freedom in this case is $k_H = 9$ and $\mathbf{P}(\chi^2 \geqslant \chi_{qH}^2) = 0.018$; small intervals are united. The hypothesis should be rejected.

Smooth the observations with the distribution law specified by a Charlier-A series and test, with the chi-square test, that the experimental data obey the resulting distribution law.

43.28 Two lots, each containing 100 items, are measured. The number of items h_{ij} with normal, underestimated and overestimated dimensions are exhibited in Table 82.

TABLE 82

	Dimension j			
	Results of measurements j			
Lot no. i	1 (*underestimated dimension*)	2 (*normal dimension*)	3 (*overestimated dimension*)	h_{i0}
1	25	50	25	100
2	52	41	7	100
h_{0j}	77	91	32	200

Using the chi-square test, determine whether the number of a lot and the character of the dimensions of the items are independent at a 5 per cent significance level.

44. DATA PROCESSING BY THE METHOD OF LEAST SQUARES

Basic Formulas

The method of least squares is applied for finding estimates of parameters appearing in a functional dependence between variables whose values are experimentally determined.

If the experiment gives $n + 1$ pairs of values (x_i, y_i), where x_i are the values of the argument and y_i are the values of the function, then the parameters of the approximating function $F(x)$ are selected to minimize the sum

$$S = \sum_{i=0}^{n} [y_i - F(x_i)]^2.$$

If the approximating function is a polynomial, that is,

$$F(x) = Q_m(x) = a_0 + a_1 x + \cdots + a_m x^m \qquad (m \leqslant n),$$

then the estimates of its coefficients $\tilde{a}_k$ are determined from a system of $m + 1$ normal equations

$$\sum_{j=0}^{m} s_{k+j} a_j \equiv s_k a_0 + s_{k+1} a_1 + \cdots + s_{k+m} a_m = v_k \qquad (k = 0, 1, 2, \ldots, m),$$

where

$$s_k = \sum_{i=0}^{n} x_i^k \qquad (k = 0, 1, 2, \ldots, 2m),$$

$$v_k = \sum_{i=0}^{n} y_i x_i^k \qquad (k = 0, 1, 2, \ldots, m).$$

If the values x_i are given without errors and the values y_i are independent and equally accurate, the estimate for the variance $\tilde{\sigma}^2$ of y_i is given by the formula

$$\tilde{\sigma}^2 = \frac{1}{n - m} S_{\min},$$

where $S_{\min}$ is the value of S, computed under the assumption that the coefficients of the polynomial $F(x) = Q_m(x)$ are replaced by their estimates that are determined from the system of normal equations.

If y_i are normally distributed, then the method given is best for finding the approximating function $F(x)$.

The estimates $\tilde{\sigma}_{a_k}$ of the variances of the coefficients $\tilde{a}_k$ and the covariances $\tilde{K}_{a_k, a_j}$ are given by the formulas

$$\tilde{\sigma}_{a_k} = M_{k,k} \tilde{\sigma}^2, \qquad \tilde{K}_{a_k, a_j} = M_{k,j} \tilde{\sigma}^2,$$

where $M_{k,j} = \Delta_{kj}/\Delta$, $\Delta = |d_{kj}|$ is the determinant of the system of normal equations of the $(m + 1)$st order,

$$d_{kj} = s_{k+j} \qquad (j, k = 0, 1, 2, \ldots, m),$$

Δ_{kj} is the cofactor of d_{kj} in the determinant Δ.

In solving the system of normal equations by the elimination method, the quantities $M_{k,j}$ may also be obtained without replacing the v_k by their numerical values. The linear combination of the v_k used to represent $\tilde{a}_k$ will have as the coefficient of v_j the desired number $M_{k,j}$.

In the particular case of a linear dependence $m = 1$, we have:

$$y = a_0 + a_1 x,$$

$$\tilde{a}_0 = \frac{s_2 v_0 - s_1 v_1}{s_2 s_0 - s_1^2}, \qquad \tilde{a}_1 = \frac{-s_1 v_0 + s_0 v_1}{s_2 s_0 - s_1^2},$$

$$\tilde{\sigma}_{a_0}^2 = \frac{s_2}{s_2 s_0 - s_1^2} \frac{S_{\min}}{n-1}, \qquad \tilde{\sigma}_{a_1}^2 = \frac{s_0}{s_2 s_0 - s_1^2} \frac{S_{\min}}{n-1},$$

$$\tilde{K}_{a_0, a_1} = -\frac{s_1}{s_2 s_0 - s_1^2} \frac{S_{\min}}{n-1}.$$

In the case in which the measurements are not equally accurate, that is, y_i have different variances σ_i^2, all the previous formulas remain valid if S, s_k and v_k are replaced by

$$S' = \sum_{i=0}^{n} p_i^2 (y_i - a_0 - a_1 x_i - \dots - a_m x_i^m),$$

$$s_k' = \sum_{i=0}^{n} p_i^2 x_i^k \qquad (k = 0, 1, 2, \dots, 2m),$$

$$v_k' = \sum_{i=0}^{n} p_i^2 y_i x_i^k \qquad (k = 0, 1, 2, \dots, m),$$

where the "weights" p_i^2 of y_i are

$$p_i^2 = \frac{A^2}{\sigma_i^2};$$

A^2 is a coefficient of proportionality.

If the "weights" p_i are known, the estimates of the variances of individual measurements y_i are computed by the formula

$$\tilde{\sigma}_i^2 = \frac{S'_{\min}}{(n-m) p_i^2}.$$

If y_i is obtained by averaging n_i equally accurate results, then the "weight" of the measurement y_i is proportional to n_i. One may take $p_i^2 = n_i$. All the formulas remain unchanged except the one for $\tilde{\sigma}_i^2$; in this case,

$$\tilde{\sigma}_i^2 = \frac{S'_{\min}}{(n-m)(n_i - 1)}.$$

The confidence intervals for the coefficients a_k for any given confidence level have the form

$$\tilde{a}_k - \gamma \tilde{\sigma}_{a_k} < a_k < \tilde{a}_k + \gamma \tilde{\sigma}_{a_k},$$

where γ is determined from Table 16T for Student's distribution for the values of α and $k = n - m$ degrees of freedom.

In the case of equally accurate measurements, the confidence interval for the standard deviation σ and the confidence level α are determined from the inequalities

$$\gamma_1\tilde{\sigma} < \sigma < \gamma_2\tilde{\sigma},$$

where γ_1 and γ_2 are found from Table 19T for a chi-square distribution with entry value α and k degrees of freedom. For the same purpose, one can use Table 18T; in this case,

$$\gamma_1 = \sqrt{\frac{n-m}{\chi_1^2}}, \qquad \gamma_2 = \sqrt{\frac{n-m}{\chi_2^2}},$$

where χ_1^2 and χ_2^2 are determined from the equations

$$\mathbf{P}(\chi^2 \leqslant \chi_1^2) = \frac{1-\alpha}{2}, \qquad \mathbf{P}(\chi^2 \geqslant \chi_2^2) = \frac{1+\alpha}{2}$$

for $k = n - m$ degrees of freedom.

The confidence limits form a strip containing the graph of the unknown correct dependence $y = F(x)$ with a given confidence level α; they are determined by the inequalities

$$Q_m(x_i) - \gamma\tilde{\sigma}_y(x_i) < y(x_i) < Q_m(x_i) + \gamma\tilde{\sigma}_y(x_i),$$

where $\tilde{\sigma}_y^2(x_i)$ is the estimate for the variance of y defined by the dependence $y = Q_m(x)$ (it depends on the random variables represented by the estimates of a_k).

In the general case, the computation of $\tilde{\sigma}_y^2(x)$ is difficult because it requires the knowledge of all the covariances k_{a_k,a_l}. For a linear dependence ($m = 1$),

$$\tilde{\sigma}_y^2(x) = \tilde{\sigma}_{a_0}^2 + \tilde{\sigma}_{a_1}^2 x^2 + 2\tilde{k}_{a_0,a_1}x.$$

The value of γ is determined from Table 16T for Student's distribution for the entry α and $k = n - m$ degrees of freedom.

In the case of equidistant values x_i of the argument, the computation of the approximating polynomial can be simplified by using the representation

$$Q_m(x) = \sum_{k=0}^{m} b_k P_{k,n}(x_i'),$$

where $P_{k,n}(x_i')$ are the orthogonal Chebyshev polynomials:

$$P_{k,n}(x') = \sum_{j=0}^{n} (-1)^j C_k^j C_{k+j}^j \frac{x'(x'-1)\cdots(x'-j+1)}{n(n-1)\cdots(n-j+1)},$$

$$x' = \frac{x - x_{\min}}{h}, \qquad h = \frac{x_{\max} - x_{\min}}{n},$$

$x_{\max}$, $x_{\min}$ are the maximal and minimal values of x_i,

$$b_k = \frac{c_k}{S_k}, \qquad c_k = \sum_{i=0}^{n} y_i P_{k,n}(x_i'), \qquad S_k = \sum_{i=0}^{n} P_{k,n}^2(x_i').$$

The estimates for the variances of the coefficients b_k are determined by the formula

$$\tilde{\sigma}_{b_k}^2 = \frac{S_{\min}}{n-m}\frac{1}{S_k}.$$

The values of the Chebyshev polynomials multiplied by $P_{k,n}(0)$ for $k = 1$ to 5, $n = 5$ to 20, $x' = 0, 1, \ldots, n$ are given in Table 30T.

If the coefficients b_k are computed from Table 30T, then for the computation of the polynomials $P_{k,n}(x')$ in the formula for $Q_m(x)$ it is also necessary to consider the coefficient $P_{k,n}(0)$ and to choose the ordinates of these polynomials from the same tables or to multiply the value of the polynomial, obtained according to the preceding formula, by $P_{k,n}(0)$.

In some cases, the approximating function is not a polynomial, but can be reduced to a polynomial by a change of variables. Examples of such change are given in Table 83.

TABLE 83

Ex. No.	*Initial function*	*The form to which it is reduced*	*Change of variables*
1	$y = Ae^{kx}$	$z = a_0 + a_1x$	$z = \ln y;\ a_0 = \ln A;\ a_1 = k$
2	$y = Bx^b$	$z = a_0 + a_1u$	$z = \log y;\ u = \log x$
3	$y = a_0 + \frac{a_1}{x}$	$y = a_0 + a_1u$	$u = \frac{1}{x}$
4	$y = a_0 + \frac{a_1}{x^b}$	$y = a_0 + a_1u$	$u = \frac{1}{x^b}$
5	$y = A \exp\left\{-\frac{(x-a)^2}{2\sigma^2}\right\}$	$z = a_0 + a_1x + a_2x^2$	$z = \log y;$ $a_0 = \log A - \frac{\log e}{2\sigma^2};$ $a_1 = \frac{a \log e}{\sigma^2};$ $a_2 = -\frac{\log e}{2\sigma^2}a^2$
6	$y = a_0 + \frac{a_1}{x} + \frac{a_2}{x^2} + \cdots$	$y = a_0 + a_1u + a_2u^2 + \cdots$	$u = \frac{1}{x}$
7	$y = a_0 + a_1x^b + a_2x^{2b} + \cdots$	$y = a_0 + a_1u + a_2u^2 \cdots$	$u = x^b$
8	$y = a_0x^{-m} + a_1x^n$	$z = a_0 + a_1u$	$z = yx^m;\ u = x^{m+n}$

If y is a function of several arguments z_i, then to obtain the linear approximating function

$$y = \alpha_0 z_0 + \alpha_1 z_1 + \cdots + \alpha_m z_m$$

corresponding to the values y_i and z_{ki} in $(n + 1)$ experiments, it is necessary to find the solutions $\tilde{\alpha}_k$ of the system of normal equations

$$s_{k0}\tilde{\alpha}_0 + s_{k1}\tilde{\alpha}_1 + \cdots + s_{km}\tilde{\alpha}_m = \beta_k \qquad (k = 0, 1, 2, \ldots, m),$$

where

$$s_{kj} = \sum_{i=0}^{n} z_{ki}z_{ji} \qquad (k, j = 0, 1, 2, \ldots, m);$$

$$\beta_k = \sum_{i=0}^{n} y_i z_{ki} \qquad (k = 0, 1, 2, \ldots, m).$$

If the values z_{ki} are known without error and the measurements of y_i are equally accurate, the estimates of the variances of α_k are determined by the formula

$$\tilde{\sigma}^2_{\alpha_k} = N_{k,k}\tilde{\sigma}^2,$$

where $\tilde{\sigma}^2 = S_{\min}/(n - m)$ and $N_{k,k}$ is the ratio of the cofactor of a diagonal element of the determinant (of the system of normal equations) to the value of the determinant itself. In solving the system without using the determinant, $N_{k,k}$ will be the solutions of this system if we replace all β_k by 1 and the other β_l by zeros.

The role of z_k can be played by any functions $f_k(x)$ of some argument x. For example, if the function y, defined in the interval $(0, 2\pi)$, is approximated by the trigonometric polynomial

$$y = \lambda_0 + \sum_{k=1}^{m} (\lambda_k \cos kx + \mu_k \sin kx),$$

then for equidistant values x_i the estimates for the coefficients λ_k and μ_k are determined by the Bessel formulas:

$$\tilde{\lambda}_0 = \frac{1}{n+1} \sum_{i=0}^{n} y_i; \qquad \tilde{\lambda}_k = \frac{2}{n+1} \sum_{i=0}^{n} y_i \cos kx_i;$$

$$\tilde{\mu}_k = \frac{2}{n+1} \sum_{i=0}^{n} y_i \sin kx_i \qquad (k = 1, 2, \ldots, m).$$

For a complex functional dependence and a sufficiently small range of variation of the arguments z_k, the computations are simplified if the function is expanded in a power series of deviations of arguments from their approximate values (for example, from their mean).

If there are errors in x_i and y_i too, and these variables obey a normal distribution, then, in the case of linear dependence

$$y = a_0 + a_1 x,$$

the estimate $\tilde{a}_1$ is the root of the quadratic equation

$$\tilde{a}_1^2 + \frac{[s_1^2 - (n+1)s_2]\dfrac{\sigma_y^2}{\sigma_x^2} - [r_1^2 - (n+1)r_2]}{s_1 r_1 - (n+1)v_1}\,\tilde{a}_1 - \frac{\sigma_y^2}{\sigma_x^2} = 0,$$

and the estimate $\tilde{a}_0$ is given by the formula

$$\tilde{a}_0 = \frac{r_1 - \tilde{a}_1 s_1}{n+1},$$

where σ_x^2, σ_y^2 are, respectively, the variances of the x_i and the y_i,

$$s_k = \sum_{i=0}^{n} x_i^k, \qquad r_k = \sum_{i=0}^{n} y_i^k \quad (k = 1, 2), \qquad v_1 = \sum_{i=0}^{n} x_i y_i.$$

Of the two roots of the quadratic equation, we select the one that better fits the conditions of the problem.

Solutions for Typical Examples

Example 44.1 In studying the influence of temperature t on the motion ω of a chronometer, the following results were obtained (Table 84).

TABLE 84

t_i, °C	5.0	9.6	16.0	19.6	24.4	29.8	34.4
ω_i	2.60	2.01	1.34	1.08	0.94	1.06	1.25

If

$$\tilde{\omega} = a_0 + a_1(t - 15) + a_2(t - 15)^2,$$

holds, where $\tilde{\omega}$ are the computed values of ω, determine the estimates for the coefficients a_k and the estimates for the standard deviations, σ of an individual measurement and $\tilde{\sigma}_{a_k}$ of the coefficients a_k. Establish the confidence intervals for a_k and for the standard deviation σ, characterizing the precision of an individual measurement for a confidence level $\alpha = 0.90$.

SOLUTION. We determine the normal equations for the coefficients a_k and $M_{k,k}$. To decrease the sizes of the coefficients of the normal equations, we introduce the variable

$$x = \frac{t - 15}{15}$$

and seek the approximating function

$$y = a_0' + a_1'x + a_2'x^2.$$

We then determine the coefficients of the normal equations s_k and v_k, as in the computations in Table 85.

TABLE 85

i	x_i^0	x_i	x_i^2	x_i^3	x_i^4	ω_i	$\omega_i x_i$	$\omega_i x_i^2$
0	1	−0.667	0.4449	−0.2967	0.1979	2.60	−1.7342	1.1567
1	1	−0.360	0.1296	−0.0467	0.0168	2.01	−0.7236	0.2605
2	1	0.067	0.0045	0.0003	0.0000	1.34	0.0898	0.0060
3	1	0.307	0.0942	0.0289	0.0089	1.08	0.3316	0.1017
4	1	0.627	0.3931	0.2465	0.1546	0.94	0.5894	0.3695
5	1	0.987	0.9742	0.9615	0.9490	1.06	1.0462	1.0327
6	1	1.293	1.6718	2.1617	2.7949	1.25	1.6162	2.0898
	$s_0=7$	$s_1=2.254$	$s_2=3.7123$	$s_3=3.0555$	$s_4=4.1221$	$v_0=10.28$	$v_1=1.2154$	$v_2=5.0169$

We obtain:

$$s_0 = 7; \quad s_1 = 2.254; \quad s_2 = 3.712; \quad s_3 = 3.056; \quad s_4 = 4.122;$$

$$v_0 = 10.28; \quad v_1 = 1.215; \quad v_2 = 5.017.$$

The system of normal equations becomes

$$7\tilde{a}'_0 + 2.254\tilde{a}'_1 + 3.712\tilde{a}'_2 = v_0,$$

$$2.254\tilde{a}'_0 + 3.712\tilde{a}'_1 + 3.056\tilde{a}'_2 = v_1,$$

$$3.712\tilde{a}'_0 + 3.056\tilde{a}'_1 + 4.122\tilde{a}'_2 = v_2.$$

Solving this system by elimination and without substituting the numerical values for v_k, we obtain:

$$\tilde{a}'_0 = 0.2869v_0 + 0.0986v_1 - 0.3314v_2,$$

$$\tilde{a}'_1 = 0.0986v_0 + 0.7248v_1 - 0.6260v_2,$$

$$\tilde{a}'_2 = -0.3314v_0 - 0.6260v_1 + 1.0051v_2.$$

Substituting the values of v_k, we find:

$$\tilde{a}'_0 = 1.404; \qquad \tilde{a}'_1 = -1.246; \qquad \tilde{a}'_2 = 0.8741.$$

$M_{k,k}$ are the coefficients of v_k in each equation for $\tilde{a}'_k$; that is,

$$M_{0,0} = 0.2869; \qquad M_{1,1} = 0.7248; \qquad M_{2,2} = 1.0051.$$

We compute the value $S_{\min}$ necessary for finding the estimates of the variance of an individual y_i and the variances of the coefficients $\tilde{a}_k$; the computations are in Table 86.

TABLE 86

i	$a'_0 + a'_1x_i$	$a'_2x_i^2$	$\tilde{\omega}_i$	ε_i	ε_i^2
0	2.2352	0.3889	2.624	−0.024	0.000576
1	1.8527	0.1133	1.966	0.044	0.001936
2	1.3207	0.0039	1.325	0.015	0.000225
3	1.0217	0.0823	1.104	−0.024	0.000576
4	0.6230	0.3436	0.967	−0.027	0.000729
5	0.1745	0.8515	1.026	0.034	0.001156
6	−0.2067	1.4613	1.255	−0.005	0.000025
				$S_{\min}$ =	0.005223

We obtain $S_{\min} = 0.005223$. Furthermore, we find

$$\tilde{\sigma}^2 = \frac{S_{\min}}{6 - 2} = 0.001306; \qquad \tilde{\sigma} = 0.03614;$$

$$\tilde{\sigma}^2_{a'_0} = M_{0,0}\tilde{\sigma}^2 = 0.0003746; \qquad \tilde{\sigma}^2_{a'_1} = 0.0009464; \qquad \tilde{\sigma}^2_{a'_2} = 0.001312;$$

$$\tilde{\sigma}_{a'_0} = 0.01936; \qquad \tilde{\sigma}_{a'_1} = 0.03076; \qquad \tilde{\sigma}_{a'_2} = 0.03623.$$

Returning to the argument t, we obtain

$$\tilde{\omega} = \tilde{a}_0 + \tilde{a}_1(t - 15) + \tilde{a}_2(t - 15)^2,$$

where

$$\tilde{a}_0 = \tilde{a}'_0 = 1.404; \qquad \tilde{a}_1 = \frac{\tilde{a}'_1}{15} = -0.08306;$$

$$\tilde{a}_2 = \frac{\tilde{a}'_2}{15^2} = 0.003885,$$

and the corresponding estimates for the standard deviations $\tilde{\sigma}_{a_k}$:

$$\tilde{\sigma}_{a_0} = \tilde{\sigma}_{a'_0} = 0.01936; \qquad \tilde{\sigma}_{a_1} = \frac{\tilde{\sigma}_{a'_1}}{15} = 0.001291;$$

$$\tilde{\sigma}_{a_2} = \frac{\tilde{\sigma}_{a'_2}}{15^2} = 0.0001610.$$

We find the confidence intervals for the coefficients a_k for a confidence level $\alpha = 0.90$. Using Table 16T, for the values of α and $k = n - m = 4$ degrees of freedom, we find

$$\gamma = 2.132.$$

The confidence intervals for a_k:

$$\tilde{a}_k - \gamma\tilde{\sigma}_{a_k} < a_k < \tilde{a}_k + \gamma\tilde{\sigma}_{a_k},$$

become

$$1.363 < a_0 < 1.446,$$

$$0.08031 < a_1 < 0.08581,$$

$$0.003542 < a_2 < 0.004228.$$

We find the confidence interval for the standard deviation σ, characterizing the precision of an individual measurement:

$$\gamma_1\tilde{\sigma} < \sigma < \gamma_2\tilde{\sigma},$$

where γ_1 and γ_2 are determined from Table 19T for $k = 4$, $\alpha = 0.90$. We have $\gamma_1 = 0.649$; $\gamma_2 = 2.37$, hence,

$$0.02345 < \sigma < 0.08565.$$

Similarly one can solve Problems 44.1 to 44.3, 44.5, 44.9, 44.10 and 44.13.

Example 44.2 The results of several equally accurate measurements of a quantity y, known to be a function of x, are given in Table 87.

TABLE 87

x	y	x	y
0.0	1.300	1.5	0.037
0.3	1.245	1.8	−0.600
0.6	1.095	2.1	−1.295
0.9	0.855	2.4	−1.767
1.2	0.514	2.7	−1.914

Select a fifth-degree polynomial that approximates the dependence of y on x in the interval [0, 2.7]. Use (the orthogonal) Chebyshev polynomials. Estimate the precision of each individual measurement as characterized by the standard deviation σ, and find the estimates of the standard deviations of the coefficients b_k for the Chebyshev polynomials $P_{k,n}(x)$.

SOLUTION. We make the change of variable $z = x/0.3$ in order to make the increase of the argument unity. We compute the quantities S_k, c_k, b_k ($k = 0, 1, \ldots, 5$) according to the formulas given in the introduction to this section. The tabulated values of the Chebyshev polynomials are taken from 30T. The computations are listed in Table 88.

TABLE 88

z	$P_{0.9}(z)$	$P_{1.9}(z)$	$P_{2.9}(z)$	$P_{3.9}(z)$	$P_{4.9}(z)$	$P_{5.9}(z)$
0	1	9	6	42	18	6
1	1	7	2	−14	−22	−14
2	1	5	−1	−35	−17	1
3	1	3	−3	−31	3	11
4	1	1	−4	−12	18	6
5	1	−1	−4	12	18	−6
6	1	−3	−3	31	3	−11
7	1	−5	−1	35	−17	−1
8	1	−7	2	14	−22	14
9	1	−9	6	−42	18	−6
	$S_0 = 10$	$S_1 = 330$	$S_2 = 132$	$S_3 = 8580$	$S_4 = 2860$	$S_5 = 780$

The computations, performed on a (keyboard) desk calculator with accumulation of the results, give:

$$S_0 = 10, \qquad S_1 = 330, \qquad S_2 = 132,$$
$$S_3 = 8580, \qquad S_4 = 2860, \qquad S_5 = 780,$$
$$c_0 = -0.530, \qquad c_1 = 66.802, \qquad c_2 = -7.497,$$
$$c_3 = -14.659, \qquad c_4 = 14.515, \qquad c_5 = -1.627.$$

For the estimates of the coefficients b_k we get:

$$\tilde{b}_0 = -0.530, \qquad \tilde{b}_1 = 0.20243, \qquad \tilde{b}_2 = -0.05680,$$
$$\tilde{b}_3 = -0.00486, \qquad \tilde{b}_4 = 0.00508, \qquad \tilde{b}_5 = -0.00209.$$

Recall that if one uses the tabulated values of the Chebyshev polynomials, the formula for the required fifth-degree polynomial has the form

$$\tilde{y} = \tilde{b}_0 P_{0.9}(z) + \tilde{b}_1 P_{1.9}(z) + \tilde{b}_2 P_{2.9}(z) + \tilde{b}_3 P_{3.9}(z) + \tilde{b}_4 P_{4.9}(z) + \tilde{b}_5 P_{5.9}(z).$$

However, if one uses the analytic formulas for the calculation of the Chebyshev polynomials, then the coefficients b_k should be replaced by $b'_k = b_k P_{k,n}(0)$, where $P_{k,n}(0)$ is the tabulated value of $P_{k,n}(z)$ for $z = 0$.

We compute the estimate $\tilde{\sigma}^2$:

$$\tilde{\sigma}^2 = \frac{S_{\min}}{n - m}, \qquad S_{\min} = \sum_{i=0}^{n} (\tilde{y}_i - y_i)^2,$$

where we use the tabulated values of the Chebyshev polynomials from Table 88 for finding the values $\tilde{y}_i$. The computation of $S_{\min}$ is indicated in Table 89.

TABLE 89

x_i	z_i	y_i	$\tilde{y}_i$	ε_i	ε_i^2
0.0	0	1.300	1.310	−0.010	0.000100
0.3	1	1.245	1.236	0.009	0.000081
0.6	2	1.095	1.098	−0.003	0.000009
0.9	3	0.855	0.868	−0.013	0.000169
1.2	4	0.514	0.514	0.000	0.000000
1.5	5	0.037	0.017	0.020	0.000400
1.8	6	−0.600	−0.602	0.002	0.000004
2.1	7	−1.295	−1.263	−0.032	0.001024
2.4	8	−1.767	−1.793	0.026	0.000676
2.7	9	−1.914	−1.908	−0.006	0.000036
				$S_{\min}$ =	0.002499

We obtain:

$$S_{\min} = 0.002499, \qquad \tilde{\sigma} = \sqrt{\frac{S_{\min}}{n - m}} = 0.02503.$$

Next, according to the formula

$$\tilde{\sigma}_{b_k} = \frac{\tilde{\sigma}}{\sqrt{S_k}},$$

we find

$$\tilde{\sigma}_{b_0} = 0.007917; \qquad \tilde{\sigma}_{b_1} = 0.001378; \qquad \tilde{\sigma}_{b_2} = 0.002179;$$

$$\tilde{\sigma}_{b_3} = 0.0002702; \qquad \tilde{\sigma}_{b_4} = 0.0004680; \qquad \tilde{\sigma}_{b_5} = 0.0008947.$$

Problems 44.4, 44.6 and 44.12 can be solved by following this solution.

Example 44.3 The readings of an aneroid barometer A and a mercury barometer B for different temperatures t are given in Table 90.

TABLE 90

i	t, °C	A, mm.	B, mm.	i	t, °C	A, mm.	B, mm.
0	10.0	749.0	744.4	5	3.8	757.5	754.0
1	6.2	746.1	741.3	6	17.1	752.4	747.8
2	6.3	756.6	752.7	7	22.2	752.5	748.6
3	5.3	758.9	754.7	8	20.8	752.2	747.7
4	4.8	751.7	747.9	9	21.0	759.5	755.6

If the dependence of B on t and A has the form

$$B = A + \alpha_0 + \alpha_1 t + \alpha_2(760 - A),$$

find estimates of the coefficients α_k, construct the confidence intervals for the coefficients α_k and for the standard deviation σ of the errors in measuring B for a confidence level $\alpha = 0.90$.

SOLUTION. Let us use the notations $z_0 = 1$, $z_i = t$, $z_2 = 760 - A$, $y = B - A$. Then, the required formula becomes

$$y = \alpha_0 z_0 + \alpha_1 z_1 + \alpha_2 z_2.$$

The initial data for these notations are represented in Table 91.

TABLE 91

i	z_0	z_1	z_2	y	$\alpha_0 + \alpha_1 z_1$	$\alpha_2 z_2$	$\tilde{y}$	$\lvert\varepsilon_i\rvert$	ε_i^2
0	1	10.0	11.0	−4.6	−3.725	−0.739	−4.46	0.14	0.0196
1	1	6.2	13.9	−4.8	−3.686	−0.934	−4.62	0.18	0.0324
2	1	6.3	3.4	−3.9	−3.687	−0.228	−3.92	0.02	0.0004
3	1	5.3	1.1	−4.2	−3.676	−0.074	−3.75	0.45	0.2025
4	1	4.8	8.3	−3.8	−3.671	−0.558	−4.23	0.43	0.1849
5	1	3.8	2.5	−3.5	−3.661	−0.168	−3.83	0.33	0.1089
6	1	17.1	7.6	−4.6	−3.799	−0.511	−4.31	0.29	0.0841
7	1	22.2	7.5	−3.9	−3.852	−0.504	−4.36	0.46	0.2116
8	1	20.8	7.8	−4.5	−3.838	−0.524	−4.36	0.14	0.0196
9	1	21.0	0.5	−3.9	−3.840	−0.034	−3.87	0.03	0.0009
								$S_{\min}$ =	0.8649

We determine the values $s_{kj} = \sum_{i=0}^{n} z_{ki} z_{ji}$ and $\beta_k = \sum_{i=0}^{n} y_i z_{ki}$ $(k, j = 0, 1, 2)$:

$$s_{00} = 10; \quad s_{01} = s_{10} = 117.5; \quad s_{02} = s_{20} = 63.6; \quad s_{11} = 1902.6;$$

$$s_{12} = s_{21} = 741.97; \quad s_{22} = 577.22; \quad \beta_0 = -41.7;$$

$$\beta_1 = 494.87; \quad \beta_2 = -276.75.$$

We write the system of normal equations, but for the time being we do not replace β_k by their numerical values:

$$10\alpha_0 + 117.5\alpha_1 + 63.6\alpha_2 = \beta_0,$$

$$117.5\alpha_0 + 1902.59\alpha_1 + 741.97\alpha_2 = \beta_1,$$

$$63.6\alpha_0 + 741.97\alpha_1 + 577.22\alpha_2 = \beta_2.$$

Solving this system by elimination, we find:

$$\alpha_0 = -0.6076\beta_0 + 0.02289\beta_1 - 0.03754\beta_2,$$

$$\alpha_1 = -0.02289\beta_0 + 0.001916\beta_1 + 0.0000591\beta_2,$$

$$\alpha_2 = -0.03754\beta_0 + 0.0000591\beta_1 + 0.005792\beta_2.$$

Setting the numerical values of β_k in these expressions, we find α_k; the coefficients of β_k in the expression for α_k are the values of $N_{k,k}$:

$$\alpha_0 = -3.621; \qquad \alpha_1 = -0.01041; \qquad \alpha_2 = -0.06719;$$

$$N_{0,0} = 0.6076; \qquad N_{1,1} = 0.001916; \qquad N_{2,2} = 0.005792.$$

Furthermore, we find: $S_{\min} = 0.8649$ (see Table 91);

$$\tilde{\sigma}^2 = 0.12356; \qquad \tilde{\sigma} = 0.3515;$$

$$\tilde{\sigma}^2_{\alpha_0} = 0.07508; \qquad \tilde{\sigma}_{\alpha_0} = 0.272;$$

$$\tilde{\sigma}^2_{\alpha_1} = 0.0002368; \qquad \tilde{\sigma}_{\alpha_1} = 0.0154;$$

$$\tilde{\sigma}^2_{\alpha_2} = 0.0007156; \qquad \tilde{\sigma}_{\alpha_2} = 0.0268.$$

We construct the confidence intervals for the coefficients α_k and for the standard deviation σ, which determines the accuracy of an individual measurement, by using Student's distribution for α_k (see Table 16T) and the chi-square distribution for σ (see Table 19T).

The number of degrees of freedom is $k = n - m = 7$ and the confidence level is $\alpha = 0.90$.

We find: $\gamma = 1.897$, $\gamma_1 = 0.705$, $\gamma_2 = 1.797$.

The confidence intervals for α_k,

$$\tilde{\alpha}_k - \gamma\tilde{\sigma}_{\alpha_k} < \alpha_k < \tilde{\alpha}_k + \gamma\tilde{\sigma}_{\alpha_k},$$

become

$$-4.141 < \alpha_0 < -3.101,$$

$$-0.0396 < \alpha_1 < 0.0188,$$

$$-0.1180 < \alpha_2 < 0.0164,$$

and for the standard deviation σ

$$\gamma_1\tilde{\sigma} < \sigma < \gamma_2\tilde{\sigma}$$

or

$$0.2478 < \sigma < 0.6316.$$

Example 44.4 Table 92 contains the values x_i, y_i and the "weights" p_i^2 that determine the accuracy in measuring y_i for a given value x_i.

TABLE 92

i	x_i	y_i	p_i^2	i	x_i	y_i	p_i^2
0	1.5	6.20	0.5	5	−0.5	4.55	1.0
1	1.1	3.45	1.0	6	−1.0	8.85	1.0
2	0.7	2.00	1.0	7	−1.5	15.70	0.5
3	0.3	1.80	1.0	8	−2.0	24.40	0.25
4	−0.1	2.40	1.0				

If y is a second-degree polynomial in x,

$$y = a_0 + a_1x + a_2x^2,$$

find the estimates for the variances of individual measurements of y_i and the variances of the coefficients a_k ($k = 0, 1, 2$). Construct the confidence limits for the unknown true relation $y = F(x)$ at a confidence level $\alpha = 0.90$.

SOLUTION. We compute the quantities s'_k and v'_k for the system of normal equations but consider the "weight" of each measurement. The computations are given in Table 93.

TABLE 93

i	p_i^2	x_i	x_i^2	x_i^3	x_i^4	y_i	y_ix_i	$y_ix_i^2$
0	0.50	1.5	2.25	3.375	5.0625	6.20	9.300	13.950
1	1.00	1.1	1.21	1.331	1.4641	3.45	3.795	4.174
2	1.00	0.7	0.49	0.343	0.2401	2.00	1.400	0.980
3	1.00	0.3	0.09	0.027	0.0081	1.80	0.540	0.162
4	1.00	−0.1	0.01	−0.001	0.0001	2.40	−0.240	0.024
5	1.00	−0.5	0.25	−0.125	0.0625	4.55	−2.275	1.138
6	1.00	−1.0	1.00	−1.000	1.0000	8.85	−8.850	8.850
7	0.50	−1.5	2.25	−3.375	5.0625	15.70	−23.550	35.325
8	0.25	−2.0	4.00	−8.000	16.0000	24.40	−48,800	97.600

We obtain:

$$s'_0 = 7.250; \qquad s'_1 = 0; \qquad s'_2 = 6.300;$$

$$s'_3 = -1.425; \qquad s'_4 = 11.837;$$

$$v'_0 = 40.100; \qquad v'_1 = -24.955; \qquad v'_2 = 64.366.$$

We write the system of normal equations:

$$\left.\begin{aligned} 7.250\alpha_0 + 0 \qquad + 6.300a_2 &= 40.100 \\ 0 \quad + 6.300a_1 - 1.425a_2 &= -24.955 \\ 6.300a_0 - 1.425a_1 + 11.837a_2 &= 64.366 \end{aligned}\right\}.$$

We find the numerical values of the determinant Δ of the system and the cofactors δ_{kj} of the elements $d_{kj} = s'_{k+j}$ of this determinant:

$$\Delta = 275.87; \quad \delta_{00} = 72.54; \quad \delta_{11} = 46.12; \quad \delta_{22} = 45.68;$$

$$\delta_{01} = \delta_{10} = -8.98; \quad \delta_{02} = \delta_{20} = -39.69; \quad \delta_{12} = \delta_{21} = 10.33.$$

We compute the estimates of the coefficients a_k:

$$\tilde{a}_k = \frac{\delta_{k0}v'_0 + \delta_{k1}v'_1 + \delta_{k2}v'_2}{\Delta},$$

and get

$$\tilde{a}_0 = 2.096; \qquad \tilde{a}_1 = -3.068; \qquad \tilde{a}_2 = 3.955.$$

We find $S_{\min}$ by performing the computations given in Table 94:

$$S_{\min} = \sum_{i=0}^{8} p_i^2[y_i - \tilde{a}_0 - \tilde{a}_1x_i - \tilde{a}_2x_i^2] = 0.2208.$$

TABLE 94

i	y_i	$\tilde{\alpha}_0 + \tilde{\alpha}_1 x_i$	$\tilde{\alpha}_2 x_i^2$	$\tilde{y}_i$	ε_i	ε_i^2
0	6.20	−2.5044	8.8945	6.390	−0.190	0.0361
1	3.45	−1.2775	4.7833	3.506	−0.056	0.0031
2	2.00	−0.0507	1.9370	1.886	0.114	0.0130
3	1.80	1.1762	0.3558	1.532	0.268	0.0718
4	2.40	2.4030	0.0395	2.442	−0.042	0.0018
5	4.55	3.6298	0.9883	4.618	−0.068	0.0046
6	8.85	5.1634	3.9531	9.116	−0.266	0.0708
7	15.70	6.6970	8.8945	15.592	0.108	0.0117
8	24.40	8.2305	15.8124	24.043	0.357	0.1274
					$S_{\min}$ =	0.2208

We compute the estimates of the variances of individual measurements $\tilde{\sigma}_i^2$ by the formula

$$\tilde{\sigma}_i^2 = \frac{S_{\min}}{n - m} \frac{1}{p_i^2},$$

and obtain:

$$\tilde{\sigma}_0^2 = \tilde{\sigma}_7^2 = 0.0368 \cdot \frac{1}{0.5} = 0.0736;$$

$$\tilde{\sigma}_1^2 = \tilde{\sigma}_2^2 = \tilde{\sigma}_3^2 = \tilde{\sigma}_4^2 = \tilde{\sigma}_5^2 = \tilde{\sigma}_6^2 = 0.0368;$$

$$\sigma_8^2 = 0.1472.$$

The estimates of the variances of the coefficients a_k and their covariances are given by the formulas

$$\tilde{\sigma}_{a_k}^2 = \frac{S_{\min}}{n - m} \frac{\delta_{kk}}{\Delta}, \qquad \tilde{K}_{a_k, a_j} = \frac{S_{\min}}{n - m} \frac{\delta_{kj}}{\Delta}.$$

We have:

$$\tilde{\sigma}_{a_0}^2 = 0.009336; \qquad \tilde{\sigma}_{a_1}^2 = 0.005936; \qquad \tilde{\sigma}_{a_2}^2 = 0.005879;$$

$$\tilde{K}_{a_0, a_1} = -0.001156; \qquad \tilde{K}_{a_0, a_2} = -0.005108; \qquad \tilde{K}_{a_1, a_2} = 0.001329.$$

We calculate the estimate of the variance $\tilde{\sigma}_y^2(x)$ of $\tilde{y}$ by the formula

$$\tilde{\sigma}_y^2(x) = \tilde{\sigma}_{a_0}^2 + \tilde{\sigma}_{a_1}^2 x^2 + \tilde{\sigma}_{a_2}^2 x^4 + 2\tilde{K}_{a_0, a_1} x + 2\tilde{K}_{a_0, a_2} x^2 + 2\tilde{K}_{a_1, a_2} x^3$$

or by

$$\tilde{\sigma}_y^2(x) = 10^{-5}(933.6 - 231.2x - 428.0x^2 + 265.8x^3 + 587.9x^4).$$

The values $\sigma_y^2(x_i)$ for all x_i are calculated in Table 95.

We construct the confidence limits for the unknown true relation $y = F(x)$:

$$\tilde{y}_i - \gamma\tilde{\sigma}_y(x_i) < y < \tilde{y}_i + \gamma\tilde{\sigma}_y(x_i),$$

where γ is determined from Table 16T for $\alpha = 0.90$ and $k = n - m = 6$ degrees of freedom:

$$\gamma = 1.943.$$

The confidence limits for y are computed as in Table 95.

Similarly one can solve Problems 44.7, 44.8 and 44.11.

TABLE 95

i	x_i	$933.0-231.2x_i$	$-428.0x_i^2$	$265.8x_i^3$	$587.9x_i^4$	$\tilde{\sigma}_y^2(x_i)$	$\tilde{\sigma}_y(x_i)$	$\tilde{y}_i$	$\tilde{y}_i-\gamma\tilde{\sigma}_y(x_i)$	$\tilde{y}_i+\gamma\tilde{\sigma}_y(x_i)$
0	1.5	586.8	−963.0	897.1	2976.2	0.03497	0.187	6.390	6.023	6.753
1	1.1	679.3	−517.9	353.8	860.7	0.01375	0.117	3.506	3.279	3.733
2	0.7	771.8	−209.7	91.2	141.2	0.00794	0.088	1.886	1.715	2.057
3	0.3	864.2	−38.5	7.8	4.8	0.00838	0.092	1.532	1.353	1.711
4	−0.1	956.7	−4.3	0.3	0.1	0.00953	0.098	2.442	2.252	2.632
5	−0.5	1049.2	−107.0	33.2	36.7	0.01012	0.101	4.618	4.422	4.814
6	−1.0	1164.8	−428.0	265.8	587.9	0.01590	0.126	9.116	8.871	9.361
7	−1.5	1280.4	−963.0	897.1	2976.2	0.04197	0.205	15.592	15.194	15.990
8	−2.0	1396.0	−1712.0	2126.4	9406.4	0.11217	0.335	24.043	23.392	24.694

Example 44.5 The values of the electric resistance of molybdenum depend on temperature T°K as shown in Table 96.

TABLE 96

T, °K	ρ, micro-ohm/cm.	T, °K	ρ, micro-ohm/cm.
2289	61.97	1489	37.72
2132	57.32	1286	32.09
1988	52.70	1178	28.94
1830	47.92		

If ρ is linearly dependent on T:

$$\rho = a_0 + a_1 T,$$

determine the coefficients a_0 and a_1 by the method of least squares. The errors in measurements of ρ and T are specified by the standard deviations $\sigma_\zeta = 0.8$ and $\sigma_T = 15°$, respectively. Find the maximal deviation of the calculated value of ρ from the experimental one.

SOLUTION. We calculate the quantities s_k, r_k $(k = 1, 2)$, v_1 as shown in Table 97.

TABLE 97

i	T_i	$T_i^2\cdot 10^{-2}$	ρ_i	ρ_i^2	$T_i\rho_i\cdot 10^{-1}$	$\tilde{\rho}_i$	ε_i
0	2,289	52,395	61.97	3,840.3	14,185	61.82	0.15
1	2,132	45,454	57.32	3,285.6	12,221	57.15	0.17
2	1,988	39,521	52.70	2,777.3	10,477	52.86	−0.16
3	1,830	33,489	47.92	2,296.3	8,769	48.15	−0.23
4	1,489	22,171	37.72	1,422.8	5,617	38.00	−0.28
5	1,286	16,538	32.09	1,029.8	4,127	31.95	0.14
6	1,178	13,877	28.94	837.5	3,409	28.73	0.21
	$s_1=12{,}192$	$s_2=22{,}344\cdot 10^3$	$r_1=318.66$	$r_2=15{,}490$	$v_1=58{,}805\cdot 10$		

We obtain:

$$s_1 = 12{,}192; \qquad s_2 = 22{,}344 \cdot 10^3;$$
$$r_1 = 318.66; \qquad r_2 = 15{,}490;$$
$$v_1 = 58{,}805 \cdot 10.$$

We write the quadratic equation for the coefficient $\tilde{a}_1$:

$$\tilde{a}_1^2 + \frac{[s_1^2 - (n+1)s_2]\dfrac{\sigma_\rho^2}{\sigma_T^2} - [r_1^2 - (n+1)r_2]}{s_1 r_1 - (n+1)v_1}\,\tilde{a}_1 - \frac{\sigma_\rho^2}{\sigma_T^2} = 0,$$

which, after the substitution of the numerical values, becomes

$$\tilde{a}_1^2 + 0.065708\tilde{a}_1 - 0.0028444 = 0.$$

Solving this equation, we find two values for $\tilde{a}_1$:

$$\tilde{a}_{11} = 0.029786; \qquad \tilde{a}_{12} = -0.095494.$$

Obviously, the negative root $\tilde{a}_{12}$ is extraneous since the data contained in Table 97 show that when T increases ρ increases. Consequently,

$$\tilde{a}_1 = 0.029786.$$

We determine the coefficient $\tilde{a}_0$ by the formula

$$\tilde{a}_0 = \frac{r_1 - \tilde{a}_1 s_1}{n+1} = -6.356.$$

We calculate the values of $\tilde{a}_0$ in Table 97:

$$\varepsilon_i = \rho_i - \tilde{\rho}_i,$$

where $\tilde{\rho}$ are the computed values of the quantity

$$\tilde{\rho} = -6.3558 + 0.029786T.$$

From the data of Table 97 we find that $|\varepsilon_{\max}| = 0.28$.
One can solve Problem 44.15 similarly.

PROBLEMS

44.1 The results of several equally accurate measurements of the depth h of penetration of a body into a barrier for different values of its specific energy E (that is, the energy per unit area) are given in Table 98.

TABLE 98

i	E_i	h_i	i	E_i	h_i	i	E_i	h_i
0	41	4	5	139	20	9	241	30
1	50	8	6	154	19	10	250	31
2	81	10	7	180	23	11	269	36
3	104	14	8	208	26	12	301	37
4	120	16						

Select a linear combination of the form

$$h = a_0 + a_1 E.$$

Determine the estimates $\tilde{\sigma}^2_{a_k}$ of the variances of the coefficients a_k and the estimate $\tilde{\sigma}^2$ of the variance determining the accuracy of an individual measurement.

44.2 Solve the preceding problem by shifting the origin of E to the arithmetic mean of E and the origin of h to a point close to the expectation of h and thereby simplify the computations.

44.3 The height h of a body in free fall at time t is determined by the formula

$$h = a_0 + a_1 t + a_2 t^2,$$

where a_0 is the height at $t = 0$, a_1 is the initial velocity of the body and a_2 is half the acceleration of gravity g.

Determine the estimates of the coefficients a_0, a_1, a_2 and estimate the accuracy of determination of the acceleration of gravity by the indicated method, by using a series of equally accurate measurements whose results appear in Table 99.

TABLE 99

t, sec.	h, cm.	t, sec.	h, cm.	t, sec.	h, cm.	t, sec.	h, cm.	t, sec.	h, cm.
$\frac{1}{30}$	11.86	$\frac{4}{30}$	26.69	$\frac{7}{30}$	51.13	$\frac{10}{30}$	85.44	$\frac{13}{30}$	129.54
$\frac{2}{30}$	15.67	$\frac{5}{30}$	33.71	$\frac{8}{30}$	61.49	$\frac{11}{30}$	99.08	$\frac{14}{30}$	146.48
$\frac{3}{30}$	20.60	$\frac{6}{30}$	41.93	$\frac{9}{30}$	72.90	$\frac{12}{30}$	113.77		

44.4 Solve the preceding problem by using (the orthgonal) Chebyshev polynomials.

44.5 Several equally accurate measurements of a quantity y at equally spaced values of the argument x give the results appearing in Table 100.

TABLE 100

x	-3	-2	-1	0	1	2	3
y	-0.71	-0.01	0.51	0.82	0.88	0.81	0.49

If y is quite accurately approximated by the second-degree polynomial

$$y = a_0 + a_1 x + a_2 x^2,$$

determine the estimates of the coefficients $\tilde{a}_k$, the variance of an individual measurement $\tilde{\sigma}^2$ and the variances $\tilde{\sigma}^2_{a_k}$ of the coefficients $\tilde{a}_k$.

44.6 The amount of wear of a cutter is determined by its thickness (in millimeters) as a function of operating time t (in hours). The results are given in Table 101.

TABLE 101

t	y	t	y	t	y
0	30.0	6	27.5	12	26.1
1	29.1	7	27.2	13	25.7
2	28.4	8	27.0	14	25.3
3	28.1	9	26.8	15	24.8
4	28.0	10	26.5	16	24.0
5	27.7	11	26.3		

Using (the orthogonal) Chebyshev polynomials, express y both as a first- and then as a third-degree polynomial of t. Considering that the results are valid, in both cases estimate the magnitude of the variance of an individual measurement and construct the confidence intervals for the standard deviation σ for a confidence level $\alpha = 0.90$.

44.7 The value of the compression of a steel bar x_i under a load y_i and the values of the variances σ_i^2, which determine the accuracy in measurements of y_i, are given in Table 102.

TABLE 102

i	0	1	2	3	4
x_i, μ	5	10	20	40	60
y_i, Kg.	51.33	78.00	144.3	263.6	375.2
σ_i^2	82.3	25.0	49.3	51.3	46.7

Find the linear dependence,

$$y = a_0 + a_1 x,$$

associated with Hooke's law. Construct the confidence intervals for the coefficients a_k ($k = 0, 1$) and also the confidence limits for the unknown correct value of the load for x ranging from 5 to 60 μ if the confidence level is $\alpha = 0.90$.

The "weights" of the measurements corresponding to each value x_i of the compression are taken inversely proportional to σ_i^2.

44.8 Table 103 contains the average values of y_i corresponding to the values x_i of the argument and also the number n_i of measurements of y for $x = x_i$.

TABLE 103

i	x_i	y_i	n_i	i	x_i	y_i	n_i
0	1	0.10	21	3	4	0.32	11
1	2	0.19	8	4	5	0.39	11
2	3	0.24	13	5	6	0.48	10

Construct the approximating second-degree polynomial and determine the estimates of the standard deviations $\tilde{\sigma}_{a_k}$ of the coefficients $\tilde{a}_k$.

44.9 The net cost (in dollars) of one copy of a book as a function of the number (in thousands of copies) in a given printing is characterized by the data accumulated by the publisher over several years (Table 104).

TABLE 104

x	y	x	y	x	y
1	10.15	10	2.11	100	1.21
2	5.52	20	1.62	200	1.15
3	4.08	30	1.41		
5	2.85	50	1.30		

Select the coefficients for a hyperbolic dependence of the form

$$y = a_0 + \frac{a_1}{x}$$

and construct the confidence intervals for the coefficients ($k = 0, 1$) and also for the quantity y for different values of x_i if the confidence level is $\alpha = 0.90$.

44.10 A condenser is initially charged to a voltage U after which it is discharged through a resistance. The voltage U is rounded-off to the nearest multiple of 5 volts at different times. The results of several measurements appear in Table 105.

TABLE 105

i	t_i, sec.	U_i, v.	i	t_i, sec.	U_i, v.	i	t_i, sec.	U_i, v.
0	0	100	4	4	30	8	8	10
1	1	75	5	5	20	9	9	5
2	2	55	6	6	15	10	10	5
3	3	40	7	7	10			

It is known that the dependence of U on t has the form

$$U = U_0 e^{-at}.$$

Select the coefficients U_0 and a and construct the confidence intervals for U_0 and a for a confidence level $\alpha = 0.90$.

44.11 The following data obtained from an aerodynamical test of a model airplane (see Table 106) express the dependence of the angle of inclination δ_B (of the elevator ensuring a rectilinear horizontal flight) on the velocity v of the air stream:

$$\delta_B = a_0 + \frac{a_1}{v^2}.$$

TABLE 106

i	v_i, m./sec.	δ_{Bi}	n_i	i	v_i, m./sec.	δ_{Bi}	n_i
0	80	$-3°\,44'$	8	5	140	$-0°\,38'$	6
1	90	$-2°\,58'$	12	6	160	$-0°\,07'$	9
2	100	$-2°\,16'$	11	7	180	$0°\,10'$	12
3	110	$-1°\,39'$	9	8	200	$0°\,35'$	10
4	120	$-1°\,21'$	14				

Find the estimates of the coefficients a_0 and a_1 and their standard deviations. The n_i denote the number of measurements for a given value of the velocity v_i.

44.12 The results of several measurements of the dimension x of a lot of items are divided into intervals and the frequencies p_i^* in Table 107 are computed for them.

TABLE 107

The limits of the interval of x_i	p_i^*	*The limits of the interval of* x_i	p_i^*	*The limits of the interval of* x_i	p_i^*
50 to 60	0.00333	100 to 110	0.18667	140 to 150	0.06333
60 to 70	0.00667	110 to 120	0.20333	150 to 160	0.05333
70 to 80	0.03000	120 to 130	0.16333	160 to 170	0.01333
80 to 90	0.07667	130 to 140	0.08333	170 to 180	0.00667
90 to 100	0.11000				

If the values of p_i^* refer to the midpoints of the intervals x_i, select, by the method of least squares, the parameters for the relation

$$p = p_0 \exp\left\{-\frac{(x - \bar{x})^2}{2\sigma^2}\right\}$$

that approximates the experimental distribution. Apply (the orthogonal) Chebyshev polynomials. Test whether the resulting dependence obeys a normal distribution law for x; that is, whether the following equation holds:

$$p_0 = \frac{10}{\sqrt{2\pi}\,\sigma}.$$

44.13 Table 108 contains the measured values of some quantity y as a function of time t (for a 20 hour period).

TABLE 108

t	y	t	y	t	y
0.00	−25	0.35	26	0.70	−16
0.05	−26	0.40	32	0.75	3
0.10	−4	0.45	40	0.80	−21
0.15	7	0.50	32	0.85	−22
0.20	6	0.55	21	0.90	−29
0.25	13	0.60	11	0.95	−32
0.30	−30	0.65	−5		

If

$$y = a \sin(\omega t - \varphi), \qquad \text{where } \omega = 360 \frac{\text{degrees}}{24 \text{ hours}},$$

determine the estimates of the parameters a and φ. Find the maximal deviation of the measured quantity y from the approximating function $\tilde{y}$.

Hint. First choose the approximate value φ' and represent y in the form

$$y = a \sin\theta + b \cos\theta,$$

where

$$\theta = \omega t - \varphi',$$

$$b = -a(\varphi - \varphi').$$

44.14 Table 109 contains the experimental data for the values of a function $y = f(x)$ with period 2π.

TABLE 109

x_i, degrees	y_i	x_i, degrees	y_i	x_i, degrees	y_i	x_i, degrees	y_i
15	1.31	105	2.12	195	2.89	285	−2.30
30	1.84	120	2.38	210	2.01	300	−2.22
45	2.33	135	2.98	225	0.92	315	−1.57
60	2.21	150	3.44	240	−0.24	330	−1.03
75	2.24	165	3.51	255	−1.23	345	−0.01
90	2.39	180	3.33	270	−1.98	360	−0.82

Find the representation of this function by the polynomial

$$\tilde{y} = \tilde{a}_0 + \tilde{a}_1 \cos x + \tilde{b}_1 \sin x + \tilde{a}_2 \cos 2x + \tilde{b}_2 \sin 2x$$

and the maximal deviation of the measured quantity y from the approximating function $\tilde{y}$.

44.15 Table 110 contains the levels x and y of the water in a river at points A and B, respectively (B is 50 km. downstream from A). These levels are measured at noon during the first 15 days of April.

TABLE 110

i	0	1	2	3	4	5	6	7	8	9	10	11	12	13	14
x_i, m.	12.1	11.2	9.8	10.4	9.2	8.5	8.8	7.4	6.6	7.0	6.4	6.0	6.5	5.8	5.4
y_i, m.	10.5	9.3	8.3	9.6	8.6	7.1	6.9	5.8	5.2	5.0	5.1	4.6	5.0	4.4	3.9

If the relation

$$y = a_0 + a_1 x,$$

holds, determine the estimates of the coefficients $\tilde{a}_0$ and $\tilde{a}_1$ and the maximal deviation y_i from the calculated values $\tilde{y}_i$ if it is known that the errors in measurements of x and y are characterized by standard deviations $\sigma_x = \sigma_y = 0.5$ m.

45. STATISTICAL METHODS FOR QUALITY CONTROL

Basic Notions

Quality control methods permit us to regulate product quality by testing. A lot of items is sampled according to a scheme guaranteed to reject a good lot with probability α ("supplier's risk") and to accept a defective lot with probability β ("consumer's risk").

A lot is considered good if the parameter that characterizes its quality does not exceed a certain limiting value and defective if this parameter has a value not smaller than another limiting value. This quality parameter can be the number l of defective items in the lot (with the limits l_0 and $l_1 > l_0$), the average value of ξ or λ (with the limits ξ_0 and $\xi_1 > \xi_0$ or λ_0 and $\lambda_1 > \lambda_0$), or (for the homogeneity control of the production) the variance of the parameter in the lot (with the limits σ_0^2 and $\sigma_1^2 > \sigma_0^2$). In the case in which the quality of a lot improves with the increase of the parameter, the corresponding inequalities are reversed.

There are different methods of control: single sampling, double sampling and sequential analysis. The determination of the size of the sample and the criteria of acceptance or rejection of a lot according to given values of α and β constitutes *planning*.

In the case of *single sampling*, one determines the sample size n_0 and the acceptance number ν; if the value of the controlled parameter is $\leqslant \nu$ in the sample, then the lot is accepted, if it is $> \nu$, then the lot is rejected.

If one controls the number (proportion) of defective items in a sample of size n_0, the total number of defective items in the lot being L and the size of the lot being N, then

$$\alpha = \mathbf{P}(M > \nu \mid L = l_0) = 1 - \sum_{m=0}^{\nu} \frac{C_{l_0}^m C_{N-l_0}^{n_0-m}}{C_N^{n_0}},$$

$$\beta = \mathbf{P}(M \leqslant \nu \mid L = l_1) = \sum_{m=0}^{\nu} \frac{C_{l_1}^m C_{N-l_1}^{n_0-m}}{C_N^{n_0}},$$

where the values C_n^m can be taken from Table 1T or computed with the aid of Table 2T.

For $n_0 \leqslant 0.1N$, it is possible to pass approximately to a binomial distribution law

$$\alpha = 1 - \sum_{m=0}^{\nu} C_{n_0}^m p_0^m (1 - p_0)^{n_0 - m} = 1 - P(p_0, n_0, \nu),$$

$$\beta = \sum_{m=0}^{\nu} C_{n_0}^m p_1^m (1 - p_1)^{n_0 - m} = P(p_1, n_0, \nu),$$

where $p_0 = l_0/N$, $p_1 = l_1/N$, and the values of $P(p, n, d)$ can be taken from Table 4T or computed with the aid of Tables 2T and 3T.

Moreover, if $p_0 < 0.1$, $p_1 < 0.1$, then letting $a_0 = n_0 p_0$, $a_1 = n_0 p_1$ (passing to the Poisson distribution law), we obtain

$$\alpha = \sum_{m=\nu+1}^{\infty} \frac{a_0^m}{m!} e^{-a_0} = 1 - \mathbf{P}(\chi^2 \geqslant \chi_{q0}^2),$$

$$\beta = 1 - \sum_{m=\nu+1}^{\infty} \frac{a_1^m}{m!} e^{-a_1} = \mathbf{P}(\chi^2 \geqslant \chi_{q1}^2),$$

where

$$\chi_{q0}^2 = 2a_0, \qquad \chi_{q1}^2 = 2a_1, \qquad \sum_{m=\nu+1}^{\infty} \frac{a^m}{m!} e^{-a}$$

are given in Table 7T, and the probabilities $\mathbf{P}(\chi^2 > \chi_q^2)$ can be obtained from Table 17T for $k = 2(\nu + 1)$ degrees of freedom.

If $50 \leqslant n_0 \leqslant 0.1N$, $n_0 p_0 \geqslant 4$, then one may use the more convenient formulas:

$$\alpha = \frac{1}{2} - \frac{1}{2}\Phi\left(\frac{\nu - n_0 p_0 + 0.5}{\sqrt{n_0 p_0 (1 - p_0)}}\right),$$

$$\beta = \frac{1}{2} - \frac{1}{2}\Phi\left(\frac{n_0 p_1 - \nu - 0.5}{\sqrt{n_0 p_1 (1 - p_1)}}\right),$$

where $\Phi(z)$ is the Laplace function (see Table 8T).

If one controls the average value $\tilde{x} = (1/n_0) \sum_{i=1}^{n} x_i$ of the parameter in a sample, and the value of the parameter x_i of one item obeys a normal distribution with known variance σ^2, then

$$\alpha = \frac{1}{2} - \frac{1}{2}\Phi\left(\frac{\nu - \xi_0}{\sigma/\sqrt{n_0}}\right), \qquad \beta = \frac{1}{2} - \frac{1}{2}\Phi\left(\frac{\xi_1 - \nu}{\sigma/\sqrt{n_0}}\right).$$

For $\xi_0 > \xi_1$, the lot is accepted if $\tilde{x} \geqslant \nu$; it is rejected if $\tilde{x} < \nu$, and in the formulas for α and β the minus sign is replaced by plus sign.

If the controlled parameter has the probability density

$$f(x) = \lambda e^{-\lambda x},$$

then

$$\alpha = 1 - \mathbf{P}(\chi^2 \geqslant \chi_{q0}^2),$$

$$\beta = \mathbf{P}(\chi^2 \geqslant \chi_{q1}^2),$$

where $\chi_{q0}^2 = 2n_0\lambda_0\nu$, $\chi_{q1}^2 = 2n_0\lambda_1\nu$, and the probability $\mathbf{P}(\chi^2 \geqslant \chi_q^2)$ is determined by Table 17T for $k = 2n_0$ degrees of freedom. If $n_0 > 15$, then approximately

$$\mathbf{P}(\chi^2 \geqslant \chi_q^2) = \frac{1}{2} - \frac{1}{4}\Phi\left(\frac{\chi_q^2 - 2n_0}{2\sqrt{n_0}}\right).$$

If one controls the product homogeneity and the quality parameter is normal, then

$$\alpha = 1 - \mathbf{P}(\tilde{\sigma} \leqslant q_0\sigma_0), \qquad \beta = \mathbf{P}(\tilde{\sigma} \leqslant q_1\sigma_1),$$

where $q_0 = \nu/\sigma_0$, $q_1 = \nu/\sigma_1$, $\tilde{\sigma}^2 = (1/n_0)\sum_{i=1}^{n}(x_i - \bar{x})^2$ if the expectation $\bar{x}$ of the parameter is known or

$$\tilde{\sigma}^2 = \frac{1}{n_0 - 1}\sum_{i=1}^{n}\left(x_i - \frac{1}{n_0}\sum_{j=1}^{n} x_j\right)^2$$

if $\bar{x}$ is unknown, and the probabilities $\mathbf{P}(\tilde{\sigma} \leqslant q\sigma)$ are calculated from Table 22T for $k = n_0$ degrees of freedom if x is known and for $k = n_0 - 1$ if $\bar{x}$ is unknown.

In the case of a *double sampling*, one determines the sizes n_1 of the first and n_2 of the second samples and the acceptance numbers ν_1, ν_2, ν_3 (usually $\nu_1 < [n_1/(n_1 + n_2)]\nu_3 < \nu_2$). If in the first sample the controlled parameter is $\leqslant \nu_1$, then the lot is accepted; if the controlled parameter is $> \nu_2$, then the lot is rejected; in the other cases the second sample is taken. If the value of the controlled parameter found for the sample of size $(n_1 + n_2)$ is $\leqslant \nu_3$, then the lot is accepted and otherwise, it is rejected.

If one controls by the number of defective items in a sample, then

$$\alpha = 1 - \sum_{m_1=0}^{\nu_2}\frac{C_{l_0}^{m_1}C_{N-l_0}^{n_1-m_1}}{C_N^{n_1}} + \sum_{m_1=\nu_1+1}^{\nu_2}\left[\frac{C_{l_0}^{m_1}C_{N-l_0}^{n_1-m_1}}{C_N^{n_1}}\left(1 - \sum_{m_2=0}^{\nu_3-m_1}\frac{C_{l_0-m_1}^{m_2}C_{N-l_0-n_1+m_1}^{n_2-m_2}}{C_{N-n_1}^{n_2}}\right)\right],$$

$$\beta = \sum_{m_1=0}^{\nu_1}\frac{C_{l_1}^{m_1}C_{N-l_1}^{n_1-m_1}}{C_N^{n_1}} + \sum_{m_1=\nu_1+1}^{\nu_2}\left[\frac{C_{l_1}^{m_1}C_{N-l_1}^{n_1-m_1}}{C_N^{n_1}}\sum_{m_2=0}^{\nu_3-m_1}\frac{C_{l_1-n_1}^{m_2}C_{N-l_1-n_1+m_1}^{n_2-m_2}}{C_{N-n_1}^{n_2}}\right].$$

As in the case of single sampling, in the presence of certain relations between the numbers n_1, n_2, N, l_0, l_1 an approximate passage is possible from a hypergeometric distribution to a binomial, normal or Poisson distribution law.

If one controls by the average value $\tilde{x}$ of the parameter in a sample, then for a normal distribution of the parameter of one item with given variance σ^2 in the particular case when $n_1 = n_2 = n$, $\nu_1 = \nu_3 = \nu$, $\nu_2 = \infty$, we have

$$\alpha = 1 - p_1 - 0.5(p_2 - p_1^2), \qquad \beta = p_3 + 0.5(p_4 - p_3^2),$$

where

$$p_1 = 0.5 + 0.5\Phi\left(\frac{\nu - \xi_0}{\sigma/\sqrt{n}}\right), \qquad p_2 = 0.5 + 0.5\Phi\left(\frac{\nu - \xi_0}{\sigma/\sqrt{2n}}\right),$$

$$p_3 = 0.5 + 0.5\Phi\left(\frac{\nu - \xi_1}{\sigma/\sqrt{n}}\right), \qquad p_4 = 0.5 + 0.5\Phi\left(\frac{\nu - \xi_1}{\sigma/\sqrt{2n}}\right).$$

For $\xi_0 > \xi_1$, the inequality signs appearing in the conditions of acceptance and rejection are reversed and, in the formulas for p_1, p_2, p_3, p_4, the plus sign appearing in front of the second term is replaced by a minus sign.

If one controls by $\tilde{x}$ and the probability density of the parameter X for one item is exponential: $f(x) = \lambda e^{-\lambda x}$, $n_1 = n_2 = n$, $\nu_1 = \nu_3 = \nu$, $\nu_2 = \infty$, then

$$\alpha = 1 - p_1 - 0.5(p_2 - p_1^2), \qquad \beta = p_3 + 0.5(p_4 - p_3^2),$$

where

$$p_1 = 1 - \mathbf{P}(\chi^2 \geqslant \chi^2_{q0}), \qquad p_2 = 1 - \mathbf{P}(\chi^2 \geqslant \chi^2_{q0}),$$

$$p_3 = 1 - \mathbf{P}(\chi^2 \geqslant \chi^2_{q1}), \qquad p_4 = 1 - \mathbf{P}(\chi^2 \geqslant \chi^2_{q1}),$$

$\chi^2_{q0} = 2n\lambda_0\nu$, $\chi^2_{q1} = 2n\lambda_1\nu$, and the probabilities $\mathbf{P}(\chi^2 \geqslant \chi^2_q)$ are computed according to Table 17T for $k = 2n$ degrees of freedom (for p_1 and p_3) and $k = 4n$ (for p_2 and p_4).

If one controls the homogeneity of the production when the controlled parameter is normally distributed, $n_1 = n_2 = n$, $\nu_1 = \nu_3 = \nu$, $\nu_2 = \infty$, then

$$\alpha = 1 - p_1 - 0.5(p_2 - p_1^2), \qquad \beta = p_2 + 0.5(p_4 - p_3^2),$$

where p_1, p_2, p_3, p_4 are determined from Table 22T for q and k, and $q = q_0$ for p_1 and p_2, $q = q_1$ for p_3 and p_4; for a known $\bar{x}$, $k = n$ for p_1 and p_3, $k = 2n$ for p_2 and p_4; for an unknown $\bar{x}$, $k = n - 1$ for p_1 and p_3, $k = 2(n - 1)$ for p_2 and p_4.

In the *sequential Wald analysis* for a variable sample size n and a random value of the controlled parameter in the sample, the likelihood coefficient γ is computed and the control lasts until γ leaves the limits of the interval (B, A), where $B = \beta/(1 - \alpha)$, $A = (1 - \beta)/\alpha$; if $\gamma \leqslant B$, then the lot is accepted, if $\gamma \geqslant A$, the lot is rejected and for $B < \gamma < A$ the tests continue.

If one controls by means of m defective items in a sample, then

$$\gamma = \gamma(n, m) = \frac{C_{l_1}^m C_{N-l_1}^{n-m}}{C_{l_0}^m C_{N-l_0}^{n-m}}.$$

For $n \leqslant 0.1N$, a formula valid for a binomial distribution is useful:

$$\gamma(n, m) = \frac{p_1^m(1 - p_1)^{n-m}}{p_0^m(1 - p_0)^{n-m}},$$

where

$$p_0 = \frac{l_0}{N}, \qquad p_1 = \frac{l_1}{N}.$$

In this case, the lot is accepted if $m \leqslant h_1 + nh_3$, the lot is rejected if $m \geqslant h_2 + nh_3$ and the tests continue if $h_1 + nh_3 < m < h_2 + nh_3$, where

$$h_1 = \frac{\log B}{\log \frac{p_1}{p_0} + \log \frac{1 - p_0}{1 - p_1}}, \qquad h_2 = \frac{\log A}{\log \frac{p_1}{p_0} + \log \frac{1 - p_0}{1 - p_1}},$$

$$h_3 = \frac{\log \frac{1 - p_0}{1 - p_1}}{\log \frac{p_1}{p_0} + \log \frac{1 - p_0}{1 - p_1}}.$$

In Figure 37 the strip II gives the range of values for n and m for which the tests are continued, I being the acceptance range, and III being the rejection range.

If $n \leqslant 0.1N$, $p_1 < 0.1$, then

$$\gamma(n, m) = \frac{a_1^m e^{-a_1}}{a_0^m e^{-a_0}},$$

where $a_0 = np_0$, $a_1 = np_1$. For the most part, the conditions for sequential control and the graphical method remain unchanged, but in the present case

$$h_1 = \frac{\log B}{\log \frac{p_1}{p_0}}, \qquad h_2 = \frac{\log A}{\log \frac{p_1}{p_0}},$$

$$h_3 = \frac{0.4343(p_1 - p_0)}{\log \frac{p_1}{p_0}}.$$

If the binomial distribution law is acceptable, the expectation of the sample size is determined by the formulas

$$\mathbf{M}[n \mid p_0] = \frac{(1 - \alpha) \log B + \alpha \log A}{p_0 \log \frac{p_1}{p_0} - (1 - p_0) \log \frac{1 - p_0}{1 - p_1}},$$

$$\mathbf{M}[n \mid p_1] = \frac{\beta \log B + (1 - \beta) \log A}{p_1 \log \frac{p_1}{p_0} - (1 - p_1) \log \frac{1 - p_0}{1 - p_1}}.$$

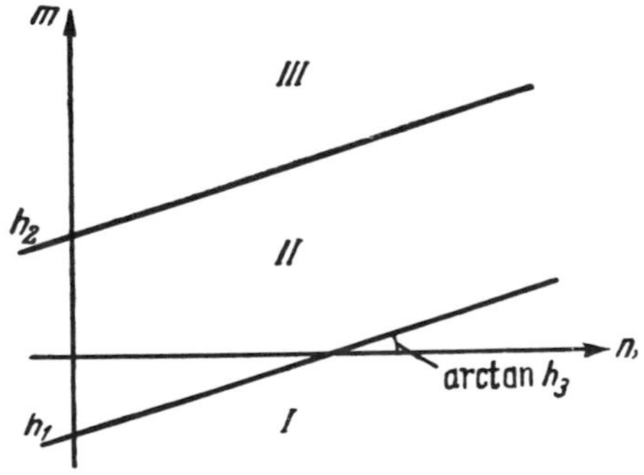

FIGURE 37

The expectation of the sample size becomes maximal when the number of defective items in the lot is $l = Nh_3$:

$$\mathbf{M}[n]_{\max} = -\frac{\log B \log A}{\log \dfrac{p_1}{p_0} \log \dfrac{1 - p_0}{1 - p_1}}, \qquad \text{where } p_0 = \frac{l_0}{N},\ p_1 = \frac{l_1}{N}.$$

If one controls by the average value $\tilde{x}$ of the parameter in the sample and the parameter of one item is a normal random variable with known variance σ^2, then

$$\gamma = \gamma(n, \tilde{x}) = \exp\left\{-\frac{1}{2\sigma^2} \sum_{i=1}^{n} [(x_i - \xi_1)^2 - (x_i - \xi_0)^2]\right\}.$$

The lot is accepted if $n\tilde{x} \leqslant h_1 + h_3 n$, the lot is rejected if $n\tilde{x} \geqslant h_2 + h_3 n$ and the tests are continued if $h_1 + nh_3 < n\tilde{x} < h_2 + nh_3$, where

$$h_1 = 2.303 \frac{\sigma^2}{\xi_1 - \xi_0} \log B; \qquad h_2 = 2.303 \frac{\sigma^2}{\xi_1 - \xi_0} \log A; \qquad h_3 = \frac{\xi_0 + \xi_1}{2}.$$

The method of control in the present case can also be graphically represented as in Figure 37 if $n\tilde{x}$ is used in place of m on the y-axis. For $\xi_0 > \xi_1$, we shall have $h_1 > 0$, $h_2 < 0$ and the inequalities in the acceptance and rejection conditions change their signs.

The expected number of tests is determined by the formulas:

$$\mathbf{M}[n \mid \xi_0] = \frac{h_2 + (1 - \alpha)(h_1 - h_2)}{\xi_0 - h_3},$$

$$\mathbf{M}[n \mid \xi_1] = \frac{h_2 + \beta(h_1 - h_2)}{\xi_1 - h_3},$$

$$\mathbf{M}[n]_{\max} = -\frac{h_1 h_2}{\sigma^2}.$$

If the parameter of an individual item has the probability density $f(x) = \lambda e^{-\lambda x}$, then

$$\gamma(n, \tilde{x}) = \frac{\lambda_1^n}{\lambda_0^n} e^{-(\lambda_1 - \lambda_0)n\tilde{x}}.$$

The lot is accepted if $n\tilde{x} \geqslant h_1 + nh_3$, it is rejected if $n\tilde{x} \leqslant h_2 + nh_3$, and the tests are continued if $h_1 + nh_3 > n\tilde{x} > h_2 + nh_3$, where

$$h_1 = -2.303 \frac{\log B}{\lambda_1 - \lambda_0}; \qquad h_2 = -2.303 \frac{\log A}{\lambda_1 - \lambda_0};$$

$$h_3 = 2.303 \frac{\log \dfrac{\lambda_1}{\lambda_0}}{\lambda_1 - \lambda_0}.$$

The graphical representation of the method of control differs from that represented in Figure 37 only because in the present case I represents the rejection region and III represents the acceptance region. The expected number of tests is computed by the formulas

$$\mathbf{M}[n \mid \lambda_0] = \frac{(1-\alpha)\log B + \alpha \log A}{\log \dfrac{\lambda_1}{\lambda_0} - 0.4343 \dfrac{\lambda_1 - \lambda_0}{\lambda_0}},$$

$$\mathbf{M}[n \mid \lambda_1] = \frac{\beta \log B + (1-\beta)\log A}{\log \dfrac{\lambda_1}{\lambda_0} - 0.4343 \dfrac{\lambda_1 - \lambda_0}{\lambda_1}},$$

$$\mathbf{M}[n]_{\max} = -\frac{h_1 h_2}{h_3^2}.$$

If the production is checked for homogeneity (normal distribution law), then

$$\gamma = \gamma(n, \tilde{\sigma}) = \frac{\sigma_0^n}{\sigma_1^n} \exp\left\{-\frac{n}{2}\left(\frac{\tilde{\sigma}^2}{\sigma_1^2} - \frac{\tilde{\sigma}^2}{\sigma_0^2}\right)\right\}.$$

The lot is accepted (for a known $\bar{x}$) if $n\tilde{\sigma}^2 \leqslant h_1 + nh_3$, it is rejected if $n\tilde{\sigma}^2 \geqslant h_2 + nh_3$ and the tests are continued if $h_1 + nh_3 < n\tilde{\sigma}^2 < h_2 + nh_3$, where

$$h_1 = \frac{4.606 \log B}{\dfrac{1}{\sigma_0^2} - \dfrac{1}{\sigma_1^2}}, \qquad h_2 = \frac{4.606 \log A}{\dfrac{1}{\sigma_0^2} - \dfrac{1}{\sigma_1^2}}, \qquad h_3 = \frac{2.303 \log \dfrac{\sigma_1^2}{\sigma_0^2}}{\dfrac{1}{\sigma_0^2} - \dfrac{1}{\sigma_1^2}}.$$

The graphical representation is analogous to Figure 37 with the values of $n\tilde{\sigma}^2$ on the y-axis.

If $\bar{x}$ is unknown, then whenever n appears in the formulas it should be replaced by $(n - 1)$.

The expected numbers of tests are

$$\mathbf{M}[n \mid \sigma_0] = \frac{h_2 + (1-\alpha)(h_1 - h_2)}{\sigma_0^2 - h_3}, \qquad \mathbf{M}[n \mid \sigma_1] = \frac{h_2 + \beta(h_1 - h_2)}{\sigma_1^2 - h_3},$$

$$\mathbf{M}[n]_{\max} = -\frac{h_1 h_2}{2h_3^2}.$$

If the total number of defects of the items belonging to the sample is checked and the number of defects of one item obeys a Poisson law with parameter a, then all the preceding formulas are applicable for the Poisson distribution if we replace:

m by $n\tilde{x}$, p_0 and p_1 by a_0 and a_1, a_0 and a_1 by na_0 and na_1, χ_{q0}^2 by $2na_0$ and χ_{q1}^2 by $2na_1$, where n is the size of the sample.

For $n \geqslant 50$, $na \geqslant 4$, it is possible to pass to a normal distribution

$$\sum_{m=\nu+1}^{\infty} \frac{a^m e^{-a}}{m!} \approx \frac{1}{2} + \frac{1}{2}\Phi\left(\frac{na - \nu - 0.5}{\sqrt{na}}\right).$$

To determine the probability that the number of tests is $n < n_g$ in a sequential analysis when $\alpha \ll \beta$ or $\beta \ll \alpha$, one may apply Wald's distribution

$$\mathbf{P}(y < y_g) = W_c(y_g) = \sqrt{\frac{c}{2\pi}} \int_0^{y_g} y^{-3/2} \exp\left\{-\frac{c}{2}\left(y + \frac{1}{y} - 2\right)\right\} dy,$$

where y is the ratio of the number of tests (n) to the expectation of n for some value of the control parameter of the lot (l, ξ, λ), $y_g = y|_{n=n_g}$ and the parameter c of Wald's distribution is determined by the following formulas:

(a) for a binomial distribution of the proportion of the defective product,

$$c = K\frac{\left|p \log\frac{p_1}{p_0} - (1-p)\log\frac{1-p_0}{1-p_1}\right|}{p(1-p)\left(\log\frac{p_1}{p_0} + \log\frac{1-p_0}{1-p_1}\right)}, \qquad p = \frac{l}{N};$$

(b) for a normal distribution of the product parameter,

$$c = K\frac{\left|\tilde{x} - \frac{\xi_0 + \xi_1}{2}\right|}{\xi_1 - \xi_0};$$

(c) for an exponential distribution of the product parameter,

$$c = K\frac{\left|2.303 \log\frac{\lambda_1}{\lambda_0} - \frac{\lambda_1 - \lambda_0}{\lambda}\right|}{\left(\frac{\lambda_1 - \lambda_0}{\lambda}\right)^2},$$

where

$$K = \begin{cases} 2.303|\log B| \text{ if the selected value of the parameter is } < h_3,\ \alpha \ll \beta; \\ 2.303 \log A \text{ if the selected value of the parameter is } > h_3,\ \beta \ll \alpha. \end{cases}$$

A special case of control by the number of defective products arises in reliability tests of duration t, where the time of reliable operation is assumed to obey an exponential distribution law. In this case, the probability p that an item fails during time t is given by the formula $p = 1 - e^{-\lambda t}$. All the formulas of control for the proportion of defective products in the case of a binomial distribution remain valid if one replaces p_0 by $1 - e^{-\lambda_0 t}$, p_1 by $1 - e^{-\lambda_1 t}$. If $\lambda t < 0.1$, then it is possible to pass to a Poisson distribution if, in the corresponding formulas, one replaces a_0 by $n\lambda_0 t$, a_1 by $n\lambda_1 t$, χ^2_{q0} by $2n\lambda_0 t$, χ^2_{q1} by $2n\lambda_1 t$.

The sequential analysis differs in the present case because for a fixed number n_0 of tested items, the testing time t is random. The lot is accepted if $t \geqslant t_1 + mt_3$, rejected if $t \geqslant t_2 + mt_3$ and the tests are continued if $t_1 + mt_3 > t > t_2 + mt_3$, where

$$t_1 = -2.303\frac{\log B}{n_0(\lambda_1 - \lambda_0)}; \qquad t_2 = -2.303\frac{\log A}{n_0(\lambda_1 - \lambda_0)};$$

$$t_3 = 2.303\frac{\log\frac{\lambda_1}{\lambda_0}}{n_0(\lambda_1 - \lambda_0)},$$

and m is the number of failures during time t. To plot the graph, one represents m on the x-axis and t on the y-axis.

The expectation of the testing time T for $\lambda t < 0.1$ is determined by the formulas:

$$\mathbf{M}[T \mid \lambda_0] = \frac{t_H}{n_0}\mathbf{M}[n \mid p_0], \qquad \mathbf{M}[T \mid \lambda_1] = \frac{t_H}{n_0}\mathbf{M}[n \mid p_1],$$

$$\mathbf{M}[T]_{\max} = \frac{t_H}{n_0}\mathbf{M}[n]_{\max},$$

where t_H is a number chosen to simplify the computations and $p_0 = \lambda_0 t_H$, $p_1 = \lambda_1 t_H$.

To determine the probability that the testing time $T < t_g$ if $\alpha \ll \beta$ or $\beta \ll \alpha$, one applies Wald's distribution in which one should set $y = t/\mathbf{M}[T \mid \lambda]$ and find the parameter c by the formula valid for a binomial distribution for the preceding chosen value of t_H.

Solution for Typical Examples

Example 45.1 A lot of $N = 40$ items is considered as first grade if it contains at most $l_0 = 8$ defective items. If the number of defective items exceeds $l_1 = 20$, then the lot is returned for repairs.

(a) Compute α and β by a single sampling of size $n_0 = 10$ if the acceptance number is $\nu = 3$;

(b) find α and β for a double sampling for which $n_1 = n_2 = 5$, $\nu_1 = 0$, $\nu_2 = 2$, $\nu_3 = 3$;

(c) compare the efficiency of planning by the methods of single and double samplings according to the average number of items tested in 100 identical lots;

(d) construct the sequential sampling plan for α and β obtained in (a), determine $n_{\min}$ for the lots with $L = 0$ and $L = N$.

Solution. (a) We compute α and β by the formulas

$$\alpha = 1 - \sum_{m=0}^{3} \frac{C_8^m C_{32}^{10-m}}{C_{40}^{10}}, \qquad \beta = \frac{1}{C_{40}^{10}} \sum_{m=0}^{3} C_{20}^m C_{20}^{10-m}.$$

Using Table 1T for C_n^m, we find

$$\alpha = 0.089, \qquad \beta = 0.136.$$

(b) We compute α and β by the formulas

$$\alpha = 1 - \frac{1}{C_{40}^5} \sum_{m=0}^{2} C_8^m C_{32}^{5-m} + \sum_{m_1=1}^{2} \left[\frac{C_8^{m_1} C_{32}^{5-m_1}}{C_{40}^5}\left(1 - \sum_{m_2=0}^{3-m_1} \frac{C_{8-m_1}^{m_2} C_{27+m_1}^{5-m_2}}{C_{35}^5}\right)\right],$$

$$\beta = \frac{C_{20}^0 C_{20}^5}{C_{40}^5} + \sum_{m_1=1}^{2} \left[\frac{C_{20}^{m_1} C_{20}^{5-m_1}}{C_{40}^5} \sum_{m_2=0}^{3-m_1} \frac{C_{20-m_1}^{m_2} C_{15+m_1}^{5-m_2}}{C_{35}^5}\right],$$

and obtain

$$\alpha = 0.105, \qquad \beta = 0.134.$$

(c) The probability that a first-grade lot in the case of double sampling will be accepted after the first sampling of five items is

$$\mathbf{P}(m_1 \leqslant \nu_1) = \mathbf{P}(m_1 = 0) = \frac{C_8^0 C_{32}^5}{C_{40}^5} = 0.306.$$

The expectation of the number of lots accepted after the first sampling from a total number of 100 lots is

$$100 \cdot 0.306 = 30.6 \text{ lots};$$

for the remaining 69.4 lots a second sampling is necessary. The average number of items used in double sampling is

$$30.6 \cdot 5 + 69.4 \cdot 10 = 847.$$

In the method of single sampling, the number of items used is

$$100 \cdot 10 = 1000.$$

In comparing the efficiency of the control methods, we have neglected the differences between the values of α and β obtained by single and double sampling.

(d) For $\alpha = 0.089$ and $\beta = 0.136$ the plan of sequential analysis is the following:

$$B = \frac{\beta}{1 - \alpha} = 0.149, \qquad \log B = -0.826,$$

$$A = \frac{1 - \beta}{\alpha} = 9.71, \qquad \log A = 0.987.$$

To determine $n_{\min}$ when all the items of the lot are nondefective, we compute the successive values of $\log \gamma(n; 0)$ by the formulas

$$\log \gamma(1; 0) = \log (N - l_1)! + \log (N - l_0 + 1)! - \log (N - l_0)! - \log (N - l_1 + 1)!,$$

$$\log \gamma(n + 1; 0) = \log \gamma(n; 0) - \log (N - l_0 - n)! + \log (N - l_1 - n)!$$

We have:

$$\begin{aligned} \log \gamma(1; 0) &= 0.7959; & \log \gamma(2; 0) &= 0.5833; \\ \log \gamma(3; 0) &= 0.3614; & \log \gamma(4; 0) &= 0.1295; \\ \log \gamma(5; 0) &= -0.1136; & \log \gamma(6; 0) &= -0.3688; \\ \log \gamma(7; 0) &= -0.6377; & \log \gamma(8; 0) &= -0.9217. \end{aligned}$$

Since the inequality $\log \gamma(n; 0) < \log B$ is satisfied only if $n \geqslant 8$, it follows that $n_{\min} = 8$.

For a lot consisting of defective items, $n = m$. We find $\log \gamma(1; 1) = 0.3979$. For successive values of n, we make use of the formula

$$\log \gamma(n + 1; m + 1) = \log \gamma(n; m) + \log (l_1 - m) - \log (l_0 - m).$$

We obtain $\log \gamma(2; 2) = 0.8316$; $\log \gamma(3; 3) = 1.3087 > \log A = 0.987$; consequently, in this case $n_{\min} = 3$.

Similarly one can solve Problem 45.1.

Example 45.2 A large lot of tubes ($N > 10{,}000$) is checked. If the proportion of defective tubes is $p \leqslant p_0 = 0.02$, the lot is considered good; if $p \geqslant p_1 = 0.10$, the lot is considered defective. Using the binomial and Poisson distribution laws (confirm their applicability):

(a) compute α and β for a single sampling (single control) if $n = 47$, $\nu = 2$;

(b) compute α and β for a double sampling (double control) taking $n_1 = n_2 = 25$, $\nu_1 = 0$, $\nu_2 = 2$, $\nu_3 = 2$;

(c) compare the efficiency of the single and double controls by the number of items tested per 100 lots;

(d) construct the plan of sequential control, plot the graph and determine $n_{\min}$ for the lot with $p = 0$, $p = 1$, compute the expectation for the number of tests in the case of sequential control.

SOLUTION. (a) In the case of binomial distribution,

$$\alpha = 1 - \sum_{m=0}^{2} C_{47}^{m} 0.02^{m} 0.98^{47-m}, \qquad \beta = \sum_{m=0}^{2} C_{47}^{m} 0.10^{m} 0.90^{47-m}.$$

Using Table 4T for the binomial distribution function and interpolating between $n = 40$ and $n = 50$, we get $\alpha = 0.0686$, $\beta = 0.1350$.

In the case of a Poisson distribution law, computing $\alpha_0 = n_0 p_0 = 0.94$, $a_1 = n_0 p_1 = 4.7$, we obtain

$$\alpha = \sum_{m=3}^{\infty} \frac{0.94^{m} e^{-0.94}}{m!}, \qquad \beta = 1 - \sum_{m=3}^{\infty} \frac{4.7^{m} e^{-4.7}}{m!}.$$

Using Table 7T, which contains the total probabilities for a Poisson distribution, we find (interpolating with respect to a)

$$\alpha = 0.0698, \qquad \beta = 0.159.$$

(b) For a binomial distribution law, using Table 1T and 4T, we find

$$\alpha = 1 - \sum_{m_1=0}^{2} C_{25}^{m_1} 0.02^{m_1} 0.98^{25-m_1} + \sum_{m_1=1}^{2} \left[C_{25}^{m_1} 0.02^{m_1} 0.98^{25-m_1} \left(1 - \sum_{m_2=0}^{2-m_1} C_{25}^{m_2} 0.02^{m_2} 0.98^{25-m_2} \right) \right] = 0.0704,$$

$$\beta = C_{25}^{0} 0.1^{0} 0.9^{25} + \sum_{m_1=1}^{2} \left[C_{25}^{m_1} 0.1^{m_1} 0.9^{25-m_1} \left(\sum_{m_2=0}^{2-m_1} C_{25}^{m_2} 0.1^{m_2} 0.9^{25-m_2} \right) \right] = 0.1450.$$

In the case of a Poisson distribution law, using Tables 6T and 7T and computing $\alpha_{01} = 0.5$, $a_{02} = 0.5$, $a_{11} = 2.5$, $a_{21} = 2.5$, we obtain

$$\alpha = \sum_{m_1=3}^{\infty} \frac{0.5^{m_1} e^{-0.5}}{m_1!} + \sum_{m_1=1}^{2} \left[\frac{0.5^{m_1} e^{-0.5}}{m_1!} \left(\sum_{m_2=3-m_1}^{\infty} \frac{0.5^{m_2} e^{-0.5}}{m_2!} \right) \right] = 0.0715,$$

$$\beta = 1 - \sum_{m_1=1}^{\infty} \frac{2.5^{m_1} e^{-0.25}}{m_1!} + \sum_{m_1=1}^{2} \left[\frac{2.5^{m_1} e^{-2.5}}{m_1!} \left(1 - \sum_{m_2=3-m_1}^{\infty} \frac{2.5^{m_2} e^{-2.5}}{m_2!} \right) \right] = 0.1935.$$

The essential difference between the values of β computed with the aid of binomial and Poisson distributions is explained by the large value of $p_1 = 0.10$.

(c) The probability of acceptance of a good lot ($p \leqslant 0.02$) after the first sampling in the case of double control (we compare the results of the binomial distribution) is

$$\mathbf{P}(m_1 \leqslant \nu_1) = \mathbf{P}(m_1 = 0) = C_{25}^{0} 0.02^0 \cdot 0.98^{25} = 0.6035.$$

The average number of good lots accepted after the first sampling from the total number of 100 lots is

$$100 \cdot 0.6035 = 60.35.$$

For the remaining 39.65 lots, a second sampling will be necessary. The average expenditure in tubes for a double control of 100 lots is equal to

$$60.35 \cdot 25 + 39.64 \cdot 50 = 3497.$$

In a defective lot, the probability of rejection after the first sampling in the case of double control is

$$\mathbf{P}(m_1 > \nu_2) = \mathbf{P}(m_1 > 2) = 1 - \sum_{m_1=0}^{2} C_{25}^{m_1} 0.1^{m_1} 0.9^{25-m_1} = 0.4629.$$

The average number of lots rejected after the first sampling from a total of 100 lots is

$$100 \cdot 0.4629 = 46.29.$$

For the remaining 53.71 lots a second sampling will be necessary. The average expenditure in tubes for a double control of 100 lots will be

$$46.29 \cdot 25 + 53.71 \cdot 50 = 3843.$$

For a single control, in all cases

$$100 \cdot 50 = 5000 \text{ tubes}$$

will be consumed.

(d) For $\alpha = 0.0686$, $\beta = 0.1350$ for a sequential control, using a binomial distribution we get:

$$B = 0.1450, \qquad \log B = -0.8388, \qquad A = 1.261, \qquad \log A = 1.1007.$$

Furthermore, $h_1 = -1.140$, $h_2 = 1.496$, $h_3 = 0.0503$ (Figure 38). We find $n_{\min}$ for a good lot for $p = 0$:

$$0 = h_1 + n_{\min} h_3, \qquad n_{\min} = -\frac{h_1}{h_3} = \frac{1.140}{0.0503} = 22.7 \approx 23;$$

for a defective lot when $p = 1$:

$$n_{\min} = h_2 + n_{\min} h_3,$$

$$n_{\min} = \frac{h_2}{1 - h_3} = \frac{1.496}{0.9497} = 1.5 \approx 2.$$

We determine the average numbers of tests for different p:

$$\mathbf{M}[n \mid 0.02] = 31.7; \qquad \mathbf{M}[n \mid 0.10] = 22.9; \qquad \mathbf{M}[n]_{\max} = 35.7.$$

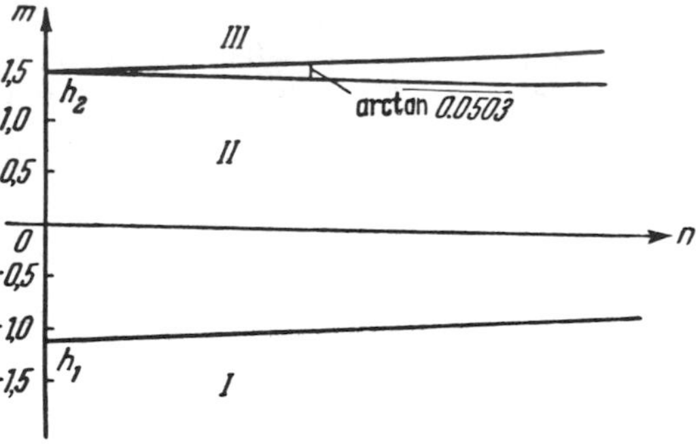

FIGURE 38

Problems 45.2 to 45.5, 45.7, 45.8 and 45.10 can be solved by following this solution.

Example 45.3 A large lot of resistors, for which the time of reliable operation obeys an exponential distribution, is subjected to reliability tests. If the failure parameter $\lambda \leqslant \lambda_0 = 2 \cdot 10^{-6}$ hours^{-1}, the lot is considered good; if $\lambda \geqslant \lambda_1 = 1 \cdot 10^{-5}$ hours^{-1}, the lot is considered defective. Assuming that $\lambda t_0 < 0.1$, where t_0 is a fixed testing time for each item in a sample of size n_0, determine for $\alpha = 0.005$, $\beta = 0.08$, the value of n_0. Use the method of single sampling for different t_0, find ν with the condition that $t_0 = 1000$ hours and also construct the plan of sequential control in the case $n = n_0$ for $t_0 = 1000$ hours. Compute $t_{\min}$ for a good lot and a defective one and $\mathbf{M}[T \mid \lambda]$, $\mathbf{P}(t < 1000)$, $\mathbf{P}(t < 500)$.

SOLUTION. The size n_0 of the sample and the acceptance number ν are determined by noting that $\lambda t_0 < 0.1$, which permits use of the Poisson distribution and furthermore, permits passing from a Poisson distribution to a chi-square distribution. We compute the quotient $\lambda_0/\lambda_1 = 0.2$. Next, from Table 18T we find the values χ_{q0}^2 for the entry quantities $\mathbf{P}(\chi^2 \geqslant \chi_{q0}^2) = 1 - \alpha = 0.995$ and k; χ_{q1}^2 for $\mathbf{P}(\chi^2 \geqslant \chi_{q1}^2) = \beta = 0.08$ and k.

By the method of sampling, we establish that

for $k = 15$

$$\chi_{q0}^2 = 4.48, \qquad \chi_{q1}^2 = 23.22, \qquad \frac{\chi_{q0}^2}{\chi_{q1}^2} = 0.1930;$$

for $k = 16$

$$\chi_{q0}^2 = 5.10, \qquad \chi_{q1}^2 = 24.48, \qquad \frac{\chi_{q0}^2}{\chi_{q1}^2} = 0.2041.$$

Interpolating with respect to $\chi_{q0}^2/\chi_{q1}^2 = 0.2$, we find: $k = 15.63$, $\chi_{q0}^2 = 4.87$, $\chi_{q1}^2 = 23.99$. We compute $\nu = (k/2) - 1 = 6.815$; we take $\nu = 6$, $2n_0\lambda_0 t_0 = 4.87$, hence, it follows that $n_0 t_0 = 4.87/2 \cdot 0.000002 = 1.218 \cdot 10^{-6}$. The condition $\lambda t_0 < 0.1$ leads to

$$t_0 < 0.1/0.00001 = 10{,}000 \text{ hours} \quad (\text{since } \lambda_1 = 0.00001).$$

Taking different values $t_0 < 10{,}000$, we obtain the corresponding values of n_0 given in Table 111.

TABLE 111

t_0 in hours	100	500	1000	2500	5000
n_0	12,180	2436	1218	487	244

We compute B, A, t_1, t_2 for the method of sequential analysis: $B = 0.08041$, $\ln B = -2.5211$, $A = 184$, $\ln A = 5.2161$. Taking $n_0 = 1218$, we have

$t_1 = 258.7$ hours;

$t_2 = -535.3$ hours,

$t_3 = 165.2$ hours (Figure 39).

The minimal testing time in the case when $m = 0$ for a good lot is $t_{\min} = 258.7$ hours; for a defective lot $t_{\min} = -535.3 + 165.2m > 0$; $m = 3.24 \approx 4$; for $m = 4$, $t_{\min} = 125.5$ hours. If for $t < 125.5$ hours $m \geqslant 4$, then the lot is rejected.

To compute the average testing time for $n = n_0 = 1218$, we take $t_H = t_0 = 1000$ hours. Then

$$p_0 = \lambda_0 t_H = 0.002;$$
$$p_1 = \lambda_1 t_H = 0.010;$$
$$\lambda^* t_H = \frac{t_H}{n_0 t_3} = 0.00497.$$

Furthermore, we find

$$\mathbf{M}[n \mid p_0] = 505, \qquad \mathbf{M}[n \mid p_1] = 572, \qquad \mathbf{M}[n]_{\max} = 1001;$$

FIGURE 39

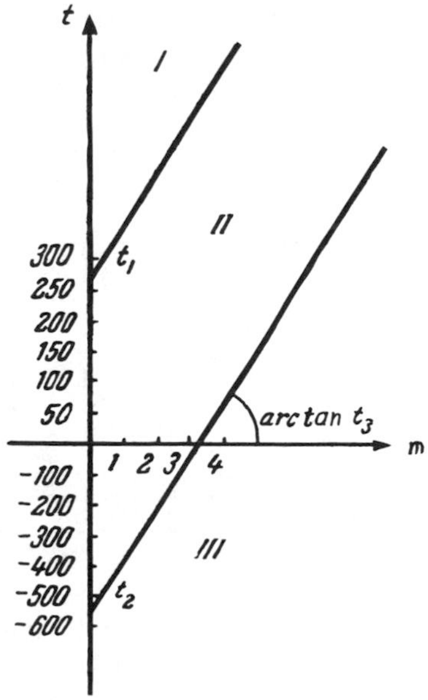

then we compute

$$\mathbf{M}[T \mid \lambda_0] = 415 \text{ hours}, \qquad \mathbf{M}[T \mid \lambda_1] = 470 \text{ hours},$$
$$\mathbf{M}[T]_{\max} = 821 \text{ hours}.$$

We find the probability that the testing time for a fixed number of items $n = n_0 = 1218$ is less than 1000 hours and 500 hours. Therefore, for $t_H = 1000$ hours, we compute the value of the parameter c of Wald's distribution and the value of

$$y = \frac{n_0}{\mathbf{M}[n \mid p_0]} = \frac{t_H}{\mathbf{M}[T \mid \lambda]}$$

with the condition that $p_0 = \lambda_0 t_0 = 0.002$; $p_1 = \lambda_1 t_0 = 0.01$. Taking $p = p_0$ since $\alpha \ll \beta$, we obtain $c = 2.37$, $y = 1000/415 = 2.406$. We find that (see Table 26T)

$$\mathbf{P}(T < 1000) = \mathbf{P}(n < 1218) = W_c(y) = 0.9599.$$

For $\gamma = 0.5$, we have

$$y = 1.203, \qquad \mathbf{P}(T < 500) = 0.725.$$

One can solve Problem 45.9 similarly.

Example 45.4 The quality of the disks produced on a flat-grinding machine is determined by the number of spots on a disk. If the average number of spots per 10 disks is at most one, then the disks are considered to be of good quality; if the average number is greater than five, then the disks are defective. A sample of 40 disks is selected from a large lot ($N > 1000$). Assuming that the number of spots on a disk obeys a Poisson distribution law:

(a) determine α and β for $\nu = 9$;

(b) for these α and β construct the plan of sequential control, compute $n_{\min}$ for a good lot and a defective one and find the values of $\mathbf{M}[n \mid a]$;

(c) test a concrete sample, whose data appear in Table 112, by the methods of single and sequential control.

TABLE 112

n	x_n	n	x_n	n	x_n	n	x_n	n	x_n
1	0	9	1	17	2	25	4	33	4
2	1	10	1	18	2	26	4	34	4
3	1	11	1	19	3	27	4	35	5
4	1	12	1	20	3	28	4	36	5
5	1	13	2	21	3	29	4	37	6
6	1	14	2	22	3	30	4	38	6
7	1	15	2	23	3	31	4	39	7
8	1	16	2	24	4	32	4	40	7

SOLUTION. (a) Using the Poisson distribution, we have $a_0 = 0.1$, $a_1 = 0.5$, $na_0 = 4$, $na_1 = 20$. Using Table 7T for the total probabilities of x_n occurrences of spots on disks in the sample, we find

$$\alpha = \sum_{x_n=10}^{\infty} \frac{4^{x_n}e^{-4}}{x_n!} = 0.00813, \qquad \beta = 1 - \sum_{x_n=10}^{\infty} \frac{20^{x_n}e^{-20}}{x_n!} 0.00500.$$

(b) For $\alpha = 0.0081$, $\beta = 0.0050$, the characteristics of the sequential control (Figure 40) are:

$$B = 0.005041; \quad \log B = -2.298; \quad A = 122.8; \quad \log A = 2.089,$$

$$h_1 = \frac{\log B}{\log \frac{a_1}{a_0}} = -3.29; \qquad h_2 = \frac{\log A}{\log \frac{a_1}{a_0}} = 2.99;$$

$$h_3 = \frac{0.4343(a_1 - a_0)}{\log \frac{a_1}{a_0}} = 0.248.$$

We compute $n_{\min}$:

$$\text{for} \quad x_n = 0, \qquad n_{\min} = 13.2 \approx 14$$
$$\text{for} \quad x_n = n, \qquad n_{\min} = 18.7 = 19.$$

The average number of tests in the case of sequential control is

$$\mathbf{M}[n \mid a_0] = 21.8; \qquad \mathbf{M}[n \mid a_1] = 11.8; \qquad \mathbf{M}[n]_{\max} = 39.5.$$

(c) In a sample with $n_0 = 40$, it turns out that $x_n = 7 < \nu = 9$; consequently, the lot is accepted. Applying the method of sequential control (see Figure 40), for $n = 30$ we obtain that the point with coordinates (n, m) lies below the lower line; that is, the lot should be accepted. Indeed,

$$\text{for} \quad n = 29, \qquad x_n = 4h_1 + mh_3 = 3.90; \quad x_n > h_1 + mh_3;$$
$$\text{for} \quad n = 30, \qquad x_n = 4h_1 + mh_3 = 4.15; \quad x_n < h_1 + mh_3.$$

Similarly one can solve Problem 45.11.

Example 45.5 The quality of punchings made by a horizontal forging machine is determined by the dispersion of their heights X, known to obey a

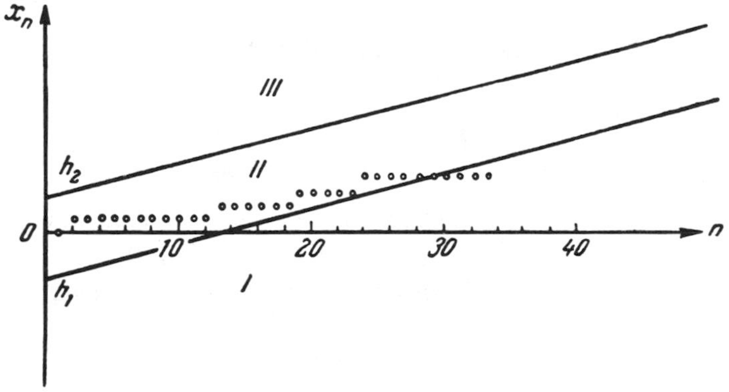

FIGURE 40

normal distribution law with expectation $\bar{x} = 32$ mm. (nominal dimension). If the standard deviation $\sigma \leqslant \sigma_0 = 0.18$ mm., the lot is considered good; if $\sigma \geqslant \sigma_1 = 0.30$ mm., the lot is defective. Find α and β for the method of single sampling if $n_0 = 39$ and $\nu = 0.22$ mm. Use the resulting values for α and β to construct a control plan by the method of sequential analysis. Compute $n_{\min}$ for a good lot and a defective one and find $\mathbf{M}[n \mid \sigma]$.

SOLUTION. We compute α and β by the formulas

$$\alpha = 1 - \mathbf{P}(\tilde{\sigma} \leqslant q_0\sigma_0), \qquad \beta = \mathbf{P}(\tilde{\sigma} \leqslant q_1\sigma_1),$$

for $k = n_0 = 39, q_0 = \nu/\sigma_0 = 1.221, q_1 = \nu/\sigma_1 = 0.733$. Interpolating according to Table 22T for the chi-square distribution, we find

$$\alpha = 0.0303; \qquad \beta = 0.0064.$$

We find the values of B, A, h_1, h_2, h_3 for the method of sequential analysis:

$$B = 0.006601; \quad \ln B = -5.021; \quad A = 30.10; \quad \ln A = 3.405;$$
$$h_1 = -0.528; \quad h_2 = 0.345; \quad h_3 = 0.0518.$$

We find $n_{\min}$. For the poorest among the good lots, $\tilde{\sigma}^2 = \sigma_0^2 = 0.0324$; $n_{\min}\sigma_0^2 = h_1 + n_{\min}h_3$; $n_{\min} = 27.2 \approx 28$.

For the best among the defective lots, $\tilde{\sigma}^2 = \sigma_1^2 = 0.0900$; $n_{\min}\sigma_1^2 = h_2 + n_{\min}h_3$; $n_{\min} = 9.3 \approx 10$.

We compute the average numbers of tests $\mathbf{M}[n \mid \sigma]$ for different σ:

$$\mathbf{M}[n \mid \sigma_0] = 25.9; \qquad \mathbf{M}[n \mid \sigma_1] = 8.8; \qquad \mathbf{M}[n]_{\max} = 34.0.$$

In a similar manner, one can solve Problem 45.12.

Example 45.6 The maximal pressure X in a powder chamber of a rocket is normally distributed with standard deviation $\sigma = 10$ kg./cm.². The rocket is considered good if $X \leqslant \xi_0 = 100$ kg./cm.²; if $X \geqslant \xi_1 = 105$ kg./cm.², the rocket is returned to the plant for adjustment. Given the values $\alpha = 0.10$ and $\beta = 0.01$, construct the plans for single control (n_0, ν) and sequential control, compute the probabilities $\mathbf{P}(n < n_0)$ and $\mathbf{P}(n < (1/2)n_0)$ that for the sequential control the average number of tests will be less than n_0 and $(1/2)n_0$, respectively.

SOLUTION. To compute the sample size n_0 and the acceptance number ν for a single control, we use the formulas

$$\Phi\left(\frac{\nu - \xi_0}{\sigma/\sqrt{n_0}}\right) = 1 - 2\alpha, \qquad \Phi\left(\frac{\xi_1 - \nu}{\sigma/\sqrt{n_0}}\right) = 1 - 2\beta.$$

Substituting the values for α and β and using Table 8T for the Laplace function, we find

$$\frac{\nu - 100}{10}\sqrt{n_0} = 1.2816, \qquad \frac{105 - \nu}{10}\sqrt{n_0} = 2.3264;$$

hence, it follows that $n_0 = 52$, $\nu = 101.8$ kg./cm.².

For the sequential control, we find that $B = 0.0111$, $\ln B = -4.500$, $A = 9.9$, $\ln A = 2.293$, $h_1 = -90$, $h_2 = 45.86$, $h_3 = 102.5$.

We determine $h_{\min}$. For the poorest among the good lots, when $\tilde{x} = \xi_0 = 100$,

$$n_{\min} \cdot 100 = -90 + n_{\min} \cdot 102.5; \qquad n_{\min} = 36;$$

for the best among the defective lots when $\tilde{x} = \xi_1 = 105$,

$$n_{\min} \cdot 105 = 45.86 + n_{\min} \cdot 102.5; \qquad n_{\min} = 18.3 \approx 19.$$

The average number of observations $\mathbf{M}[n \mid \xi]$ is equal to

$$\mathbf{M}[n \mid \xi_0] = 30.6; \qquad \mathbf{M}[n \mid \xi_1] = 17.8; \qquad \mathbf{M}[n]_{\max} = 41.3.$$

To determine the probability $\mathbf{P}(n < 52)$ since $\alpha \ll \beta$ for $\bar{x} = \xi_1 = 105$, we compute:

$$K = \ln A = 2.293; \qquad c = 1.146;$$

$$y_{11} = \frac{n_0}{\mathbf{M}[n \mid \xi_1]} = 4.031; \qquad y_{12} = \frac{1}{2} y_{11} = 2.016.$$

From Table 26T for Wald's distribution law we find that

$$\mathbf{P}(n < 52) = 0.982, \qquad \mathbf{P}(n < 26) = 0.891.$$

By following this solution one can solve Problem 45.13.

Example 45.7 The average time of operation of identical electron tubes represents $t \geqslant t_0 = 1282$ hours for a good lot and $t \leqslant t_1 = 708$ hours for a defective one. It is known that the time T of reliable operation obeys an exponential distribution law with the probability density

$$f(t) = \lambda e^{-\lambda t},$$

where the parameter λ is the intensity of failures, that is, the inverse of the mean time of operation of a tube in hours.

Determine for $\alpha = 0.001$ and $\beta = 0.01$, the size n_0 of the single sample and the acceptance number ν, construct the sequential control plan and find $n_{\min}$, $\mathbf{M}[n \mid \lambda]$, $\mathbf{P}(n < n_0)$, $\mathbf{P}(n < (1/2)n_0)$.

SOLUTION. Assuming that $n_0 > 15$ (since α and β are small), we replace the chi-square distribution, which the quantity $2\lambda n_0/\tilde{\lambda}$ obeys, by a normal distribution; i.e., we set

$$\mathbf{P}(\chi^2 \geqslant \chi_q^2) = 0.5 - 0.5\Phi\left(\frac{\chi_q^2 - 2n}{2\sqrt{n}}\right),$$

since the number of degrees of freedom is $k = 2n$. We obtain the equations

$$0.5 - 0.5\Phi\left(\frac{\chi_{q0}^2 - 2n}{2\sqrt{n}}\right) = 1 - \alpha,$$

$$0.5 - 0.5\Phi\left(\frac{\chi_{q1}^2 - 2n}{2\sqrt{n}}\right) = \beta;$$

hence, it follows, from Table 8T, that

$$\frac{\chi_{q0}^2 - 2n}{2\sqrt{n}} = -3.090, \qquad \frac{\chi_{q1}^2 - 2n}{2\sqrt{n}} = 2.324$$

or, since $\chi^2_{q0} = 2\lambda_0 n_0 \nu$, $\chi^2_{q1} = 2\lambda_1 n_0 \nu$, $\lambda_0 = 1/t_0 = 0.00078$, $\lambda_1 = 1/t_1 = 0.001413$,

$$0.000780 - \nu = -3.090 \frac{\nu}{\sqrt{n_0}},$$

$$0.001413 - \nu = 2.324 \frac{\nu}{\sqrt{n_0}}.$$

If we solve this system of equations, we obtain

$$\nu = 0.001141, \qquad n_0 = 99.03 \approx 100.$$

Since $n_0 > 15$, the use of a normal distribution is permissible.
For the sequential control, we find that:

$$B = 0.01001; \quad \ln B = -4.604; \quad A = 990; \quad \ln A = 6.898;$$

$$h_1 = 7273; \quad h_2 = -1090\cdot 10; \quad h_3 = 938.0;$$

$$\lambda^* = \frac{1}{h_3} = 0.001066.$$

We determine $n_{\min}$. For the poorest among the good lots, $\tilde{t} = t_0 = 1282$ hours, $n_{\min} = 21.1 \approx 22$; for the best among the defective lots, $\tilde{t} = t_1 = 708$ hours, $n_{\min} = 47.4 \approx 48$.

We find the average numbers of tests for different λ:

$$\mathbf{M}[n \mid \lambda_0] = 20.7; \quad \mathbf{M}[n \mid \lambda] = 46.6; \quad \mathbf{M}[n]_{\max} = 90.0.$$

Since $\alpha \ll \beta$, we determine $K = |\ln B| = 4.604$ and, then, the parameter c of Wald's distribution: $c = 1.525$; furthermore, we find $y_{01} = 100/20.7 = 4.82$; $y_{02} = 2.41$.

From Table 26T, for $y_{01}(y_{02})$ and c, we have

$$p = \mathbf{P}(n < 100) > 0.99 \qquad (\text{for} \quad p < 0.999),$$

$$\mathbf{P}(n < 50) = 0.939.$$

Similarly Problem 45.14 can be solved.

PROBLEMS

45.1 Rods in lots of 100 are checked for their quality. If a lot contains $L \leqslant l_0 = 4$ defective items, the lot is accepted; if $L \geqslant l_1 = 28$, the lot is rejected. Find α and β for the method of single sampling if $n_0 = 22$, $\nu = 2$, and for the method of double sampling for $n_1 = n_2 = 15$, $\nu_1 = 0$, $\nu_2 = 3$, $\nu_3 = 3$; compare their efficiencies according to the average number of tests; construct the sequential analysis plan and compute the minimal number of tests for a good lot and a defective one in the case of sequential control. Use the values of α and β obtained by the method of single sampling.

45.2 In the production of large lots of ball bearings, a lot is considered good if the number of defective items does not exceed 1.5 per cent and defective if it exceeds 5 per cent. Construct and compare the efficiency of the plan of single control, for which the sample size $n_0 = 410$ and acceptance number $\nu = 10$, and the plan of double control, for which $n_1 = n_2 = 220$, $\nu_1 = 2$, $\nu_2 = 7$, $\nu_3 = 11$.

Construct the sequential control plan with α and β as found for the plan of single control. Compare the efficiencies of all three methods according to the average number of tests and compute $n_{\min}$ for a good lot and a defective one for sequential control.

45.3 A large lot of punched items is considered good if the proportion of defective items $p \leqslant p_0 = 0.10$ and defective if $p \geqslant p_1 = 0.20$. Find α and β for the control by single sampling: use sample size $n_0 = 300$ and acceptance number $\nu = 45$. For the resulting values of α and β, construct the control plan by the method of sequential analysis and compute $n_{\min}$ for a good lot and a defective one; find $\mathbf{M}[n \mid p]$ and $\mathbf{P}(n < n_0)$, $\mathbf{P}(n < (1/2)n_0)$.

Hint: Pass to the normal distribution.

45.4 For a large lot of items, construct the plan of single control (n_0, ν) that guarantees (a) a supplier's risk of 1 per cent and a consumer's risk of 2 per cent, if the lot is accepted when the proportion of defective items is $p \leqslant p_0 = 0.10$ and rejected when $p \geqslant p_1 = 0.20$ (use the normal distribution), (b) $\alpha = 0.20$, $\beta = 0.10$ for the same p_0 and p_1 applied to a Poisson distribution law. Construct the corresponding plans of sequential control and find the expectations for the number of tests.

45.5 For $\alpha = 0.05$ and $\beta = 0.10$, construct the plans of single and sequential control for quality tests of large lots of rivets. The rivets are considered defective if their diameter $X > 13.575$ mm. A lot is accepted if the proportion of defective rivets is $p \leqslant p_0 = 0.03$ and rejected if $p \geqslant p_1 = 0.08$. Compute, for a Poisson distribution, the size n_0 of the single sample and the acceptance number ν. For the same α and β, construct the plan of sequential control, compute $n_{\min}$ for a good lot and a defective one and find the average number of tests $\mathbf{M}[n \mid p]$ in a sequential control.

45.6 Rivets with diameter $X > 13.575$ mm. are considered defective. At most 5 per cent of the lots whose proportion of defective items is $p < p_0 = 0.03$ may be rejected and at most 10 per cent of lots whose proportion of defective items is $p \geqslant p_1 = 0.08$ may be accepted. Assuming that the random variable X obeys a normal distribution whose estimates of the expectation $\tilde{x}$ and variance $\tilde{\sigma}^2$ are determined on the basis of sample data, find the general formulas for the size n_0 of the single sample in dimension control and for z_0 such that the following condition is satisfied

$$\mathbf{P}(\tilde{x} + \tilde{\sigma}z_0 > l \mid p = p_0) = \alpha,$$

$$\mathbf{P}(\tilde{x} + \tilde{\sigma}z_0 > l \mid p = p_1) = 1 - \beta.$$

Compute n_0 and z_0 for the conditions of the problem.

Consider the fact that the quantity

$$u = \tilde{x} + \tilde{\sigma}z_0$$

is approximately normally distributed with parameters

$$\mathbf{M}[u] = \bar{x} + \sigma z_0, \qquad \mathbf{D}[u] = \sigma^2\left(\frac{1}{n} + \frac{z_0^2}{2k}\right),$$

where $k = n - 1$. Compare the result with that of Problem 45.5.

45.7 Using the binomial and Poisson distributions, construct the plan of double control for $n_1 = n_2 = 30$, $\nu_1 = 3$, $\nu_2 = 5$, $\nu_3 = 8$, if a lot is considered good when the proportion of defective items is $p \leqslant p_0 = 0.10$ and defective when $p \geqslant p_1 = 0.20$. For the values α and β found for the binomial distribution, construct the plans of single and sequential control, compare all three methods according to the average number of tests. For the sequential control, find $n_{\min}$ for a good lot and a defective lot and compute the expectation of the number of tests $\mathbf{M}[n \mid p]$.

45.8 Construct the control plans by the methods of single and sequential sampling for large lots of radio tubes if a lot with proportion of defective items $p \leqslant p_0 = 0.02$ is considered good and with $p \geqslant p_1 = 0.07$ is considered defective. The producer's risk is $\alpha = 0.0001$ and the consumer's risk is $\beta = 0.01$. For the plan of sequential control, determine $n_{\min}$ for a good lot and a defective one, find the average number of tests $\mathbf{M}[n \mid p]$ and the probabilities $\mathbf{P}(n \leqslant \mathbf{M}[n \mid p_0])$, $\mathbf{P}(n \leqslant 2\mathbf{M}[n \mid p_0])$.

45.9 The time of operation T (in hours) of a transformer obeys an exponential distribution with an intensity of failures λ. Assuming that $\lambda t_0 < 0.1$, construct the plans of control by single sampling and sequential analysis for $\alpha = 0.10$, $\beta = 0.10$. For the single control, find the acceptance number ν and the size n_0 of the sample if the testing period of each transformer is $t_0 = 500$, 1000, 2000, 5000 hours. (Replace the Poisson distribution by a chi-square distribution.) For the sequential control, take a fixed sample size n_0 corresponding to $t_0 = 1000$ hours and find the average testing time of each transformer $\mathbf{M}[T \mid \lambda]$. Assume that a lot of transformers is good if the intensity of failures $\lambda \leqslant \lambda_0 = 10^{-5}$ hours^{-1} and defective if $\lambda \geqslant \lambda_1 = 2.10^{-5}$ hours^{-1}.

45.10 A large lot of electrical resistors is subjected to control for $\alpha = 0.005$, $\beta = 0.08$; the lot is considered good if the proportion of defective resistors is $p \leqslant p_0 = 0.02$ and defective if $p \geqslant p_1 = 0.10$. Applying a chi-square distribution instead of a Poisson one, find the size n_0 and the acceptance number ν for the method of single sampling; construct the plan of sequential control for a good lot and a defective lot; compute the expectation of the number of tested items and the probabilities $\mathbf{P}(n < n_0)$, $\mathbf{P}(n < (1/2)n_0)$.

45.11 Before planting, lots of seed potatoes are checked for rotting centers. A lot of seed potatoes is considered good for planting if in each group of 10 potatoes there is at most one spot and bad if there are five spots or more.

Assuming that the number of spots obeys a Poisson distribution, compute α and β for the method of double sampling if $n_1 = 40$, $n_2 = 20$, $\nu_1 = 4$, $\nu_2 = 12$, $\nu_3 = 14$. For the resulting values of α and β, construct the plans of single and sequential control. Compare the efficiencies of all three methods according to the mean expenditures of seed potatoes necessary to test 100 lots.

45.12 The quality characteristic in a lot of electrical resistors, whose random values obey a normal distribution law with a known mean of 200 ohms, is the standard deviation σ, and the lot is accepted if $\sigma \leqslant \sigma_0 = 10$ ohms and defective if $\sigma \geqslant \sigma_1 = 20$ ohms. Construct the control plans by the method of single sampling with $n_0 = 16$, $\nu = 12.92$ and double sampling with $n_1 = n_2 = 13$, $\nu_1 = \nu_3 = 12$, $\nu_2 = \infty$. For the resulting values of α and β (in the case of single control), construct the plan of sequential control. Compare the efficiencies of all three methods of control according to the average number of tests. Compute $n_{\min}$ for the poorest among the good lots and the best among the defective lots.

45.13 Several lots of nylon are tested for strength. The strength characteristic X, measured in g./denier (specific strength of the fiber), obeys a normal distribution with standard deviation $\sigma = 0.8$ g./denier. A lot is considered good if $X \geqslant x_0 = 5.4$ g./denier and bad if $X \leqslant x_1 = 4.9$ g./denier. Construct the plan of strength control by single sampling with $n_0 = 100$ and $\nu = 5.1$. For the resulting values of α and β, construct the plan of control by the method of sequential analysis, compute the mean expenditure in fibers and the probabilities $\mathbf{P}(n < n_0)$, $\mathbf{P}(n < (1/2)n_0)$.

45.14 It is known that if the intensity of failures is $\lambda \leqslant \lambda_0 = 0.01$, then a lot of gyroscopes is considered reliable; if $\lambda \geqslant \lambda_1 = 0.02$, the lot is unreliable and should be rejected. Assuming that the time T of reliable operation obeys an exponential distribution and taking $\alpha = \beta = 0.001$, construct the plans for single (n_0, ν) and sequential controls according to the level of the parameter λ. Find the average number of tested gyroscopes $\mathbf{M}[n \mid \lambda]$ for the case of sequential control.

45.15 A large lot of condensers is being tested. The lot is considered good if the proportion of unreliable condensers is $p \leqslant p_0 = 0.01$; for $p \geqslant p_1 = 0.06$ the lot is rejected. Construct the plan of single control (n_0, ν) for the proportion of unreliable items so that $\alpha = 0.05$, $\beta = 0.05$.

To establish the reliability, each tested condenser belonging to the considered sample is subjected to a multiple sequential control for $\alpha' = 0.0001$, $\beta' = 0.0001$ and a condenser is considered reliable if the intensity of failures $\lambda \leqslant \lambda_0 = 0.0000012$ and unreliable for $\lambda \geqslant \lambda_1 = 0.0000020$ hours^{-1} (n is the number of tests used to establish the reliability of a condenser for given α' and β'). One assumes that the time of reliable operation of a condenser obeys an exponential distribution.

45.16. Construct the plans of single and sequential controls of complex electronic devices whose reliability is evaluated according to the average time $\tilde{T}$ of unfailing (reliable) operation. If $\tilde{T} \geqslant T_0 = 100$ hours, a device is considered reliable and if $T \leqslant T_1 = 50$ hours, unreliable. It is necessary that $\alpha = \beta = 0.10$. Consider that for a fixed testing time t_T a device is accepted if $t_T/m = \tilde{T} \geqslant \nu$ and rejected if $\tilde{T} < \nu$, where m is the number of failures for time t, and ν is the acceptance number in the case of single control ($n_0 = 1$; in case of failure the device is repaired and the test is continued). In this case, $t_T/\tilde{T}$ obeys approximately a

Poisson distribution. In the case of sequential control, the quantity t depends on the progress of the test.

(a) Determine the testing time t_T and the acceptance number ν for a single control.

(b) For the plan of sequential control, reduce the condition for continuation of the tests $\ln B < \ln \gamma(t, m) < \ln A$ to the form $t_1 + mt_3 > t > t_2 + mt_3$. For t_1, t_2, t_3, obtain, preliminary general formulas.

(c) In the case of sequential control, determine the minimal testing time $t_{\min}$ for the poorest of the good lots and the best of the rejected ones.

46. DETERMINATION OF PROBABILITY CHARACTERISTICS OF RANDOM FUNCTIONS FROM EXPERIMENTAL DATA

Basic Formulas

The methods of determination of the expectation, the correlation function and the distribution laws of the ordinates of a random function by processing a series of sample functions does not differ from the methods of determination of the corresponding probability characteristics of a system of random variables. In processing the sample functions of stationary random functions, instead of averaging the sample functions, one may sometimes average with respect to time; i.e., find the probability characteristics with respect to one or several sufficiently long realizations (the condition under which this is possible is called ergodicity). In this case, the estimates (approximate values) of the expectation and correlation function are determined by the formulas

$$\tilde{x} = \frac{1}{T}\int_0^T x(t)\,dt,$$

$$\tilde{K}_x(\tau) = \frac{1}{T-\tau}\int_0^{T-\tau} [x(t) - \tilde{x}][x(t+\tau) - \tilde{x}]\,dt,$$

where T is the total time of recording of the sample function. Sometimes instead of the last formula one uses the practically equivalent formula,

$$\tilde{K}_x(\tau) = \frac{1}{T-\tau}\int_0^{T-\tau} x(t)x(t+\tau)\,dt - \tilde{x}^2.$$

In the case when the expectation $\bar{x}$ is known exactly,

$$\tilde{K}_x(\tau) = \frac{1}{T-\tau}\int_0^{T-\tau} [x(t) - \bar{x}][x(t+\tau) - \bar{x}]\,dt$$

$$\approx \frac{1}{T-\tau}\int_0^{T-\tau} x(t)x(t+\tau)\,dt - \bar{x}^2.$$

If $\tilde{x}$ and $\tilde{K}_x(\tau)$ are determined from the ordinates of a sample function of a random function at discrete time instants $t_j = (j - 1)\Delta$, the corresponding formulas become

$$\tilde{x} = \frac{1}{m}\sum_{j=1}^{m} x(t_j),$$

$$\tilde{K}_x(\tau) = \frac{1}{m - l}\sum_{j=1}^{m-l} [x(t_j) - \tilde{x}][x(t_j + \tau) - \tilde{x}]$$

or

$$\tilde{K}_x(\tau) = \frac{1}{m - l}\sum_{j=1}^{m-l} x(t_j)x(t_j + \tau) - \tilde{x}^2,$$

where $\tau = l\Delta$, $T = m\Delta$.

For normal random functions, the variances $\tilde{x}$ and $\tilde{K}_x(\tau)$ may be expressed in terms of $K_x(\tau)$. In practical computations; the unknown correlation function $K_x(\tau)$ in the formulas for $\mathbf{D}[\tilde{x}]$ and $\mathbf{D}[\tilde{K}_x(\tau)]$ is replaced by the quantity $\tilde{K}_x(\tau)$.

When one determines the value of the correlation function by processing several sample functions of different durations, one should take as the approximate value of the ordinates of $\tilde{K}_x(\tau)$ the sum of ordinates obtained by processing individual realizations whose weights are inversely proportional to the variances of these ordinates.

SOLUTION FOR TYPICAL EXAMPLES

Example 46.1 The ordinates of a stationary random function are determined by photographing the scale of the measuring instrument during equal time intervals Δ. Determine the maximal admitted value of Δ for which the increase in the variance of $\tilde{K}_x(0)$ compared with the variance obtained by processing the continuous graph of realization of a random function will be at most δ per cent if the approximate value of $\tilde{K}_x(\tau) = ae^{-\alpha|\tau|}$ and the total recording time T is $\gg 1/\alpha$. It is known that $\bar{x} = 0$ and the function $X(t)$ can be considered normal.

SOLUTION. Since $\bar{x} = 0$, by use of the continuous recording, the value of $\tilde{K}_x(0)$ is determined by the formula

$$\tilde{K}_1(0) = \frac{1}{T}\int_0^T x^2(t)\,dt.$$

For finding the variance of $\tilde{K}_1(0)$, we have

$$\mathbf{D}[\tilde{K}_1(0)] = \mathbf{M}[\tilde{K}_1^2(0)] - \{\mathbf{M}[\tilde{K}_1(0)]\}^2 = \frac{2}{T^2}\int_0^T\int_0^T K_x^2(t_2 - t_1)\,dt_1\,dt_2$$

$$\approx \frac{4}{T^2}a^2\int_0^T (T - \tau)e^{-2\alpha\tau}\,d\tau.$$

If after integration we eliminate the quantities containing the small (by assumption) factor $e^{-\alpha T}$, we get

$$\mathbf{D}[\tilde{K}_1(0)] = \frac{a^2}{T^2\alpha^2}(2\alpha T - 1).$$

If the ordinates of the random function are discrete, the value of $\tilde{K}_x(0)$ is

$$\tilde{K}_2(0) = \frac{1}{m}\sum_{j=1}^{m} x^2(j\Delta).$$

Determining the variance of $K_2(0)$, we find that

$$\begin{aligned}\mathbf{D}[\tilde{K}_2(0)] &= \frac{1}{m^2}\left\{\sum_{j=1}^{m}\sum_{l=1}^{m}\mathbf{M}[X^2(j\Delta)X^2(l\Delta)] - m^2K_x^2(0)\right\}\\ &= \frac{2}{m^2}\sum_{j=1}^{m}\sum_{l=1}^{m}K_x^2(l\Delta - j\Delta),\end{aligned}$$

where for the calculation of the expectation one uses a property of moments of systems of normal random variables.

Using the value of $\tilde{K}_x(\tau)$, we obtain

$$\begin{aligned}\mathbf{D}[\tilde{K}_2(0)] &= \frac{2a^2}{m^2}\sum_{j=1}^{m}\sum_{l=1}^{m}e^{-2\alpha|l-m|\Delta} = \frac{4a^2}{m^2}\sum_{r=0}^{m}(m-r)e^{-2\alpha r\Delta} - \frac{2a^2}{m}\\ &= \frac{2a^2\Delta}{T^2}\,\frac{T(1-e^{-4\alpha\Delta}) - 2\Delta e^{-2\alpha\Delta}}{(1-e^{-2\alpha\Delta})^2}.\end{aligned}$$

The limiting value of Δ is found from the equation

$$\frac{\mathbf{D}[\tilde{K}_2(0)]}{\mathbf{D}[\tilde{K}_1(0)]} = 1 + 0.01\delta;$$

that is, from the equation

$$\frac{2\alpha^2\Delta[T(1-e^{-4\alpha\Delta}) - 2\Delta e^{-2\alpha\Delta}]}{(2\alpha T - 1)(1-e^{-2\alpha\Delta})^2} = 1 + 0.01\delta.$$

For $\alpha\Delta \ll 1$, we obtain approximately

$$\alpha\Delta = \frac{-5\kappa + \sqrt{25\kappa^2 + 12(1-11\kappa)}}{2(1-11\kappa)}, \qquad \kappa = \frac{2\alpha T - 1}{2\alpha T - 3}\,\frac{\delta}{100}.$$

PROBLEMS

46.1 Prove that the condition

$$\lim_{\tau\to\infty} K_x(\tau) = 0$$

is necessary in order that the function $X(t)$ be ergodic.

46.2 Verify whether the expression

$$\tilde{S}_x(\omega) = \frac{1}{T}\left|\int_0^T e^{i\omega T}x(t)\,dt\right|^2$$

may be taken as an estimate of the spectral density if $X(t)$ is a normal stationary random function ($\bar{x} = 0$) and $\int_0^\infty |K(\tau)|\,d\tau < \infty$.

46.3 To determine the estimate of the correlation function of a stationary normal stochastic process $X(t)$ ($\bar{x} = 0$) a correlator is used that operates according to the formula

$$\tilde{K}_x(\tau) = \frac{1}{T - \tau} \int_0^{T-\tau} x(t)x(t + \tau)\,dt.$$

Derive the formula for $\mathbf{D}[\tilde{K}_x(\tau)]$.

46.4 Determine the expectations and the variances of the estimates of correlation functions defined by one of the formulas

$$\tilde{K}_1(\tau) = \frac{1}{T - \tau} \int_0^{T-\tau} x(t)x(t + \tau)\,dt - (\tilde{x})^2,$$

$$\tilde{K}_2(\tau) = \frac{1}{T - \tau} \int_0^{T-\tau} [x(t) - \tilde{x}][x(t + \tau) - \tilde{x}]\,dt,$$

where $\tilde{x} = 1/(T - \tau) \int_0^{T-\tau} x(t)\,dt$, if $X(t)$ is a normal random function.

46.5 The correlation function of the stationary stochastic process $X(t)$ has the form

$$K_x(\tau) = \sigma_x^2 e^{-\alpha|\tau|}.$$

Find the variance for the estimate of the expectation defined by the formula

$$\tilde{x} = \frac{1}{T} \int_0^T x(t)\,dt.$$

46.6 The spectral density $\tilde{S}_x(\omega)$ is found by a Fourier inversion of the approximate value of the correlation function. Determine $\mathbf{D}[\tilde{S}_x(\omega)]$ as a function of ω if

$$\tilde{K}_x(\tau) = \frac{1}{T} \int_0^T x(t)x(t + \tau)\,dt, \qquad \bar{x} = 0,$$

the process is normal and, to solve the problem, one may use

$$\tilde{K}_x(\tau) = ae^{-\alpha|\tau|}(1 + \alpha|\tau|)$$

instead of $K_x(\tau)$ in the final formula.

46.7 The correlation function $K_x(\tau)$ determined from an experiment is used for finding the variance of the stationary solution of the differential equation

$$\dot{Y}(t) + 2Y(t) = X(t).$$

Determine how σ_y will change if, instead of the expression

$$\tilde{K}_x(\tau) = \sigma_x^2 e^{-0.21|\tau|}(\cos 0.75\tau + 0.28 \sin 0.75|\tau|)$$

representing a sufficiently exact approximation of $K_x(\tau)$, one uses

$$\tilde{K}_x'(\tau) = \sigma_x^2 e^{-\alpha_1|\tau|} \cos \beta_1 \tau,$$

where α_1 and β_1 are chosen such that the position of the first zero and the ordinate of the first minimum of the expression of $\tilde{K}_x'(\tau)$ coincide with the corresponding quantities for $K_x(\tau)$.

46.8 An approximate value of $K_x(\tau)$ isused to find $\mathbf{D}[Y(t)]$, where

$$Y(t) = \frac{dX(t)}{dt}.$$

Determine how σ_y will change if instead of the expression

$$\tilde{K}_x(\tau) = \sigma_x^2 e^{-0.10|\tau|}\left(\cos 0.7\tau + \frac{1}{7}\sin 0.7|\tau|\right),$$

which approximates quite accurately the expression $K_x(\tau)$, one uses

$$\tilde{K}'_x(\tau) = \sigma_x^2 e^{-\alpha^2\tau^2}\cos\beta\tau,$$

where α and β are chosen such that the position of the first zeros and the value of the first minimum of the functions $\tilde{K}_x(\tau)$ and $\tilde{K}'_x(\tau)$ coincide.

46.9 The correlation function for the heel angle of a ship can be represented approximately in the form

$$\tilde{K}_\theta(\tau) = ae^{-\alpha|\tau|}\left(\cos\beta\tau + \frac{\alpha}{\beta}\sin\beta|\tau|\right),$$

where $a = 36$ deg.^{2}, $\alpha = 0.05$ sec.$^{-1}$ and $\beta = 0.75$ sec.$^{-1}$.

Determine $\mathbf{D}[\tilde{K}_\theta(\tau)]$ for $\tau = 0$ and $\tau = 3$ sec. if $\Theta(t)$ is a normal random function and $\tilde{K}_\theta(\tau)$ is obtained by processing the recorded rolling of the ship during time $T = 20$ minutes.

46.10 The ordinate of the estimate of the correlation function for $\tau = 0$ is 100 cm.^{2}, and for $\tau = \tau_1 = 4.19$ sec. Its modulus attains a maximum, corresponding to a negative value of 41.5 cm.^{2}. According to these data, select the analytic expression for $\tilde{K}(\tau)$:

(a) in the form $\tilde{K}(\tau) = \sigma^2 e^{-\alpha|\tau|}(\cos\beta\tau + (\alpha/\beta)\sin\beta|\tau|)$,
(b) in the form $\tilde{K}(\tau) = \sigma^2 e^{-\alpha|\tau|}\cos\beta\tau$.

Determine the difference in the values of the first zeros of the functions $\tilde{K}(\tau)$ in these two cases.

46.11 Determine $\mathbf{D}[\tilde{K}_\theta(\tau)]$ for $\tau = 0$, 2.09, 4.18 and 16.72 sec. if

$$\tilde{K}(_\theta\tau) = \frac{1}{T-\tau}\int_0^{T-\tau}\theta(t)\theta(t+\tau)\,dt,$$

$$K_\theta(\tau) = ae^{-\alpha|\tau|}\cos\beta\tau,$$

where $a = 25$ deg.^{2}, $\alpha = 0.12$ sec.$^{-1}$, $\beta = 0.75$ sec.$^{-1}$ and $\Theta(t)$ is a normal random function, $\bar{\theta} = 0$. To determine $K_\theta(\tau)$, one uses a 10 m. recording of $\Theta(t)$, where 1 cm. of the graph along the time axis corresponds to 1 sec.

46.12 The graph of a sample function of the random function $X(t)$ is recorded on a paper tape by using a conducting compound passing at constant speed between two contacts, one shifted with respect to the other by τ seconds along the time axis. The contacts are connected to a relay system so that the relay turns on a stop watch when the ordinates of the sample function at the points where the contacts are located have the same sign and turns it off otherwise. Show that if

$\bar{x} = 0$ and $X(t)$ is a normal stationary random function, the estimate of its normalized correlation function can be determined by the formula

$$\tilde{k}(\tau) = \cos \pi\left(1 - \frac{t_1}{t}\right),$$

where t_1 is the total reading of the stop watch and t is the total time the tape moves.

46.13 Under the assumptions of the preceding problem, determine $\mathbf{D}[\tilde{k}(5)]$ if for the determination of $\tilde{k}(5)$ one uses the graph of the sample function corresponding to the recording time $T = 10$ minutes,

$$k(\tau) = e^{-\alpha|\tau|}, \qquad \alpha = 0.2 \text{ sec.}^{-1}.$$

46.14 As a result of processing three sample functions of a single stationary random function $X(t)$ for durations T_1, T_2 and T_3, three graphs of estimates of the correlation function were obtained. Assuming that the process is normal, derive the formula for finding the ordinates of the estimate of the correlation function $\tilde{K}_x(\tau)$. Use all the experimental data with the condition that the variance of the error is minimal if for each sample function the estimate of the correlation function is given by the formula

$$\tilde{K}_j(\tau) = \frac{1}{T_j}\int_0^{T_j} x(t)x(t + \tau)\, dt, \qquad j = 1, 2, 3 \quad (\bar{x} = 0).$$

46.15 Determine variance of the estimate for the correlation function of a normal stochastic process with zero expectation if to find $\tilde{K}_x(\tau)$ one takes the ordinates of the sample function of the random function during equal time intervals Δ, the duration of recording is $T = m\Delta$ and in the final formula $K_x(\tau)$ may be replaced by $\tilde{K}_x(\tau)$.

46.16 The ordinates of a random function are determined by photographing the scale of an instrument during equal time intervals $\Delta = 1$ sec. Determine the ratio of $\mathbf{D}[\tilde{K}(0)]$ to the variance obtained by processing the continuous graph of the sample function if

$$K(\tau) = ae^{-0.5|\tau|},$$

(τ is expressed in seconds), the process is normal and the observation time $T = 5$ minutes.

46.17 An approximate determination of the ordinates of a sample function of a stationary random function $X(t)$ with zero expectation and a known correlation function $K_x(\tau)$ is given by the formula

$$X(t) = \sum_{j=0}^{m}\left(A_j \cos\frac{2\pi jT}{T} + B_j \sin\frac{2\pi jT}{T}\right)\alpha_j,$$

where A_j, B_j are mutually independent random variables with unit variances and zero expectations and T is a known number. Determine the constants α_j so that

$$\varepsilon \equiv \int_0^T [K_x(\tau) - \tilde{K}_x(\tau)]^2\, d\tau = \min.,$$

where $\tilde{K}_x(\tau)$ is the correlation function corresponding to the preceding approximate expression for $X(t)$. Determine the magnitude of ε for optimal values of the constants.

46.18 To decrease the influence of the random vibration of the frame of a mirror-galvanometer used to measure a weak current, the readings are recorded during $T = 10$ sec. and the value j of the average recorded ordinate is considered to be the required intensity of the current. Find the mean error of the result if the vibration of the frame is described by the correlation function of the intensity of current $J(t)$:

$$K(\tau) = ae^{-\alpha|\tau|},$$

where

$$a = 10^{-16}A^2, \qquad \alpha = 10^{-1}\text{ sec.}^{-1}.$$

ANSWERS AND SOLUTIONS

I RANDOM EVENTS

1. RELATIONS AMONG RANDOM EVENTS

1.1 By definition $A \cup A = A$, $AA = A$. **1.2** The event A is a particular case of B. **1.3** $B = A_6$, $C = A_5$.

1.4 (a) A certain event U, (b) an impossible event V.

1.5 (a) At least one book is taken, (b) at least one volume from each of the three complete works is taken, (c) one book from the first work or three books from the second, or one from the first and three from the second, (d) two volumes from the first and second works are taken, (e) at least one volume from the third work and one volume from the first work and three from the second, or one from the second and three from the first.

1.6 The selected number ends with 5.

1.7 $\bar{A}$ means that all items are good, $\bar{B}$ means that one or none of them is defective.

1.8 Using the properties of events ($B \cup B = B$, $BB = B$, $B \cup \bar{B} = U$, $BU = B$, $B\bar{B} = V$, $B \cup V = B$), we get $A = BC$.

1.9 (a) A means reaching the interior of the region S_A, $\bar{A}$ means hitting the exterior of S_A. Then $A \cup B = U$; that is, $A = V$, $B = U$. (b) AB means reaching the region S_{AB} common to S_A and S_B; $\bar{A}$ means falling outside S_A. Then $AB = V$; that is, $A = U$, $B = V$. (c) AB means reaching the common region S_{AB}; $A \cup B$ means hitting $S_{A \cup B}$; $S_{AB} = S_{A \cup B}$ only if $S_A = S_B$; that is, $A = B$.

1.10 $X = \bar{B}$. **1.11** Use the equalities $\bar{A} = \bar{A}B \cup \bar{A}\bar{B}$, $\bar{B} = A\bar{B} \cup \bar{A}\bar{B}$.

1.12 The equivalence is shown by passing to the complementary events. The equalities are proved by passage from n to $n + 1$.

1.13 No, since $\overline{A \cup B} = \overline{AB}$. **1.14** Use the equality $\overline{A \cup B} = \overline{AB}$.

1.15 C means a tie. **1.16** $C = A(B_1 \cup B_2)$, $\bar{C} = A \cup \bar{B}_1\bar{B}_2$.

1.17 $D = A(B_1 \cup B_2 \cup B_3 \cup B_4)(C_1 \cup C_2)$, $\bar{D} = \bar{A} \cup \bar{B}_1\bar{B}_2\bar{B}_3\bar{B}_4 \cup \bar{C}_1\bar{C}_2$.

1.18 $C = (A_1 \cup A_2)(B_1B_2 \cup B_1B_3 \cup B_2B_3)$.

2. A DIRECT METHOD FOR EVALUATING PROBABILITIES

2.1 $p = rm/n$. **2.2** 4/9. **2.3** $p = 0.25$ since the first card may belong to any suit. **2.4** $1/6^5 \approx 0.00013$. **2.5** 23/240.

2.6 The succession of draws under such conditions is immaterial and therefore $p = 2/9$.

2.7 One may consider that for control the items are taken from the total lot; $p = (n - k)/(n + m - k)$.

2.8 One may consider one-digit numbers. (a) 0.2, (b) 0.4, (c) 0.04.

2.9 (a) $N = a + 10b$. This condition is satisfied only if a is even and $a + b$ is divisible by 9, $p = 1/18$, (b) $N = a + 10b + 100c$. This number should be divisible by 4 and by 9; that is, $a + b + c$ is divisible by 9, $a + 2b$ is divisible by 4 ($m = 22$), $p = 11/360$.

2.10 $\dfrac{10\cdot 9\cdot 8\cdot 7\cdot 6}{10^5 - 1} \approx 0.302.$ **2.11** $\dfrac{8\cdot 7!\cdot 3!}{10!} = \dfrac{1}{15}\cdot$ **2.12** $\dfrac{C_5^2}{C_8^2} = \dfrac{5}{14}\cdot$

2.13 0.3. **2.14** (a) $\dfrac{5}{9}$, (b) $\dfrac{2}{9}$, (c) $\dfrac{7}{9}\cdot$ **2.15** $p = \dfrac{C_n^s C_m^{k-s}}{C_{n+m}^k}\cdot$

2.16 $p_k = \dfrac{C_5^k}{C_{90}^k}$ $(k = 1, 2, 3, 4, 5)$, $p_1 = 0.0556$, $p_2 = 0.0025$, $p_3 = 0.85\cdot 10^{-4}$, $p_4 = 0.2\cdot 10^{-5}$, $p_5 = 0.2\cdot 10^{-7}$.

2.17 (a) $\dfrac{C_2^1 C_{2n-2}^{n-1}}{C_{2n}^n} = \dfrac{n}{2n-1}$, (b) $2\dfrac{C_2^2 C_{2n-2}^{n-2}}{C_{2n}^n} = \dfrac{n-1}{2n-1}\cdot$

2.18 $p = \dfrac{C_{n+k-m}^{n-m}}{C_{n+k}^n}\cdot$ **2.19** $p = \dfrac{C_4^1 C_4^1 C_4^1}{C_{52}^3} = 0.0029.$

2.20 $n = C_{36}^3 = 7140$. The favorable combinations: (a) (7, 7, 7); (b) (9, 9, 3), (9, 6, 6); (c) (2, 8, 11), (2, 9, 10), (3, 7, 11), (3, 8, 10), (4, 6, 11), (4, 7, 10), (4, 8, 9), (6, 7, 8) and, therefore, $m = 4 + 2\cdot 4\cdot C_4^2 + 4^3\cdot 8 = 564$; $p = 0.079$.

2.21 (a) $p = 1 - \dfrac{C_5^1 C_3^1 C_2^1}{C_{10}^3} = 0.75$, (b) $p = \dfrac{C_5^1 C_3^2 + C_5^2 C_2^1}{C_{10}^3} = \dfrac{7}{24}\cdot$

2.22 It is necessary to get $n - m$ nickels from $2n$ buyers. The number of possible cases is C_{2n}^{n-m}; $p = 1 - (N/C_{2n}^{n-m})$, where N is the number of cases when it is impossible to sell $2n$ tickets, $N = \sum_{i=1}^{n-m} N_i$, $N_1 = C_{2n-(2m+1)}^{n-m}$ is the number of cases in which the first nickel came from the $(2m + 2)$nd buyer, $N_2 = C_{2n-(2m+3)}^{n-m-1}$ is the number of cases in which the first nickel came not later than from the $(2m + 1)$st buyer, and the second nickel from the $(2m + 4)$th buyer and so on;

$$p = 1 - \frac{1}{C_{2n}^{n-m}} \sum_{i=1}^{n-m} C_{2i-1}^{i}.$$

3. GEOMETRIC PROBABILITIES

3.1 $p = 1 - \dfrac{l}{L}\cdot$ **3.2** $p = \dfrac{3}{9.5} \approx 0.316.$ **3.3** $p = 1 - \dfrac{\sqrt{3}}{2} \approx 0.134.$

3.4 Construction: AB is a segment of length $2h$, C is the center of the disk. AD and BE are tangents to the disk, located on one side of the line AC. The triangles ADC and BEC coincide by rotation with angle $\varphi = \angle DCE$; therefore, $\angle ACB = \varphi$, $h = l \tan(\varphi/2)$; $p = (1/\pi)\arctan(h/l)$.

3.5 $p = 1 - \left(1 - \dfrac{2r + d}{a}\right)\left(1 - \dfrac{2r + d}{b}\right)\cdot$

3.6 (a) 0.0185, (b) $p = \dfrac{160 + 25\pi}{1000\pi} = 0.076.$

3.7 (a) 0.16, (b) 0.6.

3.8 x is the distance from the shore to the boat and y (with the corresponding sign) from the boat to the course of the ship. Possible values: $x \leqslant 1\cdot v$; for $y < 0$,

$x + y \leqslant 1 \cdot v$, for $y < 0$ $|y| \leqslant x$ (v is the speed of the boat, $1 = 1$ hour). The favorable values: $|y| \leqslant (1/3)v$; $p = 5/9$.

3.9 $k(2 - k)$.

3.10 $x = AL$, $y = AM$. Possible values: $0 \leqslant x + y \leqslant l$. The favorable values: $|y - x| \leqslant x$, $p = 0.75$.

3.11 Two segments x, y. Possible values: $0 \leqslant x + y \leqslant l$. Favorable values: $x \leqslant l/2$, $y \leqslant l/2$, $x + y \geqslant l/2$; $p = 1/4$.

3.12 Two arcs x, y. Possible values: $0 \leqslant (x + y) \leqslant 2\pi R$. Favorable values: $x \leqslant \pi R$, $y \leqslant \pi R$, $x + y \geqslant \pi R$; $p = 1/4$.

3.13 Segments x, y, z. Possible values: $0 \leqslant (x, y, z) \leqslant l$. Favorable values: $x + y \geqslant z$, $x + z \geqslant y$, $y + z \geqslant x$; $p = 1/2$.

3.14 $AM = x$, $MN = y$. Possible values: $0 \leqslant x + y \leqslant l$. Favorable values: $x \leqslant a$, $y \leqslant a$, $x + y \geqslant l - a$. For $l/3 \leqslant a \leqslant l/2$, $p = [1 - (3a/l)]^2$; for $l/2 \leqslant a \leqslant l$, $p = 1 - 3[1 - (a/l)]^2$.

3.15 x is an arbitrary instant, $0 \leqslant x \leqslant 12$ minutes. The instants of arrival of a bus belonging to line A: $x = 0, 4, 8$; the instants of arrival of a bus of line B: y, $y + 6$, where $0 \leqslant y \leqslant 4$. (a) Favorable values: for $0 < y \leqslant 2$, we have $y < x \leqslant 4$, $6 + y \leqslant x \leqslant 12$; for $y > 2$, we have $y < x < 8$ or $y + 6 < x < 12$; $p = 2/3$. (b) Favorable values: $2 \leqslant x \leqslant 4$, $6 \leqslant x \leqslant 8$, $10 \leqslant x \leqslant 12$, $4 + y \leqslant x \leqslant 6 + y$; for $y < 2$ we have $0 < x \leqslant y$ and for $y > 2$, $y - 2 \leqslant x \leqslant y$; $p = 2/3$.

3.16 x, y are the times of arrival of the ships. Possible values: $0 \leqslant x \leqslant 24$, $0 \leqslant y \leqslant 24$. Favorable values: $y - x \leqslant 1$, $x - y \leqslant 2$; $p = 0.121$.

3.17 $p = 1 - \left(1 - \dfrac{t}{T}\right)^2.$

3.18 x is the distance from the shore to the first ship, and y the distance to the second ship. Possible values: $0 \leqslant (x, y) \leqslant L$. The favorable region $|x - y| \leqslant d\sqrt{1 + (v_2/v_1)^2}$ is obtained by passage to the relative motion (the first ship remains fixed and the second ship moves with speed $v = v_2 - v_1$); for $L \geqslant d\sqrt{1 + (v_2/v_1)^2}$, $p = 1 - [1 - (d/L)\sqrt{1 + (v_2/v_1)^2}]^2$; for $L \leqslant d\sqrt{1 + (v_2/v_1)^2}$, $p = 1$.

3.19 (a) $p = 1 - (19/20)^2 = 0.0975$, (b) x, y, z are the coordinates of the inflection points. Possible values: $0 \leqslant (x, y, z) \leqslant 200$. Favorable values: $|x - y| \leqslant 10$, $|x - z| \leqslant 10$, $|y - z| \leqslant 10$; $p = 1 - (180/200)^3 = 0.271$.

3.20 $p = \dfrac{2\pi R^2(1 - \cos\alpha)}{4\pi R^2} = \sin^2\dfrac{\alpha}{2}.$

3.21 $p = \left\{R^2 \displaystyle\int_0^{2\pi}\int_{\pi/6}^{\pi/3} \cos\varphi \, d\varphi \, d\psi\right\} : \left\{2R^2 \displaystyle\int_0^{2\pi}\int_0^{\pi/3} \cos\varphi \, d\varphi \, d\psi\right\} = 0.21.$

3.22 x is the distance from the midpoint of the needle to the nearest line and φ is the angle made by the line with the needle. Possible values: $0 \leqslant x \leqslant L/2$, $0 \leqslant \varphi \leqslant \pi$. Favorable values: $x \leqslant (l/2)\sin\varphi$, $p = 2l/L\pi$.

3.23 Possible values: $|a| \leqslant n$, $|b| \leqslant m$. (a) Favorable values: $b \leqslant a^2$.
For $m \geqslant n^2$,

$$p = \frac{1}{2} + \frac{1}{2nm}\int_0^n a^2\,da = \frac{1}{2} + \frac{n^2}{6m}.$$

For $m \leqslant n^2$,

$$p = 1 - \frac{1}{2nm}\int_0^m \sqrt{b}\,db = 1 - \frac{\sqrt{m}}{3n}.$$

The roots will be positive if $a \leqslant 0$, $b \geqslant 0$. For $m \geqslant n^2$, $p = n^2/12m$; for $m \leqslant n^2$, $p = 1/4 - \sqrt{m}/6n$. (b) The roots of the equation will be real if $b^2 + a^3 \leqslant 0$. The region for favorable values of the coefficients: $a \leqslant 0$, $b^2 \leqslant -a^3$.

For $n^3 \leqslant m^2$,

$$p = \frac{1}{2nm}\int_0^n a^{3/2}\,da = \frac{n^{3/2}}{5m}.$$

For $n^3 \geqslant m^2$,

$$p = \frac{1}{2} - \frac{1}{2nm}\int_0^m b^{2/3}\,db = \frac{1}{2}\left(1 - 0.6\,\frac{m^{2/3}}{n}\right).$$

3.24 Let A and B be the positions of the moving point and the center of the circle, u and v their velocity vectors and r the distance AB. From the point B we construct a circle of radius R. We consider that $\beta > 0$ if the vector v lies to the left of the line AB, $-\pi \leqslant \beta \leqslant \pi$. From the point A we construct tangents to the circle of radius R. The point A reaches the interior of the circle if the relative velocity vector falls into the resulting sector whose angle is 2ε, $\varepsilon = \arcsin(R/r)$. From A we construct the vector $-v$. Let O be the endpoint of this vector. From O we draw a circle whose radius coincides in magnitude with the velocity of the point A. The point A will lie in the circle only if the vector $u - v$ lies in the sector. Let $u > v$. Then the required probability will be (Figure 41) $p = \alpha/2\pi$. To determine α, we set $\delta = \angle OCA$, $x = \angle OCD$, $y = \angle ODC$, $\gamma = \angle ADO$. Then $\alpha = 2\varepsilon + \delta - \gamma$. Using the equalities

$$\frac{\sin\gamma}{v} = \frac{\sin(\beta - \varepsilon)}{u} \quad \text{and} \quad \frac{\sin\delta}{v} = \frac{\sin(\beta + \varepsilon)}{u},$$

we obtain

$$p = \frac{1}{2\pi}\left\{2\varepsilon + \arcsin\left[\frac{v}{u}\sin(\beta + \varepsilon)\right] - \arcsin\left[\frac{v}{u}\sin(\beta - \varepsilon)\right]\right\}.$$

The present formula is valid for any β. For $v > u$, the problem may be solved similarly, but in this case one should consider several cases: (1) $|\beta| \geqslant \varepsilon + (\pi/2)$, $p = 0$. (2) $(\pi/2) + \varepsilon \leqslant |\beta| \geqslant \varepsilon$: (a) for $u \leqslant v\sin(|\beta| - \varepsilon)$, we shall have $p = 0$, (b) for $v\sin(|\beta| - \varepsilon) \leqslant u \leqslant v\sin(|\beta| + \varepsilon)$, we have

$$p = \frac{1}{\pi}\arccos\left[\frac{v}{u}\sin(|\beta| - \varepsilon)\right],$$

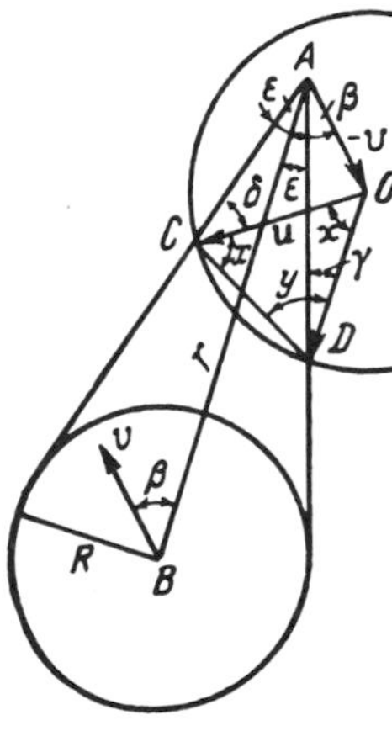

FIGURE 41

(c) for $u > v \sin(|\beta| + \varepsilon)$, we shall have

$$p = \frac{1}{\pi}\left\{\arccos\left[\frac{v}{u}\sin(|\beta| - \varepsilon)\right] - \arccos\left[\frac{v}{u}\sin(|\beta| + \varepsilon)\right]\right\}.$$

(3) $|\beta| \leqslant \varepsilon$: (a) for $u \leqslant v \sin(\varepsilon - |\beta|)$, we shall have $p = 1$, (b) for

$$v \sin(\varepsilon - |\beta|) \leqslant u \leqslant v \sin(\varepsilon + |\beta|),$$

we shall have

$$p = 1 - \frac{1}{\pi}\arccos\left[\frac{v}{u}\sin(\varepsilon - |\beta|)\right],$$

(c) for $u > v \sin(\varepsilon + |\beta|)$, we shall have

$$p = 1 - \frac{1}{\pi}\left\{\arccos\left[\frac{v}{u}\sin(\varepsilon - |\beta|)\right] + \arccos\left[\frac{v}{u}\sin(\varepsilon + |\beta|)\right]\right\}.$$

4. CONDITIONAL PROBABILITY. THE MULTIPLICATION THEOREM FOR PROBABILITIES

4.1 $p = 1 - 0.3\cdot 0.2 = 0.94.$

4.2 $p = 1 - \prod_{k=1}^{n}(1 - p_k).$

4.3 $p = (1 - 0.2)^3 = 0.512.$

4.4 $0.251.$

4.5 $p = 1 - (1 - 0.3)(1 - 0.2^2) = 0.328.$

4.6 $p(1 - p)^{n-1}.$

4.7 $1 - 0.5^n \geqslant 0.9;\ n \geqslant 4.$

4.8 $1 - (1 - p)^4 = 0.5,\ p \approx 0.159.$

4.9 $p = \left(\frac{S_\Delta}{\pi R^2}\right)^4 = \frac{729}{256\pi^4} = 0.029.$

4.10 $p = \left(1 - \frac{1}{2^2}\right)\left(1 - \frac{1}{3^2}\right)\left(1 - \frac{1}{5^2}\right)\left(1 - \frac{1}{7^2}\right)\left(1 - \frac{1}{11^2}\right)\cdots = \frac{6}{\pi^2} \approx 0.608.$[1]

4.11 From the incompatibility of the events it follows that $\mathbf{P}(A \mid B) = 0$ and $\mathbf{P}(B \mid A) = 0$; that is, the events are dependent.

4.12 $p_1 p_2.$

4.13 $p = 0.7\cdot 0.9^{12} = 0.197.$

4.14 $p = 0.7^2(1 - 0.6^2) = 0.314.$

4.15 $0.75.$

4.16 $p_1 = 0.9\cdot 0.8\cdot 0.7\cdot 0.9 \approx 0.45,\ p_2 = 0.7^2\cdot 0.8 \approx 0.39.$

4.17 (a) $0{,}1 = (p_1 p_3)^n$; that is, $n = -1/(\log p_1 \log p_3)$,
(b) $p = 1 - (1 - p_1 p_3)^3(1 - p_2 p_4)^3.$

4.18 It follows from the equality $\mathbf{P}(A)\mathbf{P}(B \mid A) = \mathbf{P}(B)\mathbf{P}(A \mid B).$

4.19 $p = 2\left(\frac{k}{n}\right)^2\left[1 - \left(\frac{k}{n}\right)^2\right].$

4.20 $p = \frac{1}{6}\cdot\frac{1}{3}\cdot\frac{2}{4}\cdot\frac{1}{3}\cdot\frac{1}{2}\cdot 1 = \frac{1}{360}.$

4.21 (a) $p = 1 - \frac{9}{10}\cdot\frac{8}{9}\cdot\frac{7}{8} = 0.3$, (b) $p = 1 - \frac{4}{5}\cdot\frac{3}{4}\cdot\frac{2}{3} = 0.6.$

4.22 $p = 1 - \frac{(n - m)!\,(n - k)!}{n!\,(n - m - k)!}.$

4.23 (a) $p = 1 - \frac{39{,}997!\ 39{,}000!}{40{,}000!\ 38{,}997!} \approx 1 - \left(\frac{39}{40}\right)^3 = 0.073,$

[1] For solution see Yaglom, A. M., and Yaglom, I. M.: Challenging Mathematical Problems with Elementary Solutions. San Francisco, Holden-Day, Inc., 1964. Problem 92, p. 29 and solution to Problem 92, pp. 202–209.

(b) $0.5 \geqslant \frac{(40{,}000 - N)(39{,}999 - N)(39{,}998 - N)}{40{,}000\cdot 39{,}999\cdot 39{,}998} \approx \left(\frac{40{,}000 - N}{40{,}000}\right)^3,$

$N \geqslant 8{,}252.$

4.24 (a) $p = 1 - \frac{(100{,}000 - 170)}{100{,}000}\cdot\frac{(100{,}000 - 2\cdot 170)}{(100{,}000 - 170)} \times \cdots \times \frac{(100{,}000 - 60\cdot 170 - 10\cdot 230)}{(100{,}000 - 59\cdot 170 - 10\cdot 230)}$

$= 1 - \frac{(100{,}000 - 60\cdot 170 - 10\cdot 230)}{100{,}000} = 0.125,$

(b) $p_{\text{sup}} = 1 - \frac{(100{,}000 - 5\cdot 170 - 230)}{(100{,}000 - 5\cdot 170)}\frac{(100{,}000 - 11\cdot 170 - 2\cdot 230)}{(100{,}000 - 11\cdot 170 - 230)} \times \cdots \times \frac{(100{,}000 - 59\cdot 170 - 10\cdot 230)}{(100{,}000 - 59\cdot 170 - 9\cdot 230)} \approx 0.0246,$

(c) $p = 1 - (1 - p_{\text{sup}})(1 - p_{\text{reg}}),\ p_{\text{reg}} = 1 - \frac{1 - p}{1 - p_{\text{sup}}} = 0.1029.$

4.25 $\mathbf{P}(A) = \mathbf{P}(B) = \mathbf{P}(C) = \frac{1}{2}\cdot$

$$\mathbf{P}(A \mid B) = \mathbf{P}(B \mid A) = \mathbf{P}(C \mid A) = \mathbf{P}(A \mid C) = \mathbf{P}(B \mid C) = \mathbf{P}(C \mid B) = \frac{1}{2},$$

that is, the events are pairwise independent;

$$\mathbf{P}(A \mid BC) = \mathbf{P}(B \mid AC) = \mathbf{P}(C \mid AB) = 1,$$

that is, the events are not independent in the set.

4.26 No (see for example, Problem 4.25). **4.27** $p = n!/n^n$.

4.28 $p = 2\frac{n}{2n}\cdot\frac{n}{(2n - 1)}\cdot\frac{(n - 1)}{(2n - 2)}\frac{(n - 1)}{(2n - 3)}\cdots\frac{1}{2}\cdot 1 = \frac{2(n!)^2}{(2n)!}\cdot$

4.29 $p = \frac{C_5^1C_{10}^2}{C_{15}^3}\frac{C_4^1C_8^2}{C_{12}^3}\frac{C_3^1C_6^2}{C_9^3}\frac{C_2^1C_4^2}{C_6^3}1 = \frac{3^5 5!\,10!}{15!} = 0.081.$

4.30 $p = \frac{C_n^1C_m^1}{C_{n+m}^2}\frac{C_{n-1}^1C_{m-1}^1}{C_{n+m-2}^2}\cdots\frac{1\cdot C_{m-(n-1)}^1}{C_{m-n+2}^2} = \frac{2^n n!\,m!}{(n + m)!}\cdot$

4.31 $p = \frac{1}{n}\cdot\frac{1}{(n - 1)}\cdots\frac{1}{[n - (k - 1)]} = \frac{(n - k)!}{n!}\cdot$

4.32 $p = \frac{1}{2}\cdot\frac{3}{4}\cdot\frac{5}{6}\cdots\frac{99}{100} = \frac{100!}{2^{100}(50!)^2} \approx 0.08.$

4.33 Let $a_1, a_2, \ldots, a_n$ be the buyers who have five-dollar bills and $b_1, b_2, \ldots, b_m$ those with ten-dollar bills, and suppose that their numbers coincide with their order in the line. The event A_k means that one will have to wait for change only because of buyer b_k $(k = 1, 2, \ldots, m)$;

$$p = \prod_{k=1}^{m} \mathbf{P}(\bar{A}_k) = \frac{n}{(n + 1)}\frac{(n - 1)}{n}\cdots\frac{(n - m + 1)}{(n - m + 2)} = \frac{n - m + 1}{n + 1}\cdot$$

4.34 It may be solved as one solves Problem 4.33;

$$\mathbf{P}(A_k) = \frac{2}{n - 2k + 3}, \qquad p = \prod_{k=1}^{m} \mathbf{P}(\bar{A}_k) = \frac{n - 2m + 1}{n + 1}\cdot$$

4.35 The first ballot drawn should be cast for the first candidate. The probability of this is $n/(n + m)$. Then the ballots must follow in succession so that the number of drawn votes cast for the first candidate is always not smaller than for the second one. The probability of this event is $(n - m)/n$ (see Problem 4.33);

$$p = \frac{n}{(n+m)}\frac{(n-m)}{n} = \frac{n-m}{n+m}.$$

5. THE ADDITION THEOREM FOR PROBABILITIES

5.1 0.03. **5.2** 0.55. **5.3** $pk = \sum_{j=1}^{n} pkj$. **5.4** $2(r/R)^2$. **5.5** 11/26.

5.6 $p = 1 - \frac{1}{C_{17}^{6}}(C_{10}^{6} + C_{10}^{5}C_{5}^{1} + C_{10}^{5}C_{2}^{1} + C_{10}^{4}C_{5}^{2} + C_{10}^{4}C_{5}^{1}C_{2}^{1} + C_{10}^{3}C_{5}^{3}) \approx 0.4.$

5.7 $\mathbf{P}(A\bar{B}) = \mathbf{P}(A) - \mathbf{P}(AB)$.

5.8 $\mathbf{P}(B) = \mathbf{P}(AB) + \mathbf{P}(\bar{A}B) = [\mathbf{P}(A) + \mathbf{P}(\bar{A})]\mathbf{P}(B \mid A) = \mathbf{P}(B \mid A)$.

5.9 $\mathbf{P}(B) = \mathbf{P}(A) + \mathbf{P}(B\bar{A}) \geqslant \mathbf{P}(A)$. **5.10** 0.323. **5.11** 0.5.

5.12 npq^{m-1}. **5.13** (a) 1/3, (b) 5/6.

5.14 A means that the first ticket has equal sums, B the second ticket. (a) $\mathbf{P}(A \cup B) = 2\mathbf{P}(A) = 0.1105$; (b) $\mathbf{P}(A \cup B) = 2\mathbf{P}(A) - \mathbf{P}^2(A) = 0.1075$.

5.15 From $\mathbf{P}(A \cup B) \leqslant 1$, it follows that $\mathbf{P}(B) - \mathbf{P}(AB) \leqslant \mathbf{P}(\bar{A})$ or

$$\mathbf{P}(A \mid B) \geqslant 1 - \frac{\mathbf{P}(\bar{A})}{\mathbf{P}(B)} = \frac{a+b-1}{b}.$$

5.16 From $Z = X \cup Y$, it follows that $Z \leqslant X + |Y|$, $Z \geqslant X - |Y|$, $\mathbf{P}(Z \leqslant 11) \geqslant \mathbf{P}(X \leqslant 10$ and $|Y| \leqslant 1) = \mathbf{P}(X \leqslant 10) + \mathbf{P}(|Y| \leqslant 1) - \mathbf{P}(X \leqslant 10$ or $|Y| \leqslant 1) \geqslant 0.9 + 0.95 - 1 = 0.85$, $\mathbf{P}(Z \geqslant 9) \geqslant 0.05$, $\mathbf{P}(Z \leqslant 9) \leqslant 0.95$.

5.17 0.44 and 0.35. **5.18** $p(2 - p)$.

5.19 $p_B = 0.1 + 0.9\cdot 0.8\cdot 0.3 = 0.316$; $p_C = 0.9(0.2 + 0.8\cdot 0.7\cdot 0.4) = 0.3816$.

5.20 $p = \frac{1}{n}\frac{1}{(n-1)} + \left(1 - \frac{1}{n}\right)\frac{1}{n} = \frac{n^2 - n + 1}{n^2(n-1)}.$

5.21 $p_B \approx 0.8$, $p_C \approx 0.2$.

5.22 $G(m + n) = G(m) + [1 - G(m)]G(n \mid m)$;

$$G(n \mid m) = \frac{G(n+m) - G(m)}{1 - G(m)}.$$

5.23 $p_1 = \frac{1}{2} + \frac{1}{2^3} + \frac{1}{2^5} + \cdots = \frac{2}{3}$, $p_2 = \frac{1}{2^2} + \frac{1}{2^4} + \cdots = \frac{1}{3}$.

Another solution: $p_1 + p_2 = 1$, $p_2 = (1/2)p_1$; that is, $p_1 = 2/3$, $p_2 = 1/3$.

5.24 $p_1 + p_2 + p_3 = 1$, $p_2 = \frac{1}{2}p_1$, $p_3 = \frac{1}{2}p_2$, i.e., $p_1 = \frac{4}{7}$, $p_2 = \frac{2}{7}$, $p_3 = \frac{1}{7}$.

5.25 $p + q = 1$, $q = \frac{1}{2}p$; $p = \frac{2}{3}$.

5.26 $p_1 + p_2 = 1$, $p_1 \frac{m}{n+m} = p_2$; $p = p_1 = \frac{n+m}{n+2m}$.

5.27 p_1 is the probability of hitting for the first marksman; p_2 is the probability of hitting for the second marksman; $p_1 + p_2 = 1, 0.2p_2 = 0.8\cdot 0.3p_1$; $p = p_1 = 0.455$.

5.28 Use the condition of Problem 1.12.

5.29 If we calculate the number of identical terms, we get

$$\mathbf{P}\left(\bigcup_{k=1}^{n} A_k\right) = C_n^1\mathbf{P}(A_1) - C_n^2\mathbf{P}(A_1A_2) + C_n^3\mathbf{P}(A_1A_2A_3) - \cdots + (-1)^{n-1}\mathbf{P}\left(\prod_{k=1}^{n} A_k\right).$$

5.30 Using the equality $\prod_{k=1}^{n} A_k = \overline{\sum_{k=1}^{n} \bar{A}_k}$ from Problem 1.12 and the general formula for the probability of a sum of events, we obtain

$$\mathbf{P}\left(\prod_{k=1}^{n} A_k\right) = 1 - \left\{\sum_{k=1}^{n} \mathbf{P}(\bar{A}_k) - \sum_{k=1}^{n-1} \sum_{j=k+1}^{n} \mathbf{P}(\bar{A}_k \bar{A}_j) + \sum_{k=1}^{n-2} \sum_{j=k+1}^{n-1} \sum_{i=j+1}^{n} \mathbf{P}(\bar{A}_k \bar{A}_j \bar{A}_i) - \cdots + (-1)^{n-1} \mathbf{P}\left(\prod_{k=1}^{n} \bar{A}_k\right)\right\}.$$

However, according to Problem 1.12 we have $\prod_{k=1}^{s} \bar{A}_k = \overline{\sum_{k=1}^{s} A_k}$ and, hence, for any s, $\mathbf{P}(\prod_{k=1}^{s} \bar{A}_k) = 1 - \mathbf{P}(\sum_{k=1}^{s} A_k)$. Also considering the equality

$$1 - C_n^1 + C_n^2 - \cdots + (-1)^n = 0,$$

we get the formula indicated in the assumption of the problem.

5.31 Use the equality

$$\mathbf{P}\left(\bar{A}_0 \prod_{k=1}^{n} A_k\right) = \mathbf{P}\left(\prod_{k=1}^{n} A_k\right) - \mathbf{P}\left(\prod_{k=0}^{n} A_k\right)$$

and the formula from the condition of Problem 5.30.

5.32 $p = \sum_{k=1}^{n} \frac{(-1)^{k-1}}{k!}$.

5.33 The probability that m persons out of n will occupy their seats is $C_n^m(n-m)/n! = 1/m!$. The probability that the remaining $n - m$ persons will not sit in their seats is

$$\sum_{k=0}^{n=m} \frac{(-1)^k}{k!}; \qquad p = \frac{1}{m} \sum_{k=0}^{n-m} \frac{(-1)^k}{k!}.$$

5.34 The event A_j means that no passenger will enter the jth car,

$$\mathbf{P}(A_j) = \left(1 - \frac{1}{n}\right)^k, \qquad \mathbf{P}(A_j A_i) = \left(1 - \frac{2}{n}\right)^k, \qquad \mathbf{P}(A_j A_i A_s) = \left(1 - \frac{3}{n}\right)^k$$

and so on. Using the formula from the answer to Problem 5.29, we obtain

$$p = 1 - C_n^1\left(1 - \frac{1}{n}\right)^k + C_n^2\left(1 - \frac{2}{n}\right)^k - \cdots + (-1)^{n-1} C_n^{n-1}\left(1 - \frac{n-1}{n}\right)^k.$$

5.35 The first player wins in the following n cases: (1) in m games he loses no game, (2) in m games he loses one but wins the $(m + 1)$st game, (3) in $m + 1$ games he loses two, but wins the $(m + 2)$nd game, ..., (n) in $m + n - 2$ games he loses $n - 1$ and, then, he wins the $(m + n - 1)$st game.

$$P = p^m(1 + C_m^1 q + C_{m+1}^2 q^2 + \cdots + C_{m+n-2}^{n-1} q^{n-1}).$$

5.36 The stack is divided in the ratio p_1/p_2 of probabilities of winning for the first and second players,

$$p_1 = \frac{1}{2^m}\left(1 + \frac{1}{2} C_m^1 + \frac{1}{2^2} C_{m+1}^2 + \cdots + \frac{1}{2^{n-1}} C_{m+n-2}^{n-1}\right),$$

$$p_2 = \frac{1}{2^n}\left(1 + \frac{1}{2} C_n^1 + \frac{1}{2^2} C_{n+1}^2 + \cdots + \frac{1}{2^{m-1}} C_{m+n-2}^{m-1}\right).$$

5.37 The event A means that the first told the truth, B means that the fourth told the truth;

$$p = \mathbf{P}(A \mid B) = \frac{\mathbf{P}(A)\mathbf{P}(B \mid A)}{\mathbf{P}(B)}.$$

Let p_k be the probability that (in view of double distortions) the kth liar transmitted the correct information; $p_1 = 1/3$, $p_2 = 5/9$, $p_3 = 13/27$, $p_4 = 41/81$, $\mathbf{P}(A) = p_1$, $\mathbf{P}(B \mid A) = p_3$, $\mathbf{P}(B) = p_4$; $p = 13/41$.

5.38 We replace the convex contour by a polygon with n sides. The event A means that line A_{ij} will be crossed by the ith and jth sides; $A = \sum_{i=1}^{n} \sum_{d=i+1}^{n} A_{ij}$, $p' = \sum_{i=1}^{n} \sum_{j=i+1}^{n} p_{ij}$, where $p_{ij} = \mathbf{P}(A_{ij})$; $p' = (1/2) \sum_{k=1}^{n} p_k^*$, $p_k^* = \sum_{i=1}^{n} p_{ki} - p_{kk}$ being the probability that the parallel lines are crossed by the kth side of length l_k. From the solution of Buffon's problem 3.22, it follows that $p_k^* = 2l_k/L\pi$; $p' = (1/L\pi) \sum_{k=1}^{n} l_k$. Since this probability is independent of the number and size of the sides, we have $p = s/L\pi$.

6. THE TOTAL PROBABILITY FORMULA

6.1 $p = \frac{11}{12}\cdot\frac{1}{11} + \frac{1}{12}\cdot\frac{2}{11} = \frac{13}{132}$. **6.2** $p = \frac{3}{4}\cdot\frac{4}{9} + \frac{1}{4}\cdot\frac{2}{9} = \frac{7}{18}$.

6.3 H_1 means that among the balls drawn there are no white balls, H_2 means that one ball is white and H_3 that both are white;

$$p = \frac{1}{2}\left(\frac{m_1}{n_1 + m_1} + \frac{m_2}{n_2 + m_2}\right).$$

6.4 H_{j1} means that a white ball is drawn from the jth urn;

$$\mathbf{P}(H_{11}) = \frac{m}{m + k},$$

$$\mathbf{P}(H_{12}) = \frac{k}{m + k},$$

$$\mathbf{P}(H_{21}) = \frac{m}{(m + k)}\frac{(m + 1)}{(m + k + 1)} + \frac{k}{(m + k)}\frac{m}{(m + k + 1)} = \frac{m}{m + k},$$

$$\mathbf{P}(H_{22}) = \frac{k}{m + k}.$$

Consider

$$\mathbf{P}(H_{j1}) = \frac{m}{m + k}, \qquad \mathbf{P}(H_{j2}) = \frac{k}{m + k}.$$

Then $\mathbf{P}(H_{j+1,1}) = m/(m + k)$. Therefore $p = m/(m + k)$.

6.5 0.7. **6.6** 2/9. **6.7** 0.225. **6.8** 0.75. **6.9** 0.332.

6.10 The event A means getting a contact. The hypothesis H_k means that a contact is possible on the kth band ($k = 1, 2$). Let x be the position of the center of the hole and y the point of application of the contact.

$$\mathbf{P}(H_1) = \mathbf{P}(15 \leqslant x \leqslant 45) = 0.3, \qquad \mathbf{P}(H_2) = \mathbf{P}(60 \leqslant x \leqslant 95) = 0.35.$$

The contact is possible on the first band if for $25 \leqslant x \leqslant 35 |x - y| \leqslant 5$, for $15 \leqslant x \leqslant 25$, $20 \leqslant y \leqslant x + 5$, for $35 \leqslant x \leqslant 45 x - 5 \leqslant y \leqslant 45$. Thus $\mathbf{P}(A \mid H_1) = 1/15$. Similarly, $\mathbf{P}(A \mid H_2) = 1/14$, $p = 0.045$.

6.11 The event A means that s calls come during the time interval $2t$. The hypothesis H_k ($k = 0, 1, \ldots, s$) means that during the first interval k calls came, $\mathbf{P}(H_k) = \mathbf{P}_t(k)$. The probability that $s - k$ calls come during the second interval will be

$$\mathbf{P}(A \mid H_k) = P_t(s - k), \qquad P_{2t}(s) = \sum_{k=0}^{s} P_t(k)P_t(s - k).$$

6.12 The hypothesis H_k means that there are k defective bulbs, $\mathbf{P}(H_k) = 1/6$ ($k = 0, 1, \ldots, 5$). The event A means that all 100 bulbs are good,

$$\mathbf{P}(A \mid H_k) = \frac{C_{1000-k}^{100}}{C_{1000}^{100}} \approx 0.9^k \qquad (k = 0, 1, \ldots, 5);$$

$$p = \frac{1}{6}\sum_{k=0}^{5} \mathbf{P}(A \mid H_k) \approx 0.78.$$

6.13 The hypothesis H_k means that there are k white balls in the urn ($k = 0, 1, \ldots, n$); the event A means that a white ball will be drawn from the urn,

$$\mathbf{P}(H_k) = \frac{1}{n+1}, \quad \mathbf{P}(A \mid H_k) = \frac{k+1}{n+1}; \qquad p = \frac{n+2}{2(n+1)}.$$

6.14 The hypothesis H_k ($k = 0, 1, 2, 3$) means that k new balls are taken for the first game. The event A means that three new balls are taken for the second game,

$$\mathbf{P}(H_k) = \frac{C_9^k C_6^{3-k}}{C_{15}^3}, \quad \mathbf{P}(A \mid H_k) = \frac{C_{9-k}^3}{C_{15}^3}; \qquad p = 0.089.$$

6.15 $p = \dfrac{1}{14C_7^5}(9C_4^2 + 8C_3^2C_4^3 + 7C_3^1) = 0.58.$

6.16 $p = \dfrac{25}{30}\cdot\dfrac{24}{29} + \left(\dfrac{25}{30}\cdot\dfrac{5}{29} + \dfrac{5}{30}\cdot\dfrac{25}{29}\right)\cdot\dfrac{24}{28} = \dfrac{190}{203}.$

6.17 $\mathbf{P}(A) = \mathbf{P}(AB) + \mathbf{P}(A\bar{B}) = \mathbf{P}(B)\mathbf{P}(A \mid B) + \mathbf{P}(\bar{B})\mathbf{P}(A \mid \bar{B})$.
The equality is valid only in several particular cases: (a) $A = V$, (b) $B = U$ (c) $B = A$, (d) $B = \bar{A}$, (e) $B = V$, where U denotes a certain event and V an impossible one.

6.18 By the formula from Example 6.2, it follows that $m \approx 13$, $p \approx 0.67$.

6.19 In the first region there are eight helicopters, $p \approx 0.74$.

7. COMPUTATION OF THE PROBABILITIES OF HYPOTHESES AFTER A TRIAL (BAYES' FORMULA)

7.1 $p = \dfrac{0.1\cdot 5/6}{0.9\cdot 1/2 + 0.1\cdot 5/6} = \dfrac{5}{32}.$ **7.2** $p = 1 : \left[1 + \dfrac{k_2m_2(m_1 + n_1)}{k_1m_1(m_2 + n_2)}\right].$

7.3 The hypothesis H_1 means that the item is a standard one and H_2 that it is nonstandard. The event A means that the item is found to be good;

$$\mathbf{P}(H_1) = 0.96, \quad \mathbf{P}(H_2) = 0.04, \quad \mathbf{P}(A \mid H_1) = 0.98, \quad \mathbf{P}(A \mid H_2) = 0.05,$$
$$\mathbf{P}(A) = 0.9428; \qquad p = \mathbf{P}(H_1 \mid A) = 0.998.$$

7.4 The hypotheses H_k ($k = 0, 1, \ldots, 5$) means that there are k defective items. The event A means that one defective item is drawn;

$$\mathbf{P}(H_k) = \frac{1}{6}, \qquad \mathbf{P}(A \mid H_k) = \frac{k}{5}, \qquad \mathbf{P}(H_k \mid A) = \frac{\mathbf{P}(H_k)\mathbf{P}(A \mid H_k)}{\mathbf{P}(A)}.$$

The most probable hypothesis is H_5; that is, there are five defective items.

7.5 $\mathbf{P}(H_0 \mid A) = \dfrac{1}{6\cdot 0.78} = 0.214$ (see Problem 6.12).

7.6 The event A denotes the win of player D; the hypothesis H_k ($k = 1, 2$) means that the opponent was player B or C;

$$\mathbf{P}(H_k) = \frac{1}{2}; \qquad \mathbf{P}(A \mid H_1) = 0.6 \times 0.3 + (1 - 0.18) \times 0.7 \times 0.5;$$
$$\mathbf{P}(A \mid H_2) = 0.2 \times 0.3 + (1 - 0.06)0.4 \times 0.7; \qquad \mathbf{P}(H_1 \mid A) = 0.59;$$
$$\mathbf{P}(H_2 \mid A) = 0.41.$$

7.7 The second group.

7.8 The event A means that two marksmen score a hit, H_k means that the kth marksman fails;

$$p = \mathbf{P}(H_3 \mid A) = \frac{6}{13}.$$

7.9 The event A means that the boar is killed by the second bullet;

$$\mathbf{P}(A) = \sum_{k=1}^{3} \mathbf{P}(H_k).$$

The hypothesis H_k means that the kth marksman hit ($k = 1, 2, 3$);

$$\mathbf{P}(H_1) = 0.048, \quad \mathbf{P}(H_2) = 0.128, \quad \mathbf{P}(H_3) = 0.288,$$
$$\mathbf{P}(H_1 \mid A) = 0.103, \quad \mathbf{P}(H_2 \mid A) = 0.277, \quad \mathbf{P}(H_3 \mid A) = 0.620.$$

7.10 The fourth part.

7.11 $p = n^k/(1 + 2^k + \cdots + n^k)$.

7.12 The events are: M_1 that the first twin is a boy, M_2 that the second is also a boy. The hypotheses are: H_1 that both are boys, H_2 that there are a boy and a girl;

$$\mathbf{P}(M_1) = a + \frac{1}{2}[1 - (a + b)]; \qquad p = \mathbf{P}(M_2 \mid M_1) = \frac{2a}{1 + a - b}.$$

7.13 A_k means that the kth child born is a boy and B_k that it is a girl ($k = 1, 2$); $\mathbf{P}(A_1A_2) + \mathbf{P}(B_1B_2) + 2\mathbf{P}(A_1B_2) = 1$, $\mathbf{P}(A_1A_2 + B_1B_2) = 4\mathbf{P}(A_1B_2)$. Therefore,

$$\mathbf{P}(A_1A_2) + \mathbf{P}(B_1B_2) = \frac{2}{3}, \quad \mathbf{P}(A_1B_2) = \frac{1}{6}, \quad \mathbf{P}(A_1A_2) = 0.51 - \frac{1}{6};$$

$$p = \mathbf{P}(A_2 \mid A_1) = \frac{103}{153}.$$

7.14 5/11.

7.15 One occurrence.

7.16 Hypothesis H_1 means that the first student is a junior and H_2 means that he is a sophomore. A denotes the event that the second student has been studying for more time than the first, B means that the second student is in the third year.

$$\mathbf{P}(H_1) = \frac{n_1}{n-1}, \quad \mathbf{P}(H_2) = \frac{n_2}{n-1}, \quad \mathbf{P}(A \mid H_1) = \frac{n_2 + n_3}{n-1}, \quad \mathbf{P}(A \mid H_2) = \frac{n_3}{n-1},$$

$$\mathbf{P}(A) = \frac{1}{(n-1)^2}[n_1(n_2 + n_3) + n_2n_3], \quad \mathbf{P}(AB) = \frac{n_3}{n-1};$$

$$p = \mathbf{P}(B \mid A) = \frac{\dfrac{1}{n_1} + \dfrac{1}{n_2}}{\dfrac{1}{n_1} + \dfrac{1}{n_2} + \dfrac{1}{n_3}}.$$

7.17 1/4 and 2/11.

7.18 The hypotheses H_k ($k = 0, 1, \ldots, 8$) mean that eight out of k items are nondefective. A denotes the event that three out of four selected items are nondefective:

$$\mathbf{P}(H_k) = \frac{1}{9}, \quad \mathbf{P}(H_j \mid A) = 0 \; (j = 0, 1, 2, 8),$$

$$\mathbf{P}(H_k \mid A) = \frac{C_k^3 C_{8-k}^1}{C_8^4} \; (k = 3, 4, 5, 6, 7), \quad \mathbf{P}(A) = \frac{1}{5};$$

$$p = \mathbf{P}(H_4 \mid A)\cdot\frac{3}{4} + \mathbf{P}(H_5 \mid A)\cdot\frac{1}{2} = \frac{3}{14}.$$

8. EVALUATION OF PROBABILITIES OF OCCURRENCE OF AN EVENT IN REPEATED INDEPENDENT TRIALS

8.1 (a) $0.9^4 = 0.656$, (b) $0.9^4 + 4\cdot 0.1\cdot 0.9^3 = 0.948$.

8.2 (a) $C_{10}^5 \dfrac{1}{2^{10}} = \dfrac{63}{256}$, (b) $1 - \dfrac{1}{2^{10}}(1 + C_{10}^1 + C_{10}^2 + C_{10}^1 + 1) = \dfrac{957}{1024}$.

8.3 (a) $p = C_{200}^{3}\cdot 0.01^{3}\cdot 0.99^{197} \approx 1.35e^{-2} = 0.18$, (b) $p \approx 0.09$.

8.4 0.17. **8.5** 0.64. **8.6** (a) 0.163, (b) 0.353.

8.7 $p = 1 - (0.8^4 + 4\cdot 0.8^3\cdot 0.2 + 5\cdot 0.8^2\cdot 0.2^2 + 2\cdot 0.8\cdot 0.2^3)\,0.7^2\cdot 0.6 = 0.718.$

8.8 $W_n = \sum_{m=0}^{n} C_n^m p^m q^{n-m}\left[1 - \left(1 - \frac{1}{\omega}\right)^m\right] = 1 - \left(1 - \frac{p}{\omega}\right)^n.$

8.9 $p = 1 - (0.7^4 + 4\cdot 0.7^3\cdot 0.3\cdot 0.4) = 0.595.$

8.10 Hypothesis H_1 means the probability of hitting in one shot is 1/2, H_2 means that this probability is 2/3. The event A means that 116 hits occurred. $\mathbf{P}(H_1 \mid A) \approx 2\mathbf{P}(H_2 \mid A)$; that is, the first hypothesis is more probable.

8.11 See Table 113.

TABLE 113

p	0.01	0.05	0.1	0.2	0.3	0.4	0.5	0.6
$R_{10;\,1}$	0.0956	0.4013	0.6513	0.8926	0.9718	0.9940	0.9990	0.9999

8.12 0.2. **8.13** 0.73.

8.14 $R_{n;\,1} \approx 1 - e^{-0.02n}$ $(n > 10)$. See Table 114.

TABLE 114

n	1	10	20	30	40	50	60	70	80	90	100
$R_{n;\,1}$	0.02	0.18	0.33	0.45	0.55	0.63	0.70	0.75	0.80	0.84	0.86

8.15 $p = 1 - 0.95^{10} = 0.4.$ **8.16** $p = 1 - 0.9^5 = 0.41.$

8.17 $p = p_{10}^3 + 3p_{10}^2(p_9 + p_8) + 3p_{10}p_9^2 = 0.0935.$

8.18 (a) $p = \sum_{k=0} P'_{3;\,k}P''_{3;\,k} = 0.311$, (b) 0.243. **8.19** 0.488.

8.20 A denotes the event that two good items are produced. The hypothesis H_k means that the kth worker produces the items $(k = 1, 2, 3)$;

$$p = \sum_{k=1}^{3} \mathbf{P}(H_k \mid A) \times \mathbf{P}(A \mid H_k) \approx 0.22.$$

8.21 (a) $p = \frac{1}{\sqrt[3]{2}} = 0.794$, (b) $3p^4 - 4p^3 + \frac{1}{2} = 0$, $p = 0.614$.

8.22 $P_{\mathrm{I}} = p^4 + C_4^1 p^4 q + C_5^2 p^4 q^2 + C_6^3 p^3 q^3 (p^2 + 2p^2 q) = 0.723$; $P_{\mathrm{II}} = 0.277.$

8.23 $p = \frac{1}{2^{2n-k}} C_{2n-k}^{n}.$ **8.24** 0.784.

8.25 The 200 w. ones ($R_{6,1} = 0.394$; $R_{10,2} = 0.117$).

8.26 0.64. **8.27** 0.2816.

8.28 $P_m = nC_{m-1}^{k-1}p^k q^{m-k}$ for $m \geqslant k$; $P_m = 0$ for $m < k$.

8.29 $p = \sum_{m=k}^{2k-1} P_m = np^k \sum_{m=k}^{2k-1} C_{m-1}^{k-1} q^{m-k}.$

8.30 We require:

$$0.1 \geqslant 0.8^n\left[1 + \frac{n}{4} + \frac{n(n-1)}{32}\right]; \qquad n \geqslant 25.$$

8.31 We require: $0.99 \cdot 5^{10} = 4^{10} + C_{10}^1 4^9 + \cdots + C_{10}^n 4^{10-n}$; $n = 5$.

8.32 $P_{4,0} = 0.3024$, $P_{4,1} = 0.4404$, $P_{4,2} = 0.2144$, $P_{4,3} = 0.0404$, $P_{4,4} = 0.0024$.

8.33 0.26. **8.34** 0.159. **8.35** 95/144. **8.36** $n = 29$.

8.37 $n \geqslant 10$. **8.38** $n \geqslant 16$. **8.39** 8. **8.40** 8.

8.41 $\mu = 4$; $p = 0.251$. **8.42** $\mu_+ = 3$, $\mu_- = 1$; $p = 32/81$.

9. THE MULTINOMIAL DISTRIBUTION. RECURSION FORMULAS. GENERATING FUNCTIONS

9.1 $p = P_{5;\,2,2,1} + 2P_{5;\,3,2,0} = 50/243$.

9.2 $p = P_{3;\,1,1,1} + P_{3;\,2,1\,0} + P_{3;\,1,2,0} = 0.245$.

9.3 (a) $p = \dfrac{9!}{(3!)^3}\cdot\dfrac{1}{3^9} = 0.085$, (b) $p = 6\,\dfrac{9!}{4!\,3!\,2!} \times \dfrac{1}{3^9} = 0.385$.

9.4 $p = \dfrac{10!}{6!\,3!}\,0.15^6 0.22^3 \cdot 0.13 = 0.13 \cdot 10^{-4}$.

9.5
$$p = 1 - 2(0.0664^4 + \tfrac{1}{2}\,0.2561^4 + 4\cdot 0.0664\cdot 0.2561^3$$
$$+ 6\cdot 0.0664^2\cdot 0.2561^2 + 4\cdot 0.2561\cdot 0.0664^3) = 0.983.$$

9.6 (a) $p = \dfrac{12!}{2^6\cdot 6^{12}} = 0.00344$, (b) $p = \dfrac{6!}{2}\cdot\dfrac{12!}{2!\,2!\,3!\,4!}\cdot\dfrac{1}{6^{12}} = 0.138$.

9.7 (a) $p_{\mathrm{I}} = \dfrac{l^{l_1} m^{m_1} n^{n_1}}{(l + m + n)^{l_1 + m_1 + n_1}}$, (b) $p = 6p_1$,

(c) $p = \dfrac{(l_1 + m_1 + n_1)!}{l_1!\,m_1!\,n_1!}\cdot\dfrac{l^{l_1} m^{m_1} n^{n_1}}{(l + m + n)^{l_1 + m_1 + n_1}}$.

9.8 $p = p_n$, $p_k = p_{k-1}\cdot\dfrac{1}{2} + (1 - p_{k-1})\,\dfrac{1}{2} = 0.5$; $p = 0.5$.

9.9 Let p_k be the probability of a tie when $2k$ resulting games have been played; $p_{k+1} = (1/2)p_k$ $(k = 0, 1, \ldots)$, $p_0 = 1$, $p_{n-1} = (1/2)^{n-1}$; $p = (1/2)p_{n-1} = 1/2^n$.

9.10 The number n should be odd. Let p_k be the probability that after $2k + 1$ games the play is not terminated; $p_0 = 1$,

$$p_k = \left(\frac{3}{4}\right)^k \left(k = 1, 2, \ldots, \frac{n-3}{2}\right); \qquad p = \frac{1}{4}\,p_{(n-3)/2} = \frac{1}{4}\left(\frac{3}{4}\right)^{(n-3)/2}.$$

9.11 Let p_k be the probability of ruin of the first player when he has k dollars. According to the formula of total probability $p_k = pp_{k+1} + qp_{k-1}$. Moreover, $p + q = 1$, $p_0 = 1$, $p_{n+m} = 0$. Consequently, $q(p_k - p_{k-1}) = p(p_{k+1} - p_k$. (1) $p = q$. Then $p_k = 1 - kc$, $c = 1/(n + m)$; that is, $p_{\mathrm{I}} = m/(n + m)$, $p_{\mathrm{II}} = n/(n + m)$. (2) $p \neq q$. Then $p_k - p_{k-1} = (p/q)^k(p_1 - 1)$. Summing these equalities from 1 to n and from 1 to $n + m$, we obtain

$$1 - p_n = (1 - p_{\mathrm{I}})\,\frac{1 - \left(\frac{q}{p}\right)^n}{1 - \frac{q}{p}}, \qquad 1 - p_{n+m} = (1 - p_{\mathrm{I}})\,\frac{1 - \left(\frac{q}{p}\right)^{n+m}}{1 - \frac{q}{p}}.$$

Thus,

$$p_{\text{I}} = \frac{1 - \left(\frac{p}{q}\right)^m}{1 - \left(\frac{p}{q}\right)^{n+m}}, \qquad p_{\text{II}} = 1 - p_{\text{I}} = \frac{1 - \left(\frac{q}{p}\right)^n}{1 - \left(\frac{q}{p}\right)^{n+m}}.$$

9.12 $P = P_m$; $P_m = 0$ for $m \leqslant n$; $P_n = 1/2^{n-1}$; $P_m = 1/2^n$ for $n < m < 2n - 1$. In the general case P_m is determined from the recurrent formula

$$P_m = \frac{1}{2} P_{m-1} + \frac{1}{2^2} P_{m-2} + \cdots + \frac{1}{2^{n-1}} P_{m-n+1},$$

which is obtained by the formula of total probability. In this case, the hypothesis H_k means that the first opponent of the winner wins k games;

$$P_{m-k} = P(H_k)\left(\frac{1}{2}\right)^{n-k} \qquad (k = 1, 2, \ldots, n-1).$$

9.13 P_k is the probability that exactly k games are necessary. For $k = 1, 2, 3, 4, 5$, $P_k = 0$, $P_6 = 2p^6 = 1/2^5$, $P_7 = 2C_6^1 p^6 q = 3/2^5$, $P_8 = 2C_7^2 p^6 q^2 = 21/2^7$, $P_9 = 7/2^5$, $P_{10} = 63/2^9$; (a) $R = \sum_{k=1}^{10} P_k = 193/256$, (b) if n is odd, then $P_n = 0$. For even n, $P_n = (1/2)p_{(n-1)/2}$, where p_k is the probability that after $2k$ games the opponents have equal numbers of points; $p_5 = C_{10}^5(1/2^{10} = 63/2^8$, $p_{k+1} = (1/2)p_k$; that is,

$$p_k = \frac{63}{2^{k+3}} \quad (k = 5, 6, \ldots), \qquad P_n = \frac{63}{2^{(n/2)+3}}.$$

9.14 Expand $(1 - u)^{-1}$ into a series and find the coefficient of u^m.

9.15 The same as in Problem 9.14.

9.16 The required probability is the constant term in the expansion of generating function

$$G(u) = \frac{1}{4^n}\left(u + 2 + \frac{1}{u}\right)^n = \frac{(1+u)^{2n}}{4^n u^n}; \qquad p = \frac{1}{4^n} C_{2n}^n.$$

9.17 The required probability is the sum of the coefficients of u raised to powers not less than m in the expansion of the function

$$G(u) = \left(\frac{1}{16}u^2 + \frac{1}{4}u + \frac{3}{8} + \frac{1}{4u} + \frac{1}{16u^2}\right)^n = \frac{(1+u)^{4n}}{(4u)^{2n}};$$

$$p = \frac{1}{4^{2n}} \sum_{k=2n+m}^{4n} C_{4n}^k.$$

For $n = m = 3$, $p = 0.073$.

9.18 The required probability is twice the sum of the coefficients of u^4 in the expansion of the function

$$G(u) = \frac{1}{5^{20}}\left(u + \frac{1}{u} + 3\right)^{20} = \frac{1}{5^{20}} \sum_{m=0}^{20} \sum_{n=0}^{20-m} \frac{20!}{m!\, n!\, (20 - m - n)!} u^{m-n} 3^{20-m-n};$$

$$p = 2\frac{20!}{5^{20}} \sum_{k=0}^{8} \frac{3^{16-2k}}{(4+k)!\, k!\, (16-2k)!} = 0.104.$$

9.19 (a) The required probability p_{champ} is the sum of the coefficients of non-negative powers of u in the expansion of the function

$$G(u) = \left(\frac{1}{4}u + \frac{1}{4u} + \frac{1}{2}\right)^{24} = \frac{(1+u)^{48}}{4^{24}u^{24}};$$

$$p_{\text{champ}} = \frac{1}{4^{24}} \sum_{k=24}^{48} C_{48}^k = \frac{1}{2 \cdot 4^{24}}(2^{48} + C_{48}^{24}) = 0.5577, \qquad p_{\text{aspirant}} = 0.4423.$$

(b) the probability of the complementary event is the sum of the coefficients of u whose powers range from -4 to 3 in the expansion of the function

$$G(u) = \frac{1}{4^{20}} \frac{(1+u)^{40}}{u^{20}}; \qquad p = 1 - \frac{1}{4^{20}} \sum_{k=16}^{23} C_{40}^{k} = 0.22.$$

9.20 (a) The required probability P_m is found with the aid of the generating function

$$G(u) = \frac{1}{6^n}(u + u^2 + \cdots + u^6) = \frac{u^n(1-u^6)^n}{6^n(1-u)^n}.$$

Using the equality $1/(1-u)^n = 1 + C_n^{n-1}u + C_{n+1}^{n-1}u^2 + \cdots$ we obtain

$$P_m = \frac{1}{6^n}(C_{m-1}^{n-1} - C_n^1 C_{m-n}^{n-1} + C_n^2 C_{m-13}^{n-1} - \cdots),$$

and the series is cut off if $m - 6k < n$; (b) $R_m = \sum_{k=n}^{m} P_k$. Using the equality $1 + C_n^{n-1} + \cdots + C_{s-1}^{n-1} = C_s^n$, we obtain

$$R_m = \frac{1}{6^n}(C_m^n - C_n^1 C_{m-6}^n + C_n^2 C_{m-12}^n - \cdots).$$

For $n = 10$, $m = 20$,

$$P_{20} = \frac{1}{6^{10}}(C_{19}^9 - C_{10}^1 C_{13}^9) = 0.0014, \qquad R_{20} = \frac{1}{6^{10}}(C_{20}^{10} - C_{10}^1 C_{14}^{10}) = 0.0029.$$

9.21 The desired probability is the coefficient of u^{21} in the expansion of the function

$$G(u) = \frac{1}{10^6}(1 + u + \cdots + u^9)^6 = \frac{1}{10^6}\left(\frac{1-u^{10}}{1-u}\right)^6$$

$$= \frac{1}{10^6}(1 - C_6^1 u^{10} + C_6^2 u^{20} - \cdots)(1 + C_6^5 u + C_7^5 u^2 + \cdots);$$

$$p = \frac{1}{10^6}(C_{26}^5 - C_6^1 C_{16}^5 + C_6^2 C_6^5) = 0.04.$$

9.22 (a) p_N is the coefficient of u^N in the expansion of the function

$$G(u) = \frac{u^n}{m^n}\left(\frac{1-u^m}{1-u}\right)^n;$$

$$p_N = \frac{1}{m^n}(C_{N-1}^{n-1} - C_n^1 C_{N-m-1}^{n-1} + C_n^2 C_{N-2m-1}^{n-1} - \cdots),$$

and the series is cut off when $N - ms < n$;

(b) $$p = 1 + p_N - \sum_{k=n}^{N} p_k = 1 + p_N - \frac{1}{m^n}(C_N^n - C_n^1 C_{N-m}^n + C_n^2 C_{N-2m}^n - \cdots)$$

(compare with Problem 9.20).

9.23 (a) $G_1(u) = \dfrac{u^{21}}{4^3}\left(\dfrac{1-u^4}{1-u}\right)^3, \quad p = \dfrac{1}{4^3}(C_6^2 - 3) = 0.1875;$

(b) $G_2(v) = \dfrac{v^{21}}{8^3}(1+v)^9, \quad p = \dfrac{1}{8^3} C_9^4 = 0.2461;$

(c) $G(u) = G_1(u) \times G_2\left(\dfrac{1}{u}\right) = \dfrac{1}{32^3}\dfrac{(1+u^2)^3(1+u)^{12}}{u^9},$

$$p = \frac{2}{32^3}(C_{12}^3 + 3C_{12}^5) = 0.1585.$$

9.24 Hypothesis H_k means that the numbers of heads for the two coins first become equal after k tosses of both coins ($k = 1, 2, \ldots, n$); the event A means that after n throws the numbers of heads become equal (previous equality is not excluded).

$$\mathbf{P}(A) = \sum_{k=1}^{n} \mathbf{P}(H_k)\mathbf{P}(A \mid H_k), \qquad p = \mathbf{P}(H_n), \qquad \mathbf{P}(A) = \mathbf{P}(A \mid H_0),$$

$$\mathbf{P}(A \mid H_k) = \frac{1}{4^{n-k}} C_{2n-2k}^{n-k}.$$

Consequently, $C_{2n}^{n} = \sum_{k=1}^{n} 4^k C_{2n-2k}^{n-k}\mathbf{P}(H_k)$. Using successful values for n, one can find $p = \mathbf{P}(H_n)$. Let $R(u) = \sum_{k=1}^{\infty} u^k \mathbf{P}(H_k)$, $Q(u) = \sum_{j=0}^{\infty} u^j p_j$, where $p_{n-j} = \mathbf{P}(A \mid H_j)$. Adding together the terms containing u^n, we obtain:

$$Q(u)R(u) = \sum_{n=1}^{\infty} u^n \sum_{k=1}^{n} p_{n-k}\mathbf{P}(H_k) = \sum_{n=1}^{\infty} u^n P_n(A) = Q(u) - 1;$$

$$Q(u) = \sum_{k=0}^{\infty} \left(\frac{u}{4}\right)^k \frac{(2k)!}{(k!)^2} = (1-u)^{-1/2};$$

$$R(u) = 1 - \sqrt{1-u} = \sum_{k=1}^{\infty} u^k \frac{(2k-2)!}{2^{2k-1}k!\,(k-1)!}; \qquad p = \frac{(2n-2)!}{2^{2n-1}(n-1)!\,n!}.$$

9.25 Let μ be the number of votes cast for a certain candidate. The probability of this is $P_\mu = C_n^\mu p^\mu q^{n-\mu}$. The probability that at most μ votes are cast for this candidate is $\alpha_\mu = \sum_{s=0}^{\mu} P_s$. The probability that among k candidates $l - 1$ receive at least μ votes, $k - l - 1$ persons get no more than μ votes and two receive μ votes each is

$$\frac{k!}{2(l-1)!\,(k-l-1)!}(1 + P_\mu - \alpha_\mu)^{l-1}\alpha_\mu^{k-l-1}P_\mu^2;$$

$$p = \frac{k!}{2(l-1)!\,(k-l-1)!}\sum_{\mu=0}^{n} P_\mu^2 \alpha_\mu^{k-l-1}(1 + P_\mu - \alpha_\mu)^{l-1}.$$

9.26 The probability of winning one point for the serving team is 2/3.

(a) $$P_k = C_{15}^1 \left(\frac{1}{3}\right)^2 \left(\frac{2}{3}\right)^{13+k} + C_{k-1}^1 C_{15}^2 \left(\frac{1}{3}\right)^4 \left(\frac{2}{3}\right)^{11+k} + C_{k-1}^2 C_{15}^3 \left(\frac{1}{3}\right)^6 \left(\frac{2}{3}\right)^{9+k} + \cdots + C_{k-1}^{k-2} C_{15}^{k-1} \left(\frac{1}{3}\right)^{2k-2} \left(\frac{2}{3}\right)^{17-k} + C_{15}^k \left(\frac{1}{3}\right)^{2k} \left(\frac{2}{3}\right)^{15-k}$$

or

$$P_k = \left(\frac{2}{3}\right)^{15} \frac{1}{6^k}(4^{k-1}C_{15}^1 + 4^{k-2}C_{k-1}^1 C_{15}^2 + 4^{k-3}C_{k-1}^2 C_{15}^3 + \cdots + 4C_{k-1}^1 C_{15}^{k-1} + C_{15}^k);$$

$$Q_k = \left(\frac{1}{3}\right)\left(\frac{2}{3}\right)^{14} \frac{1}{6^k}(4^k + 4^{k-1}C_k^1 C_{14}^1 + 4^{k-2}C_k^2 C_{14}^2 + \cdots + 4C_k^{k-1}C_{14}^{k-1} + C_{14}^k)$$

$$(k = 0, 1, \ldots, 13).$$

The numbers P_k and Q_k are given in Table 115.

TABLE 115

k	0	1	2	3	4	5	6
P_k	0.00228	0.00571	0.01047	0.01623	0.02260	0.02915	0.03546
Q_k	0.00114	0.00342	0.00695	0.01159	0.01709	0.02312	0.02929
k	7	8	9	10	11	12	13
P_k	0.04118	0.04604	0.04986	0.05254	0.05407	0.05450	0.05392
Q_k	0.03524	0.04064	0.04525	0.04890	0.05148	0.05299	0.05345

(b) $P_{\mathrm{I}} = \sum_{k=0}^{13} P_k = 0.47401, \quad Q_{\mathrm{I}} = \sum_{k=0}^{13} Q_k = 0.42056.$

(c) let α_k be the probability of scoring $14 + k$ points out of $28 + 2k$ for the first team (serving), which wins the last ball, β_k being the analogous probability for the second team;

$$\beta_0 = Q_{13}, \qquad \alpha_0 = C_{14}^1 \left(\frac{1}{3}\right)^2 \left(\frac{2}{3}\right)^{26} + C_{13}^1 C_{14}^2 \left(\frac{1}{3}\right)^4 \left(\frac{2}{3}\right)^{24} + \cdots$$
$$+ C_{13}^1 C_{14}^{13} \left(\frac{1}{3}\right)^{26} \left(\frac{2}{3}\right)^2 + \left(\frac{1}{3}\right)^{28} = 0.05198,$$

$$\alpha_{k+1} + \beta_{k+1} = \frac{1}{3}(\alpha_k + \beta_k), \qquad \alpha_{k+1} - \beta_{k+1} = -\frac{1}{9}(\alpha_k - \beta_k);$$

that is,

$$(\alpha_k + \beta_k) = \frac{1}{3_k}(\alpha_0 + \beta_0), \qquad (\alpha_k - \beta_k) = \frac{(-1)^k}{9^k}(\alpha_0 - \beta_0);$$

$$p_k = \frac{\alpha_0 + \beta_0}{3^{k+1}} + \frac{(-1)^k}{9^{k+1}}(\alpha_0 - \beta_0), \qquad q_k = \frac{\alpha_0 + \beta_0}{3^{k+1}} - \frac{(-1)^k}{9^{k+1}}(\alpha_0 - \beta_0);$$

$$p_k = \frac{0.10543}{3^{k+1}} - \frac{(-1)^k 0.00148}{9^{k+1}}, \qquad q_k = \frac{0.10543}{3^{k+1}} + \frac{(-1)^k 0.00148}{9^{k+1}};$$

(d) $P_{\mathrm{II}} = \sum_{k=0}^{\infty} p_k = 0.05257, \quad Q_{\mathrm{II}} = \sum_{k=0}^{\infty} q_k = 0.05286;$

(e) $P = P_{\mathrm{I}} + P_{\mathrm{II}} = 0.52658, \quad Q = Q_{\mathrm{I}} + Q_{\mathrm{II}} = 0.47342.$

II RANDOM VARIABLES

10. THE PROBABILITY DISTRIBUTION SERIES, THE DISTRIBUTION POLYGON AND THE DISTRIBUTION FUNCTION OF A DISCRETE RANDOM VARIABLE

10.1 See Table 116.

TABLE 116

x_i	0	1
p_i	0.7	0.3

$$F(x) = \begin{cases} 0 & \text{for } x \leqslant 0, \\ 0.7 & \text{for } 0 < x \leqslant 1, \\ 1 & \text{for } x > 1. \end{cases}$$

10.2 See Table 117.

TABLE 117

x_i	0	1	2	3
p_i	0.125	0.375	0.375	0.125

$$F(x) = \begin{cases} 0 & \text{for} \quad x \leqslant 0, \\ 0.125 & \text{for} \quad 0 < x \leqslant 1, \\ 0.500 & \text{for} \quad 1 < x \leqslant 2, \\ 0.875 & \text{for} \quad 2 < x \leqslant 3, \\ 1 & \text{for} \quad x > 3. \end{cases}$$

10.3 See Table 118.

TABLE 118

x_i	1	2	3	4	5
p_i	0.1	0.09	0.081	0.0729	0.6561

10.4 (a) $\mathbf{P}(X = m) = q^{m-1}p = 1/2^m$, (b) one experiment.

10.5 X_1 is the random number of throws for the basketball player who starts the throws and X_2 is the same for the second player;

$$\left.\begin{aligned} \mathbf{P}(X_1 = m) &= 0.6^{m-1} \cdot 0.4^m \\ \mathbf{P}(X_2 = m) &= 0.6^{m+1} \cdot 0.4^{m-1} \end{aligned}\right\} \text{ for all } m \geqslant 1.$$

10.6 See Table 119.

TABLE 119

x_i	−3	3	8	9	14	15	19	20	25	30
p_i	0.008	0.036	0.060	0.054	0.180	0.027	0.150	0.135	0.225	0.125

10.7 $\mathbf{P}(X = m) = q^{m-4}p = 1/2^{m-3}$ for all $m \geqslant 4$, since the minimal random number of inclusions is four and occurs if the first device included ceases to operate.

10.8 (a) $\mathbf{P}(X = m) = \begin{cases} q^{n-1} & \text{for} \quad m = 0, \\ q^{n-m-1} & \text{for} \quad 0 < m \leqslant n-1; \end{cases}$

(b) $\mathbf{P}(X = m) = \begin{cases} pq^{m-1} & \text{for} \quad 1 \leqslant m \leqslant n-1, \\ q^{n-1} & \text{for} \quad m = n. \end{cases}$

10.9 $\mathbf{P}(X = m) = C_n^m p^m q^{n-m}$ for all $0 \leqslant m \leqslant n$.

10.10 $\mathbf{P}(X = m) = 1 - 2 \cdot 0.25^m$ for all $m \geqslant 1$.

10.11 $\mathbf{P}(X = k) = (1 - p/\omega)^{k-1}p/\omega$ for all $k \geqslant 1$.

10.12 $\mathbf{P}(X = m) = (np)^m/m!\, e^{-np}$ for all $m \geqslant 0$.

10.13 See Table 120.

TABLE 120

z_i	0	$\frac{1}{4}$	$\frac{2}{3}$	$\frac{3}{2}$	4	∞
p_i	$\frac{1}{32}$	$\frac{5}{32}$	$\frac{10}{32}$	$\frac{10}{32}$	$\frac{5}{32}$	$\frac{1}{32}$

10.14 See Table 121.

TABLE 121

x_i	0	1	2	3	4	5	6	7	8	9	10	11	12	13
$10^3 p_i$	1	3	6	10	15	21	28	36	45	55	63	69	73	75
x_i	14	15	16	17	18	19	20	21	22	23	24	25	26	27
$10^3 p_i$	75	73	69	63	55	45	36	28	21	15	10	6	3	1

11. THE DISTRIBUTION FUNCTION AND THE PROBABILITY DENSITY FUNCTION OF A CONTINUOUS RANDOM VARIABLE

11.1 $F(x) = \begin{cases} 1 & \text{if } x \text{ belongs to } (0, 1), \\ 0 & \text{if } x \text{ does not belong to } (0, 1). \end{cases}$

11.2 $f(x) = \dfrac{1}{\sqrt{2\pi}} e^{-x^2/2}$. **11.3** $2^{1/\alpha} x_0$.

11.4 (a) $p = \dfrac{1}{e}$, (b) $f(x) = \dfrac{1}{T} e^{-t/T}$.

11.5 (a) σ, (b) $\sigma\sqrt{\dfrac{2 \log 2}{\log e}} \approx 1.18\sigma$, (c) $f(x) = \dfrac{x}{\sigma^2} e^{-x^2/2\sigma^2}$.

11.6 (a) $f(x) = \dfrac{m}{x_0} x^{m-1} e^{-x^m/x_0}$ $(x \geqslant 0)$, (b) $x_p = \{-x_0 \ln (1 - p)\}^{1/m}$,

(c) $\left(\dfrac{m-1}{m} x_0\right)^{1/m}$

11.7 (a) 10, (b) $F(x) = \dfrac{2}{\sqrt{2\pi}} \displaystyle\int_0^{t_B} e^{-t^2/2}\, dt$, where $t_B = \dfrac{\log x - \log x_0}{\sigma}$.

11.8 (a) $a = \dfrac{1}{2}$, $b = \dfrac{1}{\pi}$; (b) $F(x) = \dfrac{a}{\pi(x^2 + a^2)}$;

(c) $\mathbf{P}(\alpha < X < \beta) = \dfrac{1}{\pi} \arctan \dfrac{a(\beta - \alpha)}{a^2 + \alpha\beta}$.

11.9 $a = \dfrac{1}{\sqrt{\pi}}$. **11.10** (a) $F(x) = \dfrac{1}{2} + \dfrac{1}{\pi} \arctan x$, (b) $\mathbf{P}(|X| < 1) = \frac{1}{2}$.

11.11 $p = \dfrac{1}{2}$. **11.12** $p = \dfrac{2}{3}$.

11.13 Introduce the random variable X denoting the time interval during which a tube ceases to operate. Write the differential equation for $F(x) = \mathbf{P}(X < x)$, the distribution function of the random variable X. The solution of this equation for $x = l$ has the form $F(l) = 1 - e^{-kl}$.

11.4 (a) $\dfrac{z^2}{L^4}(6L^2 - 8Lz + 3z^2)$, (b) $1 - \left(\dfrac{z - x}{L - x}\right)^{h+1}$.

11.5 $f(x) = \displaystyle\sum_{i=1}^{\infty} \frac{1}{2^i} \delta(x - x_i)$.

12. NUMERICAL CHARACTERISTICS OF DISCRETE RANDOM VARIABLES

12.1 $\bar{x} - p$.

12.2 $\bar{x}_a = 1.8$, $\bar{x}_b = 1.7$, $\bar{x}_B = 2.0$; the minimal number of weighings will be in the case of system (b).

12.3 $\mathbf{M}[X] = 2$, $\mathbf{D}[X] = 1.1$.

12.4 To prove this it is necessary to compute $\mathbf{M}[X] = dG(u)/du|_{u=1}$, where $G(u) = (q_1 + p_1u)(q_2 + p_2u)(q_3 + p_3u)$.

12.5 We form the generating function $G(u) = (q + pu)^n$; $\mathbf{M}[X] = G'(1) = np$.

12.6 $\dfrac{2}{N}\sum_{i=1}^{n} m_i k_i$.

12.7 For the first, 7/11; for the second, $-7/11$ coins; that is, the game is lost by the second player.

12.8 Consider a, b and c as the expected wins of players A, B and C under the assumption that A wins from B. For these quantities there obtain $a = (m/2) + (b/2)$, $c = a/2$, $b = c/2$, forming a system of equations for the unknowns a, b and c. Solving the system, we obtain $a = (4/7)m$, $b = (1/7)m$, $c = (2/7)m$. In the second case, we obtain for the players A, B and C, $(5/14)m$, $(5/14)m$, $(2/7)m$, respectively.

12.9
$$\mathbf{M}[A] = \frac{2}{2^2} + \left(\frac{4}{2^4} + \frac{5}{2^5}\right) + \left(\frac{7}{2^7} + \frac{8}{2^8}\right) + \cdots$$
$$= \sum_{m=2}^{\infty} \frac{m}{2^m} - 3\sum_{m=1}^{\infty} \frac{m}{2^{3m}} = \frac{3}{2} - \frac{24}{49} = 1\,\frac{1}{98};$$
$$\mathbf{M}(C) = \frac{3}{2^2} + \frac{6}{2^5} + \frac{9}{2^8} + \cdots = \frac{3}{4}\sum_{m=0}^{\infty} \frac{m+1}{8^m} = \frac{3}{4}\,\frac{1}{\left(1 - \frac{1}{8}\right)^2} = \frac{48}{49}.$$

12.10 $\mathbf{M}[X] = p\sum_{k=1}^{\infty} k(1-p)^{k-1} = \dfrac{1}{p}$.

12.11 $\mathbf{M}[X] = p\sum_{m=4}^{\infty} m(1-p)^{m-4} = 4 + \dfrac{1-p}{p} = 3 + \dfrac{1}{p} = 8$.

12.12 $\mathbf{M}[X] = k/p$; $\mathbf{D}[X] = [k(1-p)/p]$. The series
$$S = \sum_{m=k}^{\infty} \frac{m!}{(m-k)!}\, q^{m-k}$$
is summed with the aid of the formula
$$S = \frac{d^k}{dq^k}\sum_{m=0}^{\infty} q_m = \frac{k}{(1-q)^{k+1}},$$
where $q = 1 - p$.

12.13 (a) $\mathbf{M}[m] = \omega$, where $\omega = 1/(1 - e^{-\alpha})$, (b) $\mathbf{M}[m] = \omega + 1$. For summation of the series, we use the formulas
$$\sum_{m=1}^{\infty} me^{-\alpha m} = -\frac{d}{d\alpha}\sum_{m=0}^{\infty} e^{-\alpha m} = -\frac{d}{d\alpha}\left(\frac{1}{1 - e^{-\alpha}}\right).$$

12.14 $\mathbf{M}[X] = 1/[p_1 + p_2p_3(1 - p_1)] = 4.55$, where $p_1 = 0.18$, $p_3 = p_2 = 0.22$.

12.15 $\mathbf{M}[X] = 4(2/3)$. **12.16** $\mathbf{M}[n] = n + m\sum_{k=1}^{n}(1/k)$.

12.17 Find the maximum of the variance as a function of the probability of occurrence of an event.

12.18 $\mu_3 = np(1 - p)(1 - 2p)$ vanishes at $p = 0$, $p = 0.5$, and $p = 1$.

12.19 Treat the variance as a function of probability of occurrence of an event.

12.20 In both cases the expected number of black balls in the second urn is 5, and of white balls is $4 + 1/2^{10}$ in the first case and $4 + e^{-5}$ in the second case.

12.21 Two dollars. **12.22** For $p < 3/4$.

12.23 $\mathbf{M}[X] = [(n^2 - 1)/3n]a$. For finding the probabilities $p_k = \mathbf{P}(X = ka)$ that the random length of transition equals ka, use the formula of total probabilities and take as hypothesis A_i the fact that the worker is at the ith machine.

12.24 $q = 0.9$; $P_{10} = 1 - q^{10} \approx 0.651$. **12.25** $\mathbf{M}[X] = 3/2$.

12.26 $\mathbf{M}[X] = \sum_{m=1}^{\infty} (1/n^2) = \pi^2/6$. **12.27** $y = 1/2p$; $y = 6.5$ dollars.

12.28 $\mathbf{M}[X] = n/m$; $\mathbf{D}[X] = n(m + n)/m^2$.

12.29 $$X_k = \frac{M + M_1}{N + N_1} N + \frac{MN_1 - NM_1}{N + N_1}\left(1 - \frac{1}{N} + \frac{1}{N_1}\right)^k;$$

$\lim_{k\to\infty} X_k = [(M + M_1)/(N + N_1)]N$. Write the equation of finite differences for the expected number of white balls X_k contained in the first urn after k experiments:

$$X_{k+1} - X_k = \frac{M + M_1}{N_1} - \left(\frac{1}{N_1} + \frac{1}{N}\right)X_k.$$

12.30 $$p = C_m^k \left(\frac{1}{n}\right)^k \left(\frac{n - 1}{n}\right)^{m-k}; \; \bar{x} = \frac{m}{n}; \; \mathbf{D}[X] = \frac{m(n - 1)}{n^2}.$$

12.31 $\bar{x} = q/p$; $\mathbf{D}[X] = q^2/p^2 + q/p$, where $q = 1 - p$.

12.32 $$\mathbf{M}[X] = \sum_{n=1}^{\infty} 2np_n = \sum_{n=1}^{\infty} \frac{(2n - 2)!}{2^{2n-2}[(n - 1)!]^2} = \sum_{n=0}^{\infty} \frac{(2n)!}{2^{2n}(n!)^2} = \infty,$$

since

$$\sum_{n=0}^{\infty} \left(\frac{x}{4}\right)^n \frac{(2n)!}{(n!)^2} = (1 - x)^{-1/2}.$$

13. NUMERICAL CHARACTERISTICS OF CONTINUOUS RANDOM VARIABLES

13.1 $\mathbf{M}[X] = a$, $\mathbf{D}[X] = \dfrac{l^2}{3}$, $E = \sigma \dfrac{\sqrt{3}}{2}$.

13.2 $\mathbf{M}[X] = 0$, $\mathbf{D}[X] = \dfrac{1}{2}$.

13.3 $\mathbf{M}[X] = \dfrac{E}{\rho\sqrt{\pi}}$, $\mathbf{D}[X] = \dfrac{E^2}{\rho^2}\left(\dfrac{1}{2} - \dfrac{1}{\pi}\right)$.

13.4 $\mathbf{D}[X] = \dfrac{a^2}{2}$, $E = \dfrac{a}{\sqrt{2}}$.

13.5 $\mathbf{P}(a < \bar{a}) = 1 - e^{-\pi/4}$, $\mathbf{P}(a > \bar{a}) = e^{-\pi/4}$, $\dfrac{\mathbf{P}(a < \bar{a})}{\mathbf{P}(a > \bar{a})} = \dfrac{0.544}{0.456} = 1.19$.

13.6 $A = \dfrac{4h^3}{\sqrt{\pi}}$, $\mathbf{M}[V] = \dfrac{2}{h\sqrt{\pi}}$, $\mathbf{D}[V] = \dfrac{1}{h^2}\left(\dfrac{3}{2} - \dfrac{4}{\pi}\right)$.

13.7 $\mathbf{M}[X] = \mathbf{D}[X] = m + 1$.

13.8 $\mathbf{M}[X] = \dfrac{3}{2} x_0$, $\mathbf{D}[X] = \dfrac{3}{4} x_0^2$.

13.9 $\mathbf{M}[X] = 0$, $\mathbf{D}[X] = 2$.

13.10 $A = \dfrac{1}{\beta^{\alpha+1}\Gamma(\alpha+1)}$, $\mathbf{M}[X] = (\alpha + 1)\beta$, $\mathbf{D}[X] = \beta^2(\alpha + 1)$.

13.11 $A = \dfrac{\Gamma(a+b)}{\Gamma(a)\Gamma(b)}$, $\mathbf{M}[X] = \dfrac{a}{a+b}$, $\mathbf{D}[X] = \dfrac{ab}{(a+b)^2(a+b+1)}$.

13.12 $A = \dfrac{\Gamma\left(\dfrac{n+1}{2}\right)}{\sqrt{\pi}\,\Gamma\left(\dfrac{n}{2}\right)}$, $\mathbf{M}[X] = 0$, $\mathbf{D}[X] = \dfrac{1}{n-2}$ $(n > 2)$.

To calculate the integral $\int_0^\infty (1 + x^2)^{-(n+1)/2}\,dx$, use the change of variables $x = \sqrt{y/(1-y)}$ leading to the B-function and express the latter in terms of the Γ-function.

13.13 $A = \dfrac{1}{2^{(n-3)/2}\Gamma\left(\dfrac{n-1}{2}\right)}$, $\mathbf{M}[X] = \dfrac{\sqrt{2}\,\Gamma\left(\dfrac{n}{2}\right)}{\Gamma\left(\dfrac{n-2}{2}\right)}$, $\mathbf{D}[X] = n - 1 - \bar{x}^2$.

13.14 Use the relation

$$f(x) = \frac{dF(x)}{dx} = -\frac{d[1 - F(x)]}{dx}.$$

13.15 $\mathbf{M}[T] = 1/\gamma$. Notice that $p(t)$ is the distribution function of the random time of search (T) necessary to sight the ship.

13.16 $m(t) = m_0 e^{-pt}$. Consider the fact that the probability of decay of any fixed atom during the time interval $(t, t + \Delta t)$ is $p\Delta t$ and work out the differential equation for $m(t)$.

13.17 $T_{\text{II}} = (1/p)(\log 2)/(\log e)$. Use the solution of Problem 13.16.

13.18 $[\mathbf{P}(T < \bar{T})]/[\mathbf{P}(T > \bar{T})] = 0.79$; that is, the number of scientific workers who are older than the average age (among the scientific workers) is larger than that younger than the average age. The average age among the scientific workers is $\bar{T} = 41.25$ years.

13.19 $m_{2\nu} = \dfrac{(2\nu-1)(2\nu-3)\cdots 5\cdot 3\cdot 1}{(n-2)(n-4)\cdots(n-2\nu)}\, n^\nu$ for $n \geqslant 2\nu + 1$, $m_{2\nu+1} = 0$.

For the calculation of integrals of the form

$$\int_0^\infty x^{2\nu}\left(1 + \frac{x^2}{n}\right)^{-(n+1)} dx,$$

make the change of variables $x = \sqrt{n[y/(1-y)]}$ that leads to the B-function and express the latter in terms of the Γ-function.

13.20 $m_k = \dfrac{\Gamma(p+k)\Gamma(p+q)}{\Gamma(p)\Gamma(p+q+k)}$.

13.21 $\mathbf{M}[X] = 0$, $\mathbf{D}[X] = \dfrac{\pi^2}{12} + \dfrac{1}{2}$.

13.22 $\mu_k = \sum_{j=0}^{k} (-1)^{k-j}(\bar{x})^{k-j} m_j$, where $m_j = \mathbf{M}[X^j]$.

13.23 $m_k = \sum_{j=0}^{k} C_k^j(\bar{x})^{k-j}\mu_j$, where $\mu_j = \mathbf{M}[(X - \bar{x})^j]$.

14. POISSON'S LAW

14.1 $p = 1 - e^{-0.1} \approx 0.095.$ **14.2** $p = \frac{3^4}{4!} e^{-3} \approx 0.17.$

14.3 $p = 1 - e^{-1} \approx 0.63.$ **14.4** $p = e^{-0.5} \approx 0.61.$

14.5 (1) 0.95958, (2) 0.95963. **14.6** 0.9. **14.7** 0.143.

14.8 $p = \frac{1}{e} \sum_{m=3}^{500} \frac{1}{m!} \approx 1 - \frac{1}{e} \sum_{m=0}^{2} \frac{1}{m!} \approx 0.08.$

14.9 0.4. **14.10** $\text{Sk} = \frac{1}{\sqrt{a}}.$ **14.11** (a) $\frac{(\lambda p)^k}{k!} e^{-\lambda p}$, (b) $1 - e^{-\lambda p}.$

14.12 $\mathbf{M}[X] = \mathbf{D}[X] = (\log 2)/(\log e) MN_0/AT_\pi$. Work out the differential equation for the average number of particles at the instant t. Equate the average number of particles with half the initial number. The resulting equation enables one to find the probability of decay of a given particle; multiplying it by the number of particles, we get $\mathbf{M}[X]$.

14.13 (a) $p = \frac{n^{10}}{10!} e^{-n} \approx 1.02 \cdot 10^{-10}$, (b) $p = 1 - e^{-n} - ne^{-n} \approx 0.673,$

where

$$n = \frac{MN_0 ptS}{4\pi r^2 A} \approx 0.475.$$

14.14 Express $P_n(k_1, k_2, \ldots, k_m, k_{m+1})$ in the form

$$P_n(k_1, k_2, \ldots, k_m, k_{m+1}) = \frac{\left(1 - \frac{1}{n}\right)\left(1 - \frac{2}{n}\right)\cdots\left(1 - \frac{s-1}{n}\right)}{\left(1 - \frac{\lambda_1 + \lambda_2 + \cdots + \lambda_m}{n}\right)^s} \times \left(1 - \frac{\sum_{i=1}^m \lambda_i}{n}\right) \prod_{n=1}^{m} \frac{\lambda_i^{k_i}}{k_i!},$$

where $s = \sum_{i=1}^m k_i$. In as much as $\sum_{i=1}^n \lambda_i$ and s is finite, then

$$\lim_{n \to \infty} \frac{\left(1 - \frac{1}{n}\right)\left(1 - \frac{2}{n}\right)\cdots\left(1 - \frac{s-1}{n}\right)}{\left(1 - \frac{1}{n}\sum_{i=1}^n \lambda_i\right)^s} \left(1 - \frac{1}{n}\sum_{i=1}^n \lambda_i\right)^n = \exp\left\{-\sum_{i=1}^m \lambda_i\right\}.$$

15. THE NORMAL DISTRIBUTION LAW

15.1 $p = 0.0536.$

15.2 $p_{\text{below}} = 0.1725$, $p_{\text{inside}} = 0.4846$, $p_{\text{above}} = 0.3429.$

15.3 (a) 1372 sq. m., (b) 0.4105. **15.4** 22 measurements.

15.5 $E_x = 2\rho\sqrt{\frac{2}{3}}\, E_y \approx 0.78 E_y.$

15.6 See Table 122.

TABLE 122

x	−65	−55	−45	−35	−25	−15	−5	+5	+15	+25	+35
$10^5 F(x)$	35	350	2,150	8,865	25,000	50,000	75,000	91,135	97,850	99,650	99,965

15.7 $E \approx 39$ m. The resulting transcendental equation may be more simply solved by a graphical method.

15.8 $E_1 = \sigma\sqrt{\frac{2}{\pi}}$.

15.9 (a) 0.1587, 0.0228, 0.00135; (b) 0.3173, 0.0455, 0.0027.

15.10 $p \approx 0.089$. **15.11** $p = 0.25$. **15.12** (a) 0.5196, (b) 0.1281.

15.13 $\mathbf{M}[X] = 3$ items. **15.14** Not less than 30 μ.

15.15 ~ 8.6 km. **15.16** (a) 1.25 mm., (b) 0.73 mm.

15.17 (a) $F_a(x) = \dfrac{\left(\dfrac{x - \bar{x}}{\sigma}\right) - \Phi\left(\dfrac{b - \bar{x}}{\sigma}\right)}{1 - \Phi\left(\dfrac{b - \bar{x}}{\sigma}\right)}$ for all $x > b$,

(b) $F_6(x) = \dfrac{\Phi\left(\dfrac{x - \bar{x}}{\sigma}\right) - \Phi\left(\dfrac{a - \bar{x}}{\sigma}\right)}{\Phi\left(\dfrac{b - \bar{x}}{\sigma}\right) - \Phi\left(\dfrac{a - \bar{x}}{\sigma}\right)}$ for $a < x < b$.

15.18 $E = \rho\sqrt{\dfrac{b^2 - a^2}{\ln(b/a)}}$.

16. CHARACTERISTIC FUNCTIONS

16.1 $E(u) = q + pe^{iu}$, where $q = 1 - p$.

16.2 $E(u) = \prod_{i=1}^{n} (q_k + p_k e^{iu})$, where $p_k + q_k = 1$.

16.3 $E(u) = (q + pe^{iu})^n$; $\mathbf{M}[X] = np$, $\mathbf{D}[X] = npq$.

16.4 $E(u) = \dfrac{1}{1 + a(1 - e^{iu})}$, $\mathbf{M}[X] = a$, $\mathbf{D}[X] = a(1 + a)$.

16.5 $E(u) = \exp\{a(e^{iu} - 1)\}$, $\mathbf{M}[X] = \mathbf{D}[X] = a$.

16.6 $E(u) = \exp\left(iu\bar{x} - \dfrac{\sigma^2 u^2}{2}\right)$.

16.7 $E(u) = \dfrac{1}{1 - iu}$, $m_k = k!$

16.8 $E(u) = \dfrac{e^{iub} - e^{iua}}{iu(b - a)}$, $m_k = \dfrac{b^{k+1} - a^{k+1}}{(k + 1)(b - a)}$.

16.9 $E(u) = 1 + v\sqrt{\pi}\, e^{-v^2}[i - \Phi(v)]$, where $v = u/2h$ and

$$\Phi(v) = \frac{2}{\sqrt{\pi}} \int_0^v e^{-z^2}\, dz.$$

Integrate by parts and then use the formulas:

$$\int_0^\infty e^{-x^2} \sin 2px\, dx = \frac{\sqrt{\pi}}{2} e^{-p^2}\Phi(p), \qquad \int_0^\infty e^{-px^2} \cos qx\, dx = \frac{1}{2} e^{-q^2/4p}\sqrt{\frac{\pi}{p}}.$$ [2]

16.10 $E(u) = \left(1 - \dfrac{iu}{\alpha}\right)^{-\lambda}$, $m_k = \dfrac{\lambda(\lambda + 1)\cdots(\lambda + k - 1)}{\alpha^k}$.

---

[2] See Jahnke, E., and Emde, R.: Tables of Functions with Formulae and Curves. 4th rev. ed. New York, Dover Publications, Inc., 1945.

16.11 $E(u) = \frac{1}{\pi}\int_0^{\pi} e^{iau\cos\varphi}\,d\varphi = I_0(au)$. Pass to polar coordinates and use one of the integral representations of the Bessel function.[2]

16.12 $E(u) = \exp[ix u - a \mid u]$. By a change of variables it is reduced to the form

$$E(u) = e^{ixu}\frac{a}{\pi}\int_{-\infty}^{+\infty}\frac{e^{ixu}}{x^2 + a^2}\,dx.$$

The integral in this formula is computed with the aid of the theory of residues, for which it is necessary to consider the integral

$$\frac{a}{\pi}\oint\frac{e^{izu}}{z^2 + a^2}\,dz$$

over a closed contour. For positive u the integration is performed over the semicircle (closed by a diameter) in the upper half-plane and for negative n over a similar semicircle in the lower half-plane.

16.13 $E_y(u) = \exp\left\{iu(b + a\bar{x}) - \frac{u^2}{2}a^2\sigma^2\right\}$, $E_x(u) = \exp\left\{-\frac{u^2\sigma^2}{2}\right\}$.

16.14 $\mu_{2k} = \sigma^{2k}(2k - 1)!!$, $\mu_{2k+1} = 0$.

16.15 $f(x) = \frac{a}{\pi(a^2 + x^2)}$ (the Cauchy law).

16.16 $f_1(x) = \begin{cases} e^{-x} & \text{for } x > 0, \\ 0 & \text{for } x < 0, \end{cases}$ $\quad f_2(x) = \begin{cases} 0 & \text{for } x > 0, \\ e^{x} & \text{for } x < 0. \end{cases}$

Solve this with the aid of the theory of residues; consider separate the cases of positive and negative values of x.

16.17 $\mathbf{P}(X = k) = 2^{-k}$, where $k = 1, 2, 3, \ldots$. Expand the characteristic function in a series of powers of $(1/2)e^{iu}$ and use the analytic representation of the δ-function given in the introduction to Section 11, p. 49.

17. THE COMPUTATION OF THE TOTAL PROBABILITY AND THE PROBABILITY DENSITY IN TERMS OF CONDITIONAL PROBABILITY

17.1 $p = \frac{b}{l(\theta_2 - \theta_1)}\left(\ln\tan\frac{\theta_2}{2} - \ln\tan\frac{\theta_1}{2}\right)$.

17.2 Denoting the diameter of the circle by D and the interval between the points by l, we obtain

$$p = \frac{D(2l - D)}{l^2} = 0.4375.$$

17.3 $p = 0.15$.

17.4
$$p = \frac{1}{2}\left[1 - \hat{\Phi}\left(\frac{L - \bar{x}}{E}\right)\right] + \frac{E}{2L\rho\sqrt{\pi}} \times \left[\exp\left\{-\rho^2\frac{\bar{x}^2}{E^2}\right\} - \exp\left\{-\rho^2\frac{(L - \bar{x})^2}{E^2}\right\}\right] + \frac{\bar{x}}{2L}\left[\hat{\Phi}\left(\frac{L - \bar{x}}{E}\right) + \hat{\Phi}\left(\frac{\bar{x}}{E}\right)\right] \approx 0.67.$$

17.5 In both cases we get the same result, $p_1 = p_2 = 0.4$.

17.6 $p = 1 - \frac{1}{2l}\int_{-l}^{+l}\left\{1 - \frac{1}{2}\left[\hat{\Phi}\left(\frac{z + r}{R}\right) - \hat{\Phi}\left(\frac{z - r}{R}\right)\right]\right\}^n dz.$

17.7 $F(w) = n\int_{-\infty}^{+\infty} f(y)\left\{\int_y^{y+w} f(x)\,dx\right\}^{n-1} dy.$

17.8 $p = 1 - \dfrac{128}{45\pi^2} \approx 0.712.$

17.9 $p_i = \dfrac{r_i}{\sum_{i=1}^{n} r_i}$, where $r_i = \displaystyle\int_{-\infty}^{+\infty} f_i(x)f_p(x - x_0)\,dx.$

17.10 $f(a \mid m_0) = \dfrac{2(2a)^{m_0+1}}{(m_0 + 1)!}\, e^{-2a}.$

III SYSTEMS OF RANDOM VARIABLES

18. DISTRIBUTION LAWS AND NUMERICAL CHARACTERISTICS OF SYSTEMS OF RANDOM VARIABLES

18.1 $f(x, y) = \begin{cases} \dfrac{1}{(b - a)(d - c)} & \text{for } a \leqslant x \leqslant b,\ c \leqslant y \leqslant d, \\ 0 & \text{outside the rectangle,} \end{cases}$

$F(x, y) = F_1(x)F_2(y)$, where

$$F_1(x) = \begin{cases} 1 & \text{for } x \geqslant b, \\ \dfrac{x - a}{b - a} & \text{for } a \leqslant x \leqslant b, \\ 0 & \text{for } x \leqslant a, \end{cases} \qquad F_2(y) = \begin{cases} 1 & \text{for } y \geqslant d, \\ \dfrac{y - c}{d - c} & \text{for } c \leqslant y \leqslant d, \\ 0 & \text{for } y \leqslant c. \end{cases}$$

18.2 (a) $A = 20$, (b) $F(x, y) = \left(\dfrac{1}{\pi}\arctan\dfrac{x}{4} + \dfrac{1}{2}\right)\left(\dfrac{1}{\pi}\arctan\dfrac{y}{5} + \dfrac{1}{2}\right).$

18.3 $f(x, y, z) = abce^{-(ax+by+cz)}.$

18.4 The triangle with vertices having coordinates:

$$\left(\frac{1}{a}\ln\frac{abc}{f_0}, 0, 0\right); \quad \left(0, \frac{1}{6}\ln\frac{abc}{f_0}, 0\right); \quad \left(0, 0, \frac{1}{c}\ln\frac{abc}{f_0}\right).$$

18.5 (a) $F(i, j) = \mathbf{P}(X < i,\ Y < j) = \mathbf{P}(X \leqslant i - 1,\ Y \leqslant j - 1)$. For the values of $F(i, j)$ see Table 123.

TABLE 123

$j-1$ \ $i-1$	0	1	2	3	4	5	6
0	0.202	0.376	0.489	0.551	0.600	0.623	0.627
1	0.202	0.475	0.652	0.754	0.834	0.877	0.887
2	0.202	0.475	0.683	0.810	0.908	0.964	0.982
3	0.202	0.475	0.683	0.811	0.911	0.971	1.000

(b) $1 - \mathbf{P}(X \leqslant 6,\ Y \leqslant 1) = 1 - 0.887 = 0.113;$

(c) $\mathbf{M}[X] = 1.947$; $\mathbf{M}[Y] = 0.504$; $\|k_{ij}\| = \begin{Vmatrix} 2.610 & 0.561 \\ 0.561 & 0.548 \end{Vmatrix}$.

18.6 $F(x_1, x_2, \ldots, x_n) = \prod_{i=1}^{n} F_i(x_i) = \prod_{i=1}^{n} \int_{-\infty}^{x_i} f_i(\xi_i)\, d\xi_i.$

18.7 $P = \dfrac{f(u, v)}{f(u, v) + f(v, u)}.$

18.8 $P = f(u, v, w) : [f(u, v, w) + f(u, w, v) + f(v, u, w) + f(v, w, u) + f(w, u, v) + f(w, v, u)].$

18.9 $P = F(a_1, b_3) - F(a_1, b_5) + F(a_2, b_1) - F(a_2, b_3) + F(a_3, b_4) - F(a_3, b_2) + F(a_4, b_2) - F(a_4, b_4) + F(a_5, b_5) - F(a_5, b_1).$

18.10 $P = a^{-3} - a^{-6} - a^{-9} + a^{-12}.$

18.11

$$P = \begin{cases} \dfrac{\pi R^2}{4ab} & \text{for } 0 \leqslant R \leqslant b, \\ \dfrac{R^2}{4ab}(\pi - 2\beta + \sin 2\beta) & \text{for } b \leqslant R \leqslant a, \\ \dfrac{R^2}{4ab}(\pi - 2\alpha - 2\beta + \sin 2\alpha + \sin 2\beta) & \text{for } a \leqslant R \leqslant \sqrt{a^2 + b^2}, \\ 1 & \text{for } R \geqslant \sqrt{a^2 + b^2}, \end{cases}$$

where $\alpha = \arccos (a/R)$, $\beta = \arccos (b/R)$.

18.12 (a) $c = \dfrac{3}{\pi R^3}$, (b) $p = \dfrac{3a^2}{R^2}\left(1 - \dfrac{2a}{3R}\right)$.

18.13 (a) $r_{xy} = \begin{cases} +1 & \text{for } n/m < 0, \\ -1 & \text{for } n/m > 0, \end{cases}$ (b) $\dfrac{\sigma_x}{\sigma_y} = \left|\dfrac{n}{m}\right|$.

18.14 Consider the expectations of the squares of the expressions

$$\sigma_y(X - \bar{x}) + \sigma_x(Y - \bar{y}) \quad \text{and} \quad \sigma_y(X - \bar{x}) - \sigma_x(Y - \bar{y}).$$

18.15 Make use of the reaction $k_{xy} = \mathbf{M}[XY] - \bar{x}\bar{y}$.

18.16 $\|r_{ij}\| = \begin{Vmatrix} 1 & -0.5 & 0.5 \\ -0.5 & 1 & -0.5 \\ 0.5 & -0.5 & 1 \end{Vmatrix}$.

18.17 (a) $\mathbf{M}[X] = \alpha + \gamma = 0.5$,

(b) $\mathbf{M}[Y] = \alpha + \beta = 0.45$,

$\mathbf{D}[X] = (\alpha + \gamma)(\beta + \delta) = 0.25$;

$\mathbf{D}[Y] = (\alpha + \beta)(\gamma + \delta) = 0.2475$;

$k_{xy} = \mathbf{M}[XY] - \mathbf{M}[X]\mathbf{M}[Y] = \alpha - (\alpha + \gamma)(\alpha + \beta) = 0.175.$

18.18 $\mathbf{M}[X] = \mathbf{M}[Y] = 0$; $\|k_{ij}\| = \begin{Vmatrix} \frac{1}{2} & 0 \\ 0 & \frac{1}{2} \end{Vmatrix}$.

18.19 $f(x, y) = \cos x \cos y$; $\mathbf{M}[X] = \mathbf{M}[Y] = \dfrac{\pi}{2} - 1$;

$$\|k_{ij}\| = \begin{Vmatrix} \pi - 3 & 0 \\ 0 & \pi - 3 \end{Vmatrix}.$$

18.20 $p = \frac{2l}{\pi L}\left[1 - \sqrt{1 - \frac{L^2}{l^2}} + \frac{L}{l}\arccos\frac{L}{l}\right].$

18.21 $p = \frac{l}{\pi ab}[2(a + b) - l].$

Hint. Use the formula $\mathbf{P}(A \cup B) = \mathbf{P}(A) + \mathbf{P}(b) - \mathbf{P}(AB)$, where the event A means that the needle crosses the side a and B that it crosses side b.

19. THE NORMAL DISTRIBUTION LAW IN THE PLANE AND IN SPACE. THE MULTIDIMENSIONAL NORMAL DISTRIBUTION

19.1 $F(x, y) = \frac{1}{4}\left[1 + \hat{\Phi}\left(\frac{x - \bar{x}}{E_x}\right)\right]\left[1 + \hat{\Phi}\left(\frac{y - \bar{y}}{E_y}\right)\right].$

19.2
$$f(x, y) = \frac{1}{182\pi\sqrt{3}} \times \exp\left\{-\frac{2}{3}\left[\frac{(x - 26)^2}{196} + \frac{(x - 26)(y + 12)}{182} + \frac{(y + 12)^2}{169}\right]\right\}.$$

19.3 (a) $c = 1.39$, (b) $\|k_{ij}\| = \begin{Vmatrix} 0.132 & -0.026 \\ -0.026 & 0.105 \end{Vmatrix}$, (c) $S_{el} = 0.162$.

19.4 $f(2, 2) = \frac{1}{2\pi e^3\sqrt{2}} = 0.00560.$

19.5
$$f(x, y, z) = \frac{1}{2\pi\sqrt{230\pi}} \times \exp\left\{-\frac{1}{230}(39x^2 + 36y^2 + 26z^2 - 44xy + 36xz - 38yz)\right\};$$
$$f_{\max} = \frac{1}{2\pi\sqrt{230\pi}} = 0.00595.$$

19.6

(a)
$$\|k_{ij}^{-1}\| = \begin{Vmatrix} 2 & -1 & 0 & 0 & 0 & \cdots & 0 \\ -1 & 2 & -1 & 0 & 0 & \cdots & 0 \\ 0 & -1 & 2 & -1 & 0 & \cdots & 0 \\ 0 & 0 & -1 & 2 & -1 & \cdots & 0 \\ 0 & 0 & 0 & -1 & 2 & \cdots & \cdot \\ \cdot & \cdot & \cdot & \cdot & \cdot & \cdot & \cdot \\ 0 & 0 & 0 & 0 & 0 & \cdots & 1 \end{Vmatrix};$$

(b)
$$f(x_1, \ldots, x_n) = \frac{\exp\left\{-\frac{1}{2}\sum_{i=1}^{n}(x_i - x_{i-1})^2\right\}}{(2\pi)^{n/2}}, \text{ where } x_0 = 0.$$

19.7
$$\|k_{ij}\| = \begin{Vmatrix} 10 & 0 & 2 & 0 \\ 0 & 10 & 0 & 2 \\ 2 & 0 & 10 & 0 \\ 0 & 2 & 0 & 10 \end{Vmatrix};$$
$$f(x_1, y_1, x_2, y_2) = \frac{1}{384\pi^2}\exp\left\{-\frac{5}{96}(x_1^2 + y_1^2 + x_2^2 + y_2^2) + \frac{1}{48}(x_1x_2 + y_1y_2)\right\}.$$

19.8 $\mathbf{P}(k) = 1 - e^{-k^2/2}$. **19.9** $\mathbf{P}(k) = \hat{\Phi}(k) - \frac{2\rho k}{\sqrt{\pi}} e^{-\rho^2 k^2}$.

19.10 $\mathbf{P}(R) = e^{-\rho^2(d^2/E^2)} \int_0^{\rho^2 R^2/E^2} I_0\left(\frac{2\rho d\sqrt{t}}{E}\right) e^{-t}\, dt,$

where $I_0(x)$ is the Bessel function of an imaginary argument.

19.11 (a) $\mathbf{P}(X < Y) = \frac{1}{2}$, (b) $\mathbf{P}(X < 0,\ Y > 0) = \frac{1}{4}$.

19.12 $P = \frac{1}{4}\left[\hat{\Phi}\left(\frac{c - \bar{x}}{E_x}\right) - \hat{\Phi}\left(\frac{a - \bar{x}}{E_x}\right)\right]\left[\hat{\Phi}\left(\frac{d - \bar{y}}{E_y}\right) - \hat{\Phi}\left(\frac{b - \bar{y}}{E_y}\right)\right] = 0.0335.$

19.13 (a) $P_{\text{circ}} = 1 - e^{-\rho^2} = 0.2035$, (b) $P_{\text{sq}} = \left[\hat{\Phi}\left(\frac{\sqrt{\pi}}{2}\right)\right]^2 = 0.2030$,

(c) $P_{\text{lin}} = \hat{\Phi}\left(\frac{\sqrt{0.1\pi}}{2}\right)\hat{\Phi}\left(\frac{\sqrt{10\pi}}{2}\right) = 0.1411.$

19.14 $P = 0.5\left(1 - \exp\left\{-\rho^2 \frac{R^2}{E^2}\right\}\right)$.

19.15 $P = \frac{\alpha}{2\pi}\left(\exp\left\{-\rho^2 \frac{R_1^2}{E^2}\right\} - \exp\left\{-\rho^2 \frac{R_2^2}{E^2}\right\}\right)$.

19.16 $A = 4dk$, $\alpha = E_x\sqrt{1 + \frac{2\rho^2 d^2}{3E_x^2}}$, $\beta = E_z\sqrt{1 + \frac{2\rho^2 k^2}{3E_z^2}}$.

19.17 $P = \hat{\Phi}\left(\frac{d}{\alpha}\right)\hat{\Phi}\left(\frac{k}{\beta}\right) < \hat{\Phi}\left(\frac{d}{E_x}\right)\hat{\Phi}\left(\frac{k}{E_z}\right)$, since $\alpha > E_x$, $\beta > E_z$.

19.18 $P_{\text{sat}} = 1 - q^3 - 3q^2(1 - q) - 3q[(p_2 + p_3)^2 + 2p_2p_4] - p_2^3 = 0.379,$
$P_{\text{excel}} = p_5^3 + 3p_5^2(p_3 + p_4) + 3p_4^2 p_5 = 0.007,$
where $p_2 = 0.196$, $p_3 = 0.198$, $p_4 = 0.148$, $p_5 = 0.055$, $q = 0.403$.

19.19 $P = \frac{1}{8}[\hat{\Phi}(k)]^2$.

19.20 $P = \hat{\Phi}\left(\frac{R}{E}\right) - \frac{2\rho R}{E\sqrt{\pi}} \exp\left\{-\rho^2 \frac{R^2}{E^2}\right\} - \left[\hat{\Phi}\left(\frac{a}{2E}\right)\right]^3$.

19.21 (a) $P = 0.5\left(1 - \exp\left\{-\rho^2 \frac{R^2}{E^2}\right\}\right)\left[\hat{\Phi}\left(\frac{mh}{(m + n)B}\right) + \hat{\Phi}\left(\frac{nh}{(m + n)B}\right)\right]$,

(b) $P = \frac{h}{2H}\left(1 - \exp\left\{-\rho^2 \frac{R^2}{E^2}\right\}\right)$.

19.22 $P = 0.5\left[\hat{\Phi}\left(\frac{h}{a}\right) - \frac{hE}{\sqrt{h^2E^2 + R^2a^2}}\hat{\Phi}\left(\frac{\sqrt{h^2E^2 + R^2a^2}}{Ea}\right)\right]$.

19.23 $25(x_1 - 10)^2 + 36(x_1 - 10)(x_2 - 10) + 36(x_2 - 10)^2 = 7484.6.$

19.24 $16(x_1 - 2)^2 + 5x_2^2 + 16(x_3 + 2)^2 + 8(x_1 - 2)(x_3 + 2) = 805.1.$

19.25 $\sum_{i=1}^{n} (x_i - x_{i-1})^2 = \frac{2}{\log e}\left[5 - \frac{n}{2}\log(2\pi)\right]$.

The problem has no solution for $n > 12$.

20. DISTRIBUTION LAWS OF SUBSYSTEMS OF CONTINUOUS RANDOM VARIABLES AND CONDITIONAL DISTRIBUTION LAWS

20.1

$$f(x, y, z) = \begin{cases} \dfrac{1}{(a_2 - a_1)(b_2 - b_1)(c_2 - c_1)} & \text{for } a_1 \leqslant x \leqslant a_2,\ b_1 \leqslant y \leqslant b_2, \\ & \quad c_1 \leqslant z \leqslant c_2, \\ 0 & \text{outside the parallelepiped;} \end{cases}$$

$$f(y, z) = \begin{cases} \dfrac{1}{(b_2 - b_1)(c_2 - c_1)} & \text{for } b_1 \leqslant y \leqslant b_2,\ c_1 \leqslant z \leqslant c_2, \\ 0 & \text{outside the triangle;} \end{cases}$$

$$f(z) = \begin{cases} \dfrac{1}{c_2 - c_1} & \text{for } c_1 \leqslant z \leqslant c_2, \\ 0 & \text{outside the interval.} \end{cases}$$

The random variables X, Y, Z are independent.

20.2 For $|x| \leqslant R$, $|y| \leqslant R$,

$$f_x(x) = \frac{2\sqrt{R^2 - x^2}}{\pi R^2}, \qquad f_y(y) = \frac{2\sqrt{R^2 - y^2}}{\pi R^2},$$

$$F_x(x) = \frac{1}{\pi}\left[\arcsin\frac{x}{R} + \frac{x}{R}\sqrt{1 - \frac{x^2}{R^2}}\right] + \frac{1}{2},$$

$$F_y(y) = \frac{1}{\pi}\left[\arcsin\frac{y}{R} + \frac{y}{R}\sqrt{1 - \frac{y^2}{R^2}}\right] + \frac{1}{2};$$

X and Y are independent, since $f(x, y) \neq f_x(x)f_y(y)$.

20.3

$$f(y \mid x) = \begin{cases} \dfrac{1}{2\sqrt{R^2 - x^2}} & \text{for } |x| < R, \\ \dfrac{1}{2}[\delta(y + R) + \delta(y - R)] & \text{for } |x| = R, \\ 0 & \text{for } |x| > R, \end{cases}$$

$\delta(z)$ being the δ-function.

20.4

$$\|k_{ij}\| = \left\| \begin{matrix} \dfrac{R^2}{4} & 0 \\ 0 & \dfrac{R^2}{4} \end{matrix} \right\|;$$ X and Y are uncorrelated.

20.5 (a) $$f(x, y) = \begin{cases} \dfrac{1}{a^2} & \text{inside the square,} \\ 0 & \text{outside the square. Inside the square;} \end{cases}$$

(b) $f_x(x) = \dfrac{a\sqrt{2} - 2|x|}{a^2}$, $f_y(y) = \dfrac{a\sqrt{2} - 2|y|}{a^2}$;

(c) $f(y \mid x) = \dfrac{1}{a\sqrt{2} - 2|x|}$, $f(x \mid y) = \dfrac{1}{a\sqrt{2} - 2|y|}$;

(d) $\mathbf{D}[X] = \mathbf{D}[Y] = \dfrac{a^2}{12}$,

(e) the random variables X and Y are dependent but uncorrelated.

20.6 $f_z(z) = [3(R^2 - z^2)/4R^3]$ for $|z| < R$, $f(x, y \mid z) = 1/[\pi(R^2 - z^2)]$ for $|z| < R$.

20.7 $k = 4$, $f_x(x) = 2xe^{-x^2}$ $(x \geqslant 0)$, $f_y(y) = 2ye^{-y^2}$ $(y \geqslant 0)$, $f(x \mid y) = f_x(x)$, $f(y \mid x) = f_y(y)$, $\mathbf{M}[X] = \mathbf{M}[Y] = \sqrt{\pi}/2$, $\mathbf{D}[X] = \mathbf{D}[Y] = 1 - \pi/4$, $k_{xy} = 0$.

20.8
$$\mathbf{M}[X] = \int_{-\infty}^{\infty} \mathbf{M}[X \mid y] f_y(y)\, dy;$$
$$\mathbf{D}[X] = \int_{-\infty}^{\infty} \mathbf{D}[X \mid y] f_y(y)\, dy + \int_{-\infty}^{\infty} \{\bar{x} - \mathbf{M}[X \mid y]\}^2 f_y(y)\, dy.$$

20.9 Since $\mathbf{M}[X] = 5$, $\mathbf{M}[Y] = -2$, $\sigma_x = \sigma$, $\sigma_y = 2\sigma$, $r = -0.8$, it follows that: (a) $\mathbf{M}[X \mid y] = 5 - 0.8/2(y + 2) = 4.2 - 0.4y$, $\mathbf{M}[Y \mid x] = -2 - 0.8 \times 2(x - 5) = 6 - 1.6x$, $\sigma_{x|y} = 0.6\sigma$, $\sigma_{y|x} = 1.2\sigma$,

(b) $f_x(x) = \dfrac{1}{\sigma\sqrt{2\pi}} \exp\left\{-\dfrac{(x - 5)^2}{2\sigma^2}\right\}$, $f_y(y) = \dfrac{1}{2\sigma\sqrt{2\pi}} \exp\left\{-\dfrac{(y + 2)^2}{8\sigma^2}\right\}$,

(c) $f(x \mid y) = \dfrac{1}{0.6\sigma\sqrt{2\pi}} \exp\left\{-\dfrac{(x + 0.4y - 4.2)^2}{0.72\sigma^2}\right\}$,

$$f(y \mid x) = \frac{1}{1.2\sigma\sqrt{2\pi}} \exp\left\{-\frac{(y + 1.6x - 6)^2}{2.88\sigma^2}\right\}.$$

20.10 $f_x(x) = A\sqrt{\dfrac{\pi}{c}} \exp\left\{-\left(a - \dfrac{b^2}{4c}\right)x^2\right\}$, $f_y(y) = A\sqrt{\dfrac{\pi}{a}} \exp\left\{-\left(c - \dfrac{b^2}{4a}\right)y^2\right\}$.

For the independence of X and Y it is necessary that
$$\frac{\sqrt{ac}}{\pi A} \exp\left\{-\frac{b^2}{4}\left(\frac{x^2}{c} - \frac{4xy}{b} + \frac{y^2}{a}\right)\right\} = 1.$$
This condition is satisfied for $b = 0$. In this case $A = \sqrt{ac}/\pi$.

20.11 $k = \dfrac{3\sqrt{3}}{\pi}$, $k_{xy} = -\dfrac{1}{18}$, $f(x \mid y) = \dfrac{2}{\sqrt{\pi}} e^{-(2x + 1.5y)^2}$,

$$f(y \mid x) = \frac{3}{\sqrt{\pi}} e^{-(x + 3y)^2}.$$

20.12 (a) $f_x(x) = \dfrac{1}{40\sqrt{2\pi}} \exp\left\{-\dfrac{(x - 125)^2}{3200}\right\}$,

(b) $f_y(y) = \dfrac{1}{30\sqrt{2\pi}} \exp\left\{-\dfrac{(y + 30)^2}{1800}\right\}$,

(c) $f(x \mid 0) = \dfrac{1}{32\sqrt{2\pi}} \exp\left\{-\dfrac{(x - 149)^2}{2048}\right\}$,

(d) $f(y \mid 25) = \dfrac{1}{24\sqrt{2\pi}} \exp\left\{-\dfrac{(y + 75)^2}{1152}\right\}$.

20.13 $\mathbf{M}[X \mid y] = 0.8y + 149$, $\mathbf{M}[Y \mid x] = 0.45x - 86.25$.

20.14 $f(r) = \dfrac{2r^2}{a^3\sqrt{2\pi}} \exp\left\{-\dfrac{r^2}{2a^2}\right\}$, $\bar{r} = \mathbf{M}(R) = \dfrac{4a}{\sqrt{2\pi}}$.

20.15 $f_r(r) = \dfrac{r}{ab} \exp\left\{-\dfrac{r^2}{4}\left(\dfrac{1}{a^2} + \dfrac{1}{b^2}\right)\right\} I_0\left[\dfrac{r^2}{4}\left(\dfrac{1}{a^2} - \dfrac{1}{b^2}\right)\right]\Bigg\}$,

where $I_0(x)$ is the Bessel function of zero order of an imaginary argument;
$$f_\varphi(\varphi) = \left[2\pi ab\left(\frac{\cos^2\varphi}{a^2} + \frac{\sin^2\varphi}{b^2}\right)\right]^{-1}.$$

20.16 $f(r \mid \varphi) = r\left(\frac{\cos^2\varphi}{a^2} + \frac{\sin^2\varphi}{b^2}\right)\exp\left\{-\frac{r^2}{2}\left(\frac{\cos^2\varphi}{a^2} + \frac{\sin^2\varphi}{b^2}\right)\right\},$

$$f(\varphi \mid r) = \frac{1}{2\pi I_0\left[\frac{r^2}{4}\left(\frac{1}{a^2} - \frac{1}{b^2}\right)\right]}\exp\left\{-\frac{r^2}{4}\cos 2\varphi\left(\frac{1}{a^2} - \frac{1}{b^2}\right)\right\}.$$

20.17 (a) $f(r, \vartheta, \varphi) = \frac{r^2 \sin\vartheta}{(2\pi)^{3/2}\, abc}\exp\left\{-\frac{r^2}{2}\left(\frac{\cos^2\vartheta\cos^2\varphi}{a^2} + \frac{\cos^2\vartheta\sin^2\varphi}{b^2} + \frac{\sin^2\vartheta}{c^2}\right)\right\},$

(b) $f(r, \vartheta) = \frac{r^2\sin\vartheta}{\sqrt{2\pi}\, abc}\exp\left\{-\frac{r^2}{2}\left(\frac{\sin^2\vartheta}{c^2} + \left(\frac{1}{a^2} + \frac{1}{b^2}\right)\frac{\cos^2\vartheta}{c^2}\right)\right\} \times I_0\left[\frac{r^2\cos^2\vartheta}{4}\left(\frac{1}{a^2} - \frac{1}{b^2}\right)\right],$

$$f(\vartheta, \varphi) = \frac{\sin\vartheta}{4\pi abc}\left(\frac{\cos^2\cos^2\varphi}{a^2} + \frac{\cos^2\vartheta\sin^2\varphi}{b^2} + \frac{\sin^2\vartheta}{c^2}\right)^{-3/2},$$

(c) $f(r \mid \vartheta, \varphi) = \frac{2r^2}{\sqrt{2\pi}}\left(\frac{\cos^2\vartheta\cos^2\varphi}{a^2} + \frac{\cos^2\vartheta\sin^2\varphi}{b^2} + \frac{\sin^2\vartheta}{c^2}\right)^{3/2} \times \exp\left\{-\frac{r^2}{2}\left(\frac{\cos^2\vartheta\cos^2\varphi}{a^2} + \frac{\cos^2\vartheta\sin^2\varphi}{b^2} + \frac{\sin^2\vartheta}{c^2}\right)\right\},$

$$f(\varphi \mid r, \vartheta) = \frac{\exp\left\{-\frac{r^2}{4}\cos^2\vartheta\cos 2\varphi\left(\frac{1}{a^2} - \frac{1}{b^2}\right)\right\}}{2\pi I_0\left[\frac{r^2\cos^2\vartheta}{4}\left(\frac{1}{a^2} - \frac{1}{b^2}\right)\right]}.$$

20.18 $f_{x_1, x_2}(x_1, x_2) = \frac{1}{2\pi\sqrt{96}}\exp\left\{-\frac{1}{96}(5x_1^2 - 2x_1x_2 + 5x_2^2)\right\},$

$$f_{x_1, y_1}(x_1, y_1) = \frac{1}{20\pi}\exp\left\{-\frac{x_1^2 + y_1^2}{20}\right\}.$$

20.19 $f(x_2, y_2 \mid 0, 10) = \frac{5}{96\pi}\exp\left\{-\frac{5}{96}[x_2^2 + (y_2 - 2)^2]\right\},$

$\mathbf{M}[X_2 \mid 0, 10] = 0$, $\mathbf{M}[Y_2 \mid 0, 10] = 2$,
$\mathbf{D}[X_2 \mid 0, 10] = \mathbf{D}[Y_2 \mid 0, 10] = 9.6$.

IV NUMERICAL CHARACTERISTICS AND DISTRIBUTION LAWS OF FUNCTIONS OF RANDOM VARIABLES

21. NUMERICAL CHARACTERISTICS OF FUNCTIONS OF RANDOM VARIABLES

21.1 $4a/\pi$. **21.2** $\pi(a/2)$. **21.3** $\mathbf{M}[G] = 4.1$ g., $\mathbf{D}[G] = 0.32$ g^2.

21.4 $\mathbf{M}[\varphi] = \arctan\frac{l}{h} - \frac{h}{2l}\ln\left(1 + \frac{l^2}{h^2}\right).$

21.5 $40/\pi$ cm. **21.6** $\mathbf{M}[Y] = 1$. **21.7** 1.15 m. **21.8** $a^2/2$.

21.9 $(n - 2)pq^2$ (for $n \geqslant 3$). **21.10** $\mathbf{M}[R] = \dfrac{E\sqrt{\pi}}{2\rho}$.

21.11 $11a^2/18\pi$. **21.12** $3/\pi$. **21.13** $\lambda\bar{\omega}$. **21.14** $\sum_{k=1}^{n} p_k$.

21.15 $n\left[1 - \left(1 - \dfrac{p}{n}\right)^m\right]$. **21.16** $n[1 - (1 - p)^m]$.

21.17 $\bar{t} = T[1 - \exp\{-a(1 - e^{-a})\}]$,
$\bar{s} = kT^2[1 - 2\exp\{-a(1 - e^{-\alpha})\} + \exp\{-a(1 - e^{-2\alpha})\}]^2$.

21.18 $$n\left[1 - \left(1 - \frac{p}{n}\right)^m\right] + \sum_{k=0}^{m} (n - k)\left[1 - \left(1 - \frac{p}{n - k}\right)^m\right] P_n^m(k),$$

where $P_n^m(k)$ is the probability that after the first series of cycles exactly k units will be damaged at least once;

$$P_n^m(k) = C_n^k \sum_{i=0}^{k} (-1)^i C_k^i \left[1 - \frac{p(n - k + i)}{n}\right]^m.$$

21.19 (a) $$mp + \sum_{k=0}^{4} \{(m - 2k)p + k[1 - (1 - p)^2]\} P_n^m(k) + P_n^m(5) \times [3 - (1 - p)^2 - 2(1 - p)^3] + 2P_n^m(6) \times [1 - (1 - p)^4] + P_n^m(7)[1 - (1 - p)^8],$$

where $P_n^m(k) = C_n p^k (1 - p)^{n-k}$ for $n = m = 8$,

(b) $2mp$ for $n > 2m$.

21.20 $$\frac{a^2}{6b} \ln \frac{b + \sqrt{a^2 + b^2}}{a} + \frac{b^2}{6a} \ln \frac{a + \sqrt{a^2 + b^2}}{b} + \frac{1}{3}\sqrt{a^2 + b^2}.$$

21.21 $$\frac{b^2}{a} \ln \frac{a + \sqrt{a^2 + b^2}}{b} + \frac{a^2 - 2b^2}{3a^2}\sqrt{a^2 + b^2} + \frac{2}{3}\frac{b^3}{a^2}.$$

21.22 $\sum_{k=1}^{n} p_k$, $\sum_{k=1}^{n} p_k(1 - p_k)$. **21.23** 0.316 g. **21.24** $l/3$; $l^2/18$.

21.25 $\mathbf{M}[Z] = 5a$; $\mathbf{D}[Z] = 100a^2 + 225b^2 - 150ab$.

21.26 $\mathbf{M}[Y] = \dfrac{E_x}{\rho\sqrt{\pi}}$, $\mathbf{D}[Y] = \dfrac{E_x^2}{\rho^2}\left(\dfrac{1}{2} - \dfrac{1}{\pi}\right)$.

21.27 $\mathbf{M}[Y] = \exp\{-\bar{x}(1 - \cos b)\}\cos(\bar{x}\sin b)$,
$$\mathbf{D}[Y] = \frac{1}{2}[1 + \exp\{-\bar{x}(1 - \cos 2b)\}\cos(\bar{x}\sin 2b)] - \bar{y}^2.$$

21.28 (a) 26.7 sq. m., (b) 22.0 sq. m., (c) 10 sq. m.

21.29 $\mathbf{M}[Z] = 2a\sqrt{\dfrac{2}{\pi}} - \dfrac{E}{2\rho}\sqrt{\pi}$, $\mathbf{D}[Z] = a^2\left(3 - \dfrac{8}{\pi}\right) + \dfrac{E^2}{4\rho^2}(4 - \pi)$.

21.30 $\mathbf{M}[Z] = 5(\sqrt{3} - 1)$, $\mathbf{D}[Z] = 7600$.

21.31 $r_{xy} = n!!/\sqrt{(2n - 2)!!}$, if n is even; $r_{xy} = 0$ if n is odd.

21.32 $\mathbf{M}[Z] = 0$, $\mathbf{D}[Z] = 2\Delta^2\sigma^2$. **21.33** $\dfrac{2a}{\pi}$, $a^2\left(\dfrac{1}{2} - \dfrac{4}{\pi^2}\right)$.

21.34 $\bar{r} = a\left(1 + \dfrac{e^2}{2}\right)$, $\mathbf{D}[R] = \dfrac{a^2e^2}{2}\left(1 - \dfrac{e^2}{2}\right)$ (where e is the eccentricity).

22. THE DISTRIBUTION LAWS OF FUNCTIONS OF RANDOM VARIABLES

22.1

$$F_y(y) = \begin{cases} F_x\left(\dfrac{y-b}{a}\right) & \text{for } a > 0, \\ 1 - F_x\left(\dfrac{y-b}{a}\right) & \text{for } a < 0. \end{cases}$$

22.2 $f_y(y) = f_x(e^y)e^y$.

22.3

$$f_z(z) = \begin{cases} \dfrac{1}{\sigma\sqrt{2\pi a z}} \exp\left\{-\dfrac{z}{2a\sigma^2}\right\} & \text{for } z > 0, \\ 0 & \text{for } z \leqslant 0. \end{cases}$$

22.4

$$f_y(y) = \begin{cases} \dfrac{2\rho}{E\sqrt{\pi}} \exp\left\{-\rho^2\left(\dfrac{y}{E}\right)^2\right\} & \text{for } y \geqslant 0, \\ 0 & \text{for } y < 0. \end{cases}$$

22.5

$$f_y(y) = \begin{cases} \dfrac{\pi}{\sin \pi y} & \text{for } \dfrac{1}{2} \leqslant y \leqslant \dfrac{2}{\pi} \arctan e, \\ 0 & \text{for } y < \dfrac{1}{2} \text{ or } y > \dfrac{2}{\pi} \arctan e. \end{cases}$$

22.6

$$f(v) = \begin{cases} \dfrac{1}{3av^{2/3}} & \text{for } 0 < v \leqslant a^3, \\ 0 & \text{for } v < 0 \text{ or } v > a^3. \end{cases}$$

22.7 $f_x(x) = \dfrac{1}{\pi} \dfrac{l}{l^2 + x^2} \quad (-\infty \leqslant x \leqslant \infty)$.

22.8

$$f_y(y) = \begin{cases} \dfrac{1}{\pi\sqrt{a^2 - y^2}} & \text{for } |y| < a \text{ (the arcsine law)} \\ 0 & \text{for } |y| \geqslant a. \end{cases}$$

22.9 (a) $f_y(y) = \dfrac{1}{3\pi[1 + (1-y)^{3/2}](1-y)^{2/3}}$,

(b) if $a > 0$, then

$$f_y(y) = \begin{cases} \dfrac{\sqrt{a}}{\pi(a + y)\sqrt{y}} & \text{for } y \geqslant 0, \\ 0 & \text{for } y < 0; \end{cases}$$

if $a < 0$, then

$$f_y(y) = \begin{cases} \dfrac{-\sqrt{a}}{\pi(a + y)\sqrt{y}} & \text{for } y \leqslant 0, \\ 0 & \text{for } y > 0; \end{cases}$$

(c)

$$f_y(y) = \begin{cases} \dfrac{1}{\pi} & \text{for } |y| \leqslant \dfrac{\pi}{2}, \\ 0 & \text{for } |y| > \dfrac{\pi}{2}. \end{cases}$$

22.10 For an odd n,

$$f_y(y) = \frac{ay^{(1/n)-1}}{\pi n(a^2 + y^{2/n})};$$

for even n,

$$f_y(y) = \begin{cases} \dfrac{2ay^{(1/n)-1}}{n\pi(a^2 + y^{2/n})} & \text{for } y > 0, \\ 0 & \text{for } y \leqslant 0. \end{cases}$$

22.11 (a) $f_y(y) = |y|e^{-y^2}$ $(-\infty \leqslant y \leqslant \infty)$,

(b) $f_y(y) = \begin{cases} 2ye^{-y^2} & \text{for } y \geqslant 0, \\ 0 & \text{for } y < 0. \end{cases}$

22.12

$$f_y(y) = \begin{cases} \dfrac{\Gamma(k + 1.5)}{\sqrt{\pi}\,\Gamma(k + 1)} \cos^{2k+1} y & \text{for } |y| \leqslant \dfrac{\pi}{2}, \\ 0 & \text{for } |y| > \dfrac{\pi}{2}. \end{cases}$$

22.13 (a) $f_y(y) = \dfrac{1}{\sqrt{2\pi}} \exp\left\{-\dfrac{y^2}{2}\right\}$, (b) $f_y(y) = \dfrac{1}{\sigma_y\sqrt{2\pi}} \exp\left\{-\dfrac{(y - \bar{y})^2}{2\sigma_y^2}\right\}$.

22.14

$$f_y(y) = \begin{cases} 1 & \text{for } 0 \leqslant y \leqslant 1, \\ 0 & \text{for } y < 0 \text{ or } y > 1. \end{cases}$$

22.16 $f_z(z) = \dfrac{1}{\sigma_z\sqrt{2\pi}} \exp\left\{-\dfrac{z^2}{2\sigma_z^2}\right\}$, where $\sigma_z^2 = \sigma_x^2 + \sigma_y^2$.

22.17 (a) $f_z(z) = \displaystyle\int_0^\infty \frac{1}{x} f\left(x, \frac{z}{x}\right) dx - \int_{-\infty}^0 \frac{1}{x} f\left(x, \frac{z}{x}\right) dx$,

(b) $f_z(z) = \dfrac{1}{2} \exp\{-|z|\}$, (c) $f_z(z) = \dfrac{1}{2\sigma_x\sigma_y} \exp\left\{-\dfrac{|z|}{\sigma_x\sigma_y}\right\}$,

(d) $f_z(z) = \dfrac{1}{\sqrt{2\pi}} \exp\left\{-\dfrac{z^2}{2}\right\}$.

22.18 (a) $f_z(z) = \displaystyle\int_0^\infty yf(zy, y)\, dy - \int_{-\infty}^0 yf(zy, y)\, dy$,

(b) $f_z(z) = \dfrac{2z}{(1 + z^2)^2}$,

(c) $f_z(z) = \dfrac{\Gamma\left(\dfrac{n+1}{2}\right)}{\Gamma\left(\dfrac{n}{2}\right)\sqrt{\pi}} (1 + z^2)^{-(n+1)/2}$ (Student's distribution law),

(d) $f_z(z) = \dfrac{\sigma_x\sigma_y\sqrt{1 - r^2}}{\pi(\sigma_y^2 z^2 - 2rz\sigma_x\sigma_y + \sigma_x^2)}$, for $r = 0$

$$f_z(z) = \frac{\dfrac{\sigma_x}{\sigma_y}}{\pi\left[z^2 + \left(\dfrac{\sigma_x}{\sigma_y}\right)^2\right]} \quad \text{(Cauchy's distribution law).}$$

22.19 (a) $$f_r(r) = r\int_{-r}^{r} \frac{1}{\sqrt{r^2 - x^2}}[f(x, \sqrt{r^2 - x^2}) + f(x, -\sqrt{r^2 - x^2})]\,dx = r\int_0^{2\pi} f(r\cos\varphi, r\sin\varphi)\,d\varphi;$$

(b) $$f_r(r) = \begin{cases} \dfrac{2r\rho^2}{E^2}\exp\left\{-\rho^2\left(\dfrac{r}{E}\right)^2\right\} & \text{for } 0 \leqslant r < \infty, \\ 0 & \text{for } r < 0; \end{cases}$$

(c) $$f_r(r) = \begin{cases} \dfrac{2r}{a^2} & \text{for } 0 \leqslant r \leqslant a, \\ 0 & \text{for } r > a \text{ or } r < 0; \end{cases}$$

(d) $$f_r(r) = \frac{r}{\sigma^2}\exp\left\{-\frac{r^2 + h^2}{2\sigma^2}\right\}I_0\left(\frac{rh}{\sigma^2}\right),$$

where $I_0(z)$ is the Bessel function of zero order of imaginary argument;

(e) $$f_r(r) = \frac{r}{\sigma_x\sigma_y}\exp\left\{-\frac{r^2(\sigma_x^2 + \sigma_y^2)}{4\sigma_x^2\sigma_y^2}\right\}I_0\left[\frac{r^2(\sigma_x^2 - \sigma_y^2)}{4\sigma_x^2\sigma_y^2}\right].$$

22.20 $U = (X - \bar{x})\cos\alpha + (Y - \bar{y})\sin\alpha$;

$V = (X - \bar{x})\sin\alpha + (Y - \bar{y})\cos\alpha$; $\tan 2\alpha = 2\dfrac{r\sigma_x\sigma_y}{\sigma_x^2 - \sigma_y^2}$;

$\sigma_u^2 = \sigma_x^2\cos^2\alpha + \sigma_y^2\sin^2\alpha + r\sigma_x\sigma_y\sin 2\alpha$;

$\sigma_v^2 = \sigma_x^2\sin^2\alpha + \sigma_y^2\cos^2\alpha - r\sigma_x\sigma_y\sin 2\alpha(\sigma_x^2 + \sigma_y^2 = \sigma_u^2 + \sigma_v^2)$.

22.21

$$f_\alpha(\alpha) = \begin{cases} \dfrac{1}{4}(2 - |\alpha|) & \text{for } |\alpha| \leqslant 2, \\ 0 & \text{for } |\alpha| > 2; \end{cases}$$

$$f_\beta(\beta) = \begin{cases} -\dfrac{1}{2}\ln|\beta| & \text{for } |\beta| \leqslant 1, \\ 0 & \text{for } |\beta| > 1. \end{cases}$$

22.22 $$f(t, \varphi) = \frac{t}{2\pi\sqrt{1 - r_{xy}^2}}\exp\left\{-\frac{t^2(1 - r_{xy}\sin 2\varphi)}{2(1 - r_{xy}^2)}\right\}.$$

For $r_{xy} = 0$, Φ is uniformly distributed in the interval $(0, 2\pi)$ and the random variable T obeys a Rayleigh distribution law.

22.23 $f(s \mid t)$ is the probability density of a normal distribution with parameters

$$\mathbf{M}[S \mid t] = \bar{s}_0 + \bar{v}_0 t + \bar{a}\frac{t^2}{2};$$

$$\mathbf{D}[S \mid t] = \mathbf{D}[S_0] + t^2\mathbf{D}[V_0] + \frac{t^4}{4}\mathbf{D}[a] + 2tk_{s_0v_0} + t^2k_{s_0a} + t^3k_{v_0a}.$$

22.24 $$f_y(y) = \frac{2\left(\dfrac{n}{2}\right)^{n/2}}{\Gamma\left(\dfrac{n}{2}\right)\sigma^n}y^{n-1}\exp\left\{-\frac{ny^2}{2\sigma^2}\right\}.$$

The characteristic function of the random variable X_j^2 if $\sigma^2 = 1$, $\bar{x}_j = 0$ is $E_{x_j}(t) = (1 - 2t)^{-1/2}$. Then, the characteristic function of the random variable $U = \sum_{j=1}^{n} X_j^2$ will be $E_u(t) = (1 - 2t)^{-n/2}$ and the probability density

$$f(U) = \frac{1}{2\pi}\int_{-\infty}^{\infty} e^{iut}(1 - 2t)^{-n/2}\,dt = \frac{1}{\Gamma\left(\frac{n}{2}\right)2^{n/2}} u^{(n/2)-1}e^{-u/2}.$$

If the random variables X_j have the same variance σ^2 and $\bar{x}_j = 0$, then the random variable

$$Y = +\sqrt{\frac{\sigma^2 U}{n}}.$$

Consequently, $f_y(y) = f_u[\psi(y)]|\psi'(y)|$, where $\psi(y) = y^2 n/\sigma^2$.

22.25 $f(r, \varphi_1, \varphi_2, \ldots, \varphi_{n-1}) = \dfrac{r^{n-1}\cos^{n-2}\varphi_1 \cos^{n-3}\varphi_2 \cdots \cos\varphi_{n-2}}{(2\pi)^{n/2}\sigma^n} e^{-r^2/2\sigma^2}.$

22.26 $f_y(y_1, y_2, \ldots, y_n) = f_x\left(\sum_{j=1}^{n} a_{1j}y_j, \ldots, \sum_{j=1}^{n} a_{nj}y_j\right)|\Delta|$, $\Delta = |a_{ij}|$.

22.27 $f(r, \vartheta) = r^2\cos\vartheta \displaystyle\int_0^{2\pi} f(r\cos\vartheta\cos\varphi, r\cos\vartheta\sin\varphi, r\sin\vartheta)\,d\varphi.$

23. THE CHARACTERISTIC FUNCTIONS OF SYSTEMS AND FUNCTIONS OF RANDOM VARIABLES

23.1 Make use of the fact that for independent random variables

$$f(x_1, x_2, \ldots, x_n) = \prod_{k=1}^{n} f_k(x_k).$$

23.2 $E_z(u) = E_{x_1,x_2,\ldots,x_n}(u, u, \ldots, u).$ **23.3** $E_y(u) = e^{iuc}\prod_{k=1}^{n} E_{x_k}(a_k u).$

23.4 $E_y(u) = (1 - 2iu\sigma_x^2)^{-1/2}$; $m_r = \mathbf{M}[Y^r] = 1\cdot 3\cdot 5\cdots(2r - 1)\sigma_x^{2r}.$

23.5 $E_y(u) = \dfrac{e^{iu(a+b)} - e^{iub}}{iua}.$

23.6 $E_y(u) = (1 + iu)^{-1}$; $m_r = \mathbf{M}[Y^r] = (-1)^r r!$

23.7 $E_y(u) = J_0(au)$, where $J_0(x) = (1/2\pi)\int_0^{2\pi} e^{ix\cos\varphi}\,d\varphi$ is the Bessel function of first kind of zero order;

$$f_y(y) = \int_{-\infty}^{\infty} J_0(au)\cos uy\,du = \frac{1}{\pi\sqrt{a^2 - y^2}}.$$

23.8 $E_{x,y}(u_1, u_2) = \exp\left\{i(\bar{x}u_1 + \bar{y}u_2) - \frac{1}{2}(\sigma_1^2 u_1^2 + 2\sigma_1\sigma_2 r u_1 u_2 + \sigma_2^2 u_2^2)\right\}.$

23.9 $E_{x_1,x_2,\ldots,x_n}(u_1, u_2, \ldots, u_n)$

$$= \exp\left\{ai\sum_{m=1}^{n} u_m - \frac{\sigma^2}{2}\sum_{m=1}^{n} u_m^2 - \alpha\sigma^2\sum_{m=1}^{n-1} u_m u_{m+1}\right\}.$$

23.10 $E_y(u) = \exp\left\{iu\dfrac{n(n + 1)}{2} - \dfrac{n(2n^2 + 1)}{6}u^2\right\}.$

23.11 $\mathbf{M}[(X_1^2 - \sigma^2)(X_2^2 - \sigma^2)] = 2k_{12}^2.$

23.12 (a) $\mathbf{M}[X_1^2X_2^2X_3^2] = 8k_{12}k_{13}k_{23} + 2\sigma^2(k_{12}^2 + k_{13}^2 + k_{23}^2) + \sigma^6$,

(b) $\mathbf{M}[(X_1^2 - \sigma^2)(X_2^2 - \sigma^2)(X_3^2 - \sigma^2)] = 8k_{12}k_{13}k_{23}$.

23.13 $\mathbf{M}[X_1X_2X_3] = 0$.

23.14 $\mathbf{M}[X_1X_2X_3X_4] = k_{12}k_{34} + k_{13}k_{24} + k_{14}k_{23}$.

23.15 For the proof, make use of the expansion of the characteristic function

$$E_{x_1,x_2,\ldots,x_n}(u_1, u_2, \ldots, u_n) = \exp\left\{-\frac{1}{2}\sum_{m=1}^{n}\sum_{l=1}^{n} k_{ml}u_mu_l\right\}$$

in an infinite power series of $u_1, u_2, \ldots, u_n$.

23.16 For the proof, use the property

$$E_{\sum_{j=1}^{n}}(u) = E(u_1, \ldots, u_n)|_{u_1=\cdots=u_n=u},$$

where $E(u_1, \ldots, u_n)$ is the characteristic function of a system of normal random variables.

23.17 $E(u_1, u_2) = (p_1e^{iu_1S_1} + q_1e^{iu_1S_2})^{N_1}(p_2e^{i(u_1+u_2)S_1} + q_2e^{i(u_1+u_2)S_2})^{N_2}$

$\times (p_3e^{i(u_1+u_2)S_1} + q_3e^{i(u_1+u_2)S_2})^{N_3}(p_4e^{iu_2S_1} + q_4e^{iu_2S_2})^{N_4}$;

$k_{xy} = -2S_1S_2(N_2p_2q_2 + N_3p_3q_3)$.

24. CONVOLUTION OF DISTRIBUTION LAWS

24.1

$$f_z(z) = \begin{cases} 0 & \text{for } z \leqslant 2a, \\ \dfrac{z - 2a}{(b - a)^2} & \text{for } 2a \leqslant z \leqslant a + b, \\ \dfrac{2b - z}{(b - a)^2} & \text{for } a + b \leqslant z \leqslant 2b, \\ 0 & \text{for } z \geqslant 2b. \end{cases}$$

24.2

$$f_z(z) = \begin{cases} 0 & \text{for } z \geqslant \bar{x} + \bar{y} + a + b, \\ \dfrac{\bar{x} + \bar{y} + a + b - z}{4ab} & \text{for } \bar{x} + \bar{y} + b - a \leqslant z \leqslant \bar{x} + \bar{y} + a + b, \\ \dfrac{1}{2b} & \text{for } \bar{x} + \bar{y} + a - b \leqslant z \leqslant \bar{x} + \bar{y} + b - a, \\ \dfrac{a + b - \bar{x} - \bar{y} + z}{4ab} & \text{for } \bar{x} + \bar{y} - a - b \leqslant z \leqslant \bar{x} + \bar{y} + a - b, \\ 0 & \text{for } z \leqslant \bar{x} + \bar{y} - a - b. \end{cases}$$

24.3 $$f_z(z) = \frac{1}{2(b - a)}\left[\Phi\left(\frac{z - a - \bar{x}}{\sigma_x}\right) - \Phi\left(\frac{z - b - \bar{x}}{\sigma_x}\right)\right],$$

where

$$\Phi(z) = \frac{2}{\sqrt{2\pi}}\int_0^z \exp\left\{-\frac{t^2}{2}\right\} dt.$$

24.4

$$f_z(z) = \begin{cases} 0 & \text{for } z \leqslant 3a, \\ \dfrac{(z-3a)^2}{2(b-a)^3} & \text{for } 3a \leqslant z \leqslant 2a+b, \\ \dfrac{(z-3a)^2 - 3[z-(b+2a)]^2}{2(b-a)^3} & \text{for } 2a+b \leqslant z \leqslant a+2b, \\ \dfrac{(3b-z)^2}{2(b-a)^3} & \text{for } a+2b \leqslant z \leqslant 3b, \\ 0 & \text{for } z \geqslant 3b. \end{cases}$$

24.5 The convolution of the normal distribution law with the uniform probability law has the probability density

$$f_z(z) = \frac{1}{4l}\left[\hat{\Phi}\left(\frac{z-2\bar{x}+l}{E}\right) - \hat{\Phi}\left(\frac{z-2\bar{x}-l}{E}\right)\right].$$

Equating the expectation and variance for $f_z(z)$ and for the probability density $f'_z(z)$ of the normal distribution law we obtain,

$$f'_z(z) = \frac{1}{\sigma_z\sqrt{2\pi}}\exp\left\{-\frac{(z-\bar{z})^2}{2\sigma_z^2}\right\},$$

where

$$\bar{z} = 2\bar{x}, \qquad \sigma_z = \sqrt{\frac{E^2}{2\rho^2} + \frac{l^2}{3}}.$$

If $\bar{x} = 0$, then the relative error of such a substitution at the point $z = 0$ is

$$\Delta\% = \frac{f_z(0) - f'_z(0)}{f_z(0)}100\% \quad \text{(Table 124)}.$$

TABLE 124

l	E	$2E$	$3E$	$4E$
$\Delta\%$	−0.30	−3.02	−9.70	−17.10

24.6
$$f_z(z) = \frac{1}{\pi}\frac{l}{1 + l^2(z-c)^2},$$

where $c = a + b$, $l = hk/(h + k)$. (For solution make use of the characteristic functions of the random variables X and Y.)

24.7 $f_z(z) = \dfrac{2}{\pi^2}\dfrac{z}{\operatorname{sh} z}$.

24.8

$$f_z(z) = \begin{cases} e^{-z/3}(1 - e^{-z/6}) & \text{for } z \geqslant 0, \\ 0 & \text{for } z < 0. \end{cases}$$

24.9 $f_r(r) = \dfrac{r}{\sqrt{\Delta}}\exp\left\{-\dfrac{(k_{11}+k_{22})r^2}{4\Delta}\right\}I_0\left[\dfrac{r^2}{4\Delta}\sqrt{(k_{22}-k_{11})^2 + 4k_{12}^2}\right]$, where $I_0(z)$ is the Bessel function of zero order;

$$\Delta = \begin{vmatrix} k_{11} & k_{12} \\ k_{21} & k_{22} \end{vmatrix};$$

$$k_{11} = \frac{1}{2\rho^2}[a_1^2 + a_2^2 + (b_2^2 - a_2^2)\sin^2\alpha];$$

$$k_{22} = \frac{1}{2\rho^2}[b_1^2 + b_2^2 + (a_2^2 - b_2^2)\sin^2\alpha];$$

$$k_{12} = \frac{1}{4\rho^2}(a_2^2 - b_2^2)\sin 2\alpha = k_{21}.$$

24.12 $\mathbf{M}[X] = \dfrac{1}{p_1} + \dfrac{1}{p_2} - 2$; $\mathbf{D}[X] = \dfrac{1}{p_1}\left(\dfrac{1}{p_1} - 1\right) + \dfrac{1}{p_2}\left(\dfrac{1}{p_2} - 1\right)$,

$$F_n(x) = \frac{1}{p_2 - p_1}\{(1 - p_1)p_2[1 - (1 - p_1)^n] - (1 - p_2)p_1[1 - (1 - p_2)^n]\}.$$

24.13 The required reserve resistance is $0.37\cdot\bar{q}_1 = 7.4$ kg.

24.14 (a) $$P = \frac{1}{4L}\int_0^L\left[1 - \hat{\Phi}\left(\frac{y - \bar{x}}{E}\right)\right]\left[1 + \hat{\Phi}\left(\frac{y - L + \bar{x}}{E}\right)\right]dy - \frac{1}{2L} \times \left\{(L - \bar{x})\hat{\Phi}\left(\frac{L - \bar{x}}{E}\right) - \bar{x}\hat{\Phi}\left(\frac{\bar{x}}{E}\right) + \frac{E}{\rho\sqrt{\pi}} \times \left[\exp\left\{-\rho^2\left(\frac{L - \bar{x}}{E}\right)^2\right\} - \exp\left\{-\rho^2\left(\frac{\bar{x}}{E}\right)^2\right\}\right]\right\},$$

(b) $$P = 0.75 - \frac{(L - \bar{x})}{2L}\hat{\Phi}\left(\frac{L - \bar{x}}{E}\right) + \frac{\bar{x}}{2L}\hat{\Phi}\left(\frac{\bar{x}}{E}\right) - \frac{E}{2L\rho\sqrt{\pi}} \times \left[\exp\left\{-\rho^2\left(\frac{L - \bar{x}}{E}\right)^2\right\} - \exp\left\{-\left(\rho\frac{\bar{x}}{E}\right)^2\right\}\right] + \frac{1}{4L} \times \int_0^L \hat{\Phi}\left(\frac{y - \bar{x}}{E}\right)\hat{\Phi}\left(\frac{y - L + \bar{x}}{E}\right)dy.$$

24.15 $\mathbf{P}(X_A > X_B) = \dfrac{1}{2}\left[1 + \hat{\Phi}\left(\dfrac{\bar{x}_A - \bar{x}_B}{\sqrt{E_A^2 + E_B^2}}\right)\right].$

24.16 $f_z(z) = \dfrac{z^{m-1}\lambda^m}{(m - 1)!}e^{-\lambda z}.$

24.17 $F_z(z) = \dfrac{1}{a - b}[a(1 - b)(1 - a^n) - b(1 - a)(1 - b^n)].$

24.18 See Table 125.

TABLE 125

z_i	0	1	2	3	4
$\mathbf{P}(Z = z_i)$	$\frac{1}{6}$	$\frac{11}{24}$	$\frac{1}{4}$	$\frac{1}{24}$	$\frac{1}{12}$

24.19 $\mathbf{P}(Z = m) = \dfrac{(2a)^m}{m!}e^{-2a}.$

24.20 The random variable Y has a binomial distribution.

24.21 $F_z(n) = \mathbf{P}(Z < n) = 1 - (n/2n - 1)$ $(n = 1, 2, \ldots)$.

25. THE LINEARIZATION OF FUNCTIONS OF RANDOM VARIABLES

25.1 $E_Q \approx 9100$ cal.

25.2 $$\mathbf{D}[\Omega] \approx \frac{\bar{p}}{16\bar{m}\bar{l}}\left[\left(\frac{\sigma_p}{\bar{p}}\right)^2 + \left(\frac{\sigma_m}{\bar{m}}\right)^2 + \left(\frac{\sigma_l}{\bar{l}}\right)^2 - \frac{2\sigma_p\sigma_m r_{pm}}{\bar{p}\bar{m}} - \frac{2\sigma_p\sigma_l r_{pl}}{\bar{p}\bar{l}} + \frac{2\sigma_m\sigma_l r_{ml}}{\bar{m}\bar{l}}\right].$$

25.3 $$E_z \approx \frac{\sqrt{\bar{r}^2E_R^2 + \left(\bar{\omega}\bar{l} - \dfrac{1}{\bar{\omega}\bar{c}}\right)^2\left[\bar{\omega}^2E_l^2 + \dfrac{E_c^2}{\bar{\omega}^2\bar{c}^4} + \left(\bar{l} + \dfrac{1}{\bar{\omega}^2\bar{c}}\right)^2E_\omega^2\right]}}{\sqrt{\bar{r}^2 + \left(\bar{\omega}\bar{l} - \dfrac{1}{\bar{\omega}\bar{c}}\right)^2}}.$$

25.4 $$\mathbf{D}[J] \approx \bar{j}^2\left[\frac{\sigma_E^2}{\bar{e}^2} + \frac{\sigma_R^2}{\left(\bar{r} + \dfrac{\bar{w}}{N}\right)^2} + \frac{\sigma_W^2}{N^2\left(\bar{r} + \dfrac{\bar{w}}{N}\right)^2}\right], \quad \text{where } \bar{j} = \frac{\bar{e}}{\bar{r} + \dfrac{\bar{w}}{N}}.$$

25.5 $E \approx 66.66$ m.; $E_y \approx 38.60$ m. **25.6** $E_{v_1} \approx 0.52$ m./sec.

25.7 For the assumed conditions the function $V_1 = -V\cos q$ cannot be linearized.

25.8 $\sigma_x \approx 23.1$ m.; $\sigma_y \approx 14.3$ m., $\sigma_z \approx 25$ m.

25.9 $\sigma_x = \sigma_y \approx 8.66$ m.; $\sigma_z \approx 7.05$ m.

25.10 $E_v \approx [U/(q + \bar{\omega})]E_\omega$. **25.11** $E_h = 43$ m.

25.12 $\sigma_z \approx 10^{-6}$. **25.13** $E_h \approx 12.98$ m.

25.14 The standard deviation of errors in determination of distance by the formula using the data of the radar station is ≈ 22.85 m.

25.15 $\bar{y} \approx \varphi(\bar{x}) + \frac{1}{2}\varphi''(\bar{x})\mathbf{D}[X]$, $\mathbf{D}[Y] \approx [\varphi'(\bar{x})]^2\mathbf{D}[X] + \frac{1}{2}[\varphi''(\bar{x})]^2\mathbf{D}^2[X]$.

25.16 $$\mathbf{M}[S] \approx \frac{ab}{2}\left(1 - \frac{\sigma_\gamma^2}{2}\right)\sin\bar{\gamma},$$
$$\mathbf{D}[S] \approx \frac{a^2b^2}{4}\left[\sigma_\gamma^2\cos^2\bar{\gamma} + \frac{\sigma_\gamma^4}{2}(1 + \cos^2\bar{\gamma}) + \frac{5}{12}\sigma_\gamma^6\cos^2\bar{\gamma}\right].$$

25.17 $$E_x = \frac{b\sqrt{\dfrac{E_a^2}{\bar{a}^2}\sin^2\bar{\alpha} + E_\alpha^2\cos^2\bar{\alpha}}}{\sqrt{\bar{a}^2 - b^2\sin^2\bar{\alpha}}}; \quad \bar{x} = \arcsin\left(\frac{b}{\bar{\alpha}}\sin\bar{\alpha}\right).$$

25.18 (a) By retaining the first two terms of the expansion in the Taylor series of the function $Y = 1/X$, we obtain $\bar{y} \approx -0.2$, $\mathbf{D}[Y] \approx 0.16$; (b) By retaining the first three terms of the expansion in the Taylor series of the function $Y = 1/X$, we obtain $\bar{y} \approx -1.00$, $\mathbf{D}[Y] \approx 1.44$.

24.19 (a) By the exact formulas
$$\bar{v} = \frac{4\pi\bar{r}}{3}(3\sigma_r^2 + \bar{r}^2), \qquad \mathbf{D}[V] = \frac{16\pi^2}{3}[3\sigma_r^6 + 12\bar{r}^2\sigma_r^4 + 3\bar{r}^4\sigma_r^2];$$
(b) according to the formulas of the linearization method
$$\bar{v} \approx \frac{4\pi\bar{r}^3}{3}, \qquad \mathbf{D}[V] \approx 16\pi^2\bar{r}^4\sigma_r^2.$$

25.20 (a) Measuring the height of the cone, we get $\mathbf{D}[V] \approx 4\pi^2$, (b) by measuring the length of the generator, we get $\mathbf{D}[V] \approx 3.57\pi^2$.

25.21 19.9 mg.

25.22 $E_g = \frac{4\pi^2 L}{T^2}\sqrt{\frac{4}{n^2T^2}(E_t^2 + 2E_{t'}^2) + \frac{E_l^2}{L^2}} = 4.67$ cm./sec.^{2}.

25.23 $\mathbf{D}[Z] \approx \frac{(1-k)^2\pi}{96(1+k)}$.

26. THE CONVOLUTION OF TWO-DIMENSIONAL AND THREE-DIMENSIONAL NORMAL DISTRIBUTION LAWS BY USE OF THE NOTION OF DEVIATION VECTORS

26.1 A normal distribution law with principal semi-axes of the unit ellipse $a = 48.4$ m., $b = 12.4$ m., making $\mathbf{c}_1$ the angles $\alpha = 19°\,40'$ and $109°\,40'$ with the deviation vectors.

26.2 For $\gamma = 0$, a degenerate normal law (deviation vector) $\sqrt{c_1^2 + c_2^2} = 50$ m. For $\gamma = 90°$, a normal distribution law with principal semi-axes of the unit ellipse $a = c_1 = 30$ m., $b = c_2 = 40$ m., coinciding with the directions of the deviation vectors.

26.3 The principal semi-axes $a = 1.2$ m., $b = 1.1$ m. make angles of 33° and 123° with the x-axis.

26.4 The principal semi-axes $a = b = 100$ m.; that is, the total dispersion is circular.

26.5 $a = 30.8$ m., $b = 26.0$ m., $\alpha = 18°\,15'$.

26.6 (a) $a = b = 25\sqrt{5}$ m., (b) $a = 68.9$ m., $b = 38.8$ m., $\alpha = 15°$.

26.7 From the system of equations for the conjugate semi-diameters m and n, $m^2 + n^2 = a^2 + b^2$, $mn = ab/(\sin\gamma)$, we find $m = 20$ m., $n = 15$ m. and

$$P = \hat{\Phi}\left(\frac{l}{m}\right)\hat{\Phi}\left(\frac{l}{n}\right) = 0.566.$$

26.8 $|\mathbf{m}| = 73.2$ m., $|\mathbf{n}| = 68.1$ m., $\varepsilon = 74°\,21'$.

26.9 (a) $f(x, y) = 1.17\cdot 10^{-5}\exp\{-7.06\cdot 10^{-2}(0.295x^2 - 0.670xy + 1.31y^2)\}$,
(b) $a = 126.5$ m., $b = 53.8$ m., $\alpha = 12°\,10'$.

26.10 $a = 880$ m., $b = 257$ m., $\alpha = 39°\,12'$.

26.11 The distribution law is defined by two error vectors (Figure 42):

$$a_1 = CC_1 = \frac{BE_\beta \sin\beta_2}{\sin^2(\beta_1 + \beta_2)}, \qquad a_2 = CC_2 = \frac{BE_\beta \sin\beta_1}{\sin^2(\beta_1 + \beta_2)},$$

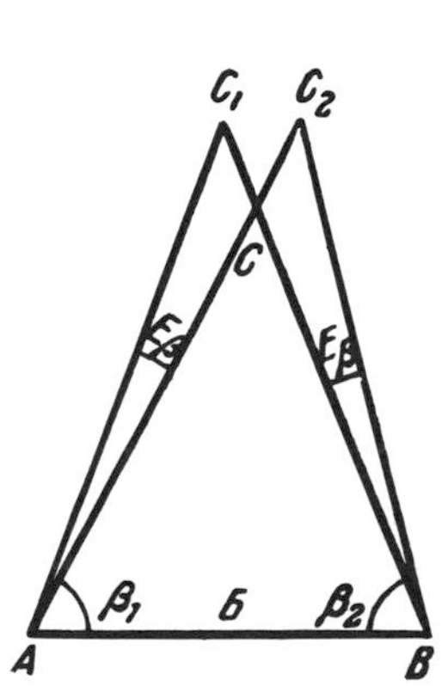

FIGURE 42

$\alpha_1 = \pi - \beta_2$, $\alpha_2 = \beta_1$ as a consequence of which

$$k_{11} = \frac{B^2 E_\beta^2}{2\rho^2 \sin^4 (\beta_1 + \beta_2)} (\sin^2 \beta_1 \cos^2 \beta_1 + \sin^2 \beta_2 \cos^2 \beta_2),$$

$$k_{22} = \frac{B^2 E_\beta^2}{2\rho^2 \sin^4 (\beta_1 + \beta_2)} (\sin^4 \beta_1 + \sin^4 \beta_2),$$

$$k_{12} = \frac{B^2 E_\beta^2}{2\rho^2 \sin^4 (\beta_1 + \beta_2)} (\sin^3 \beta_1 \cos \beta_1 - \sin^3 \beta_2 \cos \beta_2),$$

$$\tan 2\alpha = \frac{\sin^2 \beta_1 \sin 2\beta_1 - \sin^2 \beta_2 \sin 2\beta_2}{\sin^2 \beta_1 \cos 2\beta_1 + \sin^2 \beta_2 \cos 2\beta_2}.$$

26.12 $a = 18.0$ km., $b = 7.39$ km., $\alpha = 85° 36'$.

26.13 To the error vectors $\mathbf{a}_1$ and $\mathbf{a}_2$ one should add another error vector, $\mathbf{a}_3$:

$$a_3 = \frac{\sqrt{E_1^2 \sin^2 \beta_2 + E_2^2 \cos^2 \beta_2}}{\sin (\beta_1 + \beta_2)}$$

for $a_3 = \beta_0$, which gives at the point C a unit ellipse of errors with principal semi-axes $a = 41.2$ m., $b = 19.7$ m., making with the direction of the base the angles 74° 20′ and 164° 20′.

26.14 $E_v = 2.1$ m./sec., $E_q = 0.042$ rad.

26.15 $a = 156$ m., $b = 139$ m.; the principal semi-axes directed along the course of the ship.

26.16 $a = 64.0$ m., $b = c = 78.1$ m.; the semi-axis a is directed along the course of the ship.

26.17 $$f(x, y, z) = \frac{1}{120(2\pi)^{3/2}} \exp\left\{-\frac{(x-45)^2}{50} - \frac{(y-15)^2}{32} - \frac{(z+75)^2}{72}\right\}.$$

26.18 The equation of the unit ellipsoid is

$$\frac{(x-30)^2}{2100} + \frac{y^2}{1125} + \frac{z^2}{64} = 1.$$

26.19

$$\|k_{ij}\| = \begin{Vmatrix} 7421 & -2568 & -7597 \\ -2568 & 8406 & 2322 \\ -7597 & 2322 & 9672 \end{Vmatrix}.$$

26.20 $p = -1.47 \cdot 10^7$, $q = -8.9 \cdot 10^9$, $\varphi = 65° 45'$, $u_1 = 4106$, $u_2 = -622$, $u_3 = -3484$, $a = 89.3$, $b = 57.0$, $c = 19.3$, $\cos(a, x) = \pm 0.6179$, $\cos(a, y) = \mp 0.3528$, $\cos(a, z) = \mp 0.7025$.

FIGURE 43

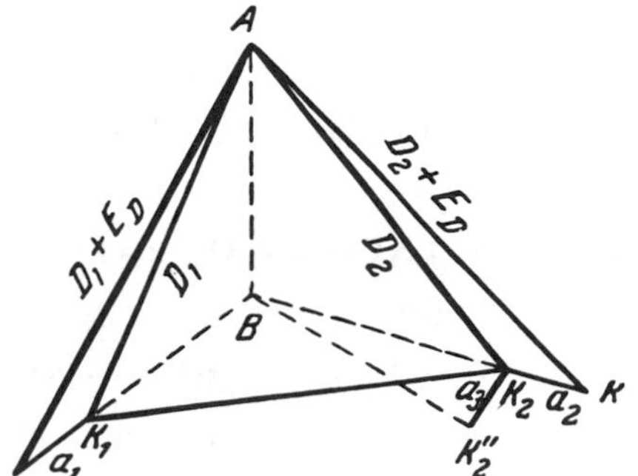

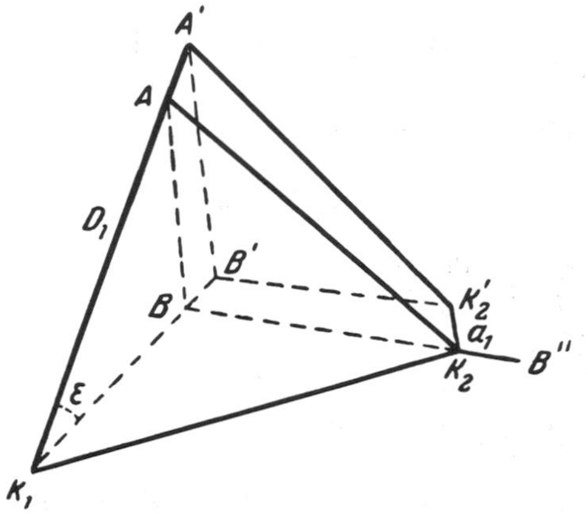

FIGURE 44

26.21 If we take as the x-axis (Figure 43) the direction BK_2 and as the y-axis the direction perpendicular to it, then by the linearization method we find three error vectors:

$$a_1 = \frac{D_1 E_D}{\sqrt{D_1^2 - H^2}}, \quad \alpha_1 = \alpha; \qquad a_2 = \frac{D_2 E_D}{\sqrt{D_2^2 - H^2}}, \quad \alpha_2 = 0;$$

$$a_3 = \sqrt{D_2^2 - H^2}\, E_\alpha, \quad \alpha_3 = 90^\circ.$$

From this we find:

$$k_{11} = \frac{E_D^2}{2\rho^2}\left(\frac{D_1^2 \cos^2\alpha}{D_1^2 - H^2} + \frac{D_2^2}{D_2^2 - H^2}\right),$$

$$k_{12} = \frac{E_D^2}{2\rho^2}\cdot\frac{D_1^2 \sin\alpha\cos\alpha}{D_1^2 - H^2},$$

$$k_{22} = \frac{E_D^2}{2\rho^2}\left[\frac{D_1^2\sin^2\alpha}{D_1^2 - H^2} + \frac{E_\alpha^2}{E_D^2}(D_2^2 - H^2)\right].$$

26.22 The error vectors $\mathbf{a}_2$ and $\mathbf{a}_3$ remain the same in magnitude and direction as in the preceding problem. The magnitude of the error vector $\mathbf{a}_1$ caused by the error in the distance D_1 and its direction $\alpha_1 = \angle K_2'K_2B''$ is determined from the formulas (Figure 44):

$$a_1 = K_2K_2' = E_2\lambda\cos\varepsilon, \qquad \sin\alpha_1 = \frac{1}{\lambda}\sin\alpha,$$

where

$$\lambda = \sqrt{1 + \frac{2D_1\sin\varepsilon\tan\varepsilon\cos\alpha}{\sqrt{D_2^2 - D_1^2\sin^2\varepsilon}} + \frac{D_1^2\sin^2\varepsilon\tan^2\varepsilon}{D_2^2 - D_1^2\sin^2\varepsilon}}.$$

V ENTROPY AND INFORMATION

27. THE ENTROPY OF RANDOM EVENTS AND VARIABLES

27.1 Since

$$H_1 - H_2 = \frac{8}{15}\log 4 - \frac{5}{15}\log 5 - \frac{3}{15}\log 3 - \log 15\left[\frac{8}{15} - \frac{5}{15} - \frac{3}{15}\right]$$
$$= -0.733 < 0,$$

the outcome of the experiment for the first urn is more certain.

27.2 $p = 1/2$.

27.3 $$H_1 = -\frac{\pi}{3\sqrt{3}} \log \frac{\pi}{3\sqrt{3}} - \left(1 - \frac{\pi}{3\sqrt{3}}\right) \log \left(1 - \frac{\pi}{3\sqrt{3}}\right)$$
$$= 0.297 \text{ decimal unit},$$
$$H_2 = -\frac{3\sqrt{3}}{4\pi} \log \frac{3\sqrt{3}}{4\pi} - \left(1 - \frac{3\sqrt{3}}{4\pi}\right) \log \left(1 - \frac{3\sqrt{3}}{4\pi}\right)$$
$$= 0.295 \text{ decimal unit};$$

that is, the uncertainties are practically the same.

27.4 (a) $H = -\cos^2 \frac{\pi}{n} \log_a \cos^2 \frac{\pi}{n} - \sin^2 \frac{\pi}{n} \log_a \sin^2 \frac{\pi}{n}$, (b) $n = 4$.

27.5 Since $\mathbf{P}(X = k) = p(1 - p)^{k-1}$, then
$$H[X] = -\frac{p \log_a p + (1 - p) \log_a (1 - p)}{p}.$$

When p decreases from 1 to 0, the entropy increases monotonically from 0 to ∞.

27.6 (a) $$H[X] = -n[p \log_a p + (1 - p) \log_a (1 - p)] - \sum_{m=1}^{n-1} C_n^m p^m (1 - p)^{n-m} \log_a C_n^m,$$

(b) $H[X] = 1.5 \log_a 2$.

27.7 (a) $\log_a (d - c)$, (b) $\log_a [\sigma_x \sqrt{2\pi e}]$, (c) $\log_a (e/c)$.

27.8 $H[X] = \log_a (0.5\sqrt{e})$.

27.9 $$H[X \mid y] = H_y[X] = \log_a (\sigma_x \sqrt{2\pi e(1 - r^2)}),$$
$$H[Y \mid x] = H_x[Y] = \log_a (\sigma_y \sqrt{2\pi e(1 - r^2)}),$$

where σ_x and σ_y are the standard deviations and r is the correlation coefficient between X and Y.

27.10 $$H[X_1, X_2, \ldots, X_n] = \int_{-\infty}^{\infty} \cdots \int_{-\infty}^{\infty} \frac{1}{\sqrt{(2\pi)^n |k|}} \times \exp\left\{-\frac{1}{2|k|} \sum_{i,j} a_{ij} x_i x_j\right\} \times \left[\frac{1}{2|k|} \sum_{i,j} a_{ij} x_i x_j \log_a e + \log_a \sqrt{(2\pi)^n |k|}\right] dx_1 \cdots dx_n$$
$$= \log_a \sqrt{(2\pi e)^n |k|},$$

where $|k|$ is the determinant of the covariance matrix.

27.11 $H_x[Y] = H[Y] - H[X] + H_y[X]$.

27.12 The uniform distribution law:
$$f(x) = \begin{cases} \dfrac{1}{b - a} & \text{for } a \leqslant x \leqslant b, \\ 0 & \text{for } x < a,\ x > b. \end{cases}$$

27.13 The exponential distribution law:
$$f(x) = \begin{cases} \dfrac{1}{\mathbf{M}[X]} \exp\left\{-\dfrac{x}{\mathbf{M}[X]}\right\} & \text{for } x \geqslant 0, \\ 0 & \text{for } x < 0. \end{cases}$$

27.14 $f(x) = \frac{1}{\sqrt{2\pi m_2}} \exp\left\{-\frac{x^2}{2m_2}\right\}.$

27.15 The normal law:

$$f(x_1, x_2, \ldots, x_n) = \frac{1}{\sqrt{(2\pi)^n |k|}} \exp\left\{-\frac{1}{2|k|} \sum_{i,j} a_{ij}(x_i - \mathbf{M}[X_i])(y_j - \mathbf{M}[Y_j])\right\}.$$

27.16 $p_{1i} = \frac{\alpha}{k}, \; p_{2j} = \frac{1-\alpha}{n}.$ **27.17** $\log_a 1050$ and $\log_a 30$.

27.18 $H[Y_1, Y_2, \ldots, Y_n] - H[X_1, X_2, \ldots, X_n]$

$$= \int_{-\infty}^{\infty} \cdots \int_{-\infty}^{\infty} f_x(x_1, x_2, \ldots, x_n) \log_a \left| I\left(\frac{\partial \varphi_k}{\partial x_j}\right)\right| dx_1 \cdots dx_n,$$

where $I(\partial\varphi_k/\partial x_j)$ is the Jacobian of the transformation from $(Y_1, Y_2, \ldots, Y_n)$ to $(X_1, X_2, \ldots, X_n)$.

27.19 (a) The logarithm of the absolute value of the determinant $|a_{kj}|$, (b) 1.85 decimal unit.

28. THE QUANTITY OF INFORMATION

28.1 (a) 5 binary units, (b) 5 binary units, (c) 3 binary units.

28.2 For a number of coins satisfying the inequality $3^{k-1} < N \leqslant 3^k$, k weighings are necessary. For $k = 5$, one may find a counterfeit coin if the total number of coins does not exceed 243.

28.3 $I = 500(-0.51 \log_2 0.51 - 0.31 \log_2 0.31 - 0.12 \log_2 0.12 - 0.06 \log_2 0.06) = 815$ binary units.

28.4 The first experiment gives the amount of information

$$I_1 = H_0 - H_1 = \log_2 N - \frac{1}{N}[k \log_2 k + (N-k)\log_2 (N-k)]$$

and the second experiment

$$I_2 = H_1 - H_2 = \frac{1}{N}[k \log_2 k + (N-k)\log_2(N-k)] - \frac{1}{N}[l \log_2 l + (k-l)\log_2(k-l) + r \log_2 r + (N-k-r)\log_2(N-k-r)].$$

28.5 The minimal number of tests is three in the sequences No. 6, No. 5 and No. 3, for example. *Hint:* Determine the amount of information given by each test and select as the first test one of those that maximizes the amount of information. Similarly, select the numbers of successive tests until the entropy of the system vanishes. To compute the amount of information, use the answer to the preceding problem.

28.6
$$\frac{I}{T} = -\frac{\sum_i \mathbf{P}(A_i)\log_2 \mathbf{P}(A_i)}{\sum_j \mathbf{P}(a_j)t_j},$$

where $\mathbf{P}(a_j) = \mathbf{P}(A_i)$ if the code a_j corresponds to the symbol A_i of the alphabet. For code No. 1,

$$\frac{I}{T} = \frac{1.782}{5.85} = 0.304 \text{ binary units/time units.}$$

For code No. 2,

$$\frac{I}{T} = \frac{1.782}{6.30} = 0.283 \text{ binary units/time units.}$$

28.7 For a more efficient code, the symbols of the code with the same serial numbers arranged in the order of their increasing durations should correspond to the symbols of the alphabet arranged in the order of decreasing probabilities; that is, the symbols d, c, b and a of the code should correspond to the symbols A_1, A_4, A_3 and A_2. The efficiency of such a code is

$$\frac{I}{T_{\min}} = \frac{1.782}{4.55} = 0.391 \text{ binary units/time units}.$$

28.8 $$l = 1 - \frac{H}{H_{\max}} = 1 + \frac{0.8 \log_2 0.8 + 0.1 \log_2 0.1 + 0.1 \log_2 0.1}{\log_2 3} = 0.42.$$

28.9 (a) See Table 126.

TABLE 126

Letter	A	B
Probabilities	0.8	0.2
Coded notations	1	0

(b) See Table 127.

TABLE 127

Letter combinations	AA	AB	BA	BB
Probabilities	0.64	0.16	0.16	0.04
Coded notations	1	01	001	000

(c) See Table 128.

TABLE 128

Letter combinations	AAA	AAB	ABA	BAA	ABB	BAB	BBA	BBB
Probabilities	0.512	0.128	0.128	0.128	0.032	0.032	0.032	0.008
Coded notations	1	011	010	001	00011	00010	00001	00000

The efficiencies of the codes are respectively:

(a) $\frac{0.722}{1} = 0.722$, (b) $\frac{1.444}{1.56} = 0.926$, (c) $\frac{2.166}{2.184} = 0.992$.

28.10 (a) $\mathbf{P}(1) = 0.8$, $\mathbf{P}(0) = 0.2$, $l_a = 1 - 0.722 = 0.278$,

(b) $\mathbf{P}(1) = \frac{0.96}{1.56} = 0.615$, $\mathbf{P}(0) = 0.385$, $l_\sigma = 1 - 0.962 = 0.038$,

(c) $\mathbf{P}(1) = \frac{1.152}{2.184} = 0.528$, $\mathbf{P}(0) = 0.472$, $l_B = 1 - 0.9977 = 0.0023$.

28.11 (a) See Tables 129 and 130.

TABLE 129

Letters	A	B	C
Probabilities	0.7	0.2	0.1
Coded notations	1	01	00

TABLE 130

Two-letter combinations	AA	AB	BA	AC	CA	BB	BC	CB	CC
Probabilities	0.49	0.14	0.14	0.07	0.07	0.04	0.02	0.02	0.01
Coded notations	1	011	010	0011	0010	0001	00001	000001	000000

(b) The efficiencies of the codes are 0.890 and 0.993, respectively. (c) The redundancies of the codes are 0.109 and 0.0007, respectively.

28.12 See Table 131.

TABLE 131

Letters	*Coded notations*	*Letters*	*Coded notations*	*Letters*	*Coded notations*	*Letters*	*Coded notations*
—	111	р	01011	я	001001	х	0000100
о	110	в	01010	ы	001000	ж	0000011
е, ё	1011	л	01001	з	000111	ю	0000010
а	1010	к	01000	ь, ъ	000110	ш	00000011
и	1001	м	00111	б	000101	ц	00000010
т	1000	д	00101	г	000100	щ	00000001
н	0111	п	001101	ч	000011	з	000000001
с	0110	у	001100	й	0000101	ф	000000000

28.13 Use the fact that the coded notation of the letter A_j will consist of k_j symbols.

28.14 In the absence of noise, the amount of information is the entropy of the input communication system:

$$I = -\mathbf{P}(A_1)\log_2 \mathbf{P}(A_1) - \mathbf{P}(A_2)\log_2 \mathbf{P}(A_2) = 1 \text{ binary unit}.$$

In the presence of noise $I = 0.919$ binary unit; it decreases by an amount equal to the magnitude of the average conditional entropy, namely

$$-\mathbf{P}(a_1)[\mathbf{P}(A_1 \mid a_1)\log_2 \mathbf{P}(A_1 \mid a_1) + \mathbf{P}(A_2 \mid a_1)\log_2 \mathbf{P}(A_2 \mid a_1)]$$
$$-\mathbf{P}(a_2)[\mathbf{P}(A_1 \mid a_2)\log_2 \mathbf{P}(A_1 \mid a_2) + \mathbf{P}(A_2 \mid a_2)\log_2 \mathbf{P}(A_2 \mid a_2)],$$

where

$$\mathbf{P}(A_j \mid a_i) = \frac{\mathbf{P}(A_j)\mathbf{P}(a_i \mid A_j)}{\mathbf{P}(a_i)}.$$

28.15 If the noise is absent, $I = H_1 = \log_2 m$; when the noise is present $I = H_1 - H_2 = \log_2 m + p \log_2 p + q \log_2 q/(m - 1)$.

28.16 $$I = \log_2 m + \sum_i \sum_j \mathbf{P}(a_i)\mathbf{P}(A_j \mid a_i) \log_2 \mathbf{P}(A_j \mid a_i),$$

where

$$\mathbf{P}(a_i) = \frac{1}{m} \sum_j p_{ij}, \qquad \mathbf{P}(A_j \mid a_i) = \frac{p_{ij}}{\sum_j p_{ij}}.$$

VI THE LIMIT THEOREMS

29. THE LAW OF LARGE NUMBERS

29.1 (a) $\mathbf{P}(|X - \bar{x}| \geqslant 4E) \leqslant 0.1375$, (b) $\mathbf{P}(|X - \bar{x}| \geqslant 3\sigma) \leqslant 1/9$.

29.2 It is proved in the same manner as one proves Chebyshev's inequality. For the proof make use of the obvious inequality

$$\int_\Omega f(x)\, dx \leqslant \int_\Omega \frac{1}{\zeta} e^{\varepsilon x - t^2} f(x)\, dx,$$

where Ω is the set of all x satisfying the condition

$$x > \frac{t^2 + \ln J}{\varepsilon}.$$

29.3 Using arguments analogous to those in the proof of the Chebyshev inequality, one obtains a chain of inequalities

$$\mathbf{P}(X \geqslant \varepsilon) \leqslant \frac{1}{e^{a\varepsilon}} \int_{e^{ax} \geqslant e^{a\varepsilon}} e^{ax}\, dF(e^{ax}) \leqslant e^{-a\varepsilon}\mathbf{M}[e^{ax}].$$

29.4 Use the Chebyshev inequality and note that $\bar{x} = m + 1$, and $\mathbf{M}[X^2] = (m + 1)(m + 2)$, hence,

$$\mathbf{P}(0 < X < 2(m + 1)) = \mathbf{P}'(|X - \bar{x}| < m + 1) > 1 - \frac{\mathbf{D}[X]}{(m + 1)^2}.$$

29.5 Denoting by X_n the random number of occurrences of the event A in n experiments, we have $\mathbf{P}(|X_n - 500| < 100) > 1 - (250/100^2) = 0.975$. Consequently, all questions may be answered "yes."

29.6 The random variables X_k are mutually independent and have equal expectations $\bar{x}_k = 0$ and variances $\mathbf{D}[X_k] = 1$, which prove that the conditions of the Chebyshev theorem are satisfied.

29.7 For $s < 1/2$, since in this case

$$\lim_{n\to\infty} \mathbf{D}\left[\frac{1}{n} \sum_{k=1}^{n} X_k\right] = \lim \frac{1}{n^2} \sum_{k=1}^{n} k^{2s} = 0.$$

29.8 $$\lim_{n\to\infty} \mathbf{D}\left[\frac{1}{n} \sum_{k=1}^{n} X_k\right] = \lim_{n\to\infty} \frac{1}{n^2} \ln (n!) = \lim_{n\to\infty} \frac{1}{n^2} \ln \{n^{n+(1/2)} e^{-n} \sqrt{2\pi}\}$$
$$= \lim_{n\to\infty} \frac{1}{n^2} \left\{\left(n + \frac{1}{2}\right) \ln n - n + \ln \sqrt{2\pi}\right\}$$
$$= \lim_{n\to\infty} \frac{\ln n}{n} = 0,$$

which proves the applicability of the law of large numbers.

29.9 (a) Not satisfied, since

$$\lim_{n\to\infty} \mathbf{D}\left[\frac{1}{n}\sum_{k=1}^{n} X_k\right] = \lim_{n\to\infty} \frac{4(4^n - 1)}{3n^2} = \infty,$$

(b) satisfied, since

$$\lim_{n\to\infty} \mathbf{D}\left\{\frac{1}{n}\sum_{k=1}^{n} X_k\right\} = \lim_{n\to\infty} \frac{1}{n} = 0,$$

(c) not satisfied, since

$$\lim_{n\to\infty} \mathbf{D}\left\{\frac{1}{n}\sum_{k=1}^{n} X_k\right\} > \lim_{n\to\infty} \frac{n(n+1)}{2n^2} = \frac{1}{2}.$$

29.10 Applicable since the inequality

$$0 \leqslant \mathbf{D}\left[\frac{1}{n}\sum_{k=1}^{n} X_k\right] < \frac{1}{n^2}\sum_{k=1}^{n} \mathbf{D}[X_k] < \frac{c}{n},$$

where c is the upper bound of $\mathbf{D}[X_k]$ for all $k = 1, 2, \ldots, n$, holds for $k_{ij} < 0$. The relation

$$\lim_{n\to\infty} \mathbf{D}\left\{\frac{1}{n}\sum_{k=1}^{n} X_k\right\} = 0,$$

follows from the inequality.

29.11 To prove this, it suffices to estimate

$$\mathbf{D}\left[\frac{1}{n}\sum_{k=1}^{n} X_k\right] = \frac{1}{n^2}\left[\sum_{k=1}^{n} \sigma_k^2 + 2\sum_{k=1}^{n-1} r_{k,k+1}\sigma_k\sigma_{k+1}\right],$$

where

$$\sigma_k^2 = \mathbf{D}[X_k], \quad \text{and} \quad r_{k,k+1} = \frac{\mathbf{M}[(X_k - \bar{x}_k)(X_{k+1} - \bar{x}_{k+1})]}{\sigma_k\sigma_{k+1}}.$$

Replacing all σ_k by their maximal value b, we obtain

$$\mathbf{D}\left[\frac{1}{n}\sum_{k=1}^{n} X_k\right] < \frac{3n-2}{n^2} b^2,$$

hence, it follows immediately that

$$\lim_{n\to\infty} \mathbf{D}\left[\frac{1}{n}\sum_{k=1}^{n} X_k\right] = 0.$$

29.12 Applicable, since all the assumptions of Khinchin's theorem are satisfied.

29.13 Consider

$$\mathbf{D}[Z_n] = \mathbf{D}\left[\frac{1}{n}\sum_{k=1}^{n} X_k\right] = \frac{1}{n^2}\left|\sum_{i=1}^{n}\sum_{j=1}^{n} r_{ij}\sigma_i\sigma_j\right| < \frac{c}{n^2}\sum_{i=1}^{n}\sum_{j=1}^{n} |r_{ij}|,$$

where σ_i is the standard deviation of the random variable X_i. Since $r_{ij} \to 0$ for $|i - j| \to \infty$, then for any $\varepsilon > 0$, one may indicate an N such that the inequality $|r_{ij}| < \varepsilon$ holds for all $|i - j| > N$. This means that in the matrix $\|r_{ij}\|$, containing n^2 elements, at most Nn elements exceed ε (these elements are replaced by unity) and the rest are less than ε. From the preceding facts, we infer the inequality

$$\left|\frac{1}{n^2}\sum_{i=1}^{n}\sum_{j=1}^{n} r_{ij}\right| < \frac{Nn}{n^2} + \frac{1}{n^2}(n^2 - Nn)\varepsilon = \varepsilon + \frac{N}{n}(1 - \varepsilon),$$

therefore, $\lim_{n\to\infty} \mathbf{D}[z_n] = 0$; this proves the theorem.

29.14 The law of large numbers cannot be applied since the series

$$\frac{6}{\pi^2}\sum_{k=1}^{\infty} \frac{(-1)^{k-1}}{k},$$

defining $\mathbf{M}[X_i]$ is not absolutely convergent.

30. THE DE MOIVRE-LAPLACE AND LYAPUNOV THEOREMS

30.1 $\mathbf{P}\left(0.2 \leqslant \frac{m}{n} < 0.4\right) = 0.97.$ **30.2** $\mathbf{P}(70 \leqslant m < 86) = 0.927.$

30.3 (a) $\mathbf{P}(m \geqslant 20) = 0.5$, (b) $\mathbf{P}(m < 28) = 0.9772$, (c) $\mathbf{P}(14 \leqslant m < 26) = 0.8664$.

30.4 In the limiting equality of the de Moivre-Laplace theorem, set

$$b = -a = \varepsilon\sqrt{\frac{n}{pq}}$$

and then make use of the integral representations of the functions $\Phi(x)$ and $\hat{\Phi}(x)$.

30.5 Because the probability of the event is unknown, the variance of the number of occurrences of the event should be taken as maximal; that is, $pq = 0.25$. In this case: (a) $n \approx 250{,}000$, (b) $n = 16{,}600$.

30.6 In the problems in which the upper limit of the permitted number of occurrences is equal to the number of experiments performed, b turns out to be so large that $\Phi(b) \approx 1$. In this case, $n \approx 108$.

30.7 $n \approx 65$. **30.8** $p = 0.943$. **30.9** 67.5.

30.10 $J = \int_0^1 x^2\,dx$ may be considered as the moment of second order of a random variable, uniformly distributed over the interval [0, 1]; then its statistical analog determined by a Monte-Carlo method will be $J_n = (1/n)\sum_{k=1}^{n} X_k^2$, where X_k are random numbers on the interval [0, 1]. With the aid of Lyapunov's theorem, we find that $\mathbf{P}(|J_{1000} - J| < 0.01) = 0.71$.

30.11 $n \approx 1.55 \cdot 10^6$. Set $J_n = (\pi/2n)\sum_{k=1}^{n} \sin X_k$, where X_k are random numbers from the interval $(0, \pi/2)$.

30.12 (a) Since the difference

$$\mathbf{P}(C) - P_n(C) = \left[\frac{m}{n} - \mathbf{P}(A)\right] \times [1 - \mathbf{P}(B \mid A)],$$

from the point of view of the law of large numbers both methods lead to correct results, (b) in the first case, 9750 experiments will be necessary and in the second case, 4500 experiments.

30.13 (a) 3100, (b) 1500.

30.14 In all three cases the limiting characteristic function equals $e^{-u^2/2}$.

30.15 $\lim_{n\to\infty} E_{Y_n}(u) = e^{-u^2/2}$.

VII THE CORRELATION THEORY OF RANDOM FUNCTIONS

31. GENERAL PROPERTIES OF CORRELATION FUNCTIONS AND DISTRIBUTION LAWS OF RANDOM FUNCTIONS

31.1 Denoting by $f(x_1, x_2 \mid t_1, t_2)$ the distribution law of second order for the random function $X(t)$, by the definition of $K_x(t_1, t_2)$ we have

$$K_x(t_1, t_2) = \int_{-\infty}^{\infty}\int_{-\infty}^{\infty} (x_1 - \bar{x}_1)(x_2 - \bar{x}_2) f(x_1, x_2 \mid t_1, t_2)\,dx_1\,dx_2.$$

Applying the Schwarz inequality, we get

$$|K_x(t_1, t_2)|^2 \leqslant \int_{-\infty}^{\infty}\int_{-\infty}^{\infty} (x_1 - \bar{x}_1)^2 f(x_1, x_2 \mid t_1, t_2)\, dx_1\, dx_2$$
$$\times \int_{-\infty}^{\infty}\int_{-\infty}^{\infty} (x_2' - \bar{x}_2)^2 f(x_1', x_2' \mid t_1, t_2)\, dx_1'\, dx_2' = \sigma^2_{x(t_1)}\sigma^2_{x(t_2)},$$

which is equivalent to the first inequality. To prove the second inequality, it suffices to consider the evident relation $\mathbf{M}\{[X(t_1) - \bar{x}(t_1)] - [X(t_2) - \bar{x}(t_2)]|^2\} \geqslant 0$.

31.2 The proof is similar to the preceding one.

31.3 It follows from the definition of the correlation function.

31.4 Since $X(t) = \sum_{j=1}^{n} \Delta_j + c$, where c is a nonrandom constant and n is the number of steps during time t, we have $\mathbf{D}[X(t)] = \mathbf{M}[n\sigma^2] = \lambda t\sigma^2$.

31.5 The correlation function $K_x(\tau)$ is the probability that an even number of sign changes will occur during time τ minus the probability of an odd number of sign changes; that is,

$$K_x(\tau) = \sum_{n=0}^{\infty} \frac{(\lambda\tau)^{2n}}{(2n)!} e^{-\lambda\tau} - \sum_{n=0}^{\infty} \frac{(\lambda\tau)^{2n+1}}{(2n+1)!} e^{-\lambda\tau} = e^{-2\lambda\tau}.$$

31.6 Since $\mathbf{M}[X(t)X(t + \tau)] \neq 0$ only if $(t, t + \tau)$ is contained in an interval between consecutive integers and, since the probability of this event is 0 if $|\tau| > 0$ and $(1 - |\tau|)$ if $|\tau| \leqslant 1$, we have for $|\tau| \leqslant 1$,

$$K_x(\tau) = (1 - |\tau|)\mathbf{M}[X^2] = (1 - |\tau|) \times \int_0^{\infty} x^2 \frac{x^{\lambda}}{\Gamma(\lambda + 1)} e^{-x}\, dx$$
$$= (\lambda + 2)(\lambda - 1)(1 - |\tau|).$$

Consequently,

$$K_x(\tau) = \begin{cases} (\lambda + 2)(\lambda + 1)(1 - |\tau|), & |\tau| \leqslant 1, \\ 0, & |\tau| > 1. \end{cases}$$

31.7 Letting $\Theta_1 = \Theta(t_1)$, $\Theta_2 = \Theta(t_1 + \tau)$, for the conditional distribution law we get

$$f(\theta_2 \mid \theta_1 = 5°) = \frac{f(\theta_1, \theta_2)}{f(\theta_1)},$$

where $f(\theta_1, \theta_2)$ is the normal distribution law of a system of random variables with correlation matrix

$$\begin{Vmatrix} K_\theta(0) & K_\theta(\tau) \\ K_\theta(\tau) & K_\theta(0) \end{Vmatrix}.$$

Substituting the data from the assumption of the problem, we get

$$P = \int_{15}^{\infty} f(\theta_2 \mid \theta_1 = 5°)\, d\theta_2 = \frac{1}{2}[1 - \Phi(2.68)] = 0.0037.$$

31.8 Denoting the heel angles at instants t and $t + \tau$ by Θ_1 and Θ_2, respectively, and their distribution law by $f(\theta_1, \theta_2)$, for the conditional distribution law of the heel angle at the instant of second measurement, we get

$$f(\theta_2 \mid -\theta_0 \leqslant \Theta_1 \leqslant \theta_0) = \frac{\int_{-\theta_0}^{\theta_0} f(\theta_1, \theta_2)\, d\theta_1}{\int_{-\theta}^{\theta_0}\int_{-\infty}^{\infty} f(\theta_1, \theta_2)\, d\theta_2\, d\theta_1}.$$

The required probability is

$$P = \frac{1}{2\sqrt{2\pi}\,\sigma_\theta \Phi\left(\frac{\theta_0}{\sigma_\theta}\right)} \int_{-\theta_0}^{\theta_0} \exp\left\{-\frac{\theta_2^2}{2\sigma_\theta^2}\right\}\left[\Phi\left(\frac{\theta_0 + k\theta_2}{\sigma_\theta\sqrt{1 - k^2}}\right) + \Phi\left(\frac{\theta_0 - k\theta_2}{\sigma_\theta\sqrt{1 - k^2}}\right)\right] d\theta_2.$$

31.9 Denoting $X_1 = \Theta(t)$, $X_2 = \dot{\Theta}(t)$, $X_3 = \Theta(t + \tau_0)$, the correlation matrix of the system X_1, X_2, X_3 becomes

$$\|k_{jl}\| = \begin{Vmatrix} K_\theta(0) & 0 & K_\theta(\tau_0) \\ 0 & -\ddot{K}_\theta(0) & -\dot{K}(\tau_0) \\ K_\theta(\tau_0) & -\dot{K}_\theta(\tau_0) & K_\theta(0) \end{Vmatrix},$$

which after numerical substitution becomes

$$\|k_{jl}\| = \begin{Vmatrix} 36 & 0 & 36e^{-0.5} \\ 0 & 36(0.25^2 + 1.57^2) & 0 \\ 36e^{-0.5} & 0 & 36 \end{Vmatrix}.$$

Determining the conditional distribution law according to the distribution law $f(x_1, x_2, x_3)$

$$f(x_3 \mid x_1 = 2, x_2 > 0) = \left. \frac{\int_0^\infty f(x_1, x_2, x_3)\, dx_2}{\int_{-\infty}^\infty \int_0^\infty f(x_1, x_2, x_3)\, dx_2\, dx_3} \right|_{x_1 = 2},$$

we obtain for the required probability,

$$P = \int_{-10}^{10} f(x_3 \mid x_1 = 2, x_2 > 0)\, dx_3 = 0.958.$$

31.10 $\bar{y}(t) = a(t)\bar{x}(t) + b(t)$; $K_y(t_1, t_2) = a^*(t_1)a(t_2)K_x(t_1, t_2)$.

31.11 $f(x)\, dx = \iint\limits_{x \leqslant a \cos\theta \leqslant x + dx} f_a(a) f_\theta(\theta)\, da\, d\theta$; $f(x) = \dfrac{1}{\sigma\sqrt{2\pi}} \exp\left\{-\dfrac{x^2}{2\sigma^2}\right\}$.

31.12 The probability that the interval T will lie between τ and $\tau + d\tau$ is the probability that there will be n points in the interval $(0, \tau)$ and one point in the interval $(\tau, \tau + d\tau)$. Since by assumption these events are independent, we have

$$\mathbf{P}(\tau \leqslant T \leqslant \tau + d\tau) = \frac{(\lambda\tau)^n}{n!} e^{-\lambda\tau}\lambda\, d\tau;$$

that is,

$$f(\tau) = \frac{\lambda^{n+1}\tau^n}{n!} e^{-\lambda\tau}.$$

31.13 $f(u) = \dfrac{1}{15.8\sqrt{2\pi}} \exp\left\{-\dfrac{u^2}{498}\right\}$.

32. LINEAR OPERATIONS WITH RANDOM FUNCTIONS

32.1 Since $\dot{K}_x(\tau)$ has no discontinuity at $\tau = 0$,

$$K_{\dot{x}}(\tau) = -\frac{d^2}{d\tau^2} K_x(\tau) = a\alpha^2 e^{-\alpha|\tau|}(1 - \alpha|\tau|).$$

32.2 $K_y(\tau) = a(\alpha^2 + \beta^2)e^{-\alpha|\tau|}\left(\cos\beta\tau - \dfrac{\alpha}{\beta}\sin\beta|\tau|\right)$,

$\mathbf{D}[Y(t)] = K_y(0) = a(\alpha^2 + \beta^2)$.

32.3 Using the definition of a mutual correlation function, we get

$$R_{x\dot{x}}(\tau) = \mathbf{M}\left\{[X^*(t) - \bar{x}^*] \frac{dX(t + \tau)}{d\tau}\right\}$$

$$= \frac{d}{d\tau} \mathbf{M}\{[X^*(t) - \bar{x}^*][X(t + \tau) - \bar{x}]\} = \frac{d}{d\tau} K_x(\tau).$$

32.4 Since any derivative of $K_x(\tau)$ is continuous at zero, $X(t)$ may be differentiated any number of times.

32.5 Twice, since $(d^2/d\tau^2)K_x(\tau)|_{\tau=0}$ and $(d^4/d\tau^4)K_x(\tau)|_{\tau=0}$ exist, $(d^5/d\tau^5)K_x(\tau)$ has a discontinuity at zero.

32.6 Only the first derivative exists since $(d^2/d\tau^2)K_x(\tau)$ exists for $\tau = 0$ and $(d^3/d\tau^3)K_x(\tau)$ has a discontinuity at this point.

32.7 $R_{\dot{x}x}(\tau) = \alpha^2\sigma_x^2(\tau - t_0)e^{-\alpha|\tau - t_0|}$.

32.8 $\mathbf{D}[Y(t)] = \sigma_x^2$, $\mathbf{D}[Z(t)] = \alpha^2\sigma_x^2$.

32.9 $K_y(\tau) = 2a^2\sigma_x^2\alpha^2 e^{-\alpha^2\tau^2}(1 - 2\alpha^2\tau^2)$.

32.10 The distribution $f(v)$ is normal with variance $\sigma_v^2 = a(\alpha^2 + \beta^2)$ and $\bar{v} = 0$, $P = 0.3085$.

32.11 $\bar{z}(t) = \bar{x}(t) + \bar{y}(t)$;

$$K_z(t_1, t_2) = K_x(t_1, t_2) + K_y(t_1, t_2) + R_{xy}(t_1, t_2) + R_{yx}(t_1, t_2).$$

32.12. $\bar{x}(t) = \sum_{j=1}^{n} \bar{x}_j(t)$; $K_x(t_1, t_2) = \sum_{j=1}^{n} K_{x_j}(t_1, t_2) + \sum_{\substack{l=1 \\ l \neq j}}^{n} \sum_{j=1}^{n} R_{x_l x_j}(t_1, t_2)$.

32.13 $K_y(\tau) = K_x(\tau) + \dfrac{d^2}{d\tau^2} K_x(\tau) + \dfrac{d^4}{d\tau^4} K_x(\tau)$.

32.14

$$K_z(\tau) = \sigma_x^2 e^{-\alpha|\tau|}\left\{1 + \alpha|\tau| + \frac{1}{3}\alpha^2\tau^2 + \frac{2\alpha^2}{3}(\alpha^2\tau^2 - \alpha|\tau| - 1) + \frac{\alpha^4}{3}(\alpha^2\tau^2 - 5\alpha|\tau| + 3)\right\}.$$

32.15 Since $K_y(t_1, t_2) = \int_0^{t_1}\int_0^{t_2} K_x(t_2' - t_1')\,dt_2'\,dt_1'$, if we let $t_2 = t_1 = t$, pass to new variables of integration, and perform the integration, we obtain

$$\mathbf{D}[Y(t)] = K_y(t, t) = 2\int_0^t (t - \tau)K_x(\tau)\,d\tau.$$

32.16 Solving the problem as we did 32.15, after transformation of the double integral we get

$$K_x(t_1, t_2) = \int_0^{t_2} (t_2 - \tau)K_y(\tau)\,d\tau + \int_0^{t_1} (t_1 - \tau)K_y(\tau)\,d\tau - \int_0^{t_2 - t_1} (t_2 - t_1 - \tau)K_y(\tau)\,d\tau.$$

32.17 $R_{xy}(t_1, t_2) = \int_0^{t_2} K_x(t_1, \xi)\,d\xi$. **32.18** $\mathbf{D}[Y(20)] = 1360$ cm.^{2}.

32.19

$$\bar{y}(t) = a_0\bar{x}(t) + a_1\frac{d\bar{x}(t)}{dt} + b_1\int_0^t e^{-\lambda t_1}\bar{x}(t_1)\,dt_1 + c,$$

$$\begin{aligned}K_y(t_1, t_2) = {}& a_0^2K_x(t_1, t_2) + a_0a_1\left[\frac{\partial K_x(t_1, t_2)}{\partial t_1} + \frac{\partial K_x(t_1, t_2)}{\partial t_2}\right] + a_1^2\frac{\partial^2 K_x(t_1, t_2)}{\partial t_1\,\partial t_2}\\ &+ a_0b_1\left[\int_0^{t_1} e^{-\lambda t_1'}K_x(t_1', t_2)\,dt_1' + \int_0^{t_2} e^{-\lambda t_2'}K_x(t_1, t_2')\,dt_2'\right]\\ &+ a_1b_1\left[\int_0^{t_1} e^{-\lambda t_1'}\frac{\partial K_x(t_1', t_2)}{\partial t_2}\,dt_1' + \int_0^{t_2} e^{-\lambda t_2'}\frac{\partial K_x(t_1, t_2')}{\partial t_1}\,dt_2'\right]\\ &+ b_1^2\int_0^{t_2}\int_0^{t_1} e^{-\lambda(t_1' + t_2')}K_x(t_1', t_2')\,dt_1'\,dt_2'.\end{aligned}$$

32.20 $R_{yz}(t_1, t_2) = ac\dfrac{\partial K_x(t_1, t_2)}{\partial t_2} + ad\dfrac{\partial^2 K_x(t_1, t_2)}{\partial t_2^2}$

$$+ bc\frac{\partial^2 K_x(t_1, t_2)}{\partial t_1\,\partial t_2} + bd\frac{\partial^3 K_x(t_1, t_2)}{\partial t_1\,\partial t_2^2}.$$

32.21 Since the variance $\mathbf{D}[\theta(t)]$ is small, $\sin\theta \approx \theta$,

$$\mathbf{D}[\Delta V(t)] = 2g^2\int_0^t (t-\tau)K_\theta(\tau)\,d\tau = \frac{2g^2}{\alpha}a\left[t - \frac{1}{\alpha}(1 - e^{-\alpha t})\right],$$

which after substitution of numerical values leads to $\sigma_{\Delta_v} = 18.6$ m./sec.

32.22 Using the definition of the correlation function as the expectation of the product of the deviations of the ordinates of a random function, and the formulas for the moments of normal random variables, we obtain

$$K_x(\tau) = a^2K_\theta^{(\text{IV})}(\tau) + b^2K_\theta(\tau) + 2c^2\ddot{K}_\theta^2(\tau) - 2ab\ddot{K}_\theta^2(\tau).$$

32.23 $K_m(\tau) = 2a^2K_\theta^2(\tau) + 2b^2K_\psi^2(\tau) - c^2K_\theta(\tau)\ddot{K}_\psi(\tau)$.

32.24 $K_y(\tau) = e^{-\alpha^2\tau^2}[1 + 2\alpha^2(1 - 2\alpha^2\tau^2)]$.

32.25 $R_{xy}(\tau) = -\dfrac{1}{3}a\alpha^2e^{-\alpha|\tau|}(1 + \alpha|\tau| - \alpha^2\tau^2)$.

32.26 $K_y(t_1, t_2) = a^*(t_1)a(t_2)K_x(\tau) + b^*(t_1)b(t_2)\dfrac{d^4K_x(\tau)}{d\tau^4}$

$$+ [a^*(t_1)b(t_2) + b^*(t_1)a(t_2)]\frac{d^2K_x(\tau)}{d\tau^2}.$$

32.27 It does not exist. **32.28** (a) Stationary, (b) nonstationary.

32.29 $\sigma_y^2 = 6.5\cdot 10^8\sigma_x^2[0.1t - 0.2 + 0.1\cos(2.48\cdot 10^{-3}t) - 8.0\sin(2.48\cdot 10^{-3}t)]$. For $t = 1$ hour, $\sigma_y \approx 1.5$ km.

32.30 $\mathbf{D}[\alpha(t)] \approx a_1t$; $\mathbf{D}[\beta(t)] \approx b_1t$;

$$a_1 = \frac{2}{\pi p}\int_0^\infty\left[(\cos\lambda\tau + 2)\frac{1}{p}k_1^2\arcsin k_{\dot\psi}(\tau) + k_2^2\frac{\cos\lambda\tau}{q}\arcsin k_{\dot\theta}(\tau)\right]d\tau;$$

$$b_1 = \frac{2}{\pi q}\int_0^\infty\left[(\cos\lambda\tau + 2)\frac{1}{q}k_2^2\arcsin k_{\dot\theta}(\tau) + k_1^2\frac{\cos\lambda\tau}{p}\arcsin k_{\dot\psi}(\tau)\right]d\tau.$$

$k_{\dot\psi}(\tau)$ and $K_{\dot\theta}(\tau)$ are the normalized correlation functions $\dot\Psi(t)$ and $\dot\Theta(t)$; $\lambda = \sqrt{pq}$.

32.31 $\mathbf{D}[Z(t)] = \displaystyle\int_0^t\int_0^t \exp\left\{a^2(\tau_1 + \tau_2) + \frac{a^4}{4}[3\varphi(\tau_1) + 3\varphi(\tau_2) - \varphi(\tau_2 - \tau_1)]\right\}$

$$\times K_x(\tau_2 - \tau_1)\,d\tau_1\,d\tau_2,$$

where

$$\varphi(\tau) = 2\int_0^\tau (\tau - \tau_1)K_y(\tau_1)\,d\tau_1.$$

33. PROBLEMS ON PASSAGES

33.1 $\bar\tau_a = 10\pi[1 - \Phi(1)]e^{1/2} = 16.45$ sec.

33.2 $\mathbf{D}[V(t)] = 0.25$ cm.²/sec.².

33.3 The number of passages (going up) beyond the level $a = 25°$ equals the number of passages going down beyond the level $a = -25°$; consequently, the required number of passages

$$2T\frac{1}{2\pi}\sqrt{\alpha^2 + \beta^2}\exp\left\{-\frac{a^2}{2b}\right\} = 11.9.$$

33.4 $\frac{\pi}{1.5}e^{0.9}\left[1 - \Phi\left(\frac{3\sqrt{5}}{5}\right)\right] = 9.91$ sec.

33.5 Starting with $t = (\sqrt{4\pi^2 p_0^2 - \alpha^2})^{-1}$.

33.6 The problem reduces to the determination of the number of passages of the random function $\dot{X}(t)$ beyond the level $\sqrt{w_0/k}$ (going up) and $-\sqrt{w_0/k}$ (going down).

Answer:
$$\frac{1}{\pi}\sqrt{\alpha^2 + \beta^2}\exp\left\{-\frac{w_0}{2ka}\right\}.$$

33.7 Since the radius of curvature is $[v/\dot{\Psi}(t)]$, the sensitive element reaches a stop when $\dot{\Psi}(t)$ leaves the limits of the strip $\pm v/R_0$, which leads to
$$\frac{1}{\pi}\sqrt{\alpha^2 + \beta^2}\exp\left\{-\frac{v^2}{2a(\alpha^2 + \beta^2)R_0^2}\right\}\text{ sec.}^{-1}.$$

33.8 For $\bar{h} \geqslant 54.5$ m. **33.9** $Q = \exp\left\{-\frac{\alpha T}{\pi}\exp\left\{-\frac{a^2}{2b}\right\}\right\}$.

33.10 Denoting by $f(x, x_1, x_2)$ the probability density of the system of normal variables $X(t)$, $\dot{X}(t)$ and $\ddot{X}(t)$, we get the required probability density,
$$f(x) = \frac{\int_{-\infty}^{0} f(x, 0, x_2)\,dx_2}{\int_{-\infty}^{\infty}\int_{-\infty}^{0} f(x, 0, x_2)\,dx_2\,dx}.$$

Considering that the correlation matrix has the form
$$\|k_{jl}\| = \left\|\begin{matrix} K_x(0) & 0 & \ddot{K}_x(0) \\ 0 & -\ddot{K}_x(0) & 0 \\ \ddot{K}_x(0) & 0 & K_x^{\mathrm{IV}}(0)\end{matrix}\right\|,$$
we find after integration that
$$f(x) = \frac{1}{\sqrt{2\pi a}}\exp\left\{-\frac{x^2}{2a}\right\}\left[1 + \Phi\left(\frac{x}{\sqrt{2a}}\right)\right].$$

33.11 $f(x) = \frac{1}{\sigma\sqrt{2\pi}}\exp\left\{-\frac{x^2}{2\sigma^2}\right\}\left[1 - \Phi\left(\frac{x}{2\sigma\sqrt{2}}\right)\right]$.

33.12 The required number equals the number of passages (from both sides) beyond the zero level; consequently,
$$\frac{T}{\pi}\sqrt{\frac{\mathbf{D}[\ddot{X}(t)]}{\mathbf{D}[\dot{X}(t)]}} = \frac{T\alpha}{\pi}\sqrt{10}.$$

33.13
$$\bar{n} = p = \frac{1}{8\pi^2\sqrt{\Delta_1 A_{55}}}\int_{-\infty}^{0}\int_{-\infty}^{0}\left[\Phi\left(\frac{A_{55}\sqrt{z_3 z_4} + A_{35}z_3 + A_{45}z_4}{\sqrt{A_{55}\Delta_2}}\right)\right.$$
$$\left. + \Phi\left(\frac{A_{55}\sqrt{z_3 z_4} - A_{35}z_3 - A_{45}z_4}{\sqrt{A_{55}\Delta_2}}\right)\right] z_3 z_4$$
$$\times \exp\left\{\frac{1}{2\Delta_2}\left[A_{33}z_3^2 + A_{44}z_4^2 + 2A_{34}z_3 z_4\right.\right.$$
$$\left.\left. - \frac{(A_{35}z_3 + A_{45}z_4)^2}{A_{35}}\right]\right\}dz_3\,dz_4,$$
where
$$\Delta_1 = k_{11}k_{22} - k_{12}^2, \qquad \Delta_2 = \left|\begin{matrix} k_{33} & k_{34} & k_{35} \\ k_{34} & k_{44} & k_{45} \\ k_{35} & k_{45} & k_{55}\end{matrix}\right|, \qquad A_{jl}\ \ (j, l = 3, 4, 5)$$

are the cofactors of the determinant Δ_2 and k_{jl} are included in the answer to Problem 33.14.

33.14 $\bar{n}$ is the probability density p of sign changes for ζ_x and ζ_y in the vicinity of the point with coordinates x, y. These are related as follows:

$$p\,dx\,dy = \mathbf{P}\left(\frac{\partial\zeta(x, y)}{\partial x} > 0;\ \frac{\partial\zeta(x + dx, y)}{\partial x} < 0;\ \frac{\partial\zeta(x, y)}{\partial y} > 0;\ \frac{\partial\zeta(x, y + dy)}{\partial y} < 0\right)$$

$$= \mathbf{P}\left(0 < \frac{\partial\zeta(x, y)}{\partial x} < -\frac{\partial^2\zeta(x, y)}{\partial x^2}\,dx;\ 0 < \frac{\partial\zeta(x, y)}{\partial y} < -\frac{\partial^2\zeta(x, y)}{\partial y^2}\,dy\right).$$

The probability $p\,dx\,dy$ can be computed if one considers that $K(\zeta, \eta)$ uniquely defines the distribution law of $\partial\zeta/\partial x$, $\partial\zeta/\partial y$, $\partial^2\zeta/\partial x^2$, $\partial^2\zeta/\partial y^2$. Performing the computations we obtain:

$$\bar{n} = p = \frac{\Delta_2}{4\pi^2\sqrt{\Delta_1\Delta_2}}\left[1 + \frac{k_{34}}{\sqrt{\Delta_2}}\left(\frac{\pi}{2} + \arctan\frac{k_{34}}{\sqrt{\Delta_2}}\right)\right],$$

where

$$\Delta_1 = k_{11}k_{22} - k_{12}^2, \qquad \Delta_2 = k_{33}k_{44} - k_{34}^2;$$

$$k_{11} = \int_{-\infty}^{\infty}\int_{-\infty}^{\infty} S_\zeta(\omega_1, \omega_2)\omega_1^2\,d\omega_1\,d\omega_2; \qquad k_{22} = \int_{-\infty}^{\infty}\int_{-\infty}^{\infty} S_\zeta(\omega_1, \omega_2)\omega_2^2\,d\omega_1\,d\omega_2;$$

$$k_{34} = \int_{-\infty}^{\infty}\int_{-\infty}^{\infty} S_\zeta(\omega_1, \omega_2)\omega_1^2\omega_2^2\,d\omega_1\,d\omega_2; \quad k_{33} = \int_{-\infty}^{\infty}\int_{-\infty}^{\infty} S_\zeta(\omega_1, \omega_2)\omega_1^4\,d\omega_1\,d\omega_2;$$

$$k_{44} = \int_{-\infty}^{\infty}\int_{-\infty}^{\infty} S_\zeta(\omega_1, \omega_2)\omega_2^4\,d\omega_1\,d\omega_2; \qquad k_{12} = \int_{-\infty}^{\infty}\int_{-\infty}^{\infty} S_\zeta(\omega_1, \omega_2)\omega_1\omega_2\,d\omega_1\,d\omega_2;$$

$$k_{35} = \int_{-\infty}^{\infty}\int_{-\infty}^{\infty} S_\zeta(\omega_1, \omega_2)\omega_1^3\omega_2\,d\omega_1\,d\omega_2; \quad k_{45} = \int_{-\infty}^{\infty}\int_{-\infty}^{\infty} S_\zeta(\omega_1, \omega_2)\omega_1\omega_2^3\,d\omega_1\,d\omega_2;$$

$$k_{55} = \int_{-\infty}^{\infty}\int_{-\infty}^{\infty} S_\zeta(\omega_1, \omega_2)\omega_1^2\omega_2^2\,d\omega_1\,d\omega_2.$$

34. SPECTRAL DECOMPOSITION OF STATIONARY RANDOM FUNCTIONS

34.1 $K(\tau) = 2a\dfrac{\sin b\tau}{\tau}$.

34.2 $K(\tau) = 2c^2(2\cos\omega_0\tau - 1)\dfrac{\sin\omega_0\tau}{\tau}$.

34.3 Denoting

$$J(\alpha, \omega) = \frac{a}{2\pi}\int_{-\infty}^{\infty} e^{-\alpha|\tau| - i\omega\tau}\,d\tau = \frac{a}{\pi}\frac{\alpha}{(\omega^2 + \alpha^2)},$$

we have

$$S(\omega) = J - \alpha\frac{\partial J}{\partial\alpha} = \frac{2\alpha}{\pi}\frac{\alpha^3}{(\omega^2 + \alpha^2)^2}.$$

34.4 $$S(\omega) = \frac{\sigma^2}{2\pi}\left(\frac{\sin\dfrac{\omega}{2}}{\dfrac{\omega}{2}}\right)^2.$$

34.5 $$S(\omega) = \frac{\alpha\sigma^2}{\pi}\frac{\omega^2 + \alpha^2 + \beta^2}{(\omega^2 + \alpha^2 + \beta^2)^2 - 4\beta^2\omega^2} = \frac{\alpha\sigma^2}{\pi}\frac{\omega^2 + \alpha^2 + \beta^2}{(\omega^2 - \alpha^2 - \beta^2)^2 + 4\alpha^2\omega^2}$$

$$= \frac{\alpha\sigma^2}{\pi}\frac{\omega^2 + \alpha^2 + \beta^2}{(\omega^2 + \alpha^2 - \beta^2)^2 + 4\alpha^2\beta^2}.$$

34.6 $S(\omega) = \frac{2\sigma^2}{\pi}\frac{\alpha(\alpha^2+\beta^2)}{(\omega^2+\alpha^2+\beta^2)^2 - 4\beta^2\omega^2} = \frac{2\sigma^2}{\pi}\frac{\alpha(\alpha^2+\beta^2)}{(\omega^2-\alpha^2-\beta^2)^2+4\alpha^2\omega^2}$

$$= \frac{2\sigma^2}{\pi}\frac{\alpha(\alpha^2+\beta^2)}{(\omega^2+\alpha^2-\beta^2)^2+4\alpha^2\beta^2}.$$

34.7 Solving this problem as we did 34.3, we get

$$S(\omega) = \frac{\sigma^2}{\pi}\frac{16a^3\omega^4}{(\omega^2+\alpha^2)^4}.$$

34.8 $S(\omega) = \dfrac{2a\alpha\omega^2}{\pi[(\omega^2+\alpha^2+\beta^2)^2 - 4\beta^2\omega^2]}.$

34.9 Two derivatives, since $S_x(\omega)$ decreases as $1/\omega^2$ when ω increases.

34.10 $S(\omega) = \dfrac{1}{\pi}\sum_{j=1}^{n}\dfrac{a_j\alpha_j}{\omega^2+\alpha_j^2}.$

34.11 $\dfrac{dS(\omega)}{d\omega} = \dfrac{2\alpha\sigma^2\omega}{\pi[(\omega^2-\alpha^2-\beta^2)^2+4\alpha^2\beta^2]^2}$

$$\times\ \{4\beta^2(\alpha^2+\beta^2) - (\alpha^2+\beta^2+\omega^2)^2\}.$$

Consequently, for $\omega = 0$ there will always be an extremum. If, for $\omega = 0$, the expression between brackets is negative, the sign of the derivative at this point changes from plus to minus; there will be one maximum at this point, and no other maxima. Thus, the condition for no maxima except at the origin is $\alpha^2 > 3\beta^2$. For

$$\alpha^2 = 3\beta^2, \qquad S(\omega) = \frac{\sigma^2\alpha}{\pi}\frac{1}{\omega^2+4\beta^2};$$

that is, $S(\omega)$ also can have only one maximum at the origin. Therefore, if $\alpha^2 \geqslant 3\beta^2$, then there exists one maximum at the origin, if $\alpha^2 < 3\beta^2$, there will be one minimum at the origin and two maxima at the points

$$\omega = \pm\omega_2, \qquad \omega_2 = \sqrt[4]{\alpha^2+\beta^2}\sqrt{2\sqrt{\beta^2} - \sqrt{\alpha^2+\beta^2}}.$$

34.12 Since

$$S_{\dot{x}}(\omega) = \frac{a^2\omega^2}{(\omega^2+\alpha^2)^2},$$

then

$$\mathbf{D}[\dot{X}(t)] = \int_{-\infty}^{\infty} S_{\dot{x}}(\omega)\,d\omega = \frac{\pi a^2}{2\alpha}.$$

34.13 Since

$$S_x(\omega) = \frac{a}{2\alpha\sqrt{\pi}}\exp\left\{-\frac{\omega^2}{4\alpha^2}\right\}, \quad \text{and} \quad R_{x\dot{x}}(\tau) = \frac{d}{d\tau}\int_{-\infty}^{\infty}\exp\{i\omega\tau\}S_x(\omega)\,d\omega,$$

then

$$S_{x\dot{x}}(\omega) = i\omega S_x(\omega) = \frac{ia\omega}{2\alpha\sqrt{\pi}}\exp\left\{-\frac{\omega^2}{4\alpha^2}\right\} = S_{x\dot{x}}(\omega).$$

34.14 Since

$$K_\Delta(\tau) = ae^{-\alpha|\tau|}[k_1^2(1+\alpha\mid\tau) + \alpha^2k_2^2(1-\alpha\mid\tau)],$$

the Fourier transform leads to

$$S_\Delta(\omega) = \frac{2a\alpha^2}{\pi(\omega^2+\alpha^2)^2}(k_1^2 + k_2^2\omega^2).$$

34.15 $R_{xy}(\tau) = K_x(\tau+\tau_0) = \displaystyle\int_{-\infty}^{\infty} e^{i\omega(\tau+\tau_0)}S_x(\omega)\,d\omega;\ \ S_{xy}(\omega) = e^{i\omega\tau_0}S_x(\omega).$

34.16 $S_{xy}(\omega) = (i\omega)^k e^{i\omega\tau_0}[S_u(\omega) + S_{vu}(\omega)].$

34.17 Since

$$K_z(\tau) = K_{\dot{x}}(\tau)K_{\dot{y}}(\tau) = a_1a_2(\alpha_1^2 + \beta_1^2)(\alpha_2^2 + \beta_2^2)e^{-(\alpha_1+\alpha_2)|\tau|}$$
$$\times \left(\cos\beta_1\tau - \frac{\alpha_1}{\beta_1}\sin\beta_1|\tau|\right)\left(\cos\beta_2\tau - \frac{\alpha_2}{\beta_2}\sin\beta_2|\tau|\right),$$

then the Fourier inversion leads to

$$S_z(\omega) = a\left\{\frac{\alpha\cos\gamma' + (\omega - \beta')\sin\gamma'}{(\omega - \beta')^2 + \alpha^2} - \frac{\alpha\cos\gamma' + (\omega + \beta')\sin\gamma'}{(\omega + \beta')^2 + \alpha^2}\right.$$
$$\left. + \frac{\alpha\cos\gamma'' + (\omega - \beta'')\sin\gamma''}{(\omega - \beta'')^2 + \alpha^2} - \frac{\alpha\cos\gamma'' + (\omega + \beta'')\sin\gamma''}{(\omega + \beta'')^2 + \alpha^2}\right\},$$

where

$$\alpha = \alpha_1 + \alpha_2, \quad \beta' = \beta_1 + \beta_2, \quad \beta'' = \beta_1 - \beta_2, \quad \gamma' = \gamma_1 + \gamma_2, \quad \gamma'' = \gamma_1 - \gamma_2,$$
$$\tan\gamma_1 = \frac{\alpha_1}{\beta_1}, \quad \tan\gamma_2 = \frac{\alpha^2}{\beta_2}, \quad a = \frac{a_1a_2\beta_1^2\beta_2^2}{4\pi\cos^3\gamma_1\cos^3\gamma_2}.$$

34.18 Since $K_z(\tau) = K_x(\tau)K_y(\tau) + \bar{x}^2K_y(\tau) + \bar{y}^2K_x(\tau)$, then

$$S_z(\omega) = \frac{a_1a_2(\alpha_1 + \alpha_2)}{\pi[\omega^2 + (\alpha_1 + \alpha_2)^2]} + \frac{\bar{x}^2a_2\alpha_2}{\pi(\omega^2 + \alpha_2^2)} + \frac{\bar{y}^2a_1\alpha_1}{\pi(\omega^2 + \alpha_1^2)}.$$

34.19 Since $K_\Delta(\tau) = K_\psi(\tau)K_\theta(\tau)$, the Fourier transform leads to

$$S_\Delta(\omega) = \frac{a_1a_2}{4\pi\cos\gamma_1\cos\gamma_2}$$
$$\times\left\{\frac{\alpha\cos\gamma' - (\omega - \beta')\sin\gamma'}{(\omega - \beta')^2 + \alpha^2} - \frac{\alpha\cos\gamma' - (\omega + \beta')\sin\gamma'}{(\omega + \beta')^2 + \alpha^2}\right.$$
$$\left. + \frac{\alpha\cos\gamma'' - (\omega - \beta'')\sin\gamma''}{(\omega - \beta'')^2 + \alpha^2} - \frac{\alpha\cos\gamma'' - (\omega + \beta'')\sin\gamma''}{(\omega + \beta'')^2 + \alpha^2}\right\},$$

where

$$\alpha = \alpha_1 + \alpha_2, \quad \beta' = \beta_1 + \beta_2, \quad \beta'' = \beta_1 - \beta_2, \quad \gamma' = \gamma_1 + \gamma_2, \quad \gamma'' = \gamma_1 - \gamma_2,$$
$$\tan\gamma_1 = \frac{\alpha_1}{\beta_1}, \quad \tan\gamma_2 = \frac{\alpha_2}{\beta_2}.$$

34.20 Applying the general formula

$$S_y(\omega) = 2\int_{-\infty}^{\infty}(\omega - \omega_1)2S_x(\omega - \omega_1)\omega_1^2S_x(\omega_1)\,d\omega_1$$

and the results of Problem 34.17, we get

$$S_y(\omega) = \frac{2a^2\alpha\beta^4}{\pi\cos^2\gamma}\left\{\frac{1}{\omega^2 + 4\alpha^2} + \frac{4(\alpha^2 + \beta^2)}{(\omega^2 + 4\alpha^2 - 4\beta^2)^2 + 16\alpha^2\beta^2}\right\}, \quad \tan\gamma = \frac{\alpha}{\beta}.$$

34.21 $S_y(\omega) = \dfrac{4a\alpha}{\pi}\left(\dfrac{a}{\omega^2 + 4\alpha^2} + \dfrac{\bar{x}^2}{\omega^2 + \alpha^2}\right).$

34.22 $S_y(\omega) = \omega^2\left(a^2\alpha\sqrt{2\pi}\exp\left\{-\dfrac{\omega^2}{4\alpha^2}\right\} + a\bar{x}\exp -\left\{\dfrac{\omega^2}{2\alpha^2}\right\}\right).$

34.23
$$S_\Delta(\omega) = S_\varphi(\omega) + \cos^4 q\int_{-\infty}^{\infty}S_\psi(\omega - \omega_1)S_\theta(\omega_1)\,d\omega_1$$
$$+ \frac{1}{8}\sin^2 2q\left[\int_{-\infty}^{\infty}S_\theta(\omega - \omega_1)S_\theta(\omega_1)\,d\omega_1 + \int_{-\infty}^{\infty}S_\psi(\omega - \omega_1)S_\psi(\omega_1)\,d\omega_1\right],$$

where $S_\varphi(\omega) = S_1(\omega)$, $S_\theta(\omega) = S_2(\omega)$, $S_\psi(\omega) = S_3(\omega)$;

$$S_j(\omega) = \frac{2a_j\alpha_j}{\pi}\,\frac{\alpha_j^2 + \beta_j^2}{(\omega^2 + \alpha_j^2 + \beta_j^2) - 4\beta_j^2\omega^2}, \qquad j = 1, 2, 3$$

and all the integrals may be computed in a finite form. Because the final result is cumbersome in the present case, it is preferable to use numerical integration methods.

34.24 Since $K_y(\tau) = 2K_x^2(\tau) + 4\bar{x}^2K_x(\tau)$, then

$$S_y(\omega) = \frac{4\sigma_x^4}{\pi}\frac{\alpha}{\omega^2 + 4\alpha^2} + 4\bar{x}_2\frac{\sigma_x^2\alpha}{\pi(\omega^2 + \alpha^2)}$$

has one maximum for $\omega = 0$.

34.25 $$S_j(\omega) = \frac{\sigma^2\alpha}{\pi}\left[\frac{1 - \cos\dfrac{\omega}{a}}{\omega^2} + \frac{1}{\omega\alpha}\left(1 - \frac{\alpha}{a}\right)\sin\frac{\omega}{a}\right],$$

where

$$\sigma^2 = \frac{\Gamma j_0^2n^2}{\pi a^2}\left\{\frac{1}{3aT}\left[1 + \frac{n(n-1)\Gamma}{2\pi}\right] - \frac{n^2\Gamma}{4\pi}\right\}, \qquad \alpha = \frac{j_0^2n^2\Gamma}{2\pi a^2T\sigma^2}\left[1 + \frac{n(n-1)\Gamma}{2\pi}\right],$$

$$T = \frac{4\pi^2}{\Omega_1\Omega_2}, \qquad a = \frac{2(\Omega_1 + \Omega_2)}{\gamma},$$

and j_0 is the intensity of photocurrent created when one hole coincides with the aperture of the diaphragm.

35. COMPUTATION OF PROBABILITY CHARACTERISTICS OF RANDOM FUNCTIONS AT THE OUTPUT OF DYNAMICAL SYSTEMS

35.1 $Y(t)$ is a stationary function; consequently,

$$S_y(\omega) = \frac{c^2}{\omega^2 + \alpha^2},$$

which after a Fourier inversion yields

$$K_y(\tau) = \frac{\pi c^2}{\alpha}e^{-\alpha|\tau|}.$$

35.2 Since $Y(t)$ is stationary, finding the expectation of both sides of the equation we obtain that $\bar{y} = (b_1/a_1)\bar{x}$. The spectral density is

$$S_y(\omega) = \frac{b_0^2\omega^2 + b_1^2}{a_0^2\omega^2 + a_1^2}S_x(\omega) = \frac{\sigma_x^2\alpha}{\pi}\frac{b_0^2\omega^2 + b_1^2}{(a_0^2\omega^2 + a_1^2)(\omega^2 + \alpha^2)},$$

which after integration between infinite limits gives

$$\mathbf{D}[Y(t)] = \frac{\sigma_x^2}{a_0a_1}\frac{a_1b_0^2\alpha + a_0b_1^2}{a_1 + a_0\alpha}.$$

35.3 $$S_u(\omega) = \frac{n^4}{g^2}\frac{\omega^2[S_{\ddot{\eta}_c}(\omega) + c^2\omega^2S_\theta(\omega)]}{(\omega^2 - n^2)^2 + 4h^2\omega^2},$$

where

$$S_{\ddot{\eta}_c}(\omega) = \frac{2a_1\alpha_1(\alpha_1^2 + \beta_1^2)}{\pi[(\omega^2 - \beta_1^2 - \alpha_1^2)^2 + 4\alpha_1^2\omega^2]}, \qquad S_\theta(\omega) = \frac{2a_2\alpha_2(\alpha_2^2 + \beta_2^2)}{\pi[(\omega^2 - \beta_2^2 - \alpha_2^2)^2 - 4\alpha_2^2\omega^2]}.$$

35.4 Since by the assumption of the problem $\alpha(t)$ can be considered stationary

$$S_\alpha(\omega) = \frac{\varepsilon^2}{\omega^2 + \varepsilon^2}S_u(\omega),$$

where $S_u(\omega)$ is obtained as in Problem 35.3. Integrating $S_\alpha(\omega)$ between infinite limits with the aid of residues, we get $\sigma_\alpha^2 = 2.13\cdot 10^{-6}$ rad.^{2}, $\sigma_\alpha = 1.46\cdot 10^{-3}$ rad.

35.5 $$S_y(\omega) = \frac{2\sigma^2\alpha(\alpha^2 + \beta^2)}{\pi[(\omega^2 - \alpha^2 - \beta^2)^2 + 4\alpha^2\omega^2]},$$

where $\alpha = h$, $\beta = \sqrt{k^2 - h^2}$, $\sigma^2 = \pi c^2/2hk^2$. Applying a Fourier inversion to $S_y(\omega)$, we get

$$K_y(\tau) = \sigma^2e^{-\alpha|\tau|}\left(\cos\beta\tau + \frac{\alpha}{\beta}\sin\beta|\tau|\right).$$

35.6 $$S_\theta(\omega) = \frac{2\sigma_\theta^2\alpha(\alpha^2+\beta^2)}{\pi[(\omega^2-\alpha^2-\beta^2)^2+4\alpha^2\omega^2]};$$

$$K_\theta(\tau) = \sigma_\theta^2 e^{-\alpha|\tau|}\left(\cos\beta\tau + \frac{\alpha}{\beta}\sin\beta|\tau|\right),$$

where

$$\sigma_\theta^2 = \frac{kT}{D}, \qquad \alpha = \frac{1}{2}\frac{r}{I}, \qquad \beta = \frac{1}{2I}\sqrt{4ID - r^2}.$$

35.7 $$S_y(\omega) = \frac{4(49\omega^6+25)}{\pi(\omega^2+1)^2(\omega^2+4)(\omega^2+9)}.$$

35.8 No, since the roots of the characteristic equation have positive real parts and, consequently, the system described by the equation is nonstationary.

35.9 Since $\zeta_c(t)$ is stationary, it follows that

$$S_{\zeta_c}(\omega) = \frac{\omega_0^4 S_x(\omega)}{|-\omega^2+2hi\omega+\omega_0^2|^2},$$

$$\mathbf{D}[\zeta_c(t)] = \frac{a\alpha(\alpha^2+\beta^2)\omega_0^4}{[(\beta_1-\beta)^2+(\alpha_1-\alpha)^2][(\beta_1+\beta)^2+(\alpha_1-\alpha)^2]}$$
$$\times\,[(\beta_1-\beta)^2+(\alpha_1+\alpha)^2][(\beta_1+\beta)^2+(\alpha_1+\alpha)^2]$$
$$\times\left\{\frac{(-\beta_1^2+\beta^2+\alpha^2+\alpha_1^2)^2+4(\alpha^2\beta_1^2-2\alpha_1^2\beta_1^2+\alpha_1^4-2\alpha^2\alpha_1^2+\alpha_1^2\beta^2)}{\alpha_1(\alpha_1^2+\beta_1^2)}\right.$$
$$\left.+\frac{(-\beta^2+\beta_1^2+\alpha_1^2+\alpha^2)^2+4(\alpha_1^2\beta^2-2\alpha^2\beta^2+\alpha^4-2\alpha_1^2\alpha^2+\alpha^2\beta_1^2)}{\alpha(\alpha^2+\beta^2)}\right\},$$

$$\alpha_1 = h, \qquad \beta_1 = \sqrt{\omega_0^2 - h^2}.$$

35.10 Letting $\omega_0 = n$, $a = 3\cdot 10^{-4}\,g^2$, we get $\mathbf{D}[\varepsilon(t)] = \mathbf{D}[\zeta_c(t)]$, where $\mathbf{D}[\zeta_c(t)]$ is mentioned in the answer to Problem 35.9. Substituting the numerical data we get $\mathbf{D}[\varepsilon(t)] = 0.06513$; $\sigma_\varepsilon = 0.255$.

35.11 The formula is a consequence of the general formula given in the introduction.

35.12 Letting $\omega_0 = k$, we obtain $\mathbf{D}[\Theta(t)] = \mathbf{D}[\zeta_c(t)]$, where $\mathbf{D}[\zeta_c(t)]$ is given in the answer to Problem 35.9.

35.13 $$S_{yx}(\omega) = \frac{k^2 S_x(\omega)}{(k^2-\omega^2)-2hi\omega},$$

$$R_{yx}(\tau) = k^2\int_{-\infty}^{\infty} e^{i\omega\tau}\frac{S_x(\omega)}{(k^2-\omega^2)-2hi\omega}\,d\omega.$$

35.14 The independent particular integrals of the homogeneous equation are e^{-t}, e^{-7t}, the weight function is $p(t) = (1/6)(e^{-t}-e^{-7t})$,

$$\frac{336}{\sqrt{\pi}}K_y(\tau) = 7\exp\left\{-\tau+\frac{1}{4\alpha^2}\right\}\left\{1+\Phi\left[\sqrt{2}\left(\alpha\tau-\frac{1}{2\alpha}\right)\right]\right\}$$
$$-\exp\left\{-7\tau+\frac{12.25}{\alpha^2}\right\}\left\{1+\Phi\left[\sqrt{2}\left(\alpha\tau-\frac{3.5}{\alpha}\right)\right]\right\}$$
$$+7\exp\left\{\tau+\frac{1}{4\alpha^2}\right\}\left\{1-\Phi\left[\sqrt{2}\left(\alpha\tau+\frac{1}{2\alpha}\right)\right]\right\}$$
$$-\exp\left\{7\tau+\frac{12.25}{\alpha^2}\right\}\left\{1-\Phi\left[\sqrt{2}\left(\alpha\tau+\frac{3.5}{\alpha}\right)\right]\right\}.$$

35.15 $$\mathbf{D}[Y(t)-Z(t)] = \mathbf{D}[Z(t)] + \int_{0-}^{\infty}\int_{0-}^{\infty} p^*(\tau_1)p(\tau_2)$$
$$\times\,K_x(\tau_2-\tau_1)\,d\tau_1\,d\tau_2 - 2\,\mathrm{Re}\int_{0-}^{\infty} p^*(\tau)R_{xz}(\tau)\,d\tau,$$

where the minus sign in the lower limits of integration means that the point 0 is included in the domain of integration.

35.16 $$\mathbf{D}[Y(t)] = \frac{\sigma_x^2}{a(a+\alpha)}\left[t^2 + \frac{2a+\alpha}{2a^2(a+\alpha)}(1-2at)\right].$$

35.17 $\bar{\alpha} = \text{const.}$, whose value may be taken zero by a proper choice of the origin;

$$\mathbf{D}[\alpha(t)] = \frac{\sigma_l^2 P^2}{H^2}\left(1+\frac{\overline{w}}{g}\right)^2 t^2 + \frac{2\sigma_l^2 P^2}{g^2 H^2}\int_0^t (t-\tau)K_w(\tau)\,d\tau.$$

35.18 Replacing $X(t)$ by its spectral decomposition, we obtain the spectral decomposition of

$$Y_1(t) = \int_{-\infty}^{\infty} \frac{1}{-\omega^2 + 2hi\omega + k^2}\left[e^{-at+i\omega t} + \frac{-(\omega+\omega_0)+(a-h)i}{2\omega_0}e^{-(h-i\omega_0)t} + \frac{-(\omega_0-\omega)-(a-h)i}{2\omega_0}e^{-(h+i\omega_0)t}\right]d\Phi(\omega),$$

where $\omega_0 = \sqrt{k^2-h^2}$. From this it follows that

$$\begin{aligned} K_{y_1}(t_1,t_2) = \int_{-\infty}^{\infty} &\frac{S_x(\omega)}{(\omega^2-k^2)^2+4h^2\omega^2} \\ \times \Big\{ &e^{-a(t_1+t_2)+i\omega(t_2-t_1)} + \frac{1}{4\omega_0^2}e^{-h(t_1-t_2)} \\ &\times [[(\omega-\omega_0)^2+(a-h)^2]e^{-i\omega_0(t_2-t_1)} + [(\omega+\omega_0)^2+(a-h)^2] \\ &\times e^{i\omega_0(t_2-t_1)} + [\omega_0^2-(\omega-ai+hi)^2] \\ &\times e^{i\omega_0(t_1+t_2)} + [\omega_0^2-(\omega+ai-hi)^2]e^{-i\omega_0(t_1+t_2)}] \\ &+ \frac{1}{2\omega_0}e^{-(a+i\omega)t_1-ht_2}[(\omega-\omega_0+ai-hi)e^{-i\omega_0t_2} \\ &\qquad + (-\omega-\omega_0+ai-hi)e^{i\omega_0t_2}] \\ &+ \frac{1}{2\omega_0}e^{-(a-i\omega)t_2-ht_1}[(\omega-\omega_0-ai+hi)e^{i\omega_0t_1} \\ &\qquad + (-\omega-\omega_0-ai+hi)e^{-i\omega_0t_1}]\Big\}\,d\omega, \end{aligned}$$

which, after we substitute the expression for $S_x(\omega)$ and integrate with the aid of residues, gives the final result in the finite form:

$$\begin{aligned} K_{y_1}(t_1,t_2) &= \sigma_x^2\alpha\left\{\frac{[e^{(\alpha-a)t_1}-M_1\alpha-N_1][e^{-(\alpha+a)t_2}+M_2\alpha-N_2]}{\alpha[(a^2+k^2-2ha+\alpha^2)^2-4\alpha^2(h-a)^2]}\right. \\ &\left.+\operatorname{Re}\frac{[e^{(\gamma-a-i\beta)t_1}+M_1(i\beta-\gamma)-N_1][e^{(i\beta-\gamma-a)t_2}-M_2(i\beta-\gamma)-N_2]}{2\gamma\beta[(\beta^2-\gamma^2+\alpha^2)+2i\beta\gamma](\beta+i\gamma)}\right\}, \end{aligned}$$

$$M_j = e^{-ht_j}\frac{\sin\beta t_j}{\beta},\qquad N_j = e^{-ht_j}\left(\cos\beta t_j + \frac{h-a}{\beta}\sin\beta t_j\right),\qquad j=1,2;$$

$$\gamma = |h-a|,\qquad \beta = \omega_0.$$

35.19 $$\begin{aligned} K_y(t_1,t_2) = \frac{a\pi}{2}&\exp\left\{\frac{1}{2}(t_1^2+t_2^2+2\alpha^2)\right\} \\ &\times\left\{[\Phi(t_1-\alpha)+\Phi(\alpha)][\Phi(t_2+\alpha)+\Phi(t_1+\alpha)]\right. \\ &\left.-\frac{4}{\sqrt{2\pi}}\int_0^{t_1}\exp\left\{-\frac{1}{2}(\xi-\alpha)^2\right\}\Phi(\xi+\alpha)\,d\xi\right\},\ t_2 \geqslant t_1.\end{aligned}$$

35.20 $\bar{y} = \frac{b}{a^2}\left(-1 + \exp\left\{\frac{a^2t^2}{2}\right\}\right);$

$$K_y(t_1, t_2) = \frac{\sigma_x^2 b^2\sqrt{2\pi}}{2\sqrt{a^2 + 2\alpha^2}} \exp\left\{\frac{a^2}{2}(t_1^2 + t_2^2)\right\} \times \int_0^{t_2} \exp\left\{-\frac{a^2(a^2 + 4\alpha^2)t^2}{2(a^2 + 2\alpha^2)}\right\} \times \left\{\Phi\left[\frac{(a^2 + 2\alpha^2)t_1 - 2\alpha^2 t}{\sqrt{a^2 + 2\alpha^2}}\right] + \Phi\left[\frac{2\alpha^2 t}{\sqrt{a^2 + 2\alpha^2}}\right]\right\} dt.$$

35.21 $\bar{y}(t) = \frac{t_0}{t} y_0 + \frac{1}{t}\int_{t_0}^{t} \bar{x}(\xi)\xi\, d\xi = 1 + \frac{t_0}{t}(y_0 - 1);$

$$K_y(t_1, t_2) = \frac{1}{t_1 t_2}\int_{t_0}^{t_1}\int_{t_0}^{t_2} K_x(\xi, \eta)\, d\xi\, d\eta = \frac{1}{t_1 t_2} \times \left\{\frac{4}{3\alpha^3}(t_1^3 - t_0^3) + \frac{1}{\alpha_2}(t_1^4 - t_0^4) + \frac{2}{5\alpha}(t_1^5 - t_0^5) - \left[\left(\frac{t_1^2}{\alpha} - \frac{2}{\alpha^2}t_1 + \frac{2}{\alpha^3}\right)e^{\alpha t_1} - \left(\frac{t_0^2}{\alpha} - \frac{2}{\alpha^2}t_0 + \frac{2}{\alpha^3}\right)e^{\alpha t_0}\right] \times \left[\left(\frac{t_1^2}{\alpha} + \frac{2}{\alpha^2}t_1 + \frac{2}{\alpha^3}\right) e^{-\alpha t_1} + \left(\frac{t_2^2}{\alpha} + \frac{2}{\alpha^2}t_2 + \frac{2}{\alpha^3}\right)e^{-\alpha t_2}\right]\right\}, \qquad t_2 \geqslant t_1.$$

35.22 $\bar{y}(t) = \frac{1}{\Delta}\sum_{k,j=1}^{n} A_{jk} y_k(t) e_j + \int_0^t p(t, \xi)\bar{x}(\xi)\, d\xi;$

$$K_y(t_1, t_2) = \frac{1}{\Delta^2}\sum_{k,j,l,m=1}^{n} A_{jm}A_{lk}y_m^*(t_1)y_k(t_2)k_{jl} + \int_0^{t_1}\int_0^{t_2} p(t_1, \xi)p(t_2, \eta)K_x(\xi, \eta)\, d\xi\, d\eta,$$

where $y_1(t), \ldots, y_n(t)$ are the independent particular integrals of the corresponding homogeneous equation,

$$\Delta = \begin{vmatrix} y_1(0) & y_2(0) & \cdots & y_n(0) \\ y_1'(0) & y_2'(0) & \cdots & y_n'(0) \\ \cdot & \cdot & \cdot & \cdot \\ y_1^{(n-1)}(0) & y_2^{(n-1)}(0) & \cdots & y_n^{(n-1)}(0) \end{vmatrix},$$

and A_{jl} are the cofactors of this determinant.

35.23 Since the solution of the system leads to

$$Y_2(t) = -2\int_0^t [e^{-(t-t_1)} - e^{-2(t-t_1)}]X(t_1)\, dt_1 + 2[Y_2(0) - Y_1(0)]e^{-t} + [2Y_1(0) - Y_2(0)]e^{-2t},$$

and

$$K_x(\tau) = 2e^{-|\tau|},$$

then

$$\mathbf{D}[Y_2(t)] = 4\left[\frac{2}{9} + (1 - 2t)e^{-2t} + \left(\frac{4}{3}t - \frac{20}{9}\right)e^{-3t} + e^{-4t}\right] + (2e^{-t} - e^{-2t})^2\mathbf{D}[Y_2(0)] + (2e^{-2t} - 2e^{-t})^2\mathbf{D}[Y_1(0)] + 2(2e^{-t} - e^{-2t})(2e^{-2t} - 2e^{-t})k_{y_1(0), y_2(0)};$$

$$\mathbf{D}[Y_2(0.5)] = 0.624.$$

35.24 $\mathbf{D}[Y_1(t)] = \frac{3}{2}e^{-4t} + \frac{4}{9}\left(-t^2 + 4t - \frac{20}{3}\right)e^{-3t}$

$$+ \left(\frac{1}{2}t^2 - 2t + \frac{5}{4}\right)e^{-2t} + \left(\frac{1}{9}t^2 - \frac{1}{6}t + \frac{23}{108}\right);$$

$$\mathbf{D}[Y_2(t)] = \frac{3}{2}e^{-4t} - \frac{8}{27}(3t^2 - 6t + 14)e^{-3t}$$

$$+ (2t^2 - 4t + 1)e^{-2t} + \left(\frac{8}{9}t^2 - \frac{20}{9}t + \frac{89}{54}\right).$$

35.25 $\mathbf{D}[Y_1(0.5)] = 0.01078$, $\mathbf{D}[Y_2(0.5)] = 0.00150$.

35.26 Since $Y(t)$ and $Z(t)$ can be assumed stationary,

$$S_y(\omega) = \frac{a^2\alpha\sigma_x^2\omega^2}{\pi b^2(\omega^2 + \alpha^2)\left(\omega^2 + \frac{1}{b^2}\right)}, \qquad S_z(\omega) = \frac{\alpha\sigma_x^2}{\pi b^2(\omega^2 + \alpha^2)\left(\omega^2 + \frac{1}{b^2}\right)},$$

which after integration leads to

$$\mathbf{D}[Y(t)] = \frac{a^2\alpha\sigma_x^2}{b(ab + 1)}, \qquad \mathbf{D}[Z(t)] = \frac{\sigma_x^2}{ab + 1}.$$

35.27 A normal law with parameters $\bar{y} = 0$, $\sigma_y = 0.78$.

35.28 $S_x(\omega) = \frac{n^4}{g^2}\left\{\omega^2 S_{\dot{\xi}_c}(\omega) + \int_{-\infty}^{\infty} \omega_1^2 S_{\dot{\eta}_c}(\omega_1)S_\varphi(\omega - \omega_1)\,d\omega_1\right.$

$$+ \rho_x^2\left[2\int_{-\infty}^{\infty} \omega_1^2(\omega - \omega_1)^2 S_\varphi(\omega - \omega_1)S_\varphi(\omega_1)\,d\omega_1\right.$$

$$+ 3\int_{-\infty}^{\infty} (\omega - \omega_1)^2 S_\psi(\omega - \omega_1)\omega_1^2 S_\psi(\omega_1)\,d\omega_1$$

$$+ \int_{-\infty}^{\infty} S_\psi(\omega - \omega_1)\omega_1^2 S_\psi(\omega_1)\,d\omega_1$$

$$\left. + 4\int_{-\infty}^{\infty} (\omega - \omega_1)\omega_1^3 S_\psi(\omega - \omega_1)S_\psi(\omega_1)\,d\omega_1\right]$$

$$+ \rho_z^2\left[\omega^4 S_\psi(\omega) + \int_{-\infty}^{\infty} (\omega - \omega_1)^4 S_\varphi(\omega - \omega_1)S_\theta(\omega_1)\,d\omega_1\right.$$

$$+ 4\int_{-\infty}^{\infty} (\omega - \omega_1)^2 S_\varphi(\omega - \omega_1)\omega_1^2 S_\theta(\omega_1)\,d\omega_1$$

$$\left.\left. + 4\int_{-\infty}^{\infty} (\omega - \omega_1)^3\omega_1 S_\varphi(\omega - \omega_1)S_\theta(\omega_1)\,d\omega_1\right]\right\};$$

$$S_y(\omega) = \frac{n^4}{g^2}\omega^2[S_{\dot{\xi}_c}(\omega) + \rho_x^2\omega^2 S_\psi(\omega)]; \qquad S_{xy}(\omega) = \frac{n^4}{g^2}\rho_x\rho_z\omega^4 S_\psi(\omega).$$

35.29 To find the asymmetry and the excess, one should determine the moments of $Y(t)$ up to and including the fourth. To find these moments it is necessary to find the expectations:

$$\mathbf{M}[X^2(t_1)X^2(t_2)], \quad \mathbf{M}[X^2(t_1)X^2(t_2)X^2(t_3)] \text{ and } \mathbf{M}[X^2(t_1)X^2(t_2)X^2(t_3)X^2(t_4)],$$

for the determination of which one should take the derivatives of corresponding orders of the characteristic function of the system of normal random variables. For example,

$$\mathbf{M}[X^2(t_1)X^2(t_2)] = \frac{\partial^4}{\partial u_1^2\,\partial u_2^2}\left\{\exp\left[-\frac{1}{2}\sum_{j,l=1}^{2} k_{jl}u_ju_l\right]\right\}\Bigg|_{u_1=u_2=0},$$

where $\|k_{jl}\|$ is the correlation matrix of the system of random variables $X(t_1)$, $\dot{X}(t_1)$, $X(t_2)$, $\dot{X}(t_2)$.

$$\mathbf{M}[X^2(t_1)X^2(t_2)] = 2K_x^2(t_2 - t_1) + K_x^2(0);$$

$$\begin{aligned}\mathbf{M}[X^2(t_1)X^2(t_2)X^2(t_3)] &= K_x^3(0) + 2K_x^2(t_2 - t_1)K_x(0) + 2K_x^2(t_3 - t_2)K_x(0) + 2K_x^2(t_3 - t_1)K_x(0) \\ &\quad + 8K_x(t_2 - t_1)K_x(t_3 - t_1)K_x(t_3 - t_2);\end{aligned}$$

$$\begin{aligned}\mathbf{M}[X^2(t_1)X^2(t_2)X^2(t_3)X^2(t_4)] &= K_x^4(0) + 2K_x^2(0)[K_x^2(t_3 - t_4) + K_x^2(t_2 - t_4) + K_x^2(t_2 - t_1) \\ &\quad + K_x^2(t_3 - t_2) + K_x^2(t_4 - t_1) + K_x^2(t_3 - t_1)] \\ &\quad + 4[K_x^2(t_2 - t_1)K_x^2(t_4 - t_3) + K_x^2(t_3 - t_1)K_x^2(t_4 - t_2) \\ &\quad + K_x^2(t_4 - t_1)K_x^2(t_3 - t_2)] \\ &\quad + 8K_x(0)[K_x(t_3 - t_2)K_x(t_4 - t_2)K_x(t_4 - t_3) \\ &\quad + K_x(t_1 - t_3)K_x(t_1 - t_4)K_x(t_4 - t_3) \\ &\quad + K_x(t_2 - t_1)K_x(t_2 - t_4)K_x(t_4 - t_1) \\ &\quad + K_x(t_3 - t_1)K_x(t_3 - t_2)K_x(t_2 - t_1)] \\ &\quad + 16[K_x(t_1 - t_2)K_x(t_1 - t_3)K_x(t_2 - t_4)K_x(t_3 - t_4) \\ &\quad + K_x(t_2 - t_1)K_x(t_1 - t_4)K_x(t_2 - t_3)K_x(t_3 - t_4) \\ &\quad + K_x(t_1 - t_3)K_x(t_1 - t_4)K_x(t_2 - t_3)K_x(t_2 - t_4)].\end{aligned}$$

Substituting the obtained expressions in the general formulas for moments of the solution of a differential equation, we get

$$\text{Sk} = \frac{2}{k + \alpha}\sqrt{2k(k + 2\alpha)}; \qquad \text{Ex} = 3\left[\frac{15k^2 + 25k\alpha + 2\alpha^2}{(k + \alpha)(3k + 2\alpha)} - 1\right].$$

35.30 For $\tau \geqslant 0$, we shall have

$$R_{yz}(\tau) = \frac{2\pi(k_1k_2c)^2}{\omega_2}\, e^{-h_2\tau}\, \frac{2\omega_2(h_1 + h_2)\cos\omega_2\tau - [\omega_2^2 - \omega_1^2 - (h_1 + h_2)^2]\sin\omega_2\tau}{[(\omega_2 - \omega_1)^2 + (h_1 + h_2)^2][(\omega_2 + \omega_1)^2 + (h_1 + h_2)^2]};$$

and for $\tau \leqslant 0$,

$$R_{yz}(\tau) = \frac{2\pi(k_1k_2c)^2}{\omega_1} \times e^{h_1\tau}\, \frac{2\omega_1(h_1 + h_2)\cos\omega_1\tau + [(\omega_2^2 - \omega_1^2) + (h_1 + h_2)^2]\sin\omega_1\tau}{[(\omega_2 - \omega_1)^2 + (h_1 + h_2)^2][(\omega_2 + \omega_1)^2 + (h_1 + h_2)^2]},$$

$$\omega_1 = \sqrt{k_1^2 - h_1^2}, \qquad \omega_2 = \sqrt{k_2^2 - h_2^2}.$$

36. OPTIMAL DYNAMICAL SYSTEMS

36.1 Determining $K_x(\tau)$ as a correlation function of a sum of correlated random functions and applying to the resulting equality a Fourier inversion, we get

$$S_x(\omega) = S_u(\omega) + S_v(\omega) + S_{uv}(\omega) + S_{uv}^*(\omega).$$

36.2 $S_{xz}(\omega) = i\omega[S_u(\omega) + S_{vu}(\omega)]$. **36.3** $L(i\omega) = i\omega e^{-i\omega\tau}$, $\mathbf{D}[\varepsilon(t)] = 0$.

36.4
$$\begin{aligned}L(i\omega) &= \frac{ia^2}{a^2(\omega^2 + \beta^2)^2 + b^2(\omega^2 + \alpha^2)^2} \\ &\quad \times \left\{\omega(\omega^2 + \beta^2)^2 e^{-i\omega\tau} - \frac{(\omega - i\alpha)^2(\omega - i\beta)^2}{2m}\right. \\ &\quad \times \left[(m - in)\left(\frac{m - in + i\beta}{m - in - i\alpha}\right)^2 e^{-(n + im)\tau}(\omega + m + in)\right. \\ &\quad + (m + in)\left(\frac{m + in - i\beta}{m + in + i\alpha}\right)^2 \\ &\quad \left.\left. \times\, e^{-(n - im)\tau}(\omega - m + in)\right]\right\},\end{aligned}$$

where

$$m = \sqrt{\frac{\sqrt{\mu^2 + \nu^2} - \mu}{2}}, \qquad n = \sqrt{\frac{\sqrt{\mu^2 + \nu^2} + \mu}{2}},$$

$$\mu = \frac{a^2\beta^2 + b^2\alpha^2}{a^2 + b^2}, \qquad \nu = \frac{ab|\beta^2 - \alpha^2|}{a^2 + b^2}.$$

36.5 $$L(i\omega) = \frac{a^2}{c^2}\frac{(\alpha + \beta)(\omega - i\beta)}{(\alpha + d)(\omega - id)},$$

where

$$c = \sqrt{a^2 + b^2}, \qquad d = \frac{1}{c}\sqrt{a^2\beta^2 + b^2\alpha^2}.$$

36.6 $$\mathbf{D}[\varepsilon(t)] = \int_{-\infty}^{\infty} |N(i\omega)|^2 S_u(\omega)\, d\omega - \int_{-\infty}^{\infty} |L(i\omega)|^2[S_u(\omega) + S_v(\omega) + S_{uv}(\omega) + S^*_{uv}(\omega)]\, d\omega.$$

36.7 $$L(i\omega) = \frac{i\alpha^2}{2mc^2}\left\{\frac{m + in}{[m + i(n + n_1)]^2 - m_1^2}\cdot\frac{\omega + m - in}{(\omega - m_1 - in_1)(\omega + m_1 - in_1)} - \frac{-m + in}{[m - i(n + n_1)]^2 - m_1^2} \times \frac{\omega - m - in}{(\omega - m_1 - in_1)(\omega + m_1 - in_1)}\right\},$$

where

$$m = \sqrt{\sqrt{\beta^4 + \gamma^4} + \beta^2}, \qquad n = \sqrt{\sqrt{\beta^4 + \gamma^4} - \beta^2},$$

$$m_1 = \sqrt{\sqrt{\beta^4 + \gamma^4 + \frac{\alpha^2}{4c^2}} + \beta^2}, \qquad n_1 = \sqrt{\sqrt{\beta^4 + \gamma^4 + \frac{\alpha^2}{4c^2}} - \beta^2};$$

$$\mathbf{D}[\varepsilon(t)] = \frac{\pi\alpha^2}{2n} - \frac{\alpha^4\pi}{2m^2c^2}\left[\frac{|A|^2}{n} + \operatorname{Im}\left(\frac{A^2}{m + in}\right)\right],$$

where

$$A = \frac{m + in}{[m^2 - m_1^2 - (n + n_1)^2] + 2im(n + n_1)}.$$

36.8 $L(i\omega) = e^{-\alpha\tau}$.

36.9 $L(i\omega) = e^{-\tau}[i\omega\tau + (1 + \tau)]$.

36.10 $$L(i\omega) = \frac{e^{-\beta\tau}}{\omega - i\alpha}\left\{\omega\left[\cos\beta\tau - \left(1 - \frac{\alpha}{\beta}\right)\sin\beta\tau\right] + i[(2\beta - \alpha)\sin\beta\tau - \alpha\cos\beta\tau]\right\};$$

$$\mathbf{D}[\varepsilon(\tau)] = \frac{\pi a^2}{2\beta}\left\{\frac{(\alpha^2 + 2\beta^2)}{2\beta^2} - e^{-2\beta\tau}\left[\cos\beta\tau - \left(1 - \frac{\alpha}{\beta}\right)\sin\beta\tau\right]^2 - \frac{\alpha^2}{2\beta^2}e^{-2\beta\tau}\left[\cos\beta\tau + \left(1 - \frac{2\beta}{\alpha}\right)\sin\beta\tau\right]^2\right\}.$$

36.11 $$L(i\omega) = \frac{a^2(\alpha + \beta)}{c^2(d + \alpha)}e^{-\alpha\tau}\frac{\omega - i\beta}{\omega - id},$$

where

$$c^2 = a^2 + b^2, \qquad d^2 = \frac{1}{c^2}(a^2\beta^2 + b^2\alpha^2).$$

36.12 $L(i\omega) = \dfrac{c^2}{a^2(\omega^2 + b^2)} \left\{(\omega^2 + \beta^2)e^{-i\omega\tau_0} - \dfrac{(b - \beta)}{(\alpha - b)}\, e^{-b\tau_0}(\omega - i\alpha)(\omega - i\beta)\right\},$

where

$$a^2 = \frac{1}{\pi}(\alpha\sigma_u^2 + \beta\sigma_v^2), \qquad b^2 = \frac{1}{a^2}\frac{\alpha\beta}{\pi}(\beta\sigma_u^2 + \alpha\sigma_v^2), \qquad c^2 = \frac{\alpha\sigma_u^2}{\pi}.$$

36.13 $L(i\omega) = e^{-\alpha\tau}\left(\cos \alpha\tau + \sin \rho\tau + i\dfrac{\omega}{\alpha} \sin \alpha\tau\right).$

36.14 $L(i\omega) = \dfrac{1}{2\beta\,(\omega - i\gamma)}\, e^{-\alpha\tau_0}\{e^{-i\beta\tau_0}[\beta - i(\alpha - \gamma)](\omega - \beta - i\alpha)$

$+ e^{i\beta\tau_0}[\beta + i(\alpha - \gamma)](\omega + \beta - i\alpha)\};$

$$\mathbf{D}[\varepsilon(t)] = [n_1^2 + n_2^2(\alpha^2 + \beta^2)]\sigma_\theta^2 - 2a^2n_2^2\pi\left[\frac{1}{\alpha}|A|^2 - \operatorname{Im}\left(\frac{A^2}{\beta + i\alpha}\right)\right],$$

where

$$a^2 = \frac{2\sigma_\theta^2}{\pi}\alpha(\alpha^2 + \beta^2)n_2^2, \qquad A = \frac{1}{2\beta}e^{-(\alpha - i\beta)\tau_0}(\beta + i\alpha - i\gamma), \qquad \gamma = \frac{n_1}{n_2}.$$

36.15 The required quantity is characterized by the standard deviation of the error of the optimal dynamical system of 1.67, 0.738, 0.0627 m./sec., respectively.

$$\sigma_\varepsilon = 2\sigma_v\sqrt{\frac{6a}{T(12 + 6\alpha T + a^2T^2)}}.$$

36.16 $\mathbf{D}[\varepsilon(t)] = 4\sigma_v^2\alpha^2 d$, where

$$d = \frac{1}{4 + 4\gamma + \gamma^2 + \dfrac{1}{12}\gamma^3}, \qquad \gamma = \alpha T,$$

which gives for σ_ε the values 1.62, 0.829, 0.0846 m./sec.

36.17 $\mathbf{D}[\varepsilon(t)] = \sigma_\theta^2(\alpha^2 + \beta^2) - \dfrac{\pi k_1^4}{a^2}\left\{\dfrac{c_1^2}{\alpha_v} + \dfrac{2|c_2|^2}{\alpha} + \dfrac{2c_1}{\beta^2 + (\alpha + \alpha_v)^2}\right.$

$\left.\times\ [\beta b' - (\alpha + \alpha_v)a']\right\},$

where

$$a^2 = \frac{\sigma_v^2\alpha_v}{\pi}; \qquad k^2 = \frac{2\sigma_\theta^2\alpha(\alpha^2 + \beta^2)}{\sigma_v^2\alpha_v}; \qquad k_1^2 = \sigma_v^2\alpha_v k^2;$$

$$c_1 = -\frac{\alpha_v}{(\alpha_v + \alpha_1)(\alpha_v + \beta_1)[\beta^2 + (\alpha_v - \alpha)^2]};$$

$$c_2 = \frac{-\alpha + i\beta}{2\beta(\beta + i\alpha + i\alpha_1)(\beta + i\alpha + i\beta_1)(\beta + i\alpha - i\alpha_v)} = a' + ib';$$

$$\alpha_1^2 = \alpha^2 - \beta^2 + \frac{k^2}{2} + \sqrt{\left(\alpha^2 - \beta^2 + \frac{k^2}{2}\right)^2 - (\alpha^2 + \beta^2)^2 - k^2\alpha_v^2};$$

$$\beta_1^2 = \alpha^2 - \beta^2 + \frac{k^2}{2} - \sqrt{\left(\alpha^2 - \beta^2 + \frac{k^2}{2}\right)^2 - (\alpha^2 + \beta^2)^2 - k^2\alpha_v^2}.$$

36.18
$$L(i\omega) = \frac{(\omega^2 + \alpha^2)^2}{4\alpha^3}$$
$$\times \left\{\frac{\beta}{\alpha + i\omega} + \frac{\gamma}{(\alpha + i\omega)^2} + \frac{\lambda_1}{i\omega} - \frac{\lambda_2}{\omega^2} - e^{-i\omega T}\right.$$
$$\left.\times \left[e^{-\alpha T}\left(\frac{\beta + \gamma T}{\alpha + i\omega} + \frac{\gamma}{(\alpha + i\omega)^2}\right) + \frac{\lambda_1 + \lambda_2 T}{i\omega} - \frac{\lambda_2}{\omega^2}\right]\right\}$$
$$- \frac{(\omega - i\alpha)^2}{4\alpha^3}[3\alpha\beta - \gamma + 2\alpha\lambda_1 - \lambda_2 - i\omega(\beta + \lambda_1)]$$
$$- \frac{(\omega + i\alpha)^2}{4\alpha^3} e^{-i\omega T}\{e^{-\alpha T}[\alpha(\beta + \gamma T) + \gamma] + 2\alpha(\lambda_1 + \lambda_2 T)$$
$$+ \lambda_2 + i\omega[(\beta + \gamma T)e^{-\alpha T} + (\lambda_1 + \lambda_2 T)]\},$$

where

$$\lambda_1 = \frac{4(\mu_1 - \beta)}{4 + \alpha T} - \frac{T}{2}\lambda_2 = -0.015202 \text{ sec.}^{-1},$$

$$\lambda_2 = -\frac{4\left[\alpha^2\mu_2 - \alpha\beta + \gamma + \dfrac{\alpha^2 T}{2}(\mu_1 - \beta)\right]}{\dfrac{1}{12}\alpha^3T^3 + \alpha^2T^2 + 4\alpha T + 4} = -0.0112 \text{ sec.}^{-1},$$

$$\mu_1 = 1, \quad \mu_2 = \tau_0, \quad \beta = (1 + \alpha\tau_0)e^{-\alpha\tau_0}, \quad \gamma = \alpha e^{-\alpha\tau_0};$$

$$\mathbf{D}[\varepsilon(t)] = \sigma_u^2\left[1 + \lambda_1\mu_1 + \lambda_2\mu_2 - \frac{\beta}{\alpha}(2\alpha\beta - 2\gamma + \alpha\lambda_1 - \lambda_2)\right.$$
$$\left. - \frac{\gamma}{\alpha^2}(\gamma + \lambda_2)\right] = 0.4525.$$

36.19 The general formula for $L(i\omega)$ is the same as in the preceding problem except that

$$\mu_1 = 0, \quad \mu_2 = 1; \quad \beta = -\alpha^2\tau_0 e^{-\alpha\tau_0}; \quad \gamma = -\alpha^2 e^{-\alpha\tau_0};$$
$$\lambda_1 = 4.58\cdot 10^{-3}; \quad \lambda_2 = -2.54\cdot 10^{-4};$$

$$\mathbf{D}[\varepsilon(t)] = \sigma_u^2\left[\alpha^2 + \lambda_2\mu_2 - \frac{\beta}{\alpha}(2\alpha\beta - 2\gamma + \alpha\lambda_1 - \lambda_2)\right.$$
$$\left. - \frac{\gamma}{\alpha^2}(\gamma + \lambda_2)\right] = 0.0110 \text{ sec.}^{-2}.$$

36.20 $l(\tau) = \delta(\tau)$, $\mathbf{D}[\varepsilon(t)] = 0$.

36.21 For the first system

$$L(i\omega) = [(\omega^2 + \alpha^2 - \beta^2)^2 + 4\alpha^2\beta^2]$$
$$\times \left\{\frac{\lambda_1}{i\omega} - \frac{\lambda_2}{\omega^2} + \frac{\lambda_3 + i\lambda_4}{2(\Omega - \omega)} + \frac{\lambda_3 - i\lambda_4}{2(\Omega + \omega)} - e^{-i\omega T}\right.$$
$$\left.\times \left[\frac{\lambda_1 + \lambda_2 T}{i\omega} - \frac{\lambda_2}{\omega^2} + \frac{\lambda_3 + i\lambda_4}{2(\Omega - \omega)} e^{i\Omega T} + \frac{\lambda_3 - i\lambda_4}{2(\Omega + \omega)} e^{-i\Omega T}\right]\right\}$$
$$- [\omega^2 - (\alpha^2 + \beta^2) - 2i\alpha\omega][2\alpha(\lambda_1 + \lambda_4) - \lambda_2 - \lambda_3\Omega - i(\lambda_1 + \lambda_4)\omega]$$
$$- e^{-i\omega T}[\omega^2 - (\alpha^2 + \beta^2) + 2i\alpha\omega][2\alpha(\lambda_1 + \lambda_2 T + \lambda_3 \sin \Omega T + \lambda_4 \cos \Omega T)$$
$$+ \lambda_2 + \lambda_3\Omega \cos \Omega T - \lambda_4\Omega \sin \Omega T + i\omega(\lambda_1 + \lambda_2 T$$
$$+ \lambda_3 \sin \Omega T + \lambda_4 \cos \Omega T)],$$

the constants λ_1, λ_2, λ_3, and λ_4 are determined from the system

$$\lambda_1 + 10\lambda_2 + 0.1244\lambda_2 + 0.9903\lambda_4 = 0.000578,$$
$$\lambda_1 + 13.4034\lambda_2 + 0.1728\lambda_3 + 0.9620\lambda_4 = 0,$$
$$\lambda_1 - 0.8752\lambda_2 + 0.1657\lambda_3 + 0.9837\lambda_4 = 0,$$
$$\lambda_1 + 10.1831\lambda_2 + 0.1236\lambda_3 + 0.9889\lambda_4 = 0.000584,$$

which has the solutions: $\lambda_1 = -0.0018$, $\lambda_2 = 0.000011$, $\lambda_3 = -0.0106$, $\lambda_4 = 0.0036$. The variance for the optimal system of first type is $\mathbf{D}[\varepsilon(t)] = 0.135 \cdot 10^{-4}$. For the second system, the form of $L(i\omega)$ remains the same but $\lambda_1 = \lambda_2 = 0$, and λ_3, λ_4 are determined from the system

$$\lambda_3 + 5.937\lambda_4 = 0,$$

$$\lambda_3 + 8.003\lambda_4 = 0.0047,$$

which leads to $\lambda_3 = -0.0136$, $\lambda_4 = 0.0023$. The variance for this system is

$$\mathbf{D}[\varepsilon(t)] = 0.266 \cdot 10^{-4}.$$

36.22 $a = e^{-\alpha\tau_2}$, $\mathbf{D}[\varepsilon(t)] = (1 - e^{-2\alpha\tau_0})\sigma_x^2$.

36.23 $a = e^{-\alpha\tau}\left(\cos\beta\tau + \dfrac{\alpha}{\beta}\sin\beta\tau\right)$; $b = \dfrac{1}{\beta}e^{-\alpha\tau}\sin\beta\tau$;

$$\mathbf{D}[\varepsilon(t)] = \sigma_x^2\left[1 - e^{-2\alpha\tau}\left(1 + 2\frac{\alpha}{\beta}\sin\beta\tau\cos\beta\tau + 2\frac{\alpha^2}{\beta^2}\sin^2\beta\tau\right)\right].$$

36.24 $a = \dfrac{\sigma_u^2}{\sigma_u^2 + \sigma_v^2}e^{-\alpha\tau_0}$, $\mathbf{D}[\varepsilon(t)] = \sigma_u^2\left(1 - \dfrac{\sigma_u^2}{\sigma_u^2 + \sigma_v^2}e^{-2\alpha\tau_0}\right)$.

36.25 $a = -\dfrac{\alpha^2 + \beta^2}{\beta}e^{-\alpha\tau_0}\sin\beta\tau_0 = -0.09721 \text{ sec.}^{-1}$,

$$b = e^{-\alpha\tau_0}\left(\cos\beta\tau_0 - \frac{\alpha}{\beta}\sin\beta\tau_0\right) = 0.9736,$$

$$c = 0, \quad \mathbf{D}[\varepsilon(t)] = 0.404 \text{ deg.}^2/\text{sec.}^2.$$

36.26 $a = e^{-\alpha\tau_0}\left(\cos\beta\tau_0 + \dfrac{\alpha}{\beta}\sin\beta\tau_0\right) = 0.99$,

$$b = \frac{1}{\beta}e^{-\alpha\tau_0}\sin\beta\tau_0 = 0.20 \text{ sec.}, \quad c = 0.$$

37. THE METHOD OF ENVELOPES

37.1
$$K_a(\tau) = \sigma_x^2\left[2\mathrm{E}(1 - q^2) - q^2\mathrm{K}(1 - q^2) - \frac{\pi}{2}\right],$$

where

$$q^2 = 1 - k^2(\tau) - r^2(\tau); \qquad k(\tau) = e^{-\alpha|\tau|},$$

$$r(\tau) = \frac{2\alpha}{\pi}\int_0^\infty \frac{\sin\omega\tau}{\omega^2 + \alpha^2}\,d\omega = \frac{1}{\pi}\left[e^{-\alpha\tau}\,\mathrm{Ei}\,(\alpha\tau) - e^{\alpha\tau}\,\mathrm{Ei}\,(-\alpha\tau)\right];$$

$\mathrm{Ei}\,(x)$ denotes the integral exponential function.

$$\mathrm{Ei}\,(x) = \int_{-\infty}^x \frac{e^u}{u}\,du.$$

37.2 Since

$$S_x(\omega) = \sigma_x^2\frac{2\alpha^3}{\pi(\omega^2 + a^2)^2},$$

we have $\omega_1 = 2\alpha/\pi$, $\omega_2 = \alpha$.

$$\mathbf{P}(\dot{\Phi} \geqslant 0) = \frac{1}{2}\left(1 + \frac{2}{\pi}\right) = 0.818; \qquad \mathbf{P}(\dot{\Phi} \leqslant 0) = \frac{1}{2}\left(1 - \frac{2}{\pi}\right) = 0.182$$

are independent of α.

37.3 $\mathbf{P}(\dot{\Phi} \geqslant 0) = \frac{1}{2}\left[1 + \frac{1}{\pi\beta}\sqrt{\alpha^2 + \beta^2}\left(\frac{\pi}{2} + \arctan\frac{\beta^2 - \alpha^2}{2\alpha\beta}\right)\right];$

$$\mathbf{P}(\dot{\Phi} \leqslant 0) = \frac{1}{2}\left[1 - \frac{1}{\pi\beta}\sqrt{\alpha^2 + \beta^2}\left(\frac{\pi}{2} + \arctan\frac{\beta^2 - \alpha^2}{2\alpha\beta}\right)\right].$$

37.4 $P = 0.5$ and are independent of α/β.

37.5 $$f(\dot{\varphi}) = \frac{\alpha^2\left(1 - \frac{4}{\pi^2}\right)}{2\left[\left(\dot{\varphi} - \frac{2\alpha}{\pi}\right)^2 + \alpha^2\left(1 - \frac{4}{\pi^2}\right)\right]^{3/2}}.$$

37.6 The phase is uniformly distributed over the interval $[0, 2\pi]$.

37.7 $$f(\dot{\varphi}) = \frac{(\alpha^2 + \beta^2)\left[1 - \frac{\alpha^2 + \beta^2}{\pi^2\beta^2}\left(\frac{\pi}{2} + \arctan\frac{\beta^2 - \alpha^2}{2\alpha\beta}\right)^2\right]}{2\left\{\left[\dot{\varphi} - \frac{\alpha^2 + \beta^2}{\pi\beta}\left(\frac{\pi}{2} + \arctan\frac{\beta^2 - \alpha^2}{2\alpha\beta}\right)\right]^2 + (\alpha^2 + \beta^2)\left[1 - \frac{\alpha^2 + \beta^2}{\pi^2\beta^2}\left(\frac{\pi}{2} + \arctan\frac{\beta^2 - \alpha^2}{2\alpha\beta}\right)^2\right]\right\}^{3/2}}.$$

37.8 $$f(a, \dot{a}) = \frac{a}{\alpha\sigma_x^3\sqrt{2\pi}\sqrt{1 - \frac{4}{\pi^2}}}\exp\left\{-\frac{1}{2\sigma_x^2}\left[a^2 + \frac{\dot{a}^2}{\alpha^2\left(1 - \frac{4}{\pi^2}\right)}\right]\right\}.$$

37.9 Since $k(\tau) = e^{-\alpha|\tau|}(1 + \alpha|\tau|)$, $k(2) = 0.982$,

$$r(2) = 2\int_0^\infty \frac{2\alpha^3}{\pi(\omega^2 + \alpha^2)}\sin 2\omega\, d\omega = \frac{1}{\pi}[1.2e^{-0.2}\,\mathrm{Ei}\,(0.2) - 0.8\,\mathrm{Ei}\,(-0.2)] = 0.122,$$

then

$$f(a_2 \mid a_1 = \sigma_x) = \frac{48.08a}{\sigma_x^2}\exp\left\{-\frac{24.04a_2^2}{\sigma_x^2} - 23.2\right\} I_0\left(47.56\frac{a_2}{\sigma_x}\right).$$

37.10 Since

$$\Delta^2 = \omega_2^2 - \omega_1^2 = (\alpha^2 + \beta^2)\left[1 - \frac{\alpha^2 + \beta^2}{\pi^2\beta^2}\left(\frac{\pi}{2} + \arctan\frac{\beta^2 - \alpha^2}{2\alpha\beta}\right)^2\right] = 0.0089,$$

$$\frac{\Delta}{\omega_2} = 0.0135 \ll 1,$$

the following formula is useful

$$f(\tau) \approx \frac{4.45\pi\cdot 10^{-3}}{\tau^2\left[\left(\frac{\pi}{\tau} - 0.693\right)^2 + 8.9\cdot 10^{-3}\right]^{3/2}}.$$

37.11 $$f(\tau) = \frac{0.041\pi}{\tau^2\left[\left(\frac{\pi}{\tau} - 0.647\right)^2 + 0.0814\right]^{3/2}}.$$

37.12 The required average number of passages equals the probability of occurrence of one passage per unit time

$$p = 2\sqrt{\frac{\omega_2^2 - \omega_1^2}{2\pi}}\,e^{-2} = 0.083\alpha \text{ sec.}^{-1}.$$

37.13 0.0424α sec.$^{-1}$.

37.14 $$f(\varphi_2 \mid \varphi_1) = \frac{q^2}{2\pi}\left\{\frac{1}{1-\kappa^2} + \frac{\kappa}{(1-\kappa^2)^{3/2}}\left[\frac{\pi}{2} + \arcsin \kappa\right]\right\},$$
where
$$q^2 = 1 - k^2(\tau) - r^2(\tau); \qquad \tau = \frac{\pi^2\beta}{\alpha^2+\beta^2}\left(\frac{\pi}{2} + \arctan\frac{\beta^2-\alpha^2}{2\alpha\beta}\right)^{-1} = 4.53 \text{ sec.};$$
$$k(\tau) = -0.95; \qquad r(\tau) = 2\int_0^\infty \frac{2\alpha(\alpha^2+\beta^2)}{\pi[(\omega^2+\alpha^2-\beta^2)^2 + 4\alpha^2\beta^2]}\sin\omega\tau\, d\tau;$$
$$\kappa = \sqrt{1-q^2}\cos(\varphi_2 - \gamma); \qquad \gamma = 179^\circ;$$
$$\mathbf{D}[X(t+\tau)] \approx \mathbf{M}[A^2]\mathbf{M}[\cos^2\Phi] - \{\mathbf{M}[A]\mathbf{M}[\cos\Phi]\}^2;$$
$$\mathbf{M}[\cos\Phi] = \int_0^{2\pi} f(\varphi_2 \mid \varphi_1)\cos\varphi_2\, d\varphi_2; \qquad \mathbf{M}[\cos^2\Phi] = \int_0^{2\pi} f(\varphi_2 \mid \varphi_1)\cos^2\varphi_2\, d\varphi_2.$$

37.15 $$R_{xy}(\tau) = 2\sigma_x^2\int_0^\infty S_x(\omega)\sin\omega\tau\, d\omega$$
$$= \frac{2\alpha(\alpha^2+\beta^2)\sigma_x^2}{\pi}\int_0^\infty \frac{\sin\omega\tau}{(\omega^2-\beta^2+\alpha^2)^2 + 4\alpha^2\beta^2}\, d\omega.$$

VIII MARKOV PROCESSES

38. MARKOV CHAINS

38.1 It follows from the equality $\mathscr{P}^{\alpha+\beta} = \mathscr{P}^\alpha\mathscr{P}^\beta$.

38.2 $p(3) = R'p(0)$, where
$$R = \mathscr{P}_1\mathscr{P}_2\mathscr{P}_3 = \|r_{ij}\|; \qquad r_1 = r_{11} = r_{22} = r_{33} = \alpha_1^3 + \alpha_2^3 + \alpha_3^3 + 6\alpha_1\alpha_2\alpha_3;$$
$$r_2 = r_{12} = r_{31} = r_{23} = 3(\alpha_1^2\alpha_2 + \alpha_2^2\alpha_3 + \alpha_1\alpha_3^2);$$
$$r_3 = r_{13} = r_{21} = r_{32} = 3(\alpha_1\alpha_2^2 + \alpha_1^2\alpha_3 + \alpha_2\alpha_3^2);$$
$$p(3) = \{r_1\alpha + r_3\beta + r_2\gamma; \quad r_2\alpha + r_1\beta + r_3\gamma; \quad r_3\alpha + r_2\beta + r_1\gamma\}.$$

38.3 States: Q_1 means that all competitions are won, Q_2 means that there is one tie, Q_3 means that a sportsman is eliminated from the competitions. By the Perron formula, $p_{21}^{(n)} = p_{31}^{(n)} = p_{32}^{(n)} = 0$, $p_{33}^{(n)} = 1$, $p_{11}^{(n)} = \alpha^n$, $p_{22}^{(n)} = \gamma^n$, $p_{23}^{(n)} = 1 - \gamma^n$, $p_{13}^{(n)} = 1 - p_{11}^{(n)} - p_{12}^{(n)}$,
$$p_{12}^{(n)} = \begin{cases} \beta\dfrac{\alpha^n - \gamma^n}{\alpha - \gamma} & \text{for } \gamma \neq \alpha, \\ n\beta\alpha^{n-1} & \text{for } \gamma = \alpha. \end{cases}$$

38.4 States: Q_1 means that the device is in good repair, Q_2 means that the blocking system is out of order, Q_3 means that the device does not operate;
$$p_{21}^{(n)} = p_{31}^{(n)} = p_{32}^{(n)} = 0, \quad p_{11}^{(n)} = (1-\alpha-\beta)^n, \quad p_{22}^{(n)} = (1-\gamma)^n, \quad p_{33}^{(n)} = 1,$$
$$p_{23}^{(n)} = 1 - (1-\gamma)^n, \quad p_{13}^{(n)} = 1 - p_{11}^{(n)} - p_{12}^{(n)},$$
$$p_{12}^{(n)} = \begin{cases} \dfrac{\alpha}{\gamma-\alpha-\beta}[(1-\alpha-\beta)^n - (1-\gamma)^n] & \text{for } \alpha+\beta \neq \gamma, \\ n\alpha(1-\gamma)^{n-1} & \text{for } \alpha+\beta = \gamma. \end{cases}$$

38.5 The state Q_j $(j = 0, 1, 2, 3)$ means that j members of a team participate in competitions. For $i < k$, $p_{ik}^{(n)} = 0$ $(i, k = 0, 1, 2, 3)$,

$$p_{00}^{(n)} = 1, \quad p_{11}^{(n)} = \alpha^n, \quad p_{22}^{(n)} = \beta^n, \quad p_{33}^{(n)} = \gamma^n, \quad p_{10}^{(n)} = 1 - \alpha^n,$$
$$p_{21}^{(n)} = \beta_1 f(\alpha, \beta), \quad p_{20}^{(n)} = 1 - p_{22}^{(n)} - p_{21}^{(n)}, \quad p_{32}^{(n)} = \gamma_1 f(\beta, \gamma),$$
$$p_{31}^{(n)} = \gamma_2 f(\alpha, \gamma) + \beta_1\gamma_1\varphi(\alpha, \beta, \gamma), \quad p_{30}^{(n)} = 1 - p_{33}^{(n)} - p_{32}^{(n)} - p_{31}^{(n)},$$

where

$$f(a, b) = \frac{a^n - b^n}{a - b}, \qquad f(a, a) = na^{n-1},$$
$$\varphi(\alpha, \beta, \gamma) = \frac{f(\alpha, \gamma) - f(\beta, \gamma)}{\alpha - \beta}, \qquad \varphi(\alpha, \alpha, \gamma) = \frac{n\alpha^{n-1}}{\alpha - \gamma} - \frac{\alpha^n - \gamma^n}{(\alpha - \gamma)^2},$$
$$\varphi(\alpha, \alpha, \alpha) = n(n-1)\alpha^{n-2}.$$

38.6 Make use of Perron's formula for single eigenvalues

$$\lambda_k = p_k \quad (k = 0, 1, \ldots, m), \qquad |\lambda\mathscr{E} - \mathscr{P}| = \prod_{\nu=0}^{m} (\lambda - p_\nu).$$

For $i > k$, $A_{ki}(\lambda) = 0$,

$$A_{kk}(\lambda) = \frac{|\lambda\mathscr{E} - \mathscr{P}|}{\lambda - p_k}.$$

For $k > i$,

$$A_{ki}(\lambda) = \frac{|\lambda\mathscr{E} - \mathscr{P}|\, D_{ki}(\lambda)}{\prod_{\nu=i}^{k} (\lambda - p_\nu)}.$$

38.7 Use Perron's formula when the eigenvalue $\lambda = p$ has multiplicity m, and the eigenvalue $\lambda = 1$ is not multiple.

$$|\lambda\mathscr{E} - \mathscr{P}| = (\lambda - p)^m(\lambda - 1); \qquad A_{m,m}(\lambda) = \frac{|\lambda\mathscr{E} - \mathscr{P}|}{\lambda - 1};$$
$$A_{kk}(\lambda) = \frac{|\lambda\mathscr{E} - \mathscr{P}|}{\lambda - p} \qquad (k = 0, 1, \ldots, m-1).$$

For $i > k$, $A_{ki}(\lambda) = 0$,

$$A_{mi}(\lambda) = \frac{|\lambda\mathscr{E} - \mathscr{P}|\, D_{mi}(\lambda)}{(\lambda - p)^{m-i}(\lambda - 1)}.$$

For $k > i$, $k \neq m$,

$$A_{ki}(\lambda) = \frac{|\lambda\mathscr{E} - \mathscr{P}|\, D_{ki}(\lambda)}{(\lambda - p)^{k-i+1}}.$$

38.8 The state Q_j means that there are j white balls in the urn after the drawings. For $j > i$, $p_{ij} = 0$; for $i \geqslant j$,

$$p_{ij} = \frac{C_i^{i-j} C_{N-i}^{m-i+j}}{C_N^m}.$$

The eigenvalues $\lambda_0 = 1$, $\lambda_k = C_{N-k}^m / C_N^m$ $(k = 1, 2, \ldots, m)$ are not multiple. The transposed matrix $\mathscr{P}'$ is upper triangular; the probabilities $p_{ij}^{(m)}$ are determined by the formulas from the hypothesis of Problem 38.6. For $N = 6$, $m = 3$,

$$p_{01}^{(n)} = p_{02}^{(n)} = p_{03}^{(n)} = p_{12}^{(n)} = p_{13}^{(n)} = p_{23}^{(n)} = 0, \quad p_{00}^{(n)} = 1, \quad p_{11}^{(n)} = \frac{1}{2^n},$$
$$p_{22}^{(n)} = \frac{1}{5^n}, \quad p_{33}^{(n)} = \frac{1}{20^n}, \quad p_{10}^{(n)} = 1 - \frac{1}{2^n}, \quad p_{20}^{(n)} = 1 - \frac{1}{2^{n-1}} + \frac{1}{5^n},$$
$$p_{21}^{(n)} = 2\left(\frac{1}{2^n} - \frac{1}{5^n}\right), \quad p_{30}^{(n)} = 1 - \frac{3}{2^n} + \frac{3}{5^n} - \frac{1}{20^n},$$
$$p_{31}^{(n)} = 3\left(\frac{1}{2^n} - \frac{2}{5^n} + \frac{1}{20^n}\right), \quad p_{32}^{(n)} = 3\left(\frac{1}{5^n} - \frac{1}{20^n}\right).$$

38.9 State Q_j means that the maximal number of points is $N + j$; $p_{ii} = i/m$, $p_{ij} = 0$ for $i > j$; $p_{ij} = 1/m$ for $i < j$ (see Example 38.1);

$$p_{ii}^{(n)} = \left(\frac{i}{m}\right)^n; \quad p_{ik}^{(n)} = 0 \quad \text{for} \quad i > k; \quad p_{ik}^{(n)} = \left(\frac{k}{m}\right)^n - \left(\frac{k-1}{m}\right)^n \quad \text{for} \quad i < k.$$

38.10 The state Q_j means that j cylinders ($j = 0, 1, \ldots, m$) remained on the segment of length L. The probability that the ball hits a cylinder is $j\alpha$, where

$$\alpha = \frac{2(r + R)}{L}; \qquad p_{j,j-1} = j\alpha, \qquad p_{jj} = 1 - j\alpha, \qquad p_{ji} = 0$$

for $i \neq j$ and $i \neq j - 1$ ($i, j = 0, 1, \ldots, m$). The eigenvalues $\lambda_k = 1 - k\alpha$ ($k = 0, 1, \ldots, m$), $p_{ik}^{(n)} = 0$ for $i < k$. For $i \geqslant k$,

$$A_{ki}(\lambda) = \alpha^{i-k} \frac{i!}{k!} \frac{|\lambda\mathscr{E} - \mathscr{P}|}{\prod_{\nu=k}^{i} (\lambda - 1 + \nu\alpha)}.$$

By Perron's formula for $i \geqslant k$, we have

$$p_{ik}^{(n)} = \alpha^{i-k} \frac{i!}{k!} \sum_{j=0}^{m} \left[\frac{\lambda^n(\lambda - \lambda_j)}{\prod_{\nu=k}^{i} (\lambda - 1 + \nu\alpha)}\right]_{\lambda=\lambda_j} = \frac{i!}{k!} \sum_{j=k}^{i} \frac{(-1)^{j-k}(1 - j\alpha)^n}{(j-k)!\,(i-j)!}$$

$$= C_i^k \sum_{\nu=0}^{i-k} (-1)^\nu C_{i-k}^\nu [1 - (k + \nu)\alpha]^n.$$

38.11 State Q_j ($j = 1, 2, \ldots, m$) means that the selected points are located in j parts of the region D; $p_{jj} = j/m$, $p_{j,j+1} = 1 - j/m$. The eigenvalues $\lambda_r = r/m$ ($r = 1, 2, \ldots, m$). From $\mathscr{P}H = HJ$, it follows that $h_{rk} = (C_{k-1}^{r-1}/C_{m-1}^{r-1})$ ($h_{1k} = 1$); from $H^{-1}\mathscr{P} = JH^{-1}$ and $H^{-1}H = \mathscr{E}$ it follows that $h_{ir}^{(-1)} = (-1)^{r-i}C_{m-i}^{r-i}C_{m-1}^{i-1}$. $\mathscr{P}^n = HJ^nH^{-1}$. $p_{ik}^{(n)} = 0$ for $i > k$ and for $i \leqslant k$,

$$p_{ik}^{(n)} = C_{m-i}^{k-i} \sum_{r=0}^{k-i} (-1)^{k-i-r}\left(\frac{r+i}{m}\right)^n C_{k-i}^r$$

(for another solution see Problem 38.10).

38.12 Set $\varepsilon = e^{2\pi i/m}$. Then

$$H = \|h_{jk}\| = \|\varepsilon^{(j-1)(k-1)}\|, \qquad H^{-1} = \|h_{jk}^{(-1)}\| = \frac{1}{m}\|\varepsilon^{-(k-1)(j-1)}\|.$$

$$\mathscr{P}^n = H\|\delta_{jk}\lambda_k^n\|H^{-1},$$

where

$$\lambda_k \sum_{r=1}^{m} \alpha_r \varepsilon^{(r-1)(k-1)} \qquad (k = 1, 2, \ldots, m).$$

$$p_{jk}^{(n)} = \frac{1}{m} \sum_{r=1}^{m} \lambda_r^n \varepsilon^{(r-1)(j-k)}, \qquad p_j^{(\infty)} = \frac{1}{m} \quad (j, k = 1, 2, \ldots, m).$$

38.13 Q_i represents the state in which the particle is at point x_i.

$$p_{i,i-1} = \frac{i}{m}, \qquad p_{i,i+1} = 1 - \frac{i}{m} \qquad (i = 0, 1, \ldots, m).$$

The matrix equation $H^{-1}\mathscr{P} = \|\delta_{ij}\lambda_i\|H^{-1}$ is equivalent to the equations

$$(1 - \xi^2)R_i^1(\xi) = m(\lambda_i - \xi)R_i(\xi), \qquad (i = 0, 1, \ldots, m),$$

where

$$Ri(\xi) = \sum_{k=0}^{m} \xi^k h_{ik}^{(-1)} = C_i(1 - \xi)^{m/2(1-\lambda i)}(1 + \xi)^{m/2(1+\lambda i)}.$$

Since $R_i(\xi)$ is a polynomial, the eigenvalues $\lambda_i = 1 - 2i/m$ ($i = 0, 1, \ldots, m$). From $HH^{-1} = \mathscr{E}$, it follows that $\xi^i = \sum_{k=0}^{m} \delta_{ik}\xi^k = \sum_{j=0}^{m} h_{ij}R_j(\xi)$. Letting

$$\zeta = \frac{1 - \xi}{1 + \xi}, \qquad C_j = 2^{-m/2},$$

we find the elements $h_{ij} = h_{ij}^{(-1)}$ of the matrices $H = H^{-1}$ are given by the expression

$$\sum_{j=0}^{m} h_{ij}\zeta^j = \sum_{j=0}^{m} h_{ij}^{(-1)}\zeta^j = 2^{-m/2}(1 - \zeta)^i(1 + \zeta)^{m-i} \qquad (i = 0, 1, \ldots, m).$$

The probabilities $p_{ik}^{(n)}$ are the elements of the matrix

$$\mathscr{P}^n = H\left\| \left(1 - \frac{2i}{m}\right)^n \delta_{ik} \right\| H.$$

38.14 Q_j describes a state in which the container of the vending machine contains j nickels.

$$p_{00} = q,\ p_{mm} = p,\ p_{j,j+1} = p,\ p_{j,j-1} = q \qquad (j = 0, 1, \ldots, m).$$

The eigenvalues are

$$\lambda_0 = 1, \qquad \lambda_k = 2\sqrt{pq}\cos\frac{k\pi}{m+1} \quad (k = 1, 2, \ldots, m).$$

$$\mathscr{P} = H\|\delta_{ik}\lambda_k\|H^{-1},$$

where

$$h_{j0} = 1 \qquad (j = 0, 1, \ldots, m),$$

$$h_{jk} = \left(\frac{q}{p}\right)^{j/2}\sin\frac{jk\pi}{m+1} - \left(\frac{q}{p}\right)^{(j+1)/2}\sin\frac{(j-1)k\pi}{m+1}$$

$$(j = 0, 1, \ldots, m; \quad k = 1, 2, \ldots, m),$$

$$h_{0k}^{(-1)} = C_1\left(\frac{p}{q}\right)^k \qquad (k = 0, 1, \ldots, m),$$

$$h_{jk}^{(-1)} = C_j\left[\left(\frac{p}{q}\right)^{k/2}\sin\frac{kj\pi}{m+1} - \left(\frac{p}{q}\right)^{(k-1)/2}\sin\frac{(k-1)j\pi}{m}\right]$$

$$(j = 1, 2, \ldots, m; \quad k = 0, 1, \ldots, m).$$

The constants C_j are determined from the condition $H^{-1}H = \mathscr{E}$:

$$C_0 = \frac{\frac{q}{p} - 1}{1 - \left(\frac{p}{q}\right)^{m+1}},$$

$$C_k = \frac{2p}{m+1}\left[1 - 2\sqrt{pq}\cos\frac{k\pi}{m+1}\right]^{-1} \qquad (k = 1, 2, \ldots, m),$$

$$p_{ik}^{(n)} = \sum_{j=0}^{m} h_{ij}\lambda_j^n h_{jk}^{(-1)} \qquad (i, k = 0, 1, \ldots, m).$$

38.15 State Q_1 means hitting the target and Q_2 means a failure;

$$p_{11} = \alpha, \quad p_{21} = \beta, \quad p_1(0) = \frac{1}{2}(\alpha + \beta), \quad p_2(0) = 1 - \frac{1}{2}(\alpha + \beta),$$

$$\{p_1(n);\ p_2(n)\} = (\mathscr{P}')^n\{p_1(0);\ p_2(0)\}.$$

The eigenvalues are: $\lambda_1 = 1$, $\lambda_2 = \alpha - \beta$. By the Lagrange-Sylvester formula for $\lambda_2 \neq 1$, we get

$$\mathscr{P}^n = \frac{1}{1 - \alpha + \beta}[\mathscr{P} - (\alpha - \beta)\mathscr{E} - (\alpha - \beta)^n(\mathscr{P} - \mathscr{E})],$$

$$p_1(n) = \frac{1}{2(1 - \alpha + \beta)}[2\beta + (1 - \alpha - \beta)(\alpha - \beta)^{n+1}].$$

If $\lambda_2 = 1$, then

$$\mathscr{P}^n = \mathscr{E}, \qquad p_1(n) = \frac{1}{2}.$$

38.16 From $\sum_{j=1}^{m} p_j^{(\infty)} = 1$, $\sum_{i=1}^{m} p_{ij} p_i^{(\infty)} = p_j^{(\infty)}$ $(j = 1, 2, \ldots, m)$, it follows that $p_j^{(\infty)} = 1/m$ $(j = 1, 2, \ldots, m)$.

38.17 Q_j describes the state in which the urn contains j white balls;

$$p_{jj} = \frac{2j(m-j)}{m^2}, \qquad p_{j,j+1} = \frac{(m-j)^2}{m^2}, \qquad p_{j,j-1} = \frac{j^2}{m^2} \qquad (j = 0, 1, \ldots, m).$$

The chain is irreducible and nonperiodic, $p_{ik}^{(\infty)} = p_k^{(\infty)}$. From the system

$$p_k^{(\infty)} = p_{k-1}^{(\infty)} p_{k-1,k} + p_k^{(\infty)} p_{kk} + p_{k+1}^{(\infty)} p_{k+1,k} \qquad (k = 0, 1, \ldots, m),$$

we get

$$p_k^{(\infty)} = (C_m^k)^2 p_0^{(\infty)}, \qquad \frac{1}{p_0^{(\infty)}} = \sum_{k=0}^{m} (C_m^k)^2 = C_{2m}^m, \qquad p_{ik}^{(\infty)} = p_k^{(\infty)} = \frac{(C_m^k)^2}{C_{2m}^m}$$

$$(k = 0, 1, \ldots, m).$$

38.18 Q_j describes the state in which the particle is located at the midpoint of the jth interval of the segment;

$$p_{11} = q, \qquad p_{mm} = p, \qquad p_{j,j+1} = p, \qquad p_{j,j-1} = q \qquad (j = 1, 2, \ldots, m).$$

The chain is irreducible and nonperiodic. The probabilities $p_k^{(\infty)}$ can be determined from the system

$$\begin{aligned} qp_1^{(\infty)} + qp_2^{(\infty)} &= p_1^{(\infty)}, \\ pp_{m-1}^{(\infty)} + pp_m^{(\infty)} &= p_m^{(\infty)}, \\ pp_{k-1}^{(\infty)} + qp_{k+1}^{(\infty)} &= p_k^{(\infty)} \qquad (k = 2, 3, \ldots, m-1). \end{aligned}$$

Then

$$p_k^{(\infty)} = \left(\frac{p}{q}\right)^{k-1} p_1^{(\infty)}, \qquad \sum_{k=1}^{m} p_k^{(\infty)} = 1.$$

For $p = q$, $p_k^{(\infty)} = 1/m$ and for $p \neq q$,

$$p_k^{(\infty)} = \frac{1 - \dfrac{p}{q}}{1 - \left(\dfrac{p}{q}\right)^m} \left(\frac{p}{q}\right)^{k-1} \qquad (k = 1, 2, \ldots, m).$$

The probabilities $p_k^{(\infty)}$ can also be obtained from $p_{ik}^{(n)}$ as $n \to \infty$ (see Problem 38.14).

38.19 The chain is irreducible and nonperiodic. From the system

$$\sum_{i=1}^{\infty} u_i p_{ij} = u_j \qquad (j = 1, 2, \ldots),$$

it follows that

$$u_j = \frac{1}{j!} u_1, \qquad u_1 = \sum_{i=1}^{\infty} \frac{i}{i+1} u_i = \sum_{i=1}^{\infty} \frac{i}{(i+1)!} u_1.$$

Since

$$\sum_{i=1}^{\infty} \frac{i}{(i+1)!} = 1,$$

there is a nonzero solution. We also have

$$\sum_{j=1}^{\infty} |u_j| = u_1 \sum_{j=1}^{\infty} \frac{1}{j!} = u_1(e-1) < \infty,$$

that is, the chain is ergodic,

$$p_j^{(\infty)} = \frac{1}{j!} p_1^{(\infty)}, \qquad \frac{1}{p_1^{(\infty)}} = e - 1, \qquad p_j^{(\infty)} = \frac{1}{(e-1)j!} \qquad (j = 1, 2, \ldots).$$

38.20 The chain is irreducible and nonperiodic. From the system

$$\sum_{i=1}^{\infty} u_i p_{ij} = u_j \qquad (j = 1, 2, \ldots),$$

it follows that $u_1 = q \sum_{i=1}^{\infty} u_i$, $u_j = u_1 p^{j-1}$. We also have

$$\sum_{j=1}^{\infty} |u_j| = u_i \sum_{j=1}^{\infty} p^{j-1} = \frac{u_1}{q} < \infty;$$

consequently, the chain is ergodic;

$$p_j^{(\infty)} = p^{j-1} p_1^{(\infty)}, \qquad p_1^{(\infty)} = q,$$

that is,

$$p_j^{(\infty)} = qp^{j-1} \qquad (j = 1, 2, \ldots).$$

38.21 The chain is irreducible and nonperiodic. From the system

$$\sum_{j=1}^{\infty} u_i p_{ij} = u_j \qquad (j = 1, 2, \ldots),$$

it follows that

$$u_j = \frac{u_1}{2(j-1)} \qquad (j = 2, 3, \ldots).$$

The series

$$\sum_{j=1}^{\infty} |u_j| = u_1\left[1 + \sum_{j=2}^{\infty} \frac{1}{2(j-1)}\right]$$

is divergent; that is, the chain is nonergodic. This is a null-regular chain for which $p_{ik}^{(\infty)} = 0\ (i, k = 1, 2, \ldots)$.

38.22 Q_j means that the particle is located at the point with coordinate $j\Delta\ (j = 1, 2, \ldots)$;
$p_{11} = 1 - \alpha,\ \ p_{j,j+1} = \alpha,\ \ p_{j+1,j} = \beta,\ \ p_{jj} = 1 - \alpha - \beta\ \ \ (j = 1, 2, \ldots)$.
The chain is irreducible and nonperiodic. From the system $\sum_{i=1}^{\infty} u_i p_{ij} = u_j$, follows that $u_k = (\alpha/\beta)^{k-1} u_1\ (k = 1, 2, \ldots)$. For $(\alpha/\beta) < 1$, we have

$$\sum_{k=1}^{\infty} |u_k| = \frac{u_1}{1 - \dfrac{\alpha}{\beta}} < \infty$$

and, consequently, the chain is ergodic;

$$p_k^{(\infty)} = \left(\frac{\alpha}{\beta}\right)^{k-1} p_1^{(\infty)}, \qquad p_1^{(\infty)} = 1 - \frac{\alpha}{\beta};$$

that is,

$$p_k^{(\infty)} = \left(\frac{\alpha}{\beta}\right)^{k-1}\left(1 - \frac{\alpha}{\beta}\right) \qquad (k = 1, 2, \ldots).$$

If $\alpha/\beta \geqslant 1$, the Markov chain is null-regular; $p_{jk}^{(\infty)} = 0\ (j, k = 1, 2, \ldots)$.

38.23 Since $W^{\infty} = 0$, $p_{*j} = \sum_{\nu=1}^{s} p_{j\nu}^{(\infty)} = 1\ (j = s + 1, s + 2, \ldots, m)$.

38.24 From the system

$$p_{*j} = \alpha \sum_{\nu=r+1}^{m} p_{*\nu} + \beta \qquad (j = r + 1, r + 2, \ldots, m),$$

we obtain

$$p_{*j} = \frac{\beta}{1 - \alpha(m - r)} \qquad (j = r + 1, r + 2, \ldots, m).$$

38.25 Q_j represents the state in which player A has j dollars ($j = 0, 1, \ldots, m$); $p_{00} = 1$, $p_{mm} = 1$, $p_{j,j+1} = p$, $p_{j,j-1} = q$ ($j = 1, 2, \ldots, m - 1$). The probabilities $p_{*j} = p_{j0}^{(\infty)}$ of ruin of player A are determined from the system

$$p_{*1} = p_{*2}p + q, \qquad p_{*,m-1} = pp_{*m-2}, \qquad p_{*j} = qp_{*j-1} + pp_{*j+1}$$
$$(j = 2, 3, \ldots, m - 2).$$

Setting $p_{*j} = a - b(q/p)^j$, we find for $p \neq q$ that

$$p_{*j} = \frac{1 - \left(\dfrac{p}{q}\right)^{m-j}}{1 - \left(\dfrac{p}{q}\right)^{m}}$$

and for $p = q$ that $p_{*j} = 1 - j/m$ ($j = 1, 2, \ldots, m - 1$). The probabilities of ruin of B are $p_{*j}(B) = 1 - p_{*j}(A)$. Another solution of this problem may be obtained from the expression for $p_{j0}^{(n)}$ as $n \to \infty$ (see Example 38.2).

38.26 $H = \|h_{jk}\| = \|\varepsilon^{(j-1)(k-1)}\|$, where $\varepsilon = e^{2\pi i/m}$. Then $\mathscr{P}H = H\|\delta_{jk}\lambda_k\|$, where $\lambda_k = \varepsilon^{k-1}$ ($k = 1, 2, \ldots, m$). Since $|\lambda_k| = 1$, the period $\kappa = m$,

$$H^{-1} = \frac{1}{m}\|\varepsilon^{-(j-1)(k-1)}\|.$$

$$p_{jk}^{(n)} = \frac{1}{m}\sum_{\nu=1}^{m} \varepsilon^{(\nu-1)(n+j-k)};$$

that is, $p_{jk}^{(n)} = 1$ if $n + j - k$ is divisible by m and $p_{jk}^{(n)} = 0$ otherwise ($j, k = 1, 2, \ldots, m$). $p_{jk}^{(mn+r)} = 1$ if $r + j - k$ is divisible by m and $p_{jk}^{(mn+r)} = 0$ otherwise ($r = 0, 1, \ldots, m - 1$).

$$\hat{p}_{jk} = \frac{1}{m}\lim_{n\to\infty}\sum_{r=0}^{m-1} p_{jk}^{(mn+r)} = \frac{1}{m} \qquad (j, k = 1, 2, \ldots, m).$$

38.27 $$\mathscr{P}^n = \begin{Vmatrix} \alpha^n & U_n \\ 0 & R^n \end{Vmatrix}; \qquad |\lambda\mathscr{E} - \mathscr{P}| = (\lambda - \alpha)(\lambda^3 - 1), \qquad \lambda_1 = 1,$$

$$\lambda_2 = \alpha, \qquad \lambda_3 = \varepsilon, \qquad \lambda_4 = \varepsilon^2,$$

where $\varepsilon = e^{2\pi i/3}$. The period $\kappa = 3$. For $j, k = 2, 3, 4$; $p_{jk}^{(n)} = 1$ if $n + j - k$ is divisible by 3 and $p_{jk}^{(n)} = 0$ otherwise. By the Perron formula,

$$p_{12}^{(n)} = \frac{1}{3} - \frac{\alpha^n(\alpha^2\beta + \alpha\delta + \gamma)}{1 - \alpha^3} + \frac{\varepsilon^{n-1}(\beta\varepsilon^2 + \delta\varepsilon + \gamma)}{(\alpha - \varepsilon)(1 - \varepsilon)^2} - \frac{\varepsilon^{2n-1}(\beta\varepsilon^4 + \delta\varepsilon^2 + \gamma)}{3(\alpha - \varepsilon^2)},$$

$$p_{13}^{(n)} = \frac{1}{3} - \frac{\alpha^n(\alpha^2\gamma + \alpha\beta + \delta)}{1 - \alpha^3} + \frac{\varepsilon^{n-1}(\gamma\varepsilon^2 + \beta\varepsilon + \delta)}{(\alpha - \varepsilon)(1 - \varepsilon)^2} - \frac{\varepsilon^{2n-1}(\gamma\varepsilon^4 + \beta\varepsilon^2 + \gamma)}{3(\alpha - \varepsilon^2)},$$

$$p_{14}^{(n)} = \frac{1}{3} - \frac{\alpha^n(\alpha^2\delta + \alpha\gamma + \beta)}{1 - \alpha^3} + \frac{\varepsilon^{n-1}(\delta\varepsilon^2 + \gamma\varepsilon + \beta)}{(\alpha - \varepsilon)(1 - \varepsilon)^2} - \frac{\varepsilon^{2n-1}(\delta\varepsilon^4 + \gamma\varepsilon^2 + \beta)}{3(\alpha - \varepsilon^2)}$$
$$= 1 - \alpha^n - p_{12}^{(n)} - p_{13}^{(n)},$$

$$\hat{p}_{jk} = \frac{1}{3}\lim_{n\to\infty}[p_{jk}^{(3n)} + p_{jk}^{(3n+1)} + p_{jk}^{(3n+2)}],$$

$$\hat{p}_{j1} = 0 \quad (j = 1, 2, 3, 4), \qquad \hat{p}_{jk} = \frac{1}{3} \quad (k = 2, 3, 4;\ j = 1, 2, 3, 4).$$

38.28 The chain is irreducible and periodic with period $\kappa = 2$. The first group consists of states with odd numbers and the second, those with even numbers. Then $\lim_{n\to\infty} p_{jk}^{(2n)} = p_k$ and $\lim_{n\to\infty} p_{jk}^{(2n+1)} = 0$ if $j + k$ is an even number, and $\lim_{n\to\infty} p_{jk}^{(2n)} = 0$, $\lim_{n\to\infty} p_{jk}^{(2n+1)} = p_k$ if $j + k$ is an odd number. The mean limiting absolute probabilities $\hat{p}_k = 1/2m$ ($k = 1, 2, \ldots, 2m$) are determined from the equality $\mathscr{P}'\hat{p} = \hat{p}$, $p_k = \kappa\hat{p}_k$.

38.29 Q_j describes the state in which the particle is at point x_j $(j = 0, 1, \ldots, m)$; $p_{01} = 1$, $p_{m,m-1} = 1$, $p_{j,j+1} = p_{j,j-1} = q$ $(j = 1, 2, \ldots, m-1)$. The chain is irreducible and periodic with period $\kappa = 2$; $q\hat{p}_1 = \hat{p}_0$, $\hat{p}_0 + q\hat{p}_2 = \hat{p}_1$, $p\hat{p}_{m-2} + \hat{p}_m = \hat{p}_{m-1}$, $p\hat{p}_{m-1} = \hat{p}_m$, $p\hat{p}_{k-1} + q\hat{p}_{k+1} = \hat{p}_k$ $(k = 1, 2, \ldots, m-1)$. For $p \neq q$, we have

$$\hat{p}_0 = \frac{1}{2}\frac{1 - \dfrac{p}{q}}{1 - \left(\dfrac{p}{q}\right)^m}, \qquad \hat{p}_m = \frac{1}{2}\left(\frac{p}{q}\right)^{m-1}\frac{1 - \dfrac{p}{q}}{1 - \left(\dfrac{p}{q}\right)^m},$$

$$\hat{p}_k = \frac{1}{2p}\left(\frac{p}{q}\right)^k \frac{1 - \dfrac{p}{q}}{1 - \left(\dfrac{p}{q}\right)^m} \qquad (k = 1, 2, \ldots, m-1).$$

For $p = q$, we have $\hat{p}_1 = \hat{p}_m = 1/2m$, $\hat{p}_k = 1/m$ $(k = 1, 2, \ldots, m-1)$.

39. THE MARKOV PROCESSES WITH A DISCRETE NUMBER OF STATES

39.1 $P_n(t) = e^{-\lambda pt}\dfrac{(\lambda pt)^n}{n!}$.

39.2 $P_n(t) = \dfrac{[(\lambda_1 + \lambda_2)t]^n}{n!} e^{-(\lambda_1+\lambda_2)t}$.

39.3

$$P_n(t) = \frac{[\Lambda(t)]^n}{n!} e^{-\Lambda(t)},$$

where $\Lambda(t) = \lambda \int_0^t [1 - F(x)]\, dx$;

$$p_n = \lim_{t\to\infty} P_n(t) = \frac{(\lambda\bar{t})^n}{n!} e^{-\lambda\bar{t}},$$

where $\bar{t} = \int_0^\infty [1 - F(x)]\, dx$ is the expected flight time of the electron.

39.4 $\rho = \sqrt{\dfrac{t}{t + \tau}}$.

39.5

$$F_n(t) = \begin{cases} 1 - \sum_{k=0}^{n-1} \dfrac{(\lambda t)^k}{k!} e^{-\lambda t} & \text{if } t \geqslant 0, \\ 0 & \text{if } t < 0; \end{cases}$$

$$f_n(t) = \begin{cases} \lambda \dfrac{(\lambda t)^{n-1}}{(n-1)!} e^{-\lambda t} & \text{if } t \geqslant 0, \\ 0 & \text{if } t < 0; \end{cases}$$

$$m_k = \frac{n(n+1)\cdots(n+k-1)}{\lambda^k}.$$

39.6 Solving the first system of equations

$$\frac{dP_{ik}(t)}{dt} = -\lambda P_{ik}(t) + \lambda P_{i,k-1}(t)$$

for initial conditions $P_{ik}(0) = \delta_{ik}$ by induction from $P_{i,k+1}(t)$ to $P_{ik}(t)$, we obtain:

$$P_{ik}(t) = \begin{cases} \dfrac{(\lambda t)^{i-k}}{(i-k)!} e^{-\lambda t} & \text{if } 0 \leqslant k \leqslant i, \\ 0 & \text{if } k > i, \end{cases}$$

39.7 For $\lambda = \mu$, the inequality

$$p_m = \frac{\dfrac{1}{m!}}{\displaystyle\sum_{n=0}^{m} \frac{1}{n!}} \leqslant 0.015$$

gives $m = 4$.

39.8 The system of equations for the limiting probabilities p_n:

$$\left.\begin{aligned} m\lambda p_0 &= \mu p_1 \\ [(m - n)\lambda + \mu]p_n &= (m - n + 1)\lambda p_{n-1} + \mu p_{n+1} \\ \mu p_m &= \lambda p_{m-1} \end{aligned}\right\},$$

has the solutions

$$p_n = \frac{m!}{(m - n)!}\left(\frac{\lambda}{\mu}\right)^n p_0,$$

where p_0 is determined by the condition $\sum_{n=0}^{m} p_n = 1$. The expected number of machines in the waiting line is

$$L_q = m - \frac{\lambda + \mu}{\lambda}(1 - p_0).$$

39.9 The system of equations for the limiting probabilities p_n is:

$$\left.\begin{aligned} m\lambda p_0 &= \mu p_1 \\ [(m - n)\lambda + n\mu]p_n &= (m - n + 1)\lambda p_{n-1} + (n + 1)\mu p_{n+1} \quad \text{for } 1 \leqslant n < r \\ [(m - n)\lambda + r\mu]p_n &= (m - n + 1)\lambda p_{n-1} + r\mu p_{n+1} \quad \text{for } r \leqslant n \leqslant m - 1 \\ r\mu p_m &= \lambda p_{m-1} \end{aligned}\right\},$$

and it has the solutions:

$$p_n = \begin{cases} \dfrac{m!}{n!\,(m - n)!}\left(\dfrac{\lambda}{\mu}\right)^n p_0 & \text{if } 1 \leqslant n \leqslant r, \\ \dfrac{m!}{r^{n-r}r!\,(m - n)!}\left(\dfrac{\lambda}{\mu}\right)^n p_0 & \text{if } n > r; \end{cases}$$

the expected number of machines in the waiting line for repairs is

$$L_q = p_0 \frac{m!}{r!} \sum_{n=r}^{m} \frac{n - r}{r^{n-r}(m - n)!}\left(\frac{\lambda}{\mu}\right)^n.$$

39.10 The probability that the computer runs is the limiting probability that there are no calls for service in the system $p_0 = e^{-\lambda/\mu}$, where μ is the average number of repairs per hour. The expected efficiency resulting from application of more reliable elements during 1000 hours of operation is

$$\begin{aligned} \delta &= \left[a + 1000c\left(1 - \exp\left\{-\frac{\lambda_A}{\mu}\right\}\right)\right] - \left[b + 1000c\left(1 - \exp\left\{-\frac{\lambda_B}{\mu}\right\}\right)\right] \\ &= 161c - (b - a). \end{aligned}$$

39.11 (a) The system of equations for the limiting probabilities

$$\left.\begin{aligned} \lambda p_0 &= \mu p_1 \\ (\lambda + k\mu)p_k &= \lambda p_{k-1} + (k + 1)\mu p_{k+1} \quad (1 \leqslant k < n) \\ (\lambda + n\mu)p_k &= \lambda p_{k-1} + n\mu p_{k+1} \quad (k \geqslant n) \end{aligned}\right\},$$

has the solutions:

$$p_k = \begin{cases} \dfrac{1}{k!}\left(\dfrac{\lambda}{\mu}\right)^k p_0 & \text{for } 1 \leqslant k \leqslant n, \\ \dfrac{1}{n!\, n^{k-n}}\left(\dfrac{\lambda}{\mu}\right)^k p_0 & \text{for } k \geqslant n, \end{cases}$$

where p_0 is the probability that all devices need no service and can be determined from the condition $\sum_{k=0}^{\infty} p_k = 1$;

$$p_0 = \left\{\sum_{k=0}^{n-1} \frac{1}{k!}\left(\frac{\lambda}{\mu}\right)^k + \frac{\mu}{(n-1)!\,(n\mu - \lambda)}\left(\frac{\lambda}{\mu}\right)^n\right\}^{-1}$$

with the condition that $\lambda < n\mu$.

(b) $p^* = \sum_{k=n}^{\infty} p_k = \dfrac{\mu p_0}{(n-1)!\,(n\mu - \lambda)}\left(\dfrac{\lambda}{\mu}\right)^n$. (c) $1 - F(t) = \sum_{k-n}^{\infty} p_k \mathbf{P}_k(T > t)$,

where $\mathbf{P}_k(T > t)$ is the probability that the waiting time in the line is longer than t if there are k calls for service in the system:

$$\mathbf{P}_k(T > t) = \sum_{j=0}^{k-n} \frac{(\mu n t)^j}{j!} e^{-\mu n t}.$$

Substituting this value, we get

$$1 - F(t) = \sum_{k=n}^{\infty} p_k \sum_{j=0}^{k-n} \frac{(\mu n t)^j}{j} e^{-\mu n t};$$

since $p_k/p_n = (\lambda/n\mu)^{k-n}$, changing the order of summation we obtain as a result,

$$1 - F(t) = p_n e^{-\mu n t} \sum_{j=0}^{\infty} \frac{(\lambda t)^j}{j!} \sum_{k=n+j}^{\infty} \left(\frac{\lambda}{n\mu}\right)^{k-n-j} = \frac{n\mu p_n}{n\mu - \lambda} e^{-(n\mu - \lambda)t}$$

and, since $p_n/p^* = 1 - (\lambda/n\mu)$, then $F(t) = 1 - p^* e^{-(n\mu - \lambda)t}$ (for $t \geqslant 0$);

$$\bar{t} = -\int_0^{\infty} t\, dF(t) = \frac{p^*}{n\mu - \lambda}.$$

(d)

$$m_1 = \sum_{k=n}^{\infty} (k - n) p_k = p_n \sum_{k=0}^{\infty} k\left(\frac{\lambda}{n\mu}\right)^k = p_n \frac{\lambda}{n\mu\left(1 - \dfrac{\lambda}{n\mu}\right)^r},$$

$$m_2 = \sum_{k=1}^{\infty} k p_k = m_1 + \frac{n p_n}{1 - \dfrac{\lambda}{n\mu}} + p_0 \sum_{k=1}^{n-1} \frac{1}{(k-1)!}\left(\frac{\lambda}{\mu}\right)^k,$$

$$m_3 = \sum_{k=0}^{n-1} (n - k) p_k = p_0 \sum_{k=0}^{n-1} \frac{n-k}{k!}\left(\frac{\lambda}{\mu}\right)^k.$$

39.12 Apply the formulas of Problem 39.11; $\bar{t} = 2/115$ hours.

39.13 Select n so that $p^* e^{-(n\mu - \lambda)} < 0.01$; $n = 4$ (see Problem 39.11).

39.14 (a) The system of equations for the limiting probabilities

$$\left.\begin{aligned} \lambda p_0 &= \mu p_1 \\ (\lambda + k\mu) p_k &= \lambda p_{k-1} + (k+1)\mu p_{k+1} \qquad (1 \leqslant k < n) \\ (\lambda + n\mu) p_k &= \lambda p_{k-1} + n\mu p_{k+1} \qquad (n \leqslant k \leqslant l - 1) \\ \lambda p_{l-1} &= n\mu p_l \end{aligned}\right\},$$

where $l = n + m$ has the solutions

$$p_k = \begin{cases} \dfrac{p^0}{k!}\left(\dfrac{\lambda}{\mu}\right)^k & \text{if } 1 \leqslant k \leqslant n, \\ \dfrac{p_0}{n!\, n^{k-n}}\left(\dfrac{\lambda}{\mu}\right)^k & \text{if } n \leqslant k \leqslant l, \end{cases}$$

where p_0 is the probability that there are no calls for service in the system;

$$p_0 = \left\{\sum_{k=0}^{n-1} \frac{1}{k!}\left(\frac{\lambda}{\mu}\right)^k + \frac{1}{n!}\left(\frac{\lambda}{\mu}\right)^n \frac{1 - \left(\dfrac{\lambda}{n\mu}\right)^{m+1}}{1 - \dfrac{\lambda}{n\mu}}\right\}^{-1}.$$

(b) the probability of refusal

$$p_l = \frac{p_0}{n!\, n^{l-n}}\left(\frac{\lambda}{\mu}\right)^l.$$

(c) the probability that all devices are busy is

$$p^* = \sum_{k=n}^{n+m} p_k = p_n \frac{1 - \left(\dfrac{\lambda}{n\mu}\right)^{m+1}}{1 - \dfrac{\lambda}{n\mu}},$$

where

$$p_n = \frac{p_0}{n!}\left(\frac{\lambda}{\mu}\right)^n.$$

(d) $$F(t) = 1 - \frac{p^* e^{-\mu t}}{1 - \left(\dfrac{\lambda}{n\mu}\right)^{m+1}} \sum_{j=0}^{m-1} \frac{(\mu n t)^j}{j!}\left[\left(\frac{\lambda}{n\mu}\right)^j - \left(\frac{\lambda}{n\mu}\right)^m\right] \quad (t \geqslant 0).$$

(e) $$m_1 = \frac{p_n}{\left(1 - \dfrac{\lambda}{n\mu}\right)^2}\left[\frac{\lambda}{n\mu} - (m+1)\left(\frac{\lambda}{n\mu}\right)^{m+1} + m\left(\frac{\lambda}{n\mu}\right)^{m+2}\right];$$

$$m_2 = m_1 + \frac{1 - \left(\dfrac{\lambda}{n\mu}\right)^{m+1}}{1 - \dfrac{\lambda}{n\mu}} n p_n + p_0 \sum_{k=1}^{n-1} \frac{1}{(k-1)!}\left(\frac{\lambda}{\mu}\right)^k;$$

$$m_3 = \sum_{k=0}^{n-1} \frac{n-k}{k!}\left(\frac{\lambda}{\mu}\right)^k p_0.$$

39.15 $p_0 = \dfrac{81}{665}$, $p = \dfrac{32}{665}$, $p^* = \dfrac{52}{133}$, $m_1 = \dfrac{264}{665}$, $m_2 = \dfrac{1550}{665} \approx 2.33$.

39.16 The system of equations for the limiting probabilities

$$\left.\begin{aligned} m\lambda p_0 &= \mu p_1 \\ [(m-n)\lambda + n\mu]p_n &= (m-n+1)\lambda p_{n-1} + (n+1)\mu p_{n+1} \\ m\mu p_m &= \lambda p_{m-1} \end{aligned}\right\},$$

has the solutions

$$p_n = C_m^n \left(\frac{\mu}{\lambda + \mu}\right)^{m-n}\left(\frac{\lambda}{\lambda + \mu}\right)^n.$$

39.17 The system of equations for the probabilities $P_n(t)$:

$$\frac{dP_n(t)}{dt} = -n\lambda P_n(t) + (n-1)\lambda P_{n-1}(t)$$

for initial conditions $P_n(0) = \delta_{n1}$ has the solution $P_n(t) = e^{-\lambda t}(1 - e^{-\lambda t})^{n-1}$.

39.18 The systems of equations

$$\left.\begin{aligned}\frac{dP_0(t)}{dt} &= \mu P_1(t),\\ \frac{dP_n(t)}{dt} &= -n(\lambda+\mu)P_n(t) + (n-1)\lambda P_{n-1}(t) + (n+1)\mu P_{n+1}(t) \quad (n \geqslant 1)\end{aligned}\right\}$$

for initial conditions $P_n(0) = \delta_{n1}$ is solved with the aid of the generating function $G(t, u) = \sum_{n=0}^{\infty} P_n(t)u^n$; $G(t, u)$ satisfies the differential equation

$$\frac{\partial G(t, u)}{\partial t} = (\lambda u - \mu)(u - 1)\frac{\partial G(t, u)}{\partial u}$$

with the initial condition $G(0, u) = u$. It has the solution

$$G(t, u) = \frac{\mu\kappa + u[1 - (\lambda + \mu)\kappa]}{1 - u\lambda\kappa}.$$

where

$$\kappa = \begin{cases} \dfrac{1 - e^{(\lambda-\mu)t}}{\mu - \lambda e^{(\lambda-\mu)t}} & \text{if } \lambda \neq \mu,\\[2ex] \dfrac{t}{1 + \mu t} & \text{if } \lambda = \mu;\end{cases}$$

thus, it follows that

$$P_0(t) = \mu\kappa, \qquad P_n(t) = (1 - \lambda\kappa)(1 - \mu\kappa)(\lambda\kappa)^{n-1} \qquad (n \geqslant 1).$$

39.19 The system of equations

$$\left.\begin{aligned}\frac{dP_0(t)}{dt} &= -\lambda_0(t)P_0(t)\\ \frac{dP_n(t)}{dt} &= -\lambda_n(t)P_n(t) + \lambda_{n-1}(t)P_{n-1}(t) \quad (n \geqslant 1)\end{aligned}\right\},$$

with the initial condition $P_n(0) = \delta_{n0}$ has the solutions: $P_0(t) = (1 + at)^{-1/a}$,

$$P_n(t) = t^n(1 + at)^{-(n+1/a)}\frac{(1 + a)(1 + 2a)\cdots[1 + (n - 1)a]}{n!}.$$

40. CONTINUOUS MARKOV PROCESSES

40.1 $a_j(t, x_1, x_2, \ldots, x_n) = \psi_j(t, x_1, x_2, \ldots, x_n)$;

$b_{jl}(t, x_1, x_2, \ldots, x_n) = \varphi_j(t, x_1, x_2, \ldots, x_n)\varphi_l(t, x_1, x_2, \ldots, x_n)$.

40.2 $a_j(t, x_1, \ldots, x_n) = \psi_j(t, x_1, \ldots, x_n)$, $j = 1, \ldots, n$;

$a_{n+1} = x_{n+2}$; $a_{n+2} = x_{n+3}$; $a_{n+3} = -\alpha^3 x_{n+1} - 3\alpha^2 x_{n+2} - 3\alpha x_{n+3}$;

$b_{n+3,n+3} = c^2$,

the remaining $b_{jl} = 0$.

40.3 $U(t) \equiv U_1(t)$ is the component of a two-dimensional Markov process for which $a_1 = x_2$,

$$a_2 = -(\alpha^2 + \beta^2)x_1 - 2\alpha x_2, \qquad b_{11} = c^2, \qquad b_{12} = -2\alpha c^2, \qquad b_{22} = 4\alpha^2 c^2.$$

40.4 $a_j(t, x_1, \ldots, x_n) = \varphi_j(t, x_1, \ldots, x_n)$; $b_{jl} = \psi_{jl}(t, x_1, \ldots, x_n)$.

40.5 The Markov process has $r + n$ dimensions;

$a_j = \varphi_j(t, x_1, \ldots, x_r)$, $j = 1, 2, \ldots, r$; $a_{r+l} = x_r + l + 1$, $l = 1, 2, \ldots, n - 1$;

$$a_{r+n} = -\sum_{j=1}^{n} c_{r+n+1-j}x_j; \qquad b_{r+p,r+q} = c_{r+p}c_{r+q}; \qquad p, q = n - m, \ldots, n,$$

the other $b_{jl} = 0$; here $c_{r+k} = \beta_{k+m-n} - \sum_{j=n-m}^{k-1} \alpha_{k-j}c_{j+r}$.

40.7 $\dot{U}_1 = \frac{1}{\mu} U_2 + \frac{1}{\mu} \sqrt{2U_2}\, \xi_1(t); \qquad \dot{U}_2 = \frac{1}{\mu} \varphi(U_1) + \frac{\kappa}{\mu} \xi_2(t),$

where $\xi_1(t)$ and $\xi_2(t)$ are mutually independent random functions with the property of "white noise."

40.8
$$f(y_1, y_2) = c \exp\left\{ -\frac{\alpha^2}{\sigma^2} \int_{x_1}^{y_1} \varphi(\eta)\, d\eta - \frac{\alpha^2}{2\sigma^2} y_2^2 \right\},$$

where c is determined from the conditions of normalization. For $\varphi(u) = \beta^2 u^3$

$$f = c_1 \exp\left\{ -\frac{\alpha^2\beta^2}{4\sigma^2} y_1^4 - \frac{\alpha^2}{2\sigma^2} y_2^2 \right\}, \qquad c_1^{-1} = \frac{2\sigma^2}{\alpha} \frac{\sqrt{\pi}}{\sqrt{\alpha\beta}} \int_{-\infty}^{\infty} e^{-\eta^4}\, d\eta.$$

40.9
$$f(y) = \frac{c}{\psi(y)} \exp\left\{ 2 \int_0^{\infty} \frac{\varphi(\eta)}{\psi(\eta)}\, d\eta \right\},$$

where c is determined from the condition $\int_{-\infty}^{\infty} f(y)\, dy = 1$.

40.10 Setting $U_1 = \zeta(t)$, $U_2 = U_1 - U$, for U_2 we find an equation that is independent of U_1. The Kolmogorov equation for U_2 will be

$$\frac{\partial f}{\partial \tau} - \frac{\partial}{\partial y_2} \left\{ \left[\frac{y_2}{RC} + \frac{1}{C} F(y_2) \right] f \right\} - \frac{1}{2} \frac{\sigma^2}{\pi RC} \frac{\partial^2 f}{\partial y_2^2} = 0$$

and its stationary solution is:

$$f(y_2) = c \exp\left\{ -\frac{\pi}{\sigma^2} y_2^2 - \frac{2\pi R}{\sigma^2} \int_0^{y_2} F(\eta)\, d\eta \right\},$$

where c is determined from the condition of normalization. The required probability density $f(y)$ is the convolution of $f(y_2)$ and the normal distribution law with zero expectation. In the particular case

$$f(y_2) = c \exp\left\{ -\frac{\pi y_2^2}{\sigma^2} - \frac{2\pi R}{\sigma^2} \frac{k}{4} y_2^2 (1 - \operatorname{sgn} y_2) \right\},$$

where

$$c_1 = \frac{2\sqrt{1 + kR}}{\sigma(1 + \sqrt{1 + kR})};$$

$$f(u) = \frac{c_1}{\sqrt{2}} \left\{ \frac{1}{\sqrt{1 + 2\pi}} \left[1 - \Phi\left(\frac{u}{\sigma\sqrt{1 + 2\pi}} \right) \right] \exp\left\{ -\frac{\pi u^2}{(2\pi + 1)\sigma^2} \right\} \right.$$
$$+ \frac{1}{\sqrt{2\pi + 2\pi kR + 1}} \left[1 + \Phi\left(\frac{u}{\sigma\sqrt{2\pi + 2\pi kR + 1}} \right) \right]$$
$$\left. \times \exp\left\{ -\frac{\pi(1 + kR)u^2}{(2\pi + 2\pi kR + 1)\sigma^2} \right\} \right\}.$$

40.11 $f(\tau, y) = \dfrac{\alpha y}{\alpha a^2 + \tau\sigma^2} \exp\left\{ -\dfrac{\alpha y^2}{2(\alpha a^2 + \tau\sigma^2)} \right\}.$

40.12 The Kolmogorov equation for $U = \exp\{-aV\}$ has the form

$$\frac{\partial f}{\partial \tau} = \frac{1}{RC} \frac{\partial}{\partial y} [(y \ln y - a i_0 R) e^{a^2\sigma^2/2}] + \frac{1}{2} \left(\frac{a i_0}{C} \right)^2 \sigma_\xi^2 \frac{\partial^2 f}{\partial y^2}.$$

The stationary solution is:

$$f(y) = N^{-1} \exp\left\{ -\frac{RC}{(a i_0 R)^2 \sigma_\xi^2} \left[y^2 \left(\ln y - \frac{1}{2} \right) - 2 a i_0 R e^{-a^2\sigma^2} y \right] \right\},$$

where

$$\sigma_\xi^2 = \frac{2}{\alpha} e^{a^2\sigma^2} [E_i(-a^2\sigma^2) - 2 \ln a\sigma - 0.57721 \cdots]$$

(compare Stratonovich, 1961, p. 243).

40.13
$$f(y) = c \exp\left\{-\frac{2}{\sigma^2}\int_0^y \varphi(\eta)\,d\eta\right\},$$
where
$$c^{-1} = \int_{-\infty}^{\infty} \exp\left\{-\frac{2}{\sigma^2}\int_0^y \varphi(\eta)\,d\eta\right\}dy.$$

40.14 The Kolmogorov equation is:
$$\frac{\partial f}{\partial \tau} + \frac{\partial}{\partial y}\{[\alpha(\tau) + \beta(\tau)y]f\} - \frac{1}{2}\frac{\partial^2}{\partial y^2}[\gamma^2(\tau)f] = 0;$$
the equation for the characteristic function $E(\tau, z)$ is:
$$\frac{\partial E}{\partial \tau} - iz\alpha(\tau)E - z\beta(\tau)\frac{\partial E}{\partial z} + \frac{\gamma^2}{2}z^2E = 0, \qquad E(\tau, z) = \exp\left\{-\frac{1}{2}\sigma_y^2z^2 + iz\bar{y}\right\},$$
$$\bar{y} = \exp\left\{\int_t^\tau \beta(\tau_1)\,d\tau_1\right\}\left\{x + \int_t^\tau \exp\left[-\int_t^{\tau_2}\beta(\tau_1)\,d\tau_1\right]\alpha(\tau_2)\,d\tau_2\right\},$$
$$\sigma_y^2 = \int_t^\tau \exp\left\{-2\int_\tau^{\tau_2}\beta(\tau_1)\,d\tau_1\right\}\gamma^2(\tau_2)\,d\tau_2.$$

40.15 The Kolmogorov equation is:
$$\frac{\partial f}{\partial \tau} - \frac{1}{T_0}\frac{\partial}{\partial y}(yf) - \frac{1}{2}\frac{i_0^2\sigma_\varepsilon^2}{T_0^2}\frac{\partial^2 f}{\partial y^2} = 0; \qquad f = \frac{1}{\sigma_y\sqrt{2\pi}}\exp\left\{-\frac{(y-\bar{y})^2}{2\sigma_y^2}\right\};$$
$$\bar{y} = x\exp\left\{-\frac{\tau - t}{T_0}\right\}; \qquad \sigma_y^2 = \frac{i_0^2\sigma_\varepsilon^2}{2T_0}\left[1 - \exp\left\{-\frac{2(\tau - t)}{T_0}\right\}\right].$$

40.16 Setting $U_1(t) = U(t)$, $U_2(t) = \dot{U}_1(t)$, we find that the coefficients of the Kolmogorov equation are:
$$a_1 = x_2; \quad a_2 = -2hx_2 - k^2x_1; \quad b_{11} = 0; \quad b_{12} = 0; \quad b_{22} = c^2;$$
$$f(\tau, y_1) = \frac{1}{\sigma_1\sqrt{2\pi}}\exp\left\{-\frac{(y_1 - \bar{y}_1)^2}{2\sigma_1^2}\right\},$$
where
$$\bar{y}_1 = xe^{-h(\tau - t)}\left[\cos\omega_0(\tau - t) + \frac{h}{\omega_0}\sin\omega_0(\tau - t)\right];$$
$$\sigma_1^2 = \frac{c^2}{4hk^2}\left\{\left(1 - \frac{k^2}{\omega_0^2}e^{-2h\tau_1}\right) + \frac{h^2}{\omega_0^2}e^{-2h\tau_1}(h\cos^2\omega_0\delta_1 - \omega_0\sin 2\omega_0\varepsilon_1)\right\};$$
$$\tau_1 = \tau - t; \qquad \omega_0 = \sqrt{k^2 - h^2}.$$

40.17 $$\frac{\partial f}{\partial \tau} - \alpha\frac{\partial}{\partial y}(f\operatorname{sgn} y) - \frac{c^2}{2}\frac{\partial^2 f}{\partial y^2} = 0.$$

40.18 $$f(t, x; \tau, y) = \left(\frac{\beta y^2}{\alpha^2}\right)^{2\mu + 1/2}\frac{2\sqrt{\beta}}{\alpha}\sum_{n=0}^{\infty} e^{-2n\beta\tau}e^{-\beta y^2/\alpha^2}\frac{n!}{\Gamma(n + 2\mu + 1)} \times L_n^{(2\mu)}\left(\frac{\beta x^2}{\alpha^2}\right)L_n^{(2\mu)}\left(\frac{\beta y^2}{\alpha^2}\right),$$
where
$$\mu = \frac{1}{2}\left(\frac{\gamma}{\alpha^2} - \frac{1}{2}\right)$$
and $L_n^{(\nu)}(x)$ are the generalized Laguerre polynomials.

40.19 $$W(T) = \int_{-\beta}^{\beta} w(\alpha T, y_1)\,dy_1;$$
$$w(\tau_1, y_1) = \sum_{j=1}^{\infty}\exp\{-\lambda_j^2\tau_1\}\exp\left\{-\frac{1}{4}y_j^2\right\}D_a(y_1)c_j,$$

where $D_a(x)$ is an even solution of the Weber equation[2] (the parabolic cylinder function):

$$\frac{d^2y}{dx^2} + \left(\frac{1}{4}x^2 - a\right)y = 0; \qquad \lambda_j^2 = a_j - 0.5;$$

a_j is a root of the equation $D_a(\beta) = 0$, $\tau_1 = a\tau$;

$$y_1 = \frac{\sqrt{2\alpha}}{c}y; \qquad \beta = \frac{\sqrt{2\alpha}}{c}u_0; \qquad c_j = \frac{\sqrt{\alpha}}{c}\frac{1}{N_j}D_{a_j}(0);$$

$$N_j = \int_{-\beta}^{\beta} \exp\left\{-\frac{1}{2}y_1^2\right\}D_{a_j}(y_1)\,dy_1.$$

40.20 $$W(T) = \int_{-\infty}^{\beta} w(\alpha T, y_1)\,dy_1;$$

$$w(\tau_1, y_1) = \sum_{j=1}^{\infty} \exp\{-\lambda_j^2\tau_1\}\exp\left\{-\frac{1}{4}y_1^2\right\}V(y_1)c_j,$$

where

$$V_a(x) = \frac{1}{\sqrt{\pi}}2^{-a/2}\left\{2^{-1/4}\Gamma\left(\frac{1}{4} - \frac{1}{2}a\right)D_a^{(1)}(x)\sin\pi\left(\frac{1}{4} + \frac{1}{2}a\right) + 2^{1/4}\Gamma\left(\frac{3}{4} - \frac{1}{2}a\right)D_a^{(2)}(x)\cos\pi\left(\frac{1}{4} + \frac{1}{2}a\right)\right\};$$

$D_a^{(1)}(x)$ and $D_a^{(2)}(x)$ are the even and odd solutions of the Weber equation[2]:

$$\frac{d^2y}{dx^2} + \left(\frac{1}{4}x^2 - a\right)y = 0;$$

a_j is the root of the equation $V_{a_j}(\beta) = 0$; $\lambda_j^2 = a_j - 0.5$; $\tau_1 = \alpha\tau$;

$$y_1 = \frac{\sqrt{2\alpha}}{c}y; \qquad \beta = \frac{\sqrt{2\alpha}}{c}u_0; \qquad c_j = \frac{\sqrt{\alpha}}{c}\frac{1}{N_j}V_{a_j}(0);$$

$$N_j = \int_{-\infty}^{\infty} \exp\left\{-\frac{1}{2}y_1^2\right\}V_{a_j}(y_1)\,dy_1.$$

IX METHODS OF DATA PROCESSING

41. DETERMINATION OF THE MOMENTS OF RANDOM VARIABLES FROM EXPERIMENTAL DATA

41.1 10.58 m. **41.2** (a) 814.87 sq. m., (b) 921.86 sq. m.

41.3 $\tilde{v} = 424.73$ m./sec., $\tilde{\sigma}_v = 8.84$ m./sec.

41.4 $\tilde{v} = 33$ m./sec., $\tilde{E}_v = 3.07$ m./sec.

41.5 $\tilde{x} = 404.85$ sq. m., $\tilde{\sigma}_x = 133$ sq. m.

41.6 For $\mathbf{P}(A) = 0.5$, $D_{\max} = 1/4n$.

41.7 $\mathbf{D}[\tilde{\sigma}_1^2] = \frac{2(n-1)}{n^2}\mathbf{D}^2[X]$, $\mathbf{D}[\tilde{\sigma}_2^2] = \frac{2}{n-1}\mathbf{D}[X]$.

[2] See Tables of Weber Parabolic Cylinder Functions in Fletcher, A., *et al.*: An Index of Mathematical Tables. Vol. II. Oxford, England, Blackwell Scientific Publications, Ltd., 1962.

41.8 $\tilde{Sk} = 0.85$, $\tilde{Ex} = 2.70$.   **41.9** $k = \dfrac{1}{2(n-1)}$.

41.10 $k = \sqrt{\dfrac{\pi}{2n(n-1)}}$.

41.11 (a) $k = \dfrac{1}{n}\sqrt{\dfrac{\pi}{2}}$, (b) $k = \dfrac{1}{\sqrt{n\left[1 + \dfrac{2}{\pi}(n-1)\right]}}$.

41.12 $A_j = \lambda/\sigma_j^2$, where λ is an arbitrary number.

41.13
$$\tilde{x} = \frac{1}{n}\sum_{k=1}^{n} x_k, \qquad \tilde{y} = \frac{1}{n}\sum_{k=1}^{n} y_k,$$

$$\tilde{E}_x = \rho k_n\sqrt{\frac{2}{n-1}\sum_{k=1}^{n}(x_k - \tilde{x})^2}, \qquad \tilde{E}_y = \rho k_n\sqrt{\frac{2}{n-1}\sum_{k=1}^{n}(y_k - \tilde{y})^2},$$

the values of k_n being given in Table 23.

41.14 $\tilde{x} = 48.31$ m., $\tilde{y} = 53.31$ m., $\tilde{E}_x = 10.75$ m., $\tilde{E}_y = 12.50$ m.

41.15
$$\tilde{x} = \frac{1}{n}\sum_{k=1}^{n} x_k, \qquad \tilde{y} = \frac{1}{n}\sum_{k=1}^{n} y_k,$$

$$\tilde{E}_\xi = \rho\sqrt{2}\sqrt{\tilde{\sigma}_x^2 \cos^2\alpha + k_{xy}\sin 2\alpha + \tilde{\sigma}_y^2 \sin^2\alpha};$$

$$\tilde{E}_\eta = \rho\sqrt{2}\sqrt{\tilde{\sigma}_x^2 \sin^2\alpha - \tilde{k}_{xy}\sin 2\alpha + \tilde{\sigma}_y^2 \cos^2\alpha},$$

where

$$\tilde{\sigma}_x^2 = \frac{1}{n-1}\sum_{k=1}^{n}(x_k - \tilde{x})^2; \qquad \tilde{\sigma}_y^2 = \frac{1}{n-1}\sum_{k=1}^{n}(y_k - \tilde{y})^2;$$

$$\tilde{k}_{xy} = \frac{1}{n-1}\sum_{k=1}^{n}(x_k - \tilde{x})(y_k - \tilde{y})$$

and angle α is determined from the equation

$$\tan 2\alpha = \frac{2k_{xy}}{\tilde{\sigma}_x^2 - \tilde{\sigma}_y^2}.$$

41.16 $\tilde{x} = 1$ m., $\tilde{y} = 40$ m., $\tilde{E}_\xi = 23$ m., $\tilde{E}_\eta = 1.07$ m.

41.17
$$k = \frac{\Gamma\left(\dfrac{n-1}{2}\right)}{\Gamma\left(\dfrac{n}{2}\right)}\sqrt{\frac{n}{2}}.$$

First, show that the probability density of the random variable $\tilde{\sigma}$ is determined by the formula

$$f_n(\tilde{\sigma}) = \frac{2}{\Gamma\left(\dfrac{n-1}{2}\right)}\left(\frac{n}{2\sigma^2}\right)^{(n-1)/2}(\tilde{\sigma})^{n-2}\exp\left\{-\frac{n(\tilde{\sigma})^2}{2\sigma^2}\right\}.$$

41.18 See Table 132.

TABLE 132

I_j	1–10	11–20	21–30	31–40	41–50	51–60	61–70	71–80	81–90	91–100
$\tilde{p}_j$	0.107	0.100	0.127	0.087	0.093	0.127	0.093	0.073	0.087	0.106
$\tilde{F}(x)$	0.107	0.207	0.334	0.421	0.514	0.641	0.734	0.807	0.894	1.0

$\tilde{x} = 48.50$, $\tilde{D}[X] = 829.18$.

41.19 See Table 133.

TABLE 133

I_j	0–3	3–6	6–9	9–12	12–15	15–18	18–21	21–24
$\tilde{p}_j$	0.000	0.002	0.006	0.040	0.070	0.114	0.156	0.164

I_j	24–27	27–30	30–33	33–36	36–39	39–42	42–45
$\tilde{p}_j$	0.180	0.122	0.108	0.030	0.004	0.004	0.000

$\tilde{x} = 22.85$, $\mathbf{D}[X] = 40.08$.

41.20 $\tilde{\sigma}_1^2$ and $\tilde{\sigma}_2^2$ are unbiased estimates of the variance

$$(\mathbf{M}[\tilde{\sigma}_2^2] = \mathbf{M}[\tilde{\sigma}_1^2] = \sigma^2); \qquad \mathbf{D}[\tilde{\sigma}_1^2] = \frac{2\sigma^4}{n-1}; \qquad \mathbf{D}[\tilde{\sigma}_2^2] = \frac{3n-4}{(n-1)^2}\sigma^4;$$

that is, $\mathbf{D}[\tilde{\sigma}_1^2] < \mathbf{D}[\tilde{\sigma}_2^2]$ (see Table 134) for any $n > 2$.

TABLE 134

n	3	5	7	10	15	∞
$\dfrac{\mathbf{D}[\tilde{\sigma}_1^2]}{\mathbf{D}[\tilde{\sigma}_2^2]}$	0.80	0.73	0.71	0.69	0.68	0.67

42. CONFIDENCE LEVELS AND CONFIDENCE INTERVALS

42.1 (92.36 m., 107.64 m.). **42.2** $\tilde{x} = 116\frac{1}{22}$ m., (115.53 m.; 116.57 m.).

42.3 0.55; 0.34. **42.4** (a) $\tilde{x} = 10.57$ m., $\tilde{\sigma}_x = 2.05$ m., (b) 0.26, (c) 0.035.

42.5 (5.249 sec., 5.751 sec.); (1.523 sec., 1.928 sec.).

42.6 (867.6 m./sec., 873.0 m./sec.). **42.7** Not less than 11 measurements.

42.8 (24,846 m., 25,154 m.), (130.7 m., 294.9 m.).

42.9 $(4.761 \cdot 10^{-10}, 4.805 \cdot 10^{-10})$, $\tilde{x} = 4.783 \cdot 10^{-10}$.

42.10 (a) (420.75 m./sec., 428.65 m./sec.), (6.69 m./sec., 12.70 m./sec.), (b) 0.61, 0.76.

42.11 Not less than three range finders.

42.12 Not less than 15 measurements.

42.13 0.44, 0.55, 0.71, 0.91. **42.14** See Table 135.

TABLE 135

n	3	5	10	25
$\sigma = 20$ m. $\tilde{\sigma} = 20$ m.	±18.98 m. ±33.72 m.	±14.71 m. ±19.05 m.	±10.40 m. ±11.59 m.	±6.58 m. ±6.84 m.

42.15 $\bar{t} = 425$ hours, (270.70 hours, 779.82 hours).

42.16 (410.21 hours, 1036.56 hours). **42.17** (50.75 hours, 85.14 hours).

42.18 (0.123, 0.459). **42.19** (0.303, 0.503), (0.276, 0.534).

42.20 (0.000, 0.149), (0.000, 0.206), (0.000, 0.369).

42.21 For marksman A (0.128, 0.872), for marksman B (0.369, 0.631).

42.22 (1.15, 3.24). **42.23** (3.721, 4.020). **42.24** (0, 4.6).

42.25

For $\alpha = 0.99$	for $\alpha = 0.95$
for r_{12} (0.42, 0.68),	for r_{12} (0.45, 0.65),
for r_{13} (0.13, 0.47),	for r_{13} (0.17, 0.43),
for r_{14} (0.21, 0.53),	for r_{14} (0.25, 0.49).

42.26 $9.82 < \bar{x} < 11.18$, $1.624 < \sigma_x < 2.632$, $70.58 < \bar{y} < 77.42$, $8.12 < \sigma_y < 13.16$, $0.369 < r_{xy} < 0.796$.

43. TESTS OF GOODNESS-OF-FIT

43.1 $\tilde{\lambda} = 0.928$, $\chi_q^2 = 2.172$, $k = 4$, $\mathbf{P}(\chi^2 \geqslant \chi_q^2) = 0.705$. The deviation is insignificant, the hypothesis on agreement of the observations with the Poisson distribution law is not contradicted.

43.2 $\tilde{\lambda} = 1.54$, $\chi_q^2 = 7.953$, $k = 6$, $\mathbf{P}(\chi^2 \geqslant \chi_q^2) = 0.246$. The deviation is insignificant.

43.3 $\tilde{x} = 5$, $p = 0.5$, $\chi_q^2 = 3.156$, $k = 9$, $\mathbf{P}(\chi^2 \geqslant \chi_q^2) = 0.944$. The hypothesis that at each shot the probability of hitting is the same is not disproved.

43.4 $\chi_q^2 = 10.32$, $k = 7$, $\mathbf{P}(\chi^2 \geqslant \chi_q^2) = 0.176$. The deviations are insignificant.

43.5 $D_{\text{hyp}} = 0.1068$, $\lambda_{\text{hyp}} = 1.068$, $\mathbf{P}(\lambda_{\text{hyp}}) = 0.202$, $D_{\text{bin}} = 0.1401$, $\lambda_{\text{bin}} = 1.401$, $\mathbf{P}(\lambda_{\text{bin}}) = 0.039$. The hypothesis that the observations agree with a hypergeometric distribution law is not disproved; the deviation of the statistical distribution from the binomial is significant and the hypothesis about the binomial distribution should be rejected.

43.6 $\tilde{x} = 11.8$ g., $\tilde{\sigma} = 4.691$ g., $k = 2$, $\chi_q^2 = 1.16$, $P(\chi^2 \geqslant \chi_q^2) = 0.568$. The hypothesis that the observations obey a normal distribution is not disproved.

43.7 $\tilde{x} = 22.85$, $\tilde{\sigma} = 6.394$, $k = 6$, $\chi_q^2 = 5.939$, $\mathbf{P}(\chi^2 \geqslant \chi_q^2) = 0.436$. The hypothesis that the statistical distribution agrees with a normal distribution is not disproved since the deviations are insignificant.

43.8 $\mathbf{M}[Z] = 4.5$, $\mathbf{D}[Z] = 8.25$, where Z is a random digit;

$$\mathbf{M}[X] = 22.5, \quad \mathbf{D}[X] = 41.23, \quad \sigma = 6.423, \quad D_0 = 0.0405,$$
$$\lambda = 0.6403, \quad \mathbf{P}(\lambda) = 0.807.$$

The hypothesis that the statistical distribution agrees with a normal distribution is not disproved.

43.9 $\chi_q^2 = 5.012$, $k = 9$, $\mathbf{P}(\chi^2 \geqslant \chi_q^2) = 0.831$. The deviations are insignificant; the hypothesis that the first 800 decimals of the number π agree with a uniform distribution law is not disproved.

43.10 $D_0 = 0.0138$, $\lambda = 0.3903$, $\mathbf{P}(\lambda) = 0.998$. The hypothesis that the first 800 decimals of π obey a uniform distribution law is not disproved.

43.11 $\chi_q^2 = 4$, $k = 9$, $\mathbf{P}(\chi^2 \geqslant \chi_q^2) = 0.91$. The hypothesis that the observations obey a uniform distribution law is not rejected.

43.12 $D_0 = 0.041$, $\lambda = 0.5021$, $\mathbf{P}(\lambda) = 0.963$. The hypothesis that the observations agree with a uniform distribution is not rejected since the deviations are insignificant.

43.13 $\chi_q^2 = 24.9$, $k = 9$, $\mathbf{P}(\chi^2 \geqslant \chi_q^2) = 0.0034$. The deviations are significant; the hypothesis that the experimental data agree with a uniform distribution should be rejected. The results of the computations contain a systematic error.

43.14 $\tilde{x} = 8.75$, $\tilde{\sigma} = 16.85$, $\chi_{qH}^2 = 11.86$, $k_H = 5$, $\mathbf{P}(\chi^2 \geqslant \chi_{qH}^2) = 0.0398$; an estimate of $\tilde{\delta} = \sqrt{6}\,\tilde{\sigma} = 41.28$ is obtained for the parameter δ of the Simpson distribution law; $\chi_{qc}^2 = 17.06$, $k_c = 5$; $\mathbf{P}(\chi^2 \geqslant \chi_q^2) = 0.00402$. The hypothesis that the observations agree with the Simpson distribution is rejected and the hypothesis that they agree with a normal distribution may be considered not rejected.

43.15 $x = \log y$, $\tilde{x} = -0.1312$, $\tilde{\sigma}_x^2 = 0.3412$, $\tilde{\sigma}_x = 0.5841$, $n = 9$, $k = 6$, $\mathbf{P}(\chi^2 \geqslant \chi_q^2) = 0.890$. The hypothesis that the experimental data obey a logarithmically normal distribution law is not disproved (the deviations are insignificant).

43.16 $\tilde{x} = 2.864, \quad \tilde{m}_2 = 11.469, \quad \mathbf{M}[X] = \nu\tilde{\sigma}, \quad \tilde{\sigma} = \sqrt{\dfrac{\tilde{m}_2}{1 + \nu^2}}$,

where ν is the root of the equation

$$T(\nu) = 0.4229; \qquad T(\nu) = \frac{\varphi(\nu) + 0.5\nu\Phi(\nu)}{\sqrt{1 + \nu^2}} = \frac{x}{2\sqrt{m_2}};$$

for $\nu = 1.2$, we have $T(\nu) = 0.4200$; for $\nu = 1.3$, $T(\nu) = 0.4241$;

$$\nu \approx 1.271, \quad \mathbf{M}[X] = 2.662, \quad \tilde{\sigma} = 2.094, \quad \chi_q^2 = 5.304, \quad k = 9,$$
$$\mathbf{P}(\chi^2 \geqslant \chi_q^2) = 0.894.$$

The hypothesis that X is the absolute value of a normally distributed variable is not disproved.

43.17 $\tilde{x} = 87.46$, $\tilde{\sigma} = 2.471$, $\tilde{\alpha} = 80.02$, $\tilde{\beta} = 94.90$, $\chi_{qH}^2 > 500$, $k_H = 7$, $\mathbf{P}(\chi^2 \geqslant \chi_{qH}^2) \approx 0$. The probability density $\Psi(x)$ for the convolution of a normal and uniform distribution has the form

$$\Psi(x) = \frac{1}{2\cdot 14.88}\left[\Phi\left(\frac{x - 80.02}{2.471}\right) + \Phi\left(\frac{94.90 - x}{2.471}\right)\right], \qquad \chi_{q\varphi}^2 = 2.949,$$

$k_\varphi = 6$, $\mathbf{P}(\chi^2 \geqslant \chi_{q\varphi}^2) = 0.814$. The hypothesis that the experimental data obey a normal distribution law is disproved. The hypothesis that the experimental data agree with the convolution of a normal distribution and a uniform one is not contradicted.

43.18 $\tilde{r} = 50.13$, $\tilde{\sigma} = \tilde{r}\sqrt{\dfrac{2}{\pi}} = 40.0$, $\chi_q^2 = 2.73$, $k = 8$, $\mathbf{P}(\chi^2 \geqslant \chi_q^2) = 0.95$.

The hypothesis that the observations agree with a Rayleigh distribution is not contradicted.

43.19 $\tilde{x} = 508.6$, $\tilde{\sigma} = 123.7$, $\chi_{qH}^2 = 2.95$, $k_H = 7$, $\mathbf{P}(\chi^2 \geqslant \chi_{qH}^2) = 0.888$. The parameter $\tilde{a}$ for a Maxwell distribution is determined from the formula

$$\tilde{a} = \frac{\tilde{x} - x_0}{1.596} = 193.4, \qquad \chi_{qM}^2 = 1.383, \qquad k_M = 7, \qquad \mathbf{P}(\chi^2 \geqslant \chi_{qM}^2) = 0.986.$$

The observations fit a Maxwell distribution better than they fit a normal distribution.

43.20 $\tilde{t} = 871.5$ hours, $\tilde{a} = 0.001148$, $k = 8$, $\chi_q^2 = 4.495$, $\mathbf{P}(\chi^2 \geqslant \chi_q^2) = 0.808$. The hypothesis that the observations agree with an exponential distribution law is not disproved (the deviations are insignificant).

43.21 $\tilde{t} = 394.5$ hours, $\tilde{\sigma} = 228.1$ hours, $\tilde{v}_m = 0.5782$, $\tilde{m} = 1.789$, $\tilde{b}_m = 0.8893$, $\chi_q^2 = 13.44$, $k = 7$, $\mathbf{P}(\chi^2 \geqslant \chi_q^2) = 0.0629$. The hypothesis on the agreement of the observations with a Weibull distribution is not disproved.

43.22 The arctan distribution law is

$$F(z) = \int_{-\infty}^{z} f(z)\,dz = \frac{1}{2} + \frac{1}{\pi}\arctan\frac{z}{2}, \qquad D_0 = 0.0195, \qquad \lambda = 0.6166,$$

$$\mathbf{P}(\lambda) = 0.842.$$

The hypothesis that the statistical distribution of variable z agrees with a Cauchy distribution and, consequently, that of the variable Y with a normal one is not disproved.

43.23 The arcsine distribution function

$$F(z) = \frac{1}{2} + \frac{1}{\pi}\arc\sin\frac{\alpha}{a}, \qquad D_0 = 0.0290, \qquad \lambda = 0.917, \qquad \mathbf{P}(\lambda) = 0.370.$$

The hypothesis that the pendulum performs harmonic oscillations is not disproved.

43.24 $\tilde{\sigma}^2 = 0.1211$, $k = 2$, $\chi_q^2 = 1.629$, $\mathbf{P}(\chi^2 \geqslant \chi_q^2) = 0.59$. The deviations are insignificant; the hypothesis that the observed values of q_i obey a chi-square distribution with $k' = 19$ degrees of freedom and, consequently, the hypothesis on the homogeneity of the series of variances are not disproved. *Hint:* The values of q_i should be arranged in their increasing order and divided into intervals so that each interval contains at least five values q_i.

43.25 $$\tilde{F}(\eta_i) = \frac{i - \dfrac{1}{2}}{n}, \qquad D_n = 0.126, \qquad \lambda = 0.797, \qquad \mathbf{P}(\lambda) = 0.549.$$

The hypothesis that the observed values obey a Student's distribution and, consequently, the hypothesis that the observed values of x_i obey a normal distribution law are not rejected.

43.26 $\tilde{x} = 115.3$, $\tilde{\sigma} = 21.43$, $\chi_{qH}^2 = 10.20$, $k_H = 10$, $\mathbf{P}(\chi^2 \geqslant \chi_{qH}^2) = 0.43$, $\tilde{\mu}_3 = 2046$, $\tilde{\mu}_4 = 6137\cdot 10^2$, $\widetilde{\mathrm{Sk}} = 0.2079$, $\widetilde{\mathrm{Ex}} = -0.0912$. The distribution function for a Charlier-A series is:

$$F(z) = 0.5 + 0.5\Phi(z) - 0.03465\varphi_2(z) - 0.0038\varphi_3(z),$$

where

$$z = \frac{x - 115.3}{21.43}, \qquad \chi_{qCh}^2 = 8.304, \qquad k_{Ch} = 8, \qquad \mathbf{P}(\chi^2 \geqslant \chi_{qCh}^2) = 0.411.$$

The hypotheses on the agreement of the observations with the normal distribution and a distribution specified by a Charlier-A series are not disproved and the latter does not improve the agreement of the observations with the theoretical distribution law.

43.27 $\tilde{\mu}_3 = -221.12$, $\tilde{\mu}_4 = 1560\cdot 10^2$, $\widetilde{\mathrm{Sk}} = -0.06961$, $\widetilde{\mathrm{Ex}} = 0.3406$. The distribution function for a Charlier-A series is

$$F(z) = 0.5 + 0.5\Phi(z) + 0.01160\varphi_2(z) + 0.01417\varphi_3(z),$$

where

$$z = \frac{c - 299773.85}{14.7}, \qquad \chi_q^2 = 17.25, \qquad k = 6, \qquad \mathbf{P}(\chi^2 \geqslant \chi_q^2) = 0.0085.$$

The deviations are significant. The hypothesis that the observations agree with a with a distribution specified by a Charlier-A series is disproved.

43.28 $\chi_q^2 = 20.48$, $k = 2$, $\mathbf{P}(\chi^2 \geqslant \chi_q^2) = 0.001$. The deviations are significant. The hypothesis on the independence of the character of the dimensions on the number of the lot is rejected. A systematic underestimate of dimensions is characteristic for the second lot.

44. DATA PROCESSING BY THE METHOD OF LEAST SQUARES

44.1 $\tilde{h} = 0.609 + 0.1242E$, $M_{0,0} = 0.3896$, $M_{1,1} = 0.00001156$, $\tilde{\sigma}^2 = 1.464$, $\tilde{\sigma}^2_{a_0} = 0.5704$, $\tilde{\sigma}^2_{a_1} = 0.0000169$.

44.2 $\tilde{h} = 0.679 + 0.124E$, $\tilde{\sigma}^2 = 1.450$, $\tilde{\sigma}^2_{a_0} = 0.5639$, $\tilde{\sigma}^2_{a_1} = 0.00001672$. The coincidence with the results of Problem 44.1 is fully satisfactory. The accuracy of the result in Problem 44.2 is higher than in Problem 44.1 since in solving 44.1 a large number of computations were performed and among them there occurred subtraction of approximately equal numbers.

44.3 $\tilde{h} = 9.14 + 65.89t + 489.28t^2$, $\tilde{\sigma}^2 = 0.001245$, $\tilde{\sigma}_g = 1.177$ cm./sec.².

44.4 $\tilde{h} = 65.021 + 5.176P_{1.13}(x)\cdot 13 + 1.087P_{2.13}(x)\cdot 13,$

where $x = 30t - 1$, or

$$\tilde{h} = 9.133 + 65.895t + 489.28t^2, \quad \tilde{\sigma}_g = 1.167 \text{ cm./sec.}^2.$$

44.5 $y = 0.8057 + 0.2004x - 0.1018x^2$, $\tilde{\sigma}^2 = 0.0002758$,

$$\tilde{\sigma}_{a_0} = 0.00009192, \quad \tilde{\sigma}_{a_1} = 0.000009848, \quad \tilde{\sigma}_{a_2} = 0.000003283.$$

44.6 $\tilde{y} = 26.97 + 0.3012P_{1.16}(t) = 29.38 - 0.3012t,$

$$\tilde{y} = 26.97 + 0.3012P_{1.16}(t) - 0.000916P_{2.16}(t) + 0.01718P_{3.16}(t)$$
$$= 29.82 - 0.7133t + 0.06782t^2 - 0.002864t^3,$$

where $P_{k,16}$ are the tabulated values of the Chebyshev polynomials. For a linear dependence $\tilde{\sigma} = 0.3048$; for $\alpha = 0.90$, we have $0.2362 < \sigma < 0.4380$. For a dependence of third degree, $\tilde{\sigma} = 0.1212$; for $\alpha = 0.90$, we have $0.0924 < \sigma < 0.1800$.

44.7 $\tilde{y} = 21.07 + 5.954x$, $\tilde{\sigma}_{a_0} = 2.90$, $\tilde{\sigma}_{a_1} = 0.0889$, $\tilde{K}_{a_0,a_1} = -0.2041$. The confidence intervals for a_k for $\alpha = 0.90$ are: $14.3 < a_0 < 27.9$, $5.75 < a_1 < 6.16$. $\tilde{\sigma}^2_{\tilde{y}}(x) = 2.900 - 0.4082x + 0.0889x^2$. The confidence limits for $y = F(x)$ for $\alpha = 0.90$ are given in Table 136.

TABLE 136

i	0	1	2	3	4
$\tilde{y}_i - \gamma\tilde{\sigma}_y(x_i)$	45.3	72.7	140.0	258.4	366.8
$\tilde{y}_i + \gamma\tilde{\sigma}_y(x_i)$	57.3	83.3	148.7	268.8	383.6

44.8 $\tilde{y} = 0.3548 + 0.06574x + 0.00130x^2$; $\tilde{\sigma}_{a_1} = 0.0147$; $\tilde{\sigma}_{a_1} = 0.0106$;

$\tilde{\sigma}_{a_2} = 0.00156$.

44.9 $$\tilde{y} = 1.1188 + \frac{8.9734}{x}, \qquad \tilde{\sigma}_{a_0} = 0.2316, \qquad \tilde{\sigma}_{a_1} = 0.6157,$$

for $\alpha = 0.95$, we have: $1.065 < a_0 < 1.172$, $8.831 < a_1 < 9.115$; $\tilde{K}_{a_0,a_1} = -0.0854$. The confidence limits for $y = F(x)$ if $\alpha = 0.95$ are given in Table 137.

TABLE 137

x_i	1	2	3	5	10	20	30	50	100	200
$y_i - \gamma\tilde{\sigma}_y(x_i)$	10.03	5.55	4.06	2.87	1.97	1.52	1.37	1.25	1.16	1.11
$y_i + \gamma\tilde{\sigma}_y(x_i)$	10.27	5.66	4.16	2.96	2.06	1.62	1.47	1.35	1.26	1.22

$$\tilde{\sigma}_y^2(x) = 0.05364 - \frac{0.1708}{x} + \frac{0.3790}{x^2}.$$

44.10 $U = 100.8e^{-0.3127t}$, $89.97 < U_0 < 112.9$, $0.2935 < a < 0.3319$.

44.11 $\delta_B = 204'.9 - \dfrac{34205}{v^2}$, $\tilde{\sigma}_{a_0} = 4'.36$, $\tilde{\sigma}_{a_1} = 504$.

44.12 $p = 0.1822 \exp\left\{-\dfrac{(x - 117.25)^2}{2 \cdot 462.91}\right\}$, $|\varepsilon_{\max}| = 0.04633$,

$$p_0' = \frac{10}{\sqrt{2\pi}\,\tilde{\sigma}} = 0.1854.$$

44.13 $\varphi' = 62°$ is chosen according to the formula $y = a' \sin(\omega t - \varphi')$, where

$$a' = \frac{|y_{0.05}| + |y_{0.45}| + |y_{0.95}|}{3} = 33, \quad \tilde{y} = 30.75 \sin(\omega t - 59°\,59'),$$

$$|\tilde{y} - y|_{\max} = 18°.4.$$

44.14 $\tilde{y} = 1.0892 - 1.2496 \cos x + 2.0802 \sin x$

$$+ 0.9795 \cos 2x + 0.4666 \sin 2x,$$

$$|\varepsilon_{\max}| = 0.24 \quad \text{for} \quad x = 120°.$$

44.15 $\tilde{y} = -3.924 + 1.306x$; $|\varepsilon_{\max}| = 1.41$.

45. STATISTICAL METHODS OF QUALITY CONTROL

45.1 For a single sample $\alpha = 0.0323$, $\beta = 0.0190$; for a double sample $\alpha = 0.0067$, $\beta = 0.0100$. The average expenditure of items for 100 lots in the case of a double sample is $48.36 \cdot 15 + 51.64 \cdot 30 = 2275$ items. The expenditure for 100 lots in the case of single sampling is 2200 items. The expenditure of items is almost the same, but in the case of double sampling the probabilities of errors in α and β are considerably smaller. $A = 30.38$, $B = 0.01963$, $\log A = 1.4825$, $\log B = -1.7069$. For a good lot if $p = 0$, $n_{\min} = 13$; $\log \gamma(12, 0) = -1.6288$, $\log \gamma(13, 0) = -1.7771$. For a defective lot when $p = 1$, $n_{\min} = 2$, $\log \gamma(1, 1) = 0.8451$, $\log \gamma(2, 2) = 1.9590$.

45.2 For a single sample $\alpha = 0.049$, $\beta = 0.009$; for a double sample $\alpha = 0.046$, $\beta = 0.008$, $A = 19.8$, $B = 0.01053$, $h_1 = -3.758$, $h_2 = 2.424$, $h_3 = 0.02915$; $\mathbf{M}[n \mid p_0] = 244.2$, $\mathbf{M}[n \mid p_1] = 113.6$, $\mathbf{M}[n]_{\max} = 321.9$. For 100 lots in the case of double sampling, the average expenditure of items is $35.1 \cdot 220 + 64.9 \cdot 440 =$ 36,278 items; in the case of single sampling, the average expenditure is 41,000 items. In the case of sequential analysis, the average expenditure for 100 good lots is not greater than 24,420 items.

45.3 The normal distribution is applicable: $\alpha = 0.0023$, $\beta = 0.0307$, $A = 415.9$, $B = 0.03077$, $h_1 = -4.295$, $h_2 = 7.439$, $h_3 = 0.1452$. For a good lot if $p = 0$, $n_{\min} = 30$; for a defective lot if $p = 1$, $n_{\min} = 9$; $\mathbf{M}[n \mid 0.10] = 94.52$, $\mathbf{M}[n \mid 0.20] = 128.9$, $\mathbf{M}[n]_{\max} = 257.4$, $c = 2.153$, $\mathbf{P}(n < 300) = 0.9842$, $\mathbf{P}(n < 150) = 0.8488$.

45.4 (a) $n_0 = 285$, $\nu = 39$ (a normal distribution is applicable); $A = 98$, $B = 0.0202$, $h_1 = -4.814$, $h_2 = 5.565$, $h_3 = 0.1452$; $\mathbf{M}[n \mid p_0] = 102.1$, $\mathbf{M}[n \mid p_1] = 101.0$; $\mathbf{M}[n]_{\max} = 219.4$; (b) $n_0 = 65$, $\nu = 8$; $A = 8$, $B = 0.2222$, $h_1 - 1.861$, $h_2 = 2.565$, $h_3 = 0.1452$; $\mathbf{M}[n \mid p_0] = 21.6$, $\mathbf{M}[n \mid p_1] = 38.6$; $\mathbf{M}[n]_{\max} = 38.6$.

45.5 Apply the passage from a Poisson distribution to a chi-square distribution: $\nu = 9$, $n_0 = 180$, $A = 18$, $B = 0.1053$, $h_1 = -2.178$, $h_2 = 2.796$, $h_3 = 0.05123$; $\mathbf{M}[n \mid p_0] = 90.86$, $\mathbf{M}[n \mid p_1] = 79.82$, $\mathbf{M}[n]_{\max} = 125.2$. For a good lot if $p = 0$, we have $n_{\min} = 43$, for a defective lot if $p = 1$, $n_{\min} = 3$.

45.6 $$z_0 = \frac{z_{1-\alpha}z_{1-p_1} + z_{1-\beta}z_{1-p_0}}{z_{1-\alpha} + z_{1-\beta}}, \qquad n_0 = \left(1 + \frac{z_0^2}{2}\right)\left(\frac{z_{1-\alpha} + z_{1-\beta}}{z_{1-p_0} - z_{1-p_1}}\right)^2,$$

where z_p are the quantiles of the normal distribution: $F(z_p) = 0.5 + 0.5\Phi(z_p) = p$, $z_{0.97} = 1.881$, $z_{0.92} = 1.405$; $z_{0.95} = 1.645$, $z_{0.90} = 1.282$, $z_0 = 1.613$, $n_0 = 87$. The single sample size in the case of magnitude control for the same α, β, p_0, p_1 is considerably smaller than in the case of control of the proportion of defectives.

45.7 In the case of a binomial distribution law (with passage to the normal distribution law) $\alpha = 0.1403$, $\beta = 0.1776$, $n_0 = 49$, $\nu = 6$, $A = 5.864$, $B = 0.2065$, $h_1 = -1.945$, $h_2 = 2.182$, $h_3 = 0.1452$, $\mathbf{M}[n \mid p_0] = 30.3$, $\mathbf{M}[n \mid p_1] = 26.4$, $\mathbf{M}[n]_{\max} = 34.2$. The average expenditure in the case of double sampling for 100 lots represents $64.34 \cdot 30 + 35.66 \cdot 60 = 4070$ items. In the case of single sampling, the expenditure of items for 100 lots is 4900 items; in the case of sequential analysis, the average expenditure for 100 good lots is not greater than 3030 items. In the case of a Poisson distribution, $\alpha = 0.1505$, $\beta = 0.2176$, $n_0 = 49$, $\nu = 6$ (passage to a chi-square distribution).

45.8 Apply the normal distribution law: $n_0 = 286$, $\nu = 15$, $A = 9900$, $B = 0.01$, $h_1 = 3.529$, $h_2 = 7.052$, $h = 0.04005$, $\mathbf{M}[n \mid 0.02] = 176.0$, $\mathbf{M}[n \mid 0.07] = 231.9$, $\mathbf{M}[n]_{\max} = 647.1$, $c = 3.608$, $\mathbf{P}(n < \mathbf{M}[n \mid 0.02]) = 0.5993$, $\mathbf{P}(n < 2\mathbf{M}[n \mid 0.02]) = 0.9476$, $\mathbf{P}(n < n_0) = 0.8860$.

45.9 For $n_0 = 925$, $\nu = 12$. For $t_0 = 1000$ hours, $A = -2.197$, $B = 2.197$, $t_1 = 237.6$, $t_2 = -237.6$, $t_3 = 74.99$; $\mathbf{M}[T \mid 10^{-5}] = 613.2$, $\mathbf{M}[T \mid 2 \cdot 10^{-5}] = 482.9$, $\mathbf{M}[T]_{\max} = 750.6$.

TABLE 138

t_0, hours	500	1000	2000	5000
n_0	1849	925	463	185

45.10 For the method of single sampling, apply the passage from a Poisson distribution to a chi-square distribution: $\nu = 6$, $n_0 = 122$, $A = 184$, $B = -0.08041$, $h_1 = -1.487$, $h_2 = 3.077$, $h_3 = 0.0503$. For a good lot, if $p = 0$, $n_{\min} = 30$; for a defective lot, if $p = 1$, $n_{\min} = 4$.

$$\mathbf{M}[n \mid 0.02] = 48.3, \quad \mathbf{M}[n \mid 0.10] = 54.6, \quad \mathbf{M}[n]_{\max} = 95.9, \quad c = 5.286;$$

$$\mathbf{P}(n < n_0) = 0.982, \qquad \mathbf{P}\left(n < \frac{1}{2}n_0\right) = 0.714.$$

45.11 For a double sample $\alpha = 0.001486$, $\beta = 0.0009152$; for a single sample $n_0 = 62$, $\nu = 13$ (the passage to the normal distribution law); $A = 671.0$, $B = 0.0009166$, $h_1 = -4.446$, $h_2 = 4.043$, $h_3 = 0.2485$, $\mathbf{M}[n \mid a_0] = 29.2$, $\mathbf{M}[n \mid a_1] = 16.0$, $\mathbf{M}[n]_{\max} = 70.7$. The average expenditure of potatoes per 100 lots in the case of double sampling is $62.88 \cdot 40 + 37.12 \cdot 60 = 4743$ items. The expenditure of potatoes per 100 lots is 6200 items. In the case of sequential analysis, the average expenditure per 100 good lots is not greater than 2920 items.

45.12 For a double sample, $\alpha = 0.0896$, $\beta = 0.0233$; for a single sample, $n_0 = 15, \nu = 12.45$; $A = 10.90, B = 0.02560, h_1 = -977.7, h_2 = 637.2, h_3 = 184.9$; $\mathbf{M}[n \mid \sigma_0] = 9.81$, $\mathbf{M}[n \mid \sigma_1] = 2.78$, $\mathbf{M}[n]_{\max} = 10$. In the case of double sampling, the average expenditure of resistors per 100 good lots is $85.66 \cdot 13 + 14.44 \cdot 26 = 1488$; in the case of single sampling, the expenditure is 1500 items; in the case of sequential analysis, the average expenditure is not larger than 981 items.

45.13 In the case of single sampling, $\alpha = 0.0000884, \beta = 0.00621, B = 0.00621$, $A = 1124 \cdot 10$, $h_1 = 6.506$, $h_2 = -11.94$, $h_3 = 5.15$; $\mathbf{M}[n \mid \xi_0] = 26.02$; $\mathbf{M}[n \mid \xi_1] = 47.32$, $\mathbf{M}[n]_{\max} = 121.4$, $c = 2.542$, $\mathbf{P}(n \leqslant 300) > 0.99(< 0.999)$; $\mathbf{P}(n \leqslant 150) = 0.9182$.

45.14 $n_0 = 86$, $\nu = 66.7$ hours, $A = 999$, $B = 0.001001$, $h_1 = 690.8$, $h_2 = -690.8$, $h_3 = 69.33$, $\lambda^* = 0.01442$, $\mathbf{M}[n \mid \lambda_0] = 22.48$, $\mathbf{M}[n \mid \lambda_1] = 35.67$, $\mathbf{M}[n]_{\max} = 99.31$.

45.15 For a single control of proportion of unreliable condensers $n_0 = 246$, $\nu = 5$. For a sequential reliability control of condensers $A = 9999$, $B = 0.0001$, $h_1 = 1152 \cdot 10^4$, $h_2 = -1152 \cdot 10^4$, $h_3 = 6384 \cdot 10^2$, $\lambda^* = 0.000001566$.

45.16 $t_T = 952.6$ hours, $\nu = 72.8$ hours, $\ln A = 2.197$, $\ln B = -2.197$,

$$t_1 = \frac{\ln AT_0T_1}{T_1 - T_0} = 219.7 \text{ hours}, \qquad t_2 = \frac{\ln BT_0T_1}{T_1 - T_0} = -219.7 \text{ hours},$$

$$t_3 = \frac{T_0 T_1 \ln \dfrac{T_1}{T_0}}{T_1 - T_0} = 69.3 \text{ hours}.$$

For the poorer of the good lots ($\tilde{T} = T_0 = 100)t_{\min} = 715.7$ hours; for the better of the defective lots ($\tilde{T} = T_1 = 50)t_{\min} = 569.2$ hours.

46. DETERMINATION OF PROBABILITY CHARACTERISTICS OF RANDOM FUNCTIONS FROM EXPERIMENTAL DATA

46.1 One should prove that if $\tilde{x} = (1/T) \int_0^T x(t)\,dt$, then $\mathbf{M}[\tilde{x}] = \bar{x}$, $\lim_{T\to\infty} \mathbf{D}[\tilde{x}] = 0$.

46.2 No, since $\lim_{T\to\infty} \mathbf{M}[\tilde{S}_x(\omega)] = S_x(\omega)$, but $\mathbf{D}[\tilde{S}_x(\omega)] = S_x^2(\omega)$ and, consequently, does not tend to zero as T increases.

46.3
$$\mathbf{D}[\tilde{K}_x(\tau)] = \frac{2}{(T-\tau)^2} \int_0^{T-\tau} (T - \tau - \tau_1) \times [K_x^2(\tau_1) + K_x(\tau_1 + \tau)K_x(\tau_1 - \tau)]\,d\tau_1.$$

46.4
$$\mathbf{M}[\tilde{K}_1(\tau)] = K(\tau) - \frac{2}{(T-\tau)^2} \int_0^{T+\tau} (T - \tau - \tau_1) K(\tau_1)\,d\tau_1;$$

$$\mathbf{M}[\tilde{K}_2(\tau)] = K(\tau) - \frac{2}{(T-\tau)^2} \int_0^{T-\tau} (T - \tau - \tau_1) K(\tau + \tau_1)\,d\tau_1;$$

$$\begin{aligned} \mathbf{D}[\tilde{K}_1(\tau)] = {} & \frac{2}{(T-\tau)^2} \int_0^{T-\tau} (T - \tau - \tau_1)[K^2(\tau_1) + K(\tau_1 + \tau)K(\tau_1 - \tau)]\,d\tau_1 \\ & + \frac{8}{(T-\tau)^4} \left[\int_0^{T-\tau} (T - \tau - \tau_1)K(\tau_1)\,d\tau_1\right]^2 - \frac{4}{(T-\tau)^3} \\ & \times \int_0^{T-\tau}\int_0^{T-\tau}\int_0^{T-\tau} K(t_3 - t_1)K(t_2 - t_1 - \tau)\,dt_1\,dt_2\,dt_3; \end{aligned}$$

$$\mathbf{D}[\tilde{K}_2(\tau)] = \frac{2}{(T-\tau)^2}\int_0^{T-\tau}(T-\tau-\tau_1) \times [K^2(\tau_1) + K(\tau_1+\tau)K(\tau_1-\tau) + K(\tau)K(\tau+\tau_1) + K(\tau)K(\tau_1-\tau)]\,d\tau_1$$
$$- \frac{2}{(T-\tau)^3}\int_0^{T-\tau}\int_0^{T-\tau}\int_0^{T-\tau}[K(t_2-t_1)K(t_3-t_1) + K(t_2-t_1+\tau) \times K(t_3-t_1-\tau)]\,dt_1\,dt_2\,dt_3$$
$$+ \frac{2}{(T-\tau)^4}\left\{\int_0^{T-\tau}(T-\tau-\tau_1)[K(\tau_1+\tau)+K(\tau_1-\tau)]\,d\tau_1\right\}^2$$
$$+ \frac{4}{(T-\tau)^4}\left\{\int_0^{T-\tau}(T-\tau-\tau_1)K(\tau_1)\,d\tau_1\right\}^2$$
$$- \frac{4}{(T-\tau)^4}\left\{\int_0^{T-\tau}(T-\tau-\tau_1)K(\tau+\tau_1)\,d\tau_1\right\}^2.$$

46.5 $\mathbf{D}[\tilde{x}] = \dfrac{2\sigma_x^2}{\alpha T}\left(1 - \dfrac{1-e^{-\alpha T}}{\alpha T}\right).$

46.6
$$\mathbf{D}[\tilde{S}(\omega)] = \frac{1}{2\pi^2T^2}\int_0^T(T-t) \times \left\{\int_0^T[K(t+\eta)+K(t-\eta)]\sin(T-\eta)\omega\,d\eta + \left|\int_{-T}^{T}e^{-i\tau\omega}K(t-\tau)\,d\tau\right|^2\right\}dt.$$

46.7 σ_y will decrease by 2 per cent. **46.8** τ_y will decrease by 3 per cent.

46.9 $\mathbf{D}[\tilde{K}_\theta(\tau)] = 22$ grad.^{4}, $\mathbf{D}[\tilde{K}_\theta(3)] = 2.8$ grad.^{4}.

46.10 The value of the first zero of the function $\tilde{K}(\tau)$ equals (a) 2.20 sec., (b) 2.30 sec.

46.11
$$\mathbf{D}[\tilde{K}_\theta(\tau)] = \frac{a^2}{2(T-\tau)} \times \left\{\frac{2\alpha^2+\beta^2}{\alpha(\alpha^2+\beta^2)} + e^{-2\alpha\tau}\left[2\tau\cos 2\beta\tau + \frac{1}{\beta}\sin 2\beta\tau + \frac{1}{\alpha} \times \cos 2\beta\tau + \frac{\alpha\cos 2\beta\tau - \beta\sin 2\beta\tau}{\alpha^2+\beta^2}\right]\right\};$$

$\mathbf{D}[\tilde{K}_\theta(0)] = 5.82$ grad.^{4}, $\mathbf{D}[\tilde{K}_\theta(2.09)] = 5.35$ grad.^{4},
$\mathbf{D}[\tilde{K}_\theta(4.18)] = 4.80$ grad.^{4}, $\mathbf{D}[\tilde{K}_\theta(16.72)] = 2.92$ grad.^{4}

and the corresponding standard deviations are 2.41, 2.32, 2.19 and 1.71 grad.^{2}.

46.12 When t increases the quotient, t_1/t converges in probability to the probability P of coincidence of the signs of the ordinates of the random functions $X(t)$ and $X(t+\tau)$, related, for a normal process, to the normalized correlation function $k(\tau)$ by $k(\tau) = \cos\pi(1-P)$, which can be proved by integrating the two-dimensional normal distribution law of the ordinates of the random function between proper limits.

46.13 Denoting by
$$Z(t) = \frac{1}{2}\left[1 + \frac{X(t)X(t+\tau)}{|X(t)X(t+\tau)|}\right]$$
and by P the probability that the signs of $X(t)$ and $X(t+\tau)$ coincide, we get $\bar{z} = P$;
$$\tilde{k}_x(\tau) = \cos\pi(1-\tilde{z}) \approx \cos\pi(1-\bar{z}) + \pi(\tilde{z}-\bar{z})\sin\pi(1-\bar{z}).$$

Consequently,

$$\mathbf{D}[\tilde{k}(\tau)] \approx \pi^2 \mathbf{D}[\tilde{z}] \sin^2 \pi(1 - P) = \pi^2[1 - k_x^2(\tau)]\mathbf{D}[\tilde{z}];$$

$$\mathbf{D}[\tilde{z}] = \frac{2}{T^2}\int_0^T (T - \tau_1)K_z(\tau_1)\,d\tau_1;$$

$$K_z(\tau) = \left\{\int_0^\infty\int_0^\infty\int_0^\infty\int_0^\infty + \int_0^\infty\int_0^\infty\int_{-\infty}^0\int_{-\infty}^0 + \int_{-\infty}^0\int_{-\infty}^0\int_{-\infty}^0\int_{-\infty}^0 + \int_{-\infty}^0\int_{-\infty}^0\int_0^\infty\int_0^\infty\right\} f(x_1, x_2, x_3, x_4)\,dx_1\,dx_2\,dx_3\,dx_4 - \bar{z}^2,$$

$f(x_1, x_2, x_3, x_4)$ being the distribution law of the system of normal variables $X(t_1)$, $X(t_1 + \tau)$, $X(t_2)$, $X(t_2 + \tau)$.

46.14 $\tilde{K}_x(\tau) = g_1\tilde{K}_1(\tau) + g_2\tilde{K}_2(\tau) + g_3\tilde{K}_3(\tau)$, where we have the approximate reaction:

$$g_j = \frac{\dfrac{1}{\sigma_j^2}}{\dfrac{1}{\sigma_1^2} + \dfrac{1}{\sigma_2^2} + \dfrac{1}{\sigma_3^2}} \quad (j = 1, 2, 3); \qquad \sigma_j^2 = \frac{2}{T_j^2}\int_0^{T_j} (T_j - \tau)\tilde{K}_j(\tau)\,d\tau.$$

For T_j exceeding considerably the damping time of $K_x(\tau)$, it is approximately true that

$$\sigma_j^2 = \frac{2}{T_j}\left(a - \frac{b}{T_j}\right),$$

where $b = \int_0^\infty \tau\tilde{K}(\tau)\,d\tau$ and $\tilde{K}(\tau)$ is a sample function.

46.15 $$\mathbf{D}[\tilde{K}_x(\tau)] = \frac{2}{(m - l)^2}\sum_{s=1}^{m-l-1} [\tilde{K}_x^2(s\Delta) + \tilde{K}_x(s\Delta + l\Delta)\tilde{K}_x(s\Delta - l\Delta)] \times (m - l - s) + \frac{1}{m - l}[\tilde{K}_x^2(0) + \tilde{K}_x^2(l\Delta)].$$

46.16 By 9 per cent.

46.17 $$\alpha_0 = \frac{1}{T}\int_0^T K_x(\tau)\,d\tau; \qquad \alpha_j = \frac{2}{T}\int_0^T K_x(\tau)\cos\frac{2\pi j\tau}{T}\,d\tau, \quad j > 0;$$

$$\Delta_{\text{opt}} = \int_0^T K_x^2(\tau)\,d\tau - T\alpha_0^2 - \frac{T}{2}\sum_{j=1}^m \alpha_j^2.$$

46.18 Since

$$\tilde{j} = \frac{1}{T}\int_0^T j(t)\,dt,$$

then

$$\mathbf{D}[\tilde{j}] = \frac{2a}{\alpha T}\left[1 - \frac{1}{\alpha T}(1 - e^{-\alpha T})\right] = \sigma_{\tilde{j}}^2 = (0.86\cdot 10^{-8})^2 A^2.$$

The mean error is $E_{\tilde{j}} = \rho\sqrt{2}\,\sigma_{\tilde{j}} = 0.58\cdot 10^{-8} A$.

SOURCES OF TABLES REFERRED TO IN THE TEXT*

1T. The binomial coefficients C_n^m: Beyer, W., pp. 339–340; Middleton, D., 1960; Kouden, D., 1961, pp. 564–567; Volodin, B. G., et al., 1962, p. 393.

2T. The factorials $n!$ or logarithms of factorials $\log n!$: Barlow, P., 1962; Beyer, W., pp. 449–450; Bronstein, I., and Semendyaev, K. A., 1964; Boev, G., 1956, pp. 350–353; Kouden, D., 1961, pp. 568–569; Segal, B. I., and Semendyaev, K. A., 1962, p. 393; Unkovskii, V. A., 1953, p. 311; Volodin, B. G., et al., 1962, p. 394.

3T. Powers of integers: Beyer, W., pp. 452–453.

4T. The binomial distribution function $\mathbf{P}(d < m + 1) = \mathbf{P}(d \leqslant m) = \sum_{k=0}^{m} C_n^k p^k (1 - p)^{n-k}$: Beyer, W., pp. 163–173; Kouden, D., 1961, pp. 573–578.

5T. The values of the gamma-function $\Gamma(x)$ or logarithms of the gamma-function $\log \Gamma(x)$: Beyer, W., p. 497; Bronstein, I., and Semendyaev, K. A., 1964; Hald, A., 1952; Middleton, D., 1960; Boev, G., 1956, p. 353; Segal, B. I., and Semendyaev, K. A., 1962, pp. 353–391; Shor, Ya., 1962, p. 528.

6T. The probabilities $P(m, a) = \dfrac{a^m}{m!} e^{-a}$ for a Poisson distribution: Beyer, W., pp. 175–187; Gnedenko, B. V.; Saaty, T., 1957; Boev, G., 1956, pp. 357–358; Dunin-Barkovskii, I. V., and Smirnov, N. V., 1955, pp. 492–494; Segal, B. I., and Semendyaev, K. A., 1962.

7T. The total probabilities $\mathbf{P}(k \geqslant m) = e^{-a} \sum_{k=m}^{\infty} a^k/k!$ for a Poisson distribution: Beyer, W., pp. 175–187.

8T. The Laplace function (the probability integral) in case of an argument expressed in terms of standard deviation $\Phi(z) = 2/\sqrt{2\pi} \int_0^z e^{-x^2/2}\,dx$: Arley, N., and Buch, K., 1950; Beyer, W., pp. 115–124; Cramér, H., 1946; Gnedenko, B. V., and Khinchin, A., 1962; Milne, W. E., 1949; Pugachev, V. S., 1965; Saaty, T., 1957; Bernstein, S., 1946, pp. 410–411.

9T. The probability density of the normal distribution $\varphi(z) = \dfrac{1}{\sqrt{2\pi}} e^{-z^2/2}$ for an argument expressed in standard deviations: Beyer, W., pp. 115–124; Gnedenko, B. V., p. 383.

* More complete information on the references is found in the Bibliography, which follows this section.

10T. The derivatives of the probability density of the normal distribution $\varphi(x)$: $\varphi_2(x) = \varphi''(x) = (x^2 - 1)\varphi(x)$; $\varphi_3(x) = \varphi'''(x) = -(x^3 - 3x)\varphi(x)$: Beyer, W., pp. 115–124.

11T. The reduced Laplace function for an argument expressed in standard deviations, $\hat{\Phi}(z) = 2\rho/\sqrt{\pi}\int_0^z e^{-\rho^2x^2}\,dx$: see 8T.

12T. The probability density of the normal distribution for an argument expressed in standard deviation, $\hat{\varphi}(z) = \rho/\sqrt{\pi}\,e^{-\rho^2z^2}$: see 9T.

13T. The function $p(z) = 2\rho/\sqrt{\pi}\int_0^z e^{-\rho^2x^2}\,dx - 2z(\rho/\sqrt{\pi})e^{-\rho^2z^2} = \hat{\Phi}(z) - 2z\hat{\varphi}(z)$: see 8T, 9T.

14T. The Student distribution law

$$\mathbf{P}(T < t) = \frac{\Gamma[(k+1)/2]}{\Gamma(k/2)\sqrt{k\pi}}\int_0^t \left(1 + \frac{x^2}{k}\right)^{-(k+1)/2} dx:$$

Beyer, W., pp. 225–226; Gnedenko, B. V.; Yaglom, A. M., and Yaglom, I. M., 1964; Volodin, B. G., et al., 1962, p. 404; Segal, B. I., and Semendyaev, K. A., 1962.

15T. The probabilities $\Phi_1(t_\alpha, k) = \mathbf{P}(|T| < t_\alpha) = 2\mathbf{P}(T < t_\alpha) - 1$ for the Student distribution law: see 14T.

16T. The values of γ associated with the confidence level $\alpha = \mathbf{P}(|T| < \gamma)$ and k degrees of freedom for the Student distribution: Arley, N., and Buch, K., 1950; Cramér, H., 1946; Laning, J. H., Jr., and Battin, R. H., 1956; Unkovskii, V. A., 1953, pp. 306–307; see also 14T.

17T. The probabilities $\mathbf{P}(\chi^2 \geqslant \chi_q^2) = \dfrac{1}{\Gamma(k/2)2^{k/2}}\displaystyle\int_{\chi_q^2}^{\infty} x^{(k/2)-1}e^{-x/2}\,dx$ for a chi-square distribution on χ_q^2 and k degrees of freedom: Beyer, W., pp. 233–239; Gnedenko, B. V.; Milne, W. E., 1949; Dunin-Barkovskii, I. V., and Smirnov, N. V., 1955, pp. 505–507.

18T. The values of χ_q^2 depending on the probability $\mathbf{P}(\chi^2 \geqslant \chi_q^2)$ and k degrees of freedom for a chi-square distribution: see 17T.

19T. The lower limit γ_1 and the upper limit γ_2 of the confidence level α and k degrees of freedom for a chi-square distribution: Laning, J. H., Jr., and Battin, R. H., 1956; Smirnov, N. V., and Dunin-Barkovskii, I. V., 1959, p. 405.

20T. The probabilities $L(q, k) = \mathbf{P}[\sqrt{k}/(1 + q) < \chi < \sqrt{k}/(1 - q)]$ for a chi-square distribution: see 22T.

21T. The probability density of a chi-square distribution

$$f(y, k) = \frac{y^{k-1}e^{-y^2/2}}{2^{(k-2)/2}\Gamma(k/2)}:$$

see 5T, 9T.

22T. The probabilities $\mathbf{P}(y \leqslant q\sqrt{k})$ for the quantity y obeying a chi-square distribution: $\mathbf{P}(y \leqslant q\sqrt{k}) = \dfrac{1}{2^{(k-2)/2}\Gamma(k/2)}\displaystyle\int_0^{q\sqrt{k}} y^{k-1}e^{-y^2/2}\,dy$: Beyer, W., pp. 233–239; Shor, Ya., 1962.

23T. The Rayleigh distribution law $\mathbf{P}(X < x) = 1 - e^{-x^2/2\sigma^2}$: Bartlett, M., 1953.

24T. The function $p(x) = 1 - e^{-\rho^2x^2}$: Bartlett, M., 1953.

25T. The probabilities

$$\mathbf{P}(D\sqrt{n} \geqslant \lambda) = \mathbf{P}(\lambda) = 1 - K(\lambda), \qquad K(\lambda) = \sum_{k=-\infty}^{\infty} (-1)^k e^{-2k^2\lambda^2}$$

for the Kolmogorov distribution law: Arley, N., and Buch, K., 1950; Gnedenko, B. V.; Milne, W. E., 1949; Dunin-Barkovskii, I. V., and Smirnov, N. V., 1955, pp. 539–540.

26T. The values of y(p-quantiles) depending on the parameter c and the Wald distribution function: $p = W_c(y) = \sqrt{\frac{c}{2\pi}}\int_0^y y^{-3/2}\exp\left\{-\frac{c}{2}\left(y + \frac{1}{y} - 2\right)\right\}dy$: Takacs, L., 1962; Basharinov, A., and Fleishman, B., 1962, pp. 338–344.

27T. Tables of random numbers: Beyer, W., pp. 341–345.

28T. The function $\eta(p) = -p \log_2 p$: Wald, A., 1947.

29T. The orthogonal Chebyshev polynomials

$$P_{k,n}(x) = \sum_{j=0}^{k} (-1)^j C_k^j C_{k+j}^j \frac{x(x-1)\cdots(x-j+1)}{n(n-1)\cdots(n-j+1)}:$$

Middleton, D., 1960.

30T. Two-sided confidence limits for the estimated parameter in the binomial distribution law: Beyer, W., 187–189.

31T. The values of $z = \tan h^{-1} r = \frac{1}{2}\ln\frac{1+r}{1-r}$: Dwight, H., 1958.

32T. The relations between the parameters b_m, v_m and m for the Weibull distribution law: Koshlyakov, N. S., Gliner, E. B., and Smirnov, M. M., 1964.

BIBLIOGRAPHY

Arley, N., and Buch, K.: *Introduction to Probability and Statistics.* New York, John Wiley and Sons, Inc., 1950.

Bachelier, L.: *Calcul des Probabilités* (*Calculus of Probabilities*). Paris, 1942.

Barlow, P.: *Barlow's Tables of Squares, Cubes, Square Roots, Cube Roots, and Reciprocals of all Integer Numbers up to 12,500.* 4th Ed. New York, Chemical Publishing Co., Inc., 1962.

Bartlett, M.: *Philosophical Magazine*, No. 44, 1953.

Basharinov, A., and Fleishman, B.: *Metody statisticheskogo posledovatel'nogo analiza i ikh priloshenYiya* (*Methods of statistical sequential analysis and their applications*). Sovetskoe Radio, 1962.

Bernstein, S.: *Teoriya Veroyatnostei* (*Probability Theory*). Gostekhizdat, 1946.

Bertrand, I.: *Calcul des Probabilités* (*Calculus of Probabilities*). Paris, 1897.

Beyer, W.: *Handbook of Tables for Probability and Statistics.* Chemical Rubber Co., Ohio.

Boev, G.: *Teoriya Veroyatnostei* (*Probability Theory*). Gostekhizdat, 1956.

Borel, E.: *Elements de la Théorie des Probabilités* (*Elements of Probability Theory*). Paris, 1924.

Bronstein, I., and Semendyaev, K. A.: *Guide Book to Mathematics for Technologists and Engineers.* New York, Pergamon Press, Inc., 1964.

Bunimovich, V.: *Fluktuatsionnye protsessy v radio-priemnykh ustroistvakh* (*Random processes in radio-reception equipment*). Sovetskoe Radio, 1951.

Cramér, H.: *Mathematical Methods of Statistics.* Princeton, N.J., Princeton University Press, 1946.

Czuber, E.: *Wahrscheinlichkeitsrechnung und ihre Anwendung auf Fehlerausgleichung Statistik und Lebensversicherung* (*Probability Theory and its Application to Error-Smoothing, Statistics and Life Insurance*). Leipzig and Berlin, 1910.

Davenport, W. B., Jr., and Root, V. L.: *Introduction to Random Signals and Noise.* New York, McGraw-Hill Book Co., Inc., 1958.

Dlin, A.: *Matematicheskaya statistika v tekhnike* (*Mathematical statistics in technology*). Sovetskaya Nauka, 1958.

Dunin-Barkovskii, I. V., and Smirnov, N. V.: *Teoriya Veroyatnostei i Matematicheskaya Statistika v Tekhnike—Obshchaya Chast* (*Probability Theory and Mathematical Statistics in Technology—General Part*). Gostekhizdat, 1955.

Dwight, H.: *Mathematical Tables of Elementary and Some Higher Order Mathematical Functions.* 3rd Rev. Ed. New York, Dover Publications, Inc., 1961.

Feller, W.: *Introduction to Probability Theory and its Applications.* New York, John Wiley and Sons, Inc., Vol. 1, 1957, Vol. 2, 1966.

Gantmakher, F. R.: *The Theory of Matrices.* New York, Chelsea Publishing Co., 1959.

Glivenko, V.: *Kurs Teorii Veroyatnostei* (*Course in Probability Theory*). GONTI, 1939.

Gnedenko, B. V.: *Theory of Probability.* New York, Chelsea Publishing Co. (4th Ed. in prep.).

Gnedenko, B. V., and Khinchin, A.: *Elementary Introduction to the Theory of Probability*, 5th Ed. New York, Dover Publications, Inc., 1962.

Goldman, S.: *Information Theory*. Englewood Cliffs, N.J., Prentice-Hall, Inc., 1953.

Goncharov, V.: *Teoriya Veroyatnostei* (*Probability Theory*). Oborongiz, 1939.

Guter, R. S., and Ovchinskii, B. V.: *Elementy Chislennogo Analiza i Matematicheskoi Obrabotki Resul'tatov Opita* (*Elements of Numeral Analysis and the Mathematical Processing of Experimental Data*). Fizmatgiz, 1962.

Gyunter, N. M., and Kuz'min, R. O.: *Sbornik Zadach po Vysshei Matematike—Ch. III* (*Collection of Problems in Higher Mathematics—Part III*). Gostekhizdat, 1951.

Hald, A.: *Statistical Theory with Engineering Applications*. New York, John Wiley and Sons, Inc., 1952.

Jahnke, E., and Emde, F.: *Tables of Functions with Formulae and Curves*. New York, Dover Publications, Inc., 1945.

Kadyrov, M.: *Tablitsy Sluchainykh Chisel* (*Table of Random Numbers*). Tashkent, 1936.

Khinchin, A.: *Raboty po Matematicheskoi Teorii Massovogo Obsluzjevaniya* (*Work in the Mathematical Theory of Mass Service* [*Queues*]). Fizmatgiz, 1963.

Koshlyakov, N. S., Gliner, E. B., and Smirnov, M. M.: *Differential Equations of Mathematical Physics*. New York, John Wiley and Sons, Inc. (Interscience), 1964.

Kotel'nikov, V.: A nomogram connecting the parameters of Weibull's distribution with probabilities. *Theory of Probability and Its Applications*, **9**: 670–674, 1964.

Kouden, D.: *Statischeskie Metody Kontrolya Kachestva* (*Statistical Methods of Quality Control*). Fizmatgiz, 1961.

Krylov, V. I.: *Approximate Calculations of Integrals*. New York, The Macmillan Co., 1962.

Laning, J. H., Jr., and Battin, R. H.: *Random Processes in Automatic Control*. New York, McGraw-Hill Book Co., Inc., 1956.

Levin, B.: *Teoriya sluchainykh protsessov i ee primenenie v radiotekhnike* (*Theory of random processes and its application to radio technology*). Sovetskoe Radio, 1957.

Linnik, Y. V.: *Method of Least Squares and Principles of the Theory of Observations*. New York, Pergamon Press, Inc., 1961.

Lukomskii, Ya.: *Teoriya Korrelyatsii i ee Primenenie k Analizu Proizvodstva* (*Correlation Theory and its Application to the Analysis of Production*). Gostekhizdat 1961.

Mesyatsev, P. P.: *Primenenie Teorii Veroyatnostei i Matematicheskoi Statistiki pri Konstruirovannii i Proizvodstve Radio-Apparatury* (*Applications of Probability Theory and Mathematical Statistics to the Construction and Production of Radios*). Voenizdat, 1958.

Middleton, D.: *Introduction to Statistical Communication Theory*. New York, McGraw-Hill Book Co., Inc., 1960.

Milne, W. E.: *Numerical Calculus*. Princeton, N.J., Princeton University Press, 1949.

Nalimov, V. V.: *Application of Mathematical Statistics to Chemical Analysis*. Reading, Mass., Addison-Wesley Publishing Co., Inc., 1963.

Pugachev, V. S.: *Theory of Random Functions*. Reading, Mass., Addison-Wesley Publishing Co., Inc., 1965.

Romanovskii, V.: *Diskretnye Tsepi Markova* (*Discrete Markov Chains*). Gostekhizdat, 1949.

Romanovskii, V.: *Matematicheskaya Statistika* (*Mathematical Statistics*). GONTI, 1938.

Rumshiskii, L. Z.: *Elements of Probability Theory*. New York, Pergamon Press, Inc., 1965.

Saaty, T.: *Resumé of useful formulas in queuing theory*. Operations Research, No. 2, 1957.

Sarymsakov, T. A.: *Osnovy Teorii Protsessov Markova* (*Basic Theory of Markov Processes*). Gostekhizdat, 1954.

Segal, B. I., and Semendyaev, K. A.: *Pyatiznachnye Matematicheskie Tablitsy* (*Five-Place Mathematical Tables*). Fizmatgiz, 1961.

Shchigolev, B. M.: *Mathematical Analysis of Observations*. New York, American Elsevier Publishing Co., Inc., 1965.

Sherstobitov, V. V., and Diner, I.: *Sbornik Zadach po Strel'be zenitoi Artilrii* (*Collection of Problems in Antiaircraft Artillery Firing*). Voenizdat, 1948.

Shor, Ya.: *Statisticheskie metody analiza i kontrolya kachestva i nadezhnosti* (*Statistical methods of analysis, quality control and safety*). Sovetskoe Radio, 1962.

Smirnov, N. V., and Dunin-Barkovskii, I. V.: *Kratkii Kurs Matematicheskoi Statistiki* (*Short Course in Mathematical Statistics*). Fizmatgiz, 1959.

Solodovnikov, V.: *Statistical Dynamics of Linear Automatic Control Systems*. Princeton, N. J., D. Van Nostrand Co., Inc., 1956.

Stratonovich, R. L.: *Izbrannye voprosy teorii fluktuatsii v radioteknike* (*Selected questions in fluctuation theory in radio technology*). Sovetskoe Radio, 1961.

Sveshnikov, A. A.: *Applied Methods of the Theory of Random Functions*. New York, Pergamon Press, Inc. (in prep.).

Takacs, L.: *Stochastic Processes, Problems and Solutions*. New York, John Wiley and Sons, Inc., 1960.

Unkovskii, V. A.: *Teoriya Veroyatnostei* (*Probability Theory*). Voenmorizdat, 1953.

Uorsing, A., and Geffner, D.: *Metody Obrabotki Eksperimental'nykh Dannykh* (*Methods for Processing Experimental Data*). IL, 1953.

Venttsel', E. S.: *Teoriya veroyatnostei* (*Probability theory*). Izd-vo Nauka, 1964.

Volodin, B. G., et al.: *Rukovodstvo Dlya Inzhenerov po Resheniyu Zadach Teorii Veroyatnostey* (*Engineer's Guide for the Solution of Problems in Probability Theory*). Sudpromgiz, 1962.

Wald, A.: *Sequential Analysis*. New York, John Wiley and Sons, Inc., 1947.

Yaglom, A. M., and Yaglom, I. M.: *Challenging Mathematical Problems with Elementary Solutions*. San Francisco, Holden-Day, Inc., 1964.

Yaglom, A. M., and Yaglom, I. M.: *Probability and Information*. New York, Dover Publications, Inc., 1962.

Yule, G. U., and Kendall, M. G.: *Introductory Theory of Statistics*. 14th Rev. Ed. New York, Hafner Publishing Co., Inc., 1958.

Index

A CATALOG OF SELECTED

DOVER BOOKS

IN SCIENCE AND MATHEMATICS

A CATALOG OF SELECTED
DOVER BOOKS
IN SCIENCE AND MATHEMATICS

QUALITATIVE THEORY OF DIFFERENTIAL EQUATIONS, V.V. Nemytskii and V.V. Stepanov. Classic graduate-level text by two prominent Soviet mathematicians covers classical differential equations as well as topological dynamics and ergodic theory. Bibliographies. 523pp. 5⅜ × 8½. 65954-2 Pa. $10.95

MATRICES AND LINEAR ALGEBRA, Hans Schneider and George Phillip Barker. Basic textbook covers theory of matrices and its applications to systems of linear equations and related topics such as determinants, eigenvalues and differential equations. Numerous exercises. 432pp. 5⅜ × 8½. 66014-1 Pa. $10.95

QUANTUM THEORY, David Bohm. This advanced undergraduate-level text presents the quantum theory in terms of qualitative and imaginative concepts, followed by specific applications worked out in mathematical detail. Preface. Index. 655pp. 5⅜ × 8½. 65969-0 Pa. $13.95

ATOMIC PHYSICS (8th edition), Max Born. Nobel laureate's lucid treatment of kinetic theory of gases, elementary particles, nuclear atom, wave-corpuscles, atomic structure and spectral lines, much mo e. Over 40 appendices, bibliography. 495pp. 5⅜ × 8½. 65984-4 Pa. $12.95

ELECTRONIC STRUCTURE AN D THE PROPERTIES OF SOLIDS: The Physics of the Chemical Bond, Walter A. Harrison. Innovative text offers basic understanding of the electronic structure of covalent and ionic solids, simple metals, transition metals and their compounds. Problems. 1980 edition. 582pp. 6⅛ × 9¼. 66021-4 Pa. $15.95

BOUNDARY VALUE PROBLEMS OF HEAT CONDUCTION, M. Necati Özisik. Systematic, comprehensive treatment of modern mathematical methods of solving problems in heat conduction and diffusion. Numerous examples and problems. Selected references. Appendices. 505pp. 5⅜ × 8½. 65990-9 Pa. $12.95

A SHORT HISTORY OF CHEMISTRY (3rd edition), J.R. Partington. Classic exposition explores origins of chemistry, alchemy, early medical chemistry, nature of atmosphere, theory of valency, laws and structure of atomic theory, much more. 428pp. 5⅜ × 8½. (Available in U.S. only) 65977-1 Pa. $10.95

A HISTORY OF ASTRONOMY, A. Pannekoek. Well-balanced, carefully reasoned study covers such topics as Ptolemaic theory, work of Copernicus, Kepler, Newton, Eddington's work on stars, much more. Illustrated. References. 521pp. 5⅜ × 8½. 65994-1 Pa. $12.95

PRINCIPLES OF METEOROLOGICAL ANALYSIS, Walter J. Saucier. Highly respected, abundantly illustrated classic reviews atmospheric variables, hydrostatics, static stability, various analyses (scalar, cross-section, isobaric, isentropic, more). For intermediate meteorology students. 454pp. 6⅛ × 9¼. 65979-8 Pa. $14.95

THE ELECTROMAGNETIC FIELD, Albert Shadowitz. Comprehensive undergraduate text covers basics of electric and magnetic fields, builds up to electromagnetic theory. Also related topics, including relativity. Over 900 problems. 768pp. 5⅜ × 8¼. 65660-8 Pa. $18.95

FOURIER SERIES, Georgi P. Tolstov. Translated by Richard A. Silverman. A valuable addition to the literature on the subject, moving clearly from subject to subject and theorem to theorem. 107 problems, answers. 336pp. 5⅜ × 8½. 63317-9 Pa. $8.95

THEORY OF ELECTROMAGNETIC WAVE PROPAGATION, Charles Herach Papas. Graduate-level study discusses the Maxwell field equations, radiation from wire antennas, the Doppler effect and more. xiii + 244pp. 5⅜ × 8½. 65678-0 Pa. $6.95

DISTRIBUTION THEORY AND TRANSFORM ANALYSIS: An Introduction to Generalized Functions, with Applications, A.H. Zemanian. Provides basics of distribution theory, describes generalized Fourier and Laplace transformations. Numerous problems. 384pp. 5⅜ × 8½. 65479-6 Pa. $9.95

THE PHYSICS OF WAVES, William C. Elmore and Mark A. Heald. Unique overview of classical wave theory. Acoustics, optics, electromagnetic radiation, more. Ideal as classroom text or for self-study. Problems. 477pp. 5⅜ × 8½. 64926-1 Pa. $12.95

CALCULUS OF VARIATIONS WITH APPLICATIONS, George M. Ewing. Applications-oriented introduction to variational theory develops insight and promotes understanding of specialized books, research papers. Suitable for advanced undergraduate/graduate students as primary, supplementary text. 352pp. 5⅜ × 8½. 64856-7 Pa. $8.95

A TREATISE ON ELECTRICITY AND MAGNETISM, James Clerk Maxwell. Important foundation work of modern physics. Brings to final form Maxwell's theory of electromagnetism and rigorously derives his general equations of field theory. 1,084pp. 5⅜ × 8½. 60636-8, 60637-6 Pa., Two-vol. set $21.90

AN INTRODUCTION TO THE CALCULUS OF VARIATIONS, Charles Fox. Graduate-level text covers variations of an integral, isoperimetrical problems, least action, special relativity, approximations, more. References. 279pp. 5⅜ × 8½. 65499-0 Pa. $7.95

HYDRODYNAMIC AND HYDROMAGNETIC STABILITY, S. Chandrasekhar. Lucid examination of the Rayleigh-Benard problem; clear coverage of the theory of instabilities causing convection. 704pp. 5⅜ × 8¼. 64071-X Pa. $14.95

CALCULUS OF VARIATIONS, Robert Weinstock. Basic introduction covering isoperimetric problems, theory of elasticity, quantum mechanics, electrostatics, etc. Exercises throughout. 326pp. 5⅜ × 8½. 63069-2 Pa. $8.95

DYNAMICS OF FLUIDS IN POROUS MEDIA, Jacob Bear. For advanced students of ground water hydrology, soil mechanics and physics, drainage and irrigation engineering and more. 335 illustrations. Exercises, with answers. 784pp. 6⅛ × 9¼. 65675-6 Pa. $19.95

GEOMETRY OF COMPLEX NUMBERS, Hans Schwerdtfeger. Illuminating, widely praised book on analytic geometry of circles, the Moebius transformation, and two-dimensional non-Euclidean geometries. 200pp. 5⅜ × 8¼. 63830-8 Pa. $8.95

MECHANICS, J.P. Den Hartog. A classic introductory text or refresher. Hundreds of applications and design problems illuminate fundamentals of trusses, loaded beams and cables, etc. 334 answered problems. 462pp. 5⅜ × 8½. 60754-2 Pa. $9.95

TOPOLOGY, John G. Hocking and Gail S. Young. Superb one-year course in classical topology. Topological spaces and functions, point-set topology, much more. Examples and problems. Bibliography. Index. 384pp. 5⅜ × 8¼. 65676-4 Pa. $9.95

STRENGTH OF MATERIALS, J.P. Den Hartog. Full, clear treatment of basic material (tension, torsion, bending, etc.) plus advanced material on engineering methods, applications. 350 answered problems. 323pp. 5⅜ × 8½. 60755-0 Pa. $8.95

ELEMENTARY CONCEPTS OF TOPOLOGY, Paul Alexandroff. Elegant, intuitive approach to topology from set-theoretic topology to Betti groups; how concepts of topology are useful in math and physics. 25 figures. 57pp. 5⅜ × 8½. 60747-X Pa. $3.50

ADVANCED STRENGTH OF MATERIALS, J.P. Den Hartog. Superbly written advanced text covers torsion, rotating disks, membrane stresses in shells, much more. Many problems and answers. 388pp. 5⅜ × 8½. 65407-9 Pa. $9.95

COMPUTABILITY AND UNSOLVABILITY, Martin Davis. Classic graduate-level introduction to theory of computability, usually referred to as theory of recurrent functions. New preface and appendix. 288pp. 5⅜ × 8½. 61471-9 Pa. $7.95

GENERAL CHEMISTRY, Linus Pauling. Revised 3rd edition of classic first-year text by Nobel laureate. Atomic and molecular structure, quantum mechanics, statistical mechanics, thermodynamics correlated with descriptive chemistry. Problems. 992pp. 5⅜ × 8½. 65622-5 Pa. $19.95

AN INTRODUCTION TO MATRICES, SETS AND GROUPS FOR SCIENCE STUDENTS, G. Stephenson. Concise, readable text introduces sets, groups, and most importantly, matrices to undergraduate students of physics, chemistry, and engineering. Problems. 164pp. 5⅜ × 8½. 65077-4 Pa. $6.95

THE HISTORICAL BACKGROUND OF CHEMISTRY, Henry M. Leicester. Evolution of ideas, not individual biography. Concentrates on formulation of a coherent set of chemical laws. 260pp. 5⅜ × 8½. 61053-5 Pa. $6.95

THE PHILOSOPHY OF MATHEMATICS: An Introductory Essay, Stephan Körner. Surveys the views of Plato, Aristotle, Leibniz & Kant concerning propositions and theories of applied and pure mathematics. Introduction. Two appendices. Index. 198pp. 5⅜ × 8½. 25048-2 Pa. $7.95

THE DEVELOPMENT OF MODERN CHEMISTRY, Aaron J. Ihde. Authoritative history of chemistry from ancient Greek theory to 20th-century innovation. Covers major chemists and their discoveries. 209 illustrations. 14 tables. Bibliographies. Indices. Appendices. 851pp. 5⅜ × 8½. 64235-6 Pa. $18.95